Second Supplements to the 2nd Edition of

RODD'S CHEMISTRY OF CARBON COMPOUNDS

ELSEVIER SCIENCE PUBLISHERS B.V.
Sara Burgerhartstraat 25
P.O. Box 211, 1000 AE Amsterdam, The Netherlands

Library of Congress Card Number 91-16129
ISBN 0-444-89873-5

This book is printed on acid-free paper.

Printed in The Netherlands

Second Supplements to the 2nd Edition of

RODD'S CHEMISTRY OF CARBON COMPOUNDS

VOLUME I

ALIPHATIC COMPOUNDS

★

VOLUME II

ALICYCLIC COMPOUNDS

★

VOLUME III

AROMATIC COMPOUNDS

★

VOLUME IV

HETEROCYCLIC COMPOUNDS

★

VOLUME V

MISCELLANEOUS

GENERAL INDEX

★

Second Supplements to the 2nd Edition of

RODD'S CHEMISTRY OF CARBON COMPOUNDS

A modern comprehensive treatise

Edited by
MALCOLM SAINSBURY
School of Chemistry, The University of Bath, Claverton Down, Bath BA2 7AY, England

Second Supplement to

VOLUME I ALIPHATIC COMPOUNDS

Part E: Trihydric Alcohols,
Their Oxidation Products and Derivatives
Part F: Penta- and Higher Polyhydric Alcohols,
Their Oxidation Products and Derivatives; Saccharides
Part G: Tetrahydric Alcohols,
Their Oxidation Products and Derivatives

ELSEVIER
Amsterdam – London – New York – Tokyo 1993

Contributors to this Volume

S.A. BOWLES

British Biotechnology Limited, Brook House, Watlington Road,
Cowley, Oxford OX4 5LY, England

B.J. COFFIN

Department of Biotechnology, South Bank Polytechnic, 103 Borough Road,
London SE1 0AA, England

R. DARCY

Department of Chemistry, University College Dublin, Belfield,
Dublin 4, Ireland

K.J. HALE

Christopher Ingold Laboratories, Chemistry Department, University College
London, 20 Gordon Street, London WC1H 0AJ, England

R.A. HILL

Chemistry Department, The University, Glasgow G12 8QQ, Scotland

P.D. JENKINS

Ferring Research Institute, 1 Venture Road, Chilworth Science Park,
Southampton SO1 7NP, England

K. McCARTHY

Department of Chemistry, University College Dublin, Belfield,
Dublin 4, Ireland

D.K. WEERASINGHE

Universal Flavor Corporation, Bio-Organic Laboratory, 5600 West Raymond
Street, Indianapolis, IN 46241-0167, U.S.A.

Preface to Volume I EFG

This volume is devoted to the chemistry of polyhydric alcohols, including saccharides. It is a vast and crucially important area, not only because of the prominent roles which polyols have in their own right, but because of the way they and their modified forms are incorporated into natural products. In setting up this volume the authors were specifically asked to be selective in the coverage of individual topics and, in my opinion, this has been achieved in style, without significant loss of any important subject material.

One reason for this selectivity was to allow space for the coverage of the applications of carbohydrates as chiral auxiliaries, templates, and starting materials in syntheses. So important is this area that many organic chemists who hitherto never ventured near carbohydrate chemistry now regard complex manipulations of sugars as 'their backyard'. These topics were not singled out in the first supplement, but have received detailed attention here.

Malcolm Sainsbury
August 1992

Contents
Volume I EFG

Aliphatic Compounds; Trihydric Alcohols, Their Oxidation Products and Derivatives

Penta- and Higher Polyhydric Alcohols, Their Oxidation Products and Derivatives; Saccharides

Tetrahydric Alcohols, Their Oxidation Products and Derivatives

Preface vii
List of common abbreviations and symbols used xv

Chapter 18. Trihydric Alcohols, Their Analogues and Derivatives and Their Oxidation Products: Trihydric Alcohols to Triketones
by B.J. COFFIN

1. Preparation of glycerol 1
2. Reactions of glycerol 3
3. Use of chiral synthons derived from glycerol in synthesis 11
   (a) Chiral γ-lactones 11
   (b) The synthesis heteroyohimbine alkaloid precursors 16
4. Glycerides 20

Chapter 19. Trihydric Alcohols and Their Oxidation Products (continued)
by B.J. COFFIN

Dihydroxycarboxylic and trihydroxycarboxylic acids and related compounds . . . 35
1. Dihydroxycarboxylic acids 35
   (i) Preparation of mevalonic acid and mevalonolactone, 36 – (ii) Biosynthesis of isoleucine and valine, 45 –
2. Amino hydroxy carboxylic acids 49
3. 2-Amino-2-alkenoic acids (didehydroaminocarboxylic acids) 50
4. Representative amino carboxylic acids and related compounds 57

Chapter 20. Trihydric Alcohols and Their Oxidation Products (continued)
by B.J. COFFIN

Hydroxy- and amino-dicarboxylic acids 81

Chapter 21. Phospholipids
by S.A. BOWLES

General references 102
1. Synthesis of phospholipids 104
   (a) Halophosphate reagents 105
   (b) Cyclic reagents 109
   (c) The cyclic enediol phosphoryl (CEP) method 109

(d) Activating reagents 111
(e) Oligonucleotide methodology 112
(f) Glycidol derivatives 116
(g) Semi-synthetic methods 118
(i) Methods of acylation, 118 – (ii) Transesterification, 122 –
2. Phospholipid analogues 123
(a) Analogues containing sulphur 123
(b) Backbone modifications 127
(c) Phosphonolipids 129
(d) Vesicles 131
(e) Membrane probes 132
(f) Electron transport 134
(g) Miscellaneous analogues 134
3. Platelet-activating factor (PAF) 136
(a) Syntheses of PAF 137
(i) Syntheses from D-mannitol, 137 – (ii) Other natural sources of chirality, 142 – (iii) Other syntheses of PAF, 142 –
(b) PAF analogues 144
(i) PAF receptor agonists, 145 – (ii) PAF receptor antagonists, 149 – (iii) Miscellaneous PAF analogues, 153 –
4. Phospholipase A_2 inhibitors 154
(a) Synthesis of PLA_2 inhibitors 156

Chapter 22. Polyhydric Alcohols and Their Oxidation Products
by P.D. JENKINS

1. Introduction 167
2. Preparation and synthesis of alditols 167
(a) Reduction 167
(b) Electrolysis 168
(c) Enzymatic 168
(d) Syntheses 169
3. Conformation, chelation and chromatography 176
(a) Conformation and stereochemistry 176
(b) Complexes and chelation 178
(c) Chromatography 181
4. Reactions and derivatives of the alditols 182
(a) Use of alditols in asymmetric synthesis 182
(i) Chiral synthetic intermediates (chirons), 182 – (ii) Reagents and catalysts, 185 –
(b) Oxidation and oxidation products 187
(c) Cyclic acetals 190
(d) Esters 193
(e) Ethers 198
5. Anhydroalditols 199
6. Deoxyalditols 202
7. Aminoalditols 203
8. Deoxynitroalditols 208
9. Sulphur containing derivatives 209

Chapter 23a. Monosaccharides: Synthesis, Chemistry, Structure and Physical Properties
by D.K. WEERASINGHE

1. Introduction 213
2. Synthesis of monosaccharides 214
   (a) Diels–Alder types 214
   (b) Oxirane ring openings 218
   (c) Wittig reactions 220
   (d) Organometallic reagents 221
   (e) Miscellaneous reactions 228
3. Amino monosaccharides 230
   (a) Oxirane ring opening 231
   (b) Condensation type reaction 234
   (c) Diels–Alder type constructions 235
   (d) Miscellaneous reactions 237
4. Monosaccharides having a phosphorus atom in the ring 243
5. Cyano monosaccharides 246
6. Halides of monosaccharides 248
7. Thio and thia monosaccharides 255
8. Carbocyclic analogues of monosaccharides 258
9. Chemistry of monosaccharides 264
10. Structure and physical properties of monosaccharides 268

Chapter 23b. Monosaccharides: Use in Synthesis as Chiral Templates
by K.J. HALE

Introduction 273
1. Asymmetric reduction 273
   (a) Asymmetric homogeneous hydrogenation with transition metal complexes of monosaccharide phosphines 273
   (i) Homogeneous hydrogenation of didehydroamino acids, 273 – (ii) Homogeneous hydrogenation of acrylic acid derivatives, 278 –
   (b) Asymmetric heterogeneous hydrogenation 280
   (c) Asymmetric hydrosilylation 281
   (d) Chiral reducing agents derived from monosaccharides 281
2. Enantioselective alkylation reactions 287
   (a) Monosaccharide derived enolates in enantioselective aldol reactions 287
   (b) Enantioselective alkylation of monosaccharide derived carbanions 289
   (c) Enantioselective alkylation with monosaccharide electrophiles 294
   (i) Monosaccharides as chiral leaving groups, 294 – (ii) Monosaccharides as chiral electrophiles, 296 –
3. Monosaccharides as chiral auxiliaries for cycloaddition reactions 300
   (a) Asymmetric Diels–Alder reactions 300
   (i) Reactions of chrial dienophiles derived from monosaccharides, 300 – (ii) Reactions of chiral dienes derived from monosaccharides, 304 –
   (b) Asymmetric [3+2] cycloaddition reactions with monosaccharide auxiliaries 310
   (c) Asymmetric [2+2] cycloadditions with monosaccharide auxiliaries 312
4. Asymmetric carbenoid additions with monosaccharide bearing copper catalysts 313

Chapter 23c. Monosaccharides: Use in the Asymmetric Synthesis of Natural Products
by K.J. HALE

Introduction 315
1. Use of monosaccharides for the assembly of natural products with carbocyclic ring systems 315
(a) Radical cyclisation strategies 315
(b) Cycloaddition strategies 325
(i) [4+2] Cycloaddition strategies, 325 – (ii) [3+2] Cycloaddition strategies, 337 – (iii) [2+2] Cycloaddition strategies, 341 – (iv) [2+1] Cycloaddition strategies, 344 –
(c) Nucleophilic cyclisation strategies for the synthesis of carbocyclic ring systems in natural products 346
(i) Intramolecular aldol reactions, 346 – (ii) Intramolecular Claisen condensation, 353 – (iii) Intramolecular Michael reactions, 357 – (iv) Intramolecular Horner–Wadsworth–Emmons reactions, 360 – (v) Intramolecular nucleophilic alkylation, 362 –
(d) Cationic cyclisation strategies 364
2. Use of monosaccharides for the assembly of molecules with heterocyclic rings 364
3. Use of monosaccharides in the total synthesis of naturally-occurring macrolides 401

Chapter 24. Disaccharides and Oligosaccharides
by R. DARCY and K. McCARTHY

Introduction 437
1. Di- and oligosaccharides 439
(a) Synthesis 439
(i) Neigbouring-group assisted procedures for the synthesis of β-glycosidic linkages (**3**), 440 – (ii) Halide-ion catalysed synthesis of glycosides (**1**) and (**2**), 443 – (iii) Synthesis of β-glycosidic linkages (**4**) and (**6**) without neighbouring-group participation, 446 – (iv) The imidate procedure: synthesis of (**1**) and (**3**), 447 – (v) The *n*-pentenyl methodology: synthesis of (**1**), (**3**) and (**2**), 448 – (vi) 1,2-halonium ion and related methods: synthesis of (**5**), (**6**) 449 – (vii) Synthesis of linkages, (**7**), (**8**), in the octulosonic and nonulosonic acid series, 452 – (viii) Synthesis of thioglycosides, 453 – (ix) Preparation by degradation of polysaccharides, 453 – (x) Enzyme-catalysed synthesis, 454 –
(b) Physical methods 458
(i) Nuclear magnetic resonance spectroscopy, 458 – (ii) Mass spectrometry, 463 – (iii) X-ray crystallography, 467 – (iv) Other spectroscopic methods, 469 –
2. Disaccharides and related compounds 470
(a) Tetrosylhexoses 470
(b) Pentosylpentoses 470
(c) Pentosylhexoses, hexosylpentoses 471
(d) Deoxyhexosylpentoses 472
(e) Deoxyhexosyldeoxyhexoses 473
(f) Deoxyhexosylhexoses 474

(g) Hexosyldeoxyhexoses 476
(h) Hexosylhexoses 477
(i) Galactosylgalactoses, 477 – (ii) Glycosylglucoses, 478 – (iii) Mannosylmannoses, 479 – (iv) Other hexosylhexoses, 479 –
(i) KDO-containing disaccharides 482
(j) Non-reducing disaccharides 482
(k) Nitrogen-containing disaccharides 484
(i) Disaccharides containing an amino-sugar and a neutral sugar residue, 484 – (ii) Disaccharides containing two amino-sugar residues, 487 –
(l) Uronic acids 488
3. Trisaccharides and related compounds 489
(a) Neutral sugar containing trisaccharides 489
(b) Deoxysugar-containing trisaccharides 492
(c) Nitrogen-containing trisaccharides 493
4. Tetrasaccharides and higher oligosaccharides 496
5. Sulphur-containing oligosaccharides 498

Chapter 25. Tetrahydric Alcohols, Their Analogues, Derivatives and Oxidation Products
by R.A. HILL

1. Tetrahydric alcohols 499
(a) Tetritols 499
(b) Derivatives of the tetritols 500
(c) Homologous tetrahydric alcohols 502
2. Trihydroxyaldehydes and trihydroxyketones 503
3. Dihydroxydiones and their derivatives 507
4. Trihydroxycarboxylic acids 508
5. Dihydroxyalkanedicarboxylic acids 510
6. Dioxodicarboxylic acids 516
7. Tetracarboxylic acids 516
(a) Alkanetetracarboxylic acids 516
(b) Alkenetetracarboxylic acids 517

Guide to the Index 519
Index 521

List of Common Abbreviations and Symbols Used

A	acid
Å	Ångström units
Ac	acetyl
a	axial
as, asymm.	asymmetrical
at.	atmosphere
B	base
Bu	butyl
b.p.	boiling point
c, C	concentration
CD	circular dichroism
conc.	concentrated
D	Debye unit, 1×10^{-18} e.s.u.
D	dissociation energy
D	dextro-rotatory; dextro configuration
d	density
dec., decomp	with decomposition
deriv.	derivative
E	energy; extinction; electromeric effect
*E*1, *E*2	uni- and bi-molecular elimination mechanisms
E1cB	unimolecular elimination in conjugate base
ESR	electron spin resonance
Et	ethyl
e	nuclear charge; equatorial
f.p.	freezing point
G	free energy
GLC	gas liquid chromatography
g	spectroscopic splitting factor, 2.0023
H	applied magnetic field; heat content
h	Planck's constant
Hz	hertz
I	spin quantum number; intensity; inductive effect
IR	infrared
J	coupling constant in NMR spectra
J	Joule
K	dissociation constant
k	Boltzmann constant; velocity constant
kcal	kilocalories
M	molecular weight; molar; mesomeric effect
Me	methyl
m	mass; mole; molecule; *meta-*
m.p.	melting point
Ms	mesyl (methanesulphonyl)

[M]	molecular rotation
N	Avogadro number; normal
NMR	nuclear magnetic resonance
NOE	Nuclear Overhauser Effect
n	normal; refractive index; principal quantum number
o	*ortho-*
ORD	optical rotatory dispersion
P	polarisation; probability; orbital state
Pr	propyl
Ph	phenyl
p	*para-*; orbital
PMR	proton magnetic resonance
R	clockwise configuration
S	counterclockwise configuration; entropy; net spin of incompleted electronic shells; orbital state
S_N1, S_N2	uni- and bi-molecular nucleophilic substitution mechanism
S_Ni	internal nucleophilic substitution mechanism
s	symmetrical; orbital
sec	secondary
soln.	solution
symm.	symmetrical
T	absolute temperature
Tosyl	*p*-toluenesulphonyl
Trityl	triphenylmethyl
t	time
temp.	temperature (in degrees centrigrade)
tert	tertiary
UV	ultraviolet
α	optical rotation (in water unless otherwise stated)
$[\alpha]$	specific optical rotation
ϵ	dielectric constant; extinction coefficient
μ	dipole moment; magnetic moment
μ_B	Bohr magneton
μg	microgram
μm	micrometer
λ	wavelength
ν	frequency; wave number
χ, χ_d, χ_μ	magnetic; diamagnetic and paramagnetic susceptibilities
(+)	dextrorotatory
(–)	laevorotatory
–	negative charge
+	positive charge

Second Supplements to the 2nd Edition of Rodd's Chemistry of Carbon Compounds, Vol. IE/F/G, edited by M. Sainsbury

Chapter 18

TRIHYDRIC ALCOHOLS, THEIR ANALOGUES AND DERIVATIVES AND THEIR OXIDATION PRODUCTS: TRIHYDRIC ALCOHOLS TO TRIKETONES.

B.J.COFFIN

Glycerol (1,2,3-trihydroxypropane, propan-1,2,3-triol), $CH_2(OH)CH(OH)CH_2OH$, continues to be a compound of importance and interest, and it and many of its related compounds are components of many substances of commercial and biological use.

(1) Preparation of glycerol

Its synthesis and production has continued to receive attention.

(1) Glycerol is made by the hydrolysis of glycerides at low temperatures by contacting their solutions with strongly basic ion exchange resins (Y.Takagi *et al.*, J. Pat., 60,163,832, 1984).

(2) Castor oil and cotton seed oil is hydrolysed with castor lipase to yield the corresponding fatty acids and glycerol (H.S.Shanker *et al.*, Front. Chem. React. Eng., 1984, 1, 499).

(3) The production of glycerol is achieved by use of the cells of *S. cerevisiae* immobilised in the sintered glass Raschig rings in a fermentation process (B.Bisping and H.J.Relim, Appl. Microbiol. Biotechnology, 1986, 23, 174).

Aureobasidium SW-45, freshly isolated from starch waste, produces erythritol and glycerol from glucose (T.Harumi *et al.*, J. Pat., 6131091, 1984).

(4) Crude animal fats, when undergoing purification by vacuum distillation rather than by conventional methods, give glycerol in high purity (C.Schrieber *et al.,* Ger. Pat., 228,973, 1985).

(5) Glycerol of high purity is recovered from high pressure cleavage products by using falling-film counter-current evaporators,

a rotational evaporator, a rectification column, coolers, phase separation apparatus and a filter (H.Gutwasser *et al.,* Ger. Pat., 228805, 1985).

(6) Glycerol and the fatty acids of C1-5 primary and secondary alcohols are made by transesterification of liquid glycerides with these alcohols by being brought into intimate contact in the presence of catalysts at about 210^{0} (S.Billenstein *et al.*, Ger. Pat., 3,421,217, 1985).

(7) Glycerol is manufactured from glycerides and ammonia using zinc salts as catalysts (H.Stuehler and K.Fischer, Ger. Pat., 880601, 1985).

(8) The epoxidation of allyl alcohol with aqueous hydrogen peroxide in the presence of a powdered heterogeneous catalyst based on an alkaline earth metal tungstate yields glycidol which is hydrolysed to glycerol (A.A.Grigor'ev *et al.*, Neftekhimiya, 1979, 19, 803).

(9) The formation of glycerol occurs when a solution of DL-glyceraldehyde is heated in the presence of hydrogen sulphide at room temperature. Also DL-glyceraldehyde and dihydroxyacetone react with hydrazine to give glycerol (D.E.Epps *et al.,* J. Mol. Evol., 1979, 14, 235).

(10) Glycerol is produced by using osmophilic yeasts with a medium containing glucose. The production of erythritol is suppressed by the addition of phosphates (N.Ramachandran and G.Sulebele, Indian J. Microbiol., 1979, 19, 136). The production of glycerol is also achieved from glucose under aerobic conditions using the cells of *P. farinosa* immobilised in calcium alginate (P.Vijaikishore and N.G.Karanth, Biotechnol. Lett., 1986, 8, 257). It is also made by using the yeast cells, *Saccharomyces cerevisiae* immobilised in calcium alginate beads and in polyacrylamide hydrazine lentils (B.Bisping and H.J.Rehm, Eur. Congr. Biotechnol., 3rd., 1984, 2, 125).

(11) Significant amounts of glycerol are obtained when the wall-less marine alga, *D. tertiolecta*, is immobilised and grown within a culture medium of 4M sodium chloride on calcium alginate beads (D.Grizeau and J.M.Navarro, Biotechnol. Lett., 1986, 8, 261).

(12) The sequential nitrosation and sodium cyanoborohydride reduction of 2-amino-2-deoxy-D-galactitol yields glycerol as the major product (S.R.Carter *et al.*, J. Chem. Soc. Perkin Trans. 1, 1985, 2775).

(13) Glycerol is produced from methanol by incubation with a

mutant MCI 1976 (FERM P-7954) of *H. polymorpha CBS 4732* (S.Chilahiro and K.Nobuo, J.Pat., 61,162,190, 1986).

(14) The synthesis of glycerol is achieved by using *Aspergillus niger* under citric acid accumulating conditions (M.Legisa and M.Mattey, Enzyme Microbiol. Technol., 1986, 8, 607).

(15) Glycols and glycerol is produced by the hydrogenolysis of alditols in the presence of a catalyst (J.Barbier *et al.*, Fr. Pat., 880304, 1986).

(16) Glycerol is manufactured from hydrogen, carbon monoxide in the presence of rhodium catalysts and trialkylamines (T.Masuda *et al.*, J. Pat., 890110, 1985).

(17) The reductive condensation of formaldehyde in the presence of ruthenium catalysts and benzimidazolium salts yields glycerol (K.Murata *et al.*, J. Pat., 880723, 1985).

(2) Reactions of glycerol

Mellitic acid is formed from glycerol when it undergoes oxidative degradation. Other condensation and cyclisation reactions occur at the same time to yield a variety of products (E.S.Rudakov *et al.*, Dokl. Akad. Nauk. SSSR., 1986, 287, 389).

Glycerols undergo *retro* Claisen condensations in the presence of hydrogen and a copper catalyst (C.Montassier *et al.*, Bull. Soc. Chim. Fr., 1989, 2, 148).

The regioselective de-*O*-benzylation of poly-*O*-benzylated polyols is achieved by use of Lewis acids, such as tin tetrachloride or titanium tetrachloride (H.Hori *et al.*, J. Org. Chem., 1989, 1346). The reaction of a bicyclophosphane with glycol, glycerol, erythritol, L-(-)-arbitol and dulcitol gives the corresponding *per*(alkoxybicyclophosphoranes) which are in tautomeric equilibrium with their phosphite forms (D.Houalla *et al.*, Tetrahedron Lett., 1985, 26, 2003):

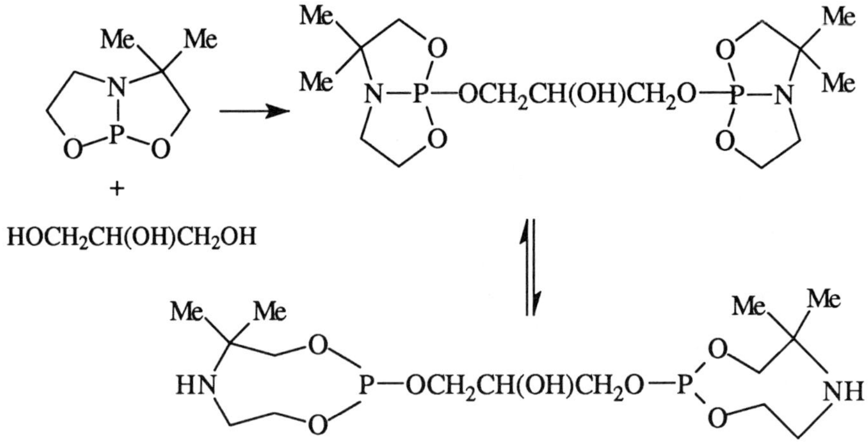

$HOCH_2CH(OH)CH_2OH$

N-Substituted-2,3-unsaturated carboxamides are prepared by the reaction of a *N*-unsubstituted-2,3-unsaturated carboxamide with glycerol under basic catalysis according to the Michael addition. Transamidination of the generated Michael adduct with a primary or secondary amino-group undergoes ammonia elimination to give the *N*-substituted carboxamide. On subsequent pyrolytic, alcoholic cleavage the *N*-substituted carboxamide is formed with a protected double bond (K.Dahmen *et al..*, Eur. Pat., 151967, 1985).

2,3-Dihydroxypropyl*trans*-9-(4-methoxy-2,3,6-trimethyl-phenyl)-3,7-dimethyl-2,4,6,8-nonatetraenoate, which has the properties of a dermatol agent, is prepared from the relevant acid chloride by stirring with glycerol in the presence of sodium glycerate or pyridine (J.Smrt, Czech. CS216,979, 1985):

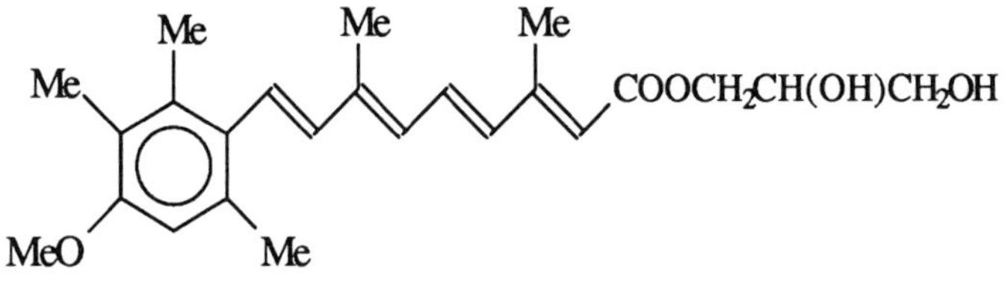

Polyglycerols are prepared by the addition of glycerol to glycidol over a sulphonated ion-exchange cation resin at 110-120° . Thus, glycerol and glycidol give a mixture of diglycerol (12.3%), triglycerol (17.1%), tetraglycerol (16.4%), pentaglycerol (20.7%) (S.Uda and E.Takemato, J. Pat., 61 140534, 1986). Glycerol undergoes polycondensation by the etherification of epichlorohydrin (G.Jakobson and W.Siemanowski, Eur. Pat. Appl., 890927). Glycerol also undergoes polycondensation using phthalic anhydride in the laboratory preparation of glyptal (Anon., Bull. Union Physiciens, 1989, 83, 662).

L-2-Glycerophosphate is formed from glycerol by use of a glycerol kinase in the presence of ATP (S.Imamura *et al.*, Patentschrift (Switz.) CH654024, 1986).

Phospholipid probes can be made by the reaction of phosphatidylcholine derivatives with glycerol and serine (Yu. Molotkovskii *et al.,* Bioorg. Khim., 1989, 15, 686).

Glycerol trivinyl ether, 1,2-*bis*(vinyloxy)propane, is prepared in a good yield by reacting ethyne and glycerol at 120 -140° and 10-12 atmos. pressure in dimethylsulphoxide in the presence of sodium hydroxide (B.A.Trofimov *et al.*, Otkrytiya Izobret, 1985, 81).

The reaction between glycerol and isobutyraldehyde in the presence of potassium sulphate in toluene at 80-135° gives a mixture containing 4-hydroxymethyl-2-isopropyl-1,3-dioxolane and 5-hydroxy-2-isopropyl-1,3-dioxane which can be used in lacquers and paints (A.I.S.Blaga *et al.*, Rom. 63,233, 1978). The reaction of bromoacetaldehyde diethyl acetal gives the corresponding 1,3-dioxolane and 1,3-dioxane moieties for use in acrylated phosphonate polymers (S.K.Sahni and I.Cabasso, J. Polym. Sci., Part A: Polym. Chem., 1988, 26, 3251).

Glycerol reacts with gallic acid to give 1-glycerol gallate which exhibits antioxidant properties (Z.Song and Z.Xiao, Linchan Huaxue Yu Gongye, 1988, 8, 9).

Acylation of a meso diol using a homochiral amine as catalyst affords a monoacyl derivative having only a modest enantiomeric excess (L.Duhamel and T.Herman, Tetrahedron Lett., 1985, 26, 3099). The two hydroxyl groups of a meso diol can, however, be more effectively differentiated by reaction with one equivalent of a homochiral acyl halide after prior treatment with dibutyltin oxide (T.Mukaiyama, I.Tomioka and M.Shimizu, Chem. Lett., 1984, 49):

MeOOC OH (i) → MeOOC SnBu$_2$ MeOOC OH MeOOC

(ii) → MeOOC OH MeOOC OCOR*

Reagents: (i) Bu_2SnO; (ii) R*COOH

The same procedure is also used to prepare homochiral derivatives from glycerol by differentiating between the two primary hydroxyl groups (T.Mukaiyama, Y.Tanabe and M.Shimizu, Chem. Lett., 1984, 401):

OH OH OH (i),(ii) → OCOR* OH OH (iii),(iv) → OH O O

Reagents: (i) Bu_2SnO; (ii) R*COOH; (iii) $Me_2C(OMe)_2/H^+$; (iv) OH^-/H_2O.

The ketalisation reaction of nonadecanone with glycerol is used in the preparation and study of PAF antagonists (C.Broquet and P.Braquet, Ger. Pat., 881222, 1986).The acid-catalysed cyclocondensation of glycerol with $MeCOCHMeCH_2SR$ (R = Et, Pr) in toluene gives dioxolanes which then undergo a Mannich reaction with diethylamine and formaldehyde to give the compound (S.S.Sabirov *et al.*, Khim-Farm.Zh.,1985, 19,1453):

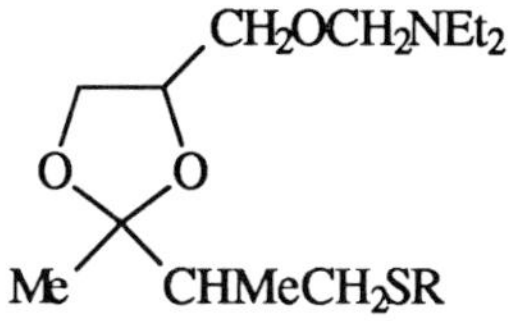

$CH_2OCH_2NEt_2$

O O

Me $CHMeCH_2SR$

Glycerol is used as the source of liquid crystalline Schiff bases using bromobenzaldehyde and amines as co-reactants (M.M.Murza and M.G.Safarov, Izv. Vyssh. Uchebn. Zaved., Khim. Khim. Tekhnol., 1988, 31, 105).

Glycerol, 1-thioglycerol and *meso*-erythritol undergo cyclo condensation with 2-hydroxy ketones to give bicyclic and tricyclic ethers (I.R.Fjeldskaar *et al.*, Acta Chem. Scand., Ser. B, 1988, B42, 280).

Allyl alcohol is formed by treating glycerol with ethyl orthoformate in the presence of benzoic acid at 160-230^{0} (I.Ya Kibina *et al.*, USSR Pat. 706394, 1979).

The Skraup reaction is an important route to quinolines; it uses glycerol as a starting material and has been used recently to obtain nitroquinolines, and from them 2-mercapto-1*H*-imidazo[4,5-*f*]quinolines (E.Surender *et al.*, Phosphorus Sulphur, 1988, 35, 267).

Glycerol can be used in the preparation of quinoline derivatives which can be used as improved chloride-sensitive fluorescent indicators for biological applications (A.S.Verkman *et al.*, Anal. Biochem., 1989, 178, 355).

The effect of the 4-substituent of the 1,3-dioxolane ring in trinitro cyclohexadienate *spiro* Meisenheimer complexes has been studied by measurement of their relative stability and regioselectivity when undergoing decomposition in acidic media. This involves the reaction of mercaptopropanediol with picryl fluoride, and the measurement of the NMR spectra of the products (V.N.Knyazev *et al.*, Zh. Org. Khim., 1988, 24, 2183).

5-Methyl-4-oxo-3,6,8-trioxabicyclo[3.2.1]octane is made by heating glycerol and 2-methoxycarbonyl-2-methyl-1,3-dioxolane in refluxing benzene in the presence of an acidic catalyst (D.L.Rakhmankulov *et al.*, USSR Pat., 722,912, 1980).

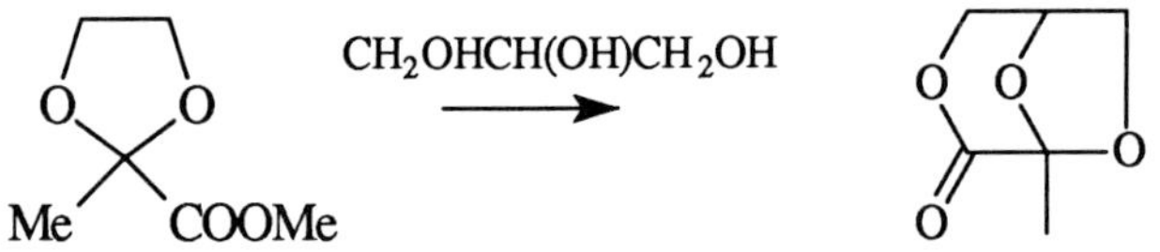

$CH_2OHCH(OH)CH_2OH$

Me COOMe

The acid-catalysed condensation of acrolein and glycerol yields *cis*-4-hydroxymethyl-2-vinyl-1,3-dioxolane (27.4%), its *trans* form (22.7%), *cis*-5-hydroxy-2-vinyl-1,3-dioxolane (22.5%) and its *trans* form (12.8%) (J.S.Shim *et al.*, Pollins, 1980, 4, 216).

Poly(oxypropylene)polyols are prepared by the polyaddition of propylene oxide to glycerol in potassium glycerate solution (H.Knopp *et al.*, Vysokomol. Soedin., Ser. A, 1980, 22, 1788).

Methylidinoglycerol is obtained as a mixture of 5-hydroxy-1,3-dioxane and 4-hydroxymethyl-1,3-dioxolane by condensing glycerol with an equimolar amount of paraformaldehyde in the presence of an acid catalyst (S.A.Calipe, Spain. Pat. 475,962, 1979).

Surface active organic boron compounds are made by the esterification of $B(OR)_3$ (where R = a lower alkyl group) or boric acid with polyhydric alcohols, followed by further esterification with C8-32 acyl halides. Thus, when one mole of boric acid is heated with two moles of glycerol in a nitrogen atmosphere at 180-210° with the loss of water it gives the boric acid triester which is then treated with octanoyl chloride to give the product (Toho Chemical Industry Co., Ltd., J. Pat., 80 15480, 1984):

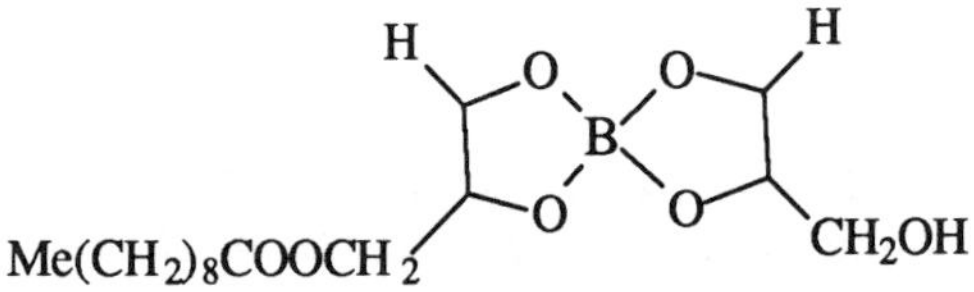

$Me(CH_2)_8COOCH_2$ CH_2OH

The reaction of glycerol mono- and dihalohydrins with *N,N*-diethyl-2,2,3,3,3-pentafluoropropylamine gives either a propyl ester or a cyclic product. Thus, 2,3-dibromo-1-propanol gives 52% of 2,3-dibromopropyl-1,2,2,2-tetrafluoropropionate, and 53% of the aminodioxane:

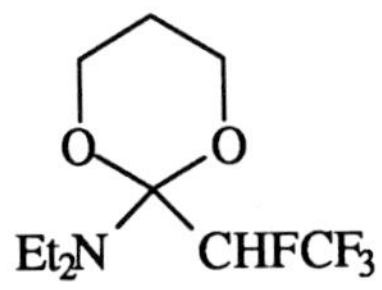

Et_2N $CHFCF_3$

(S.Watanabe *et al.,* Nippon Kagaku Kaishi, 1985, 2191).

Monoglycerides are obtained by glycerolysis of coconut oil in the presence of glycerol and sodium hydroxide as the catalyst (M.T.Tran and L.T.Tran, Tap Chi Hoa Hoc, 1985, 23, 25).

Glycerol reacts with 1,2-benzanthraquinone in 80% sulphuric acid to give mainly 7H-dibenzo[a,k,l]anthracen-7-one together with the isomeric 7H-benzo[h,i]chrysen-7-one and 13H-dibenz[a,d,e] anthracen-13-one (S.Fujisawa *et al.,* Bull. Chem. Soc. Jpn., 1985, 58, 3356):

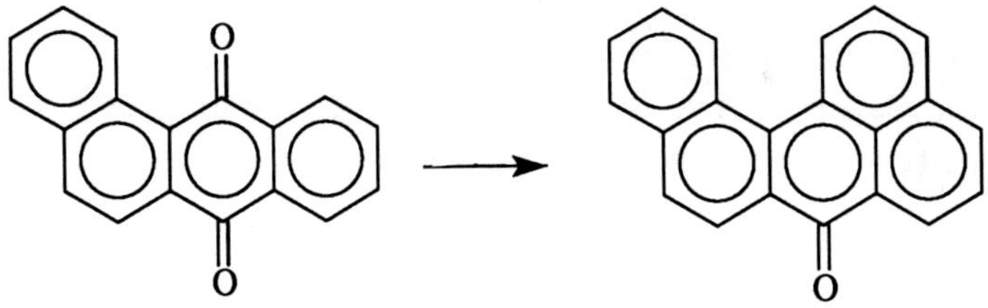

Glycerol on heating with 4-(2-methylpropyl)phenyl 2-chloropropyl ketone and thionyl chloride gives a mixture of two ketals.

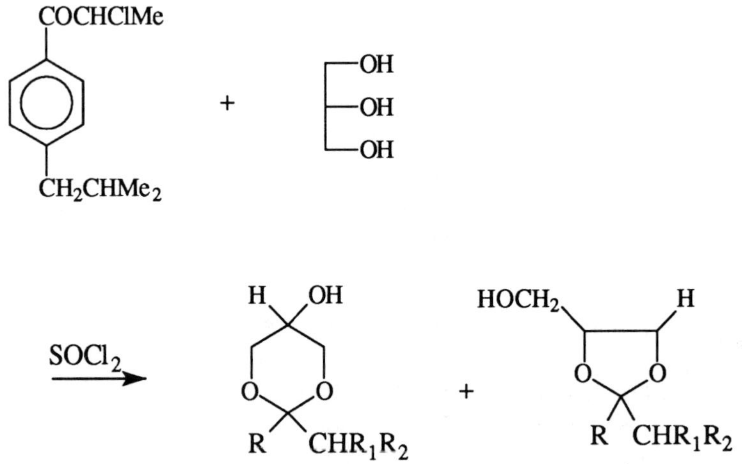

COCHClMe

CH_2CHMe_2

+ OH OH OH

$\xrightarrow{SOCl_2}$

H OH

O O

R CHR_1R_2

+

$HOCH_2$ H

O O

R CHR_1R_2

where R = 4-$Me_2CHCH_2C_6H_4$, R_1 = Cl, R_2 = Me (S.P.A.Zambon, Eur. Pat. Appl.EP 128,580, 1986). This ketal mixture, when heated in glycerol containing potassium acetate, for 17 hours at 185°, yields ibuprofen, $4Me_2CHCH_2C_6H_4CH(Me)COOH$.

The ethers, $Me(CH_2)_nOMe$, (n = 0 - 3) are reacted with glycerol and methyl oxirane to give the dioxolane derivatives:

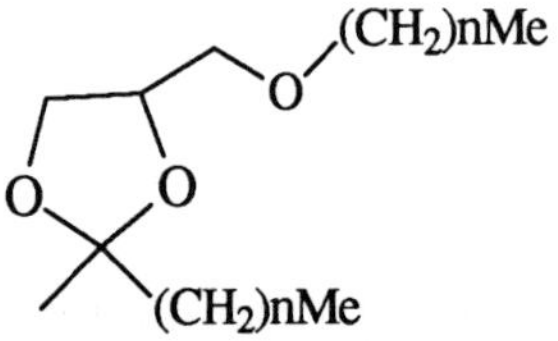

$(CH_2)nMe$

O

O O

$(CH_2)nMe$

(A.Piasecki, Pol. J. Chem., 1984, 58, 1215).

Glycerol reacts in a 1:1 mixture with phosphorus pentachloride to give the products:

$(ClCH_2)_2CHOP(O)CH_2Cl$ and $ClCH_2CHClCH_2OP(O)Cl_2$

(B.A.Arbuzov *et al*., Zh. Oshch. Khim., 1985, 55, 1866)

The reaction of glycerol with 2,4,6-tri-*tert.*-butylphenyldithiophosphorane has been reported (J.Navech *et al*., Phosphorus Sulphur, 1988, 39, 33).

The monomers and polymers of acrylated phosphonate esters are made from diethyl diethoxyethylphosphonate and glycerol, D-mannitol, D-sorbitol, pentaerythritol and dipentaerythritol (I.Cabasso *et al*., J. Polym. Sci., Part A: Polym. Chem., 1988, 26, 2997).

(3) Use of chiral synthons derived from glycerol in synthesis

Glycerol units have found use in synthesis as chiral building blocks - chirons- or chiral synthons (S.Hanessian, Total Synthesis of Natural Products: The Chiron Approach, Pergamon Press, Oxford, 1983). These are amongst the smallest chiral units and are readily available in both antipodal forms from carbohydrates. Glycerol units are accessible through asymmetric syntheses, using, for example, the Sharpless epoxidation procedure.

(a) Chiral γ-lactones

The synthesis of a number of natural products is achieved by use "glycerol units" as chiral synthons. These are obtained by the generation of γ-lactone intermediates. An example, where the chirality at C-2 in glycerol can be utilised in synthesis is in the construction of γ-lactones. These products are themselves utilised in the synthesis of several important groups of natural products including alkaloids and terpenoids. Thus, the chiral centre at C-5 in the lactone (11) derives from D-mannitol diacetonide (1) (S.Takano *et al.,* Heterocycles, 1981, 16, 381; Synthesis, 1983, 116; 1985, 503). The starting material is cleaved by treatment with aqueous periodate to glyceraldehyde acetonide (2), which is unstable. On a large scale run the acetonide (2) is not isolated, but reduced directly with sodium borohydride affording glycerol acetonide (3). Protection as the benzyl ether (4) is followed by deacetonisation which affords R-glycerol-1-benzyl ether (5).

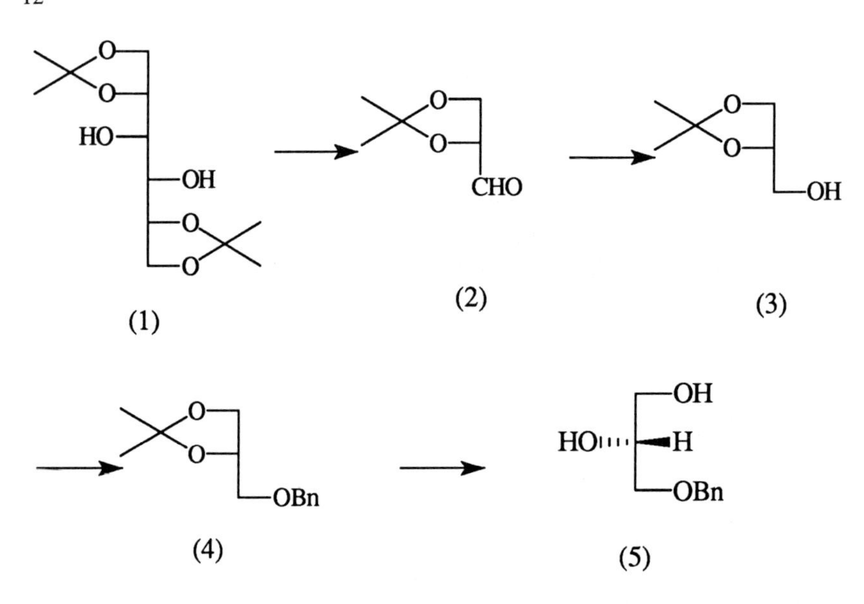

The chirality of R-glycerol benzyl ether (5) can be inversed by treating its dimesylate with potassium acetate in hot acetic anhydride solution followed by methanolysis (S.Takano *et al.,* Tetrahedron Lett., 1983, 24, 4233):

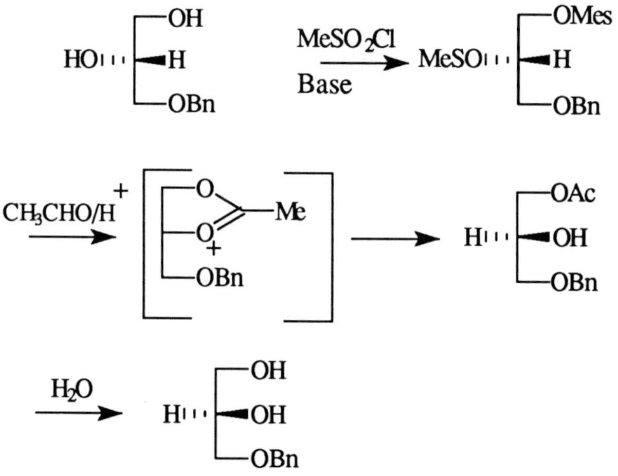

(6)

The S-glycerol benzyl ether (6), as its benzalideneacetal (7), can be brominated with *N*-bromosuccinimide and ring-opened to afford the *O*-benzoylbromohydrin (8). This yields glycidolbenzyl ether (9) when it is hydrolysed by treatment with sodium hydroxide.

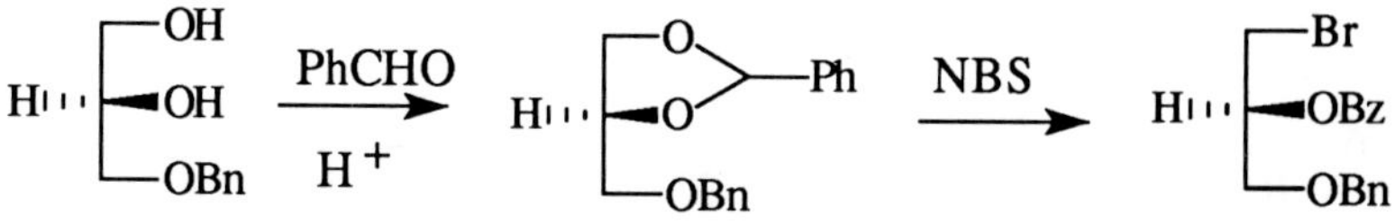

(6) (7) (8)

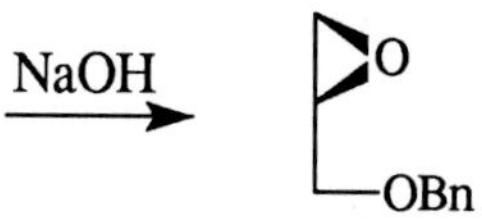

(9)

The glycidol benzyl ether (9) is then condensed with the enolate of an alkanoate ester to yield the α-substituted-γ-lactone (10) as a *syn-* and *anti-* mixture.

O

OBn

(9)

RCH_2COOR'

H

R

COOH

OH

OBn

$-H_2O$

H

R

O

O

OBn

(10)

This is then converted into either the *syn*-epimer (11) by protonation, or it may be enolised by treatment with base and alkylated at C-3. In this case the relative size of the incoming group R will dictate the stereochemical outcome of the reaction leading to (12).

R

B

O

O

OBn

(11)

(10)

B

R

O

O^-

OBn

R'X

R

R'

O

O

OBn

(12)

Clearly the R-isomer of glycerol benzyl ether (5) can be utilised to afford isomers of these lactones having the opposite stereochemistry at C-5 using similar chemical procedures.

In more recent work Takano *et al.,* (Synthesis, 1986, 403) have shown that glyceraldehyde acetal (2), when reacted with ethyl triethoxyphosphonoacetate in the presence of aqueous potassium carbonate, affords the α,β-unsaturated ester (13) which, after reduction, deacetalisation and hydrolysis, yields the lactone (14).

$(EtO)_3PCHCOOEt$

CHO

COOEt

(2) (13)

H_2/Pt H^+/H_2O

HO HO COOEt COOEt H O O OH

(14)

The inversion of chirality of this lactone is easily achieved (S.Takano *et al.*, Synthesis, 1981, 265; P.-T.Ho and N.Davies, Synthesis, 1983, 462; S.Takano *et al.*, Heterocycles, 1985, 23, 2347).

The unsaturated lactone (16) is available from the ester (13) *via* its phenylsulphide (14). This compound upon deacetalisation cyclises to the lactone (15) which is dehydrosulphurised by oxidation to the corresponding sulphoxide/sulphone and elimination of the appropriate sulphenic/suphinic acid.

(13) $\xrightarrow{PhSH}$ (14) $\xrightarrow{H^+/H_2O}$

(b) The synthesis heteroyohimbine alkaloid precursors.

Glyceraldehyde is also formally a precursor of the *cis*- and *trans*-piperidinolactones (17) and (18) which can be used in the synthesis of heteroyohomobine alkaloids (M.Uskokovic *et al.*, J. Amer. Chem. Soc., 1979, 101, 6742).

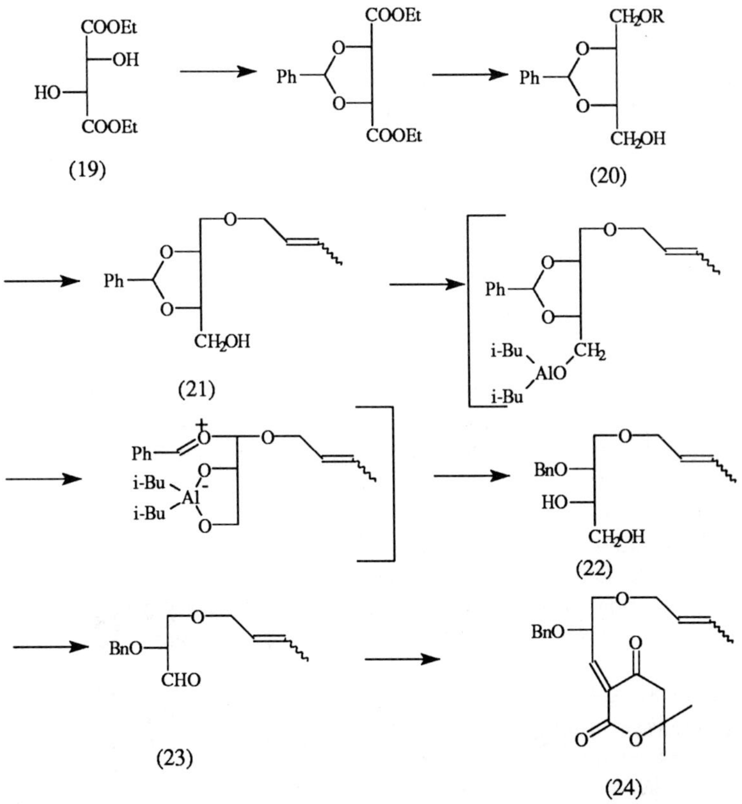

COOEt OH HO COOEt (19) → Ph O O COOEt COOEt → Ph O O CH$_2$OR CH$_2$OH (20)

→ Ph O O O CH$_2$OH (21) → [Ph O O O i-Bu i-Bu AlO CH$_2$

→ Ph $\overset{+}{O}$ O O i-Bu i-Bu Al$^-$ O] → BnO HO O CH$_2$OH (22)

→ BnO O CHO (23) → BnO O O O O (24)

Scheme 1

Glyceraldehyde itself cannot be used directly, because of its instability, but (L)-diethyl tartrate (19) is an "equivalent" which, after acetalation with benzaldehyde and reduction with lithium aluminium hydride, affords the diol (20, R = H). On reaction with crotonyl chloride the diol yields the monocrotonate (20, R = crotyl) which can be further reduced to give the butenyl ether (21). When this is heated with diisobutylaluminium hydride the reagent binds first to the free hydroxyl group allowing a selective cleavage of the acetal group to occur. The product is then the *O*-benzyldiol (22) **(Scheme 1)**.

Both *E* and *Z* isomers about the 2-butenyl group have been obtained. Cleavage to the functionalised and deprotected glyceraldehydes (23) is achieved through reactions with aqueous sodium periodate. These products condense with Meldrum's acid to give the dioxolines (24), which are processed through many further steps to give the desired piperidinolactones.

Glyceraldehyde derivatives are used in the synthesis of unsaturated trihydroxy C-18 fatty acids such as trihydroxy octadecadienoic and octadecenoic acids previously isolated from rice plants suffering from rice blast disease (H.Suemune *et al.*, Chem. Pharm. Bull., 1988, 36, 3632).

The iterative stereoselective homologation of chiral polyalkoxy aldehydes employing 2-(trimethylsilyl)thiazole as a formyl anion equivalent involves the reaction of glyceraldehyde derivatives with butyllithium in the synthesis of higher carbohydrates (A.Dondoni *et al.*, J. Org. Chem., 1989, 54, 693).

The synthesis of 1-(-D-glucopyranosyl)-1-deoxy-D-fructose, a non-metabolisable analogue of sucrose, is achieved by the condensation of glyceraldehyde derivatives with sugar ylide derivatives (M.Carcano *et al.*, J. Chem. Soc. Chem. Commun., 1989, 642).

Glyceraldehyde acetonide is used in an intramolecular Diels-Alder approach to the synthesis of nargenicins and involves the boatlike transition states in the cyclisations of substituted 1,7,9-decatrien-3-ones derived by lithiation and reaction with dihydroxy tetradecatrienoate (J.W.Coe and W.R.Roush, J. Org. Chem., 1989, 54, 915).

Methyl α-acyloxy-γ-methylene-β-tetronate is used for the synthesis of the spirotetronic acid structure in chlorothricolide involving the use of a glyceraldehyde derivative and a Diels-Alder reaction with ethylidene heptadienediol (K.Okumura *et al.*, Tetrahedron Lett., 1989, 30, 2233).

The facial selectivity of the 1,3-dipolar cycloaddition of azomethine ylide to homochiral dipolarophiles has been studied by the Wittig reaction of glyceraldehyde with methyl triphenylphosphoranylidene acetate (A.G.H.Wee, J. Chem. Soc. Perkin Trans. I, 1989, 1363).

The total synthesis of 3'-azido-3'-deoxythymidine and 3'-2',3'-dideoxyuridine has been reported involving the use of the Wittig reaction of relevant precursors with a suitable glyceraldehyde derivative derived from D-mannitol by a conventional route (C.K.Chu *et al.*, Tetrahedron Lett., 1988, 29, 5349).

The aldol reaction of iodofluoroacetate-zinc and 2,2-difluoroketene silyl acetal involves the preparation and reaction of a glyceraldehyde derivative (O.Kitagawa *et al.*, Tetrahedron Lett., 1988, 29, 1803).

The asymmetric induction in the (2,3) Wittig rearrangement of allylic ethers with a chiral substituent has been studied using glyceraldehyde allylic ether prepared asymmetrically *via* a series of conventional steps (E.Nakai and T.Nakai, Tetrahedron Lett., 1988, 29, 4587).

The observation of diastereofacial selectivity in thermal Wittig reactions has been observed, for example, in the reactions between alkoxystannes and glyceraldehyde acetal (S.V.Mortlock and E.J.Thomas, Tetrahedron Lett., 1988, 29, 2479).

A synthesis of naturally occurring 5-methylcoumarins uses the reaction of glyceraldehyde diethyl acetal silyl ether as one of the synthetic materials (C.Wolfrum and F.Bohlmann, Liebigs Ann. Chem., 1989, 295).

L-Glyceraldehyde is used as a chiral synthon in a Horner Wittig reaction in the synthesis of deltamethrin, a biologically active insecticidal pyrethrin (A.Krief and W.Dumont, Tetrahedron Lett., 1988, 29, 1083).

2-*O*-Benzylglyceraldehyde has been isolated in both its enantiomeric forms. These may be used in Wittig reactions for synthetic purposes as configurationally stable building blocks (V.Jaeger and V.Wehner, Angew. Chem., 1989, 512).

A stereocontrolled cyclopentenone synthesis involves the cyclo condensation reactions of glyceraldehyde acetonide using palladium complex catalysts (B.M.Trost *et al.*, J. Amer. Chem. Soc., 1989, 111, 7487).

R-Glyceraldehyde acetonide has been employed in the asymmetric syntheses of 3-substituted 3,4-dihydro-1(2H)isoquinolines (R.D.Clark *et al.,* J. Chem. Soc. Chem. Comm., 1989, 930).

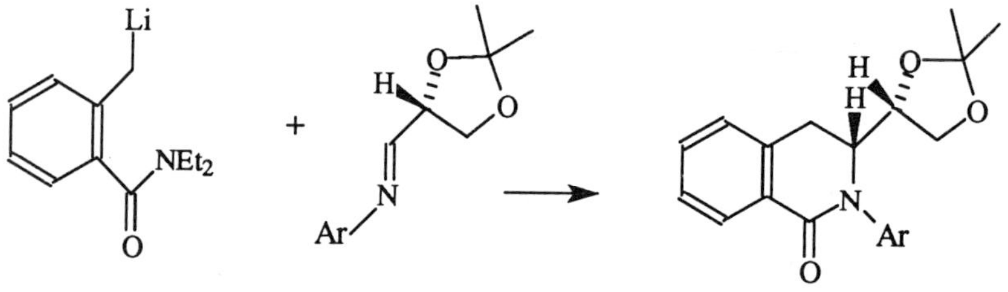

$Ar = p\text{-}MeOC_6H_4\text{-}$

The stereochemistry of the reaction can be rationalised by a Cram-type model of asymmetric induction in which the lithium atom of the amide is transferred to the N atom of the imine and is then co-ordinated to the β-oxygen atom of the acetal.

This is unusual since the 5-membered chelation of lithium is not normally observed (P.A.Wade *et al.*, J. Org. Chem., 1985, 50, 2804).

(4) Glycerides

The availability of mixed triglycerides of defined structure and configuration is important for the closer investigation of the hydrolytic action of lipases. The general synthesis of triglycerides requires that the product can be obtained in high yield and chemical purity. Any degree of substitution (one, two or three groups) should be obtainable and the positions of the substituents needs to be clearly defined. Using the standard *sn*-nomenclature, this implies enantiomeric purity of the product, since inversion at the optically active carbon exchanges positions 1 and 3.

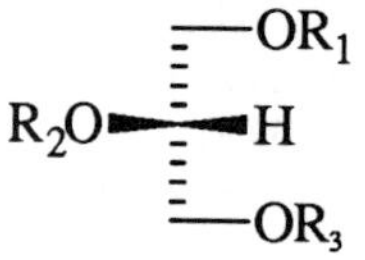

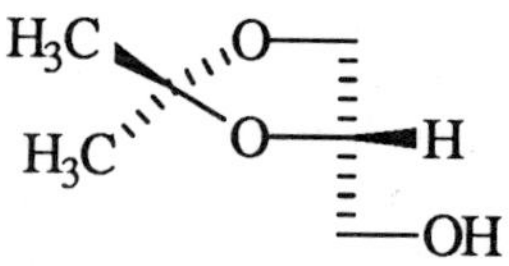

Any synthetic approach needs to start with optically pure precursors and the migration of the substituents and racemisation

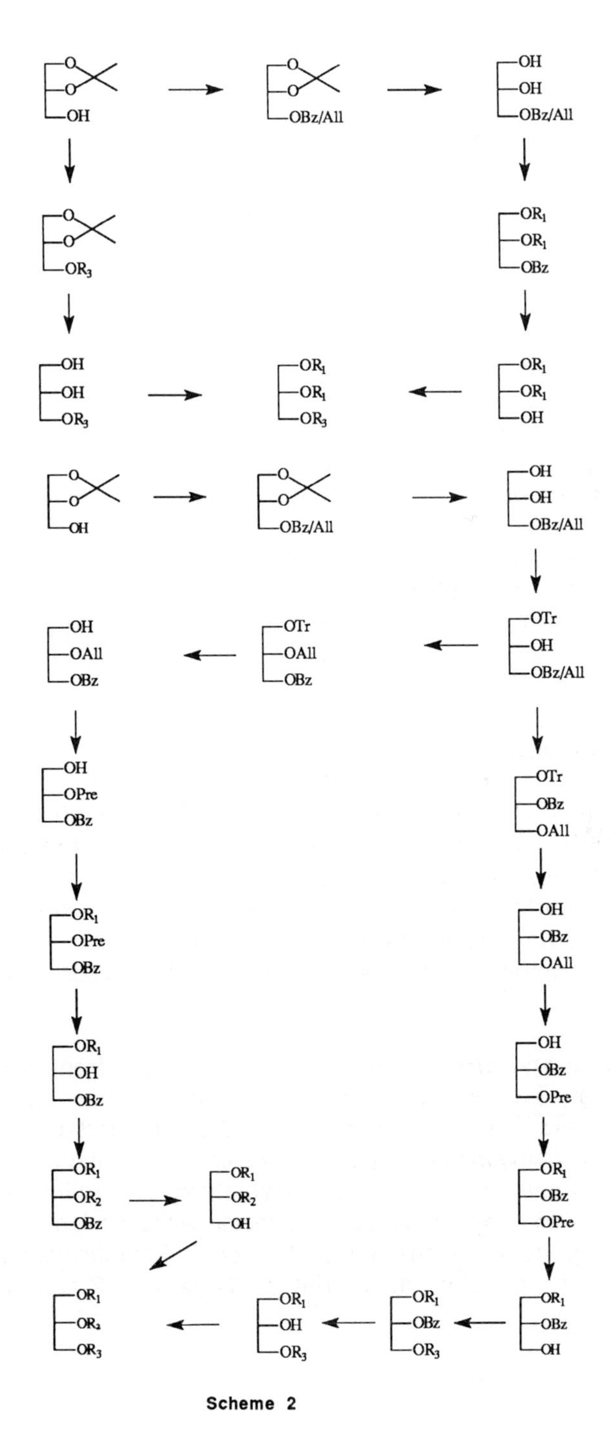

Scheme 2

Strategy for the preparation of glyceryl di- and tri-esters.
Abreviations: Bz = benzyl; All = allyl; Tr = trityl; Pre = propenyl.

needs to be rigorously avoided. The optical purity of the starting material can be ensured by using monosaccharides such as D-mannitol as the starting materials. Reaction conditions need to be chosen to avoid the possibility of racemisation. Depending on the product required, the D-mannitol is first converted into 1,2-isopropylidene-*sn*-glycerol (H.Eibl, Chem. Phys. Lipids, 1981, 28, 1; G.Hirth and W.Walther, Helv. Chim. Acta, 1985, 68, 1863), or into 3,4-isopropylidene-D-mannitol. The required glyceryl derivative can then be built up by a series of selective modification reactions and hydrolysis steps. Using this procedure it is possible to prepare combinations of mixed chain or uniform chain natural and unnatural glyceryl componds as outlined in **Scheme 2** (H.Eibl, Chem. Phys. Lipids, 1980, 26, 405).

3-Benzyl-sn-glycerol can be prepared from 1,2-isopropylidene-*sn*-glycerol by reaction with benzyl chloride in the presence of potassium *tert.*-butylate in *tert.*-butanol (H.Eibl, Chem. Phys. Lipids, 1980, 26, 405). The product is an oil, b.p. 145-147^{0}/ 1mm., [α] +6.5^{0} (M.Kates, T.H.Chan and N.Z.Stanazev, Biochemistry, 1963, 2, 394; G.W.Hirth and W.Walther, Helv. Chim. Acta, 1985, 68, 1863).

3-Allyl-sn-glycerol (b.p. 88^{0}/1mm., [α] +6.4^{0}) is similarly prepared using instead allyl bromide.

1-Trityl-2-allyl-3-benzyl-sn-glycerol is made from 3-benzyl-*sn*-glycerol by reaction with trityl chloride in the presence of triethylamine. The trityl protecting group can be removed by reaction with hydrochloric acid to provide 2-allyl-3-benzyl-*sn*-glycerol, [α] +6.0^{0}. Other milder methods for the detritylation have been reported (A.Hermetter and F.Paltauf, Chem. Phys. Lipids, 1981, 29, 191).

1-Stearoyl-2-myristoyl-sn-glycerol is obtained from 2-allyl-3-benzyl-*sn*-glycerol by firstly converting it into the 2-propenyl derivative and rearrangement into 2-propenyl-3-benzyl-*sn*-glycerol. This compound can then be esterified with stearic acid by use of 4-dimethylaminopyridine and dicyclohexylcarbodiimide. The removal of the propenyl group can be effected without migration of the stearoyl group from the position *sn*-1 to *sn*-2. Esterification at the 2-position of myristic acid yields 1-stearoyl-2-myristoyl-3-

benzyl-*sn*-glycerol which, on catalytic hydrogenolysis, loses the benzyl group to give 1-stearoyl-2-myristoyl-*sn*-glycerol.

1-Octadecyl-2-myristoyl-sn-glycerol can be obtained from 2-allyl -3-benzyl-*sn*-glycerol by reaction with octadecyl mesylate in the presence of potassium *tert.*-butyrate. After rearrangement, the propenyl group is removed, and the product is esterified and *O*-debenzylated to give the required product.

1-Stearoyl-3-myristoyl-sn-glycerol can be similarly prepared by starting from 2-benzyl-3-allyl-*sn*-glycerol.

1-Octadecyl-3-myristoyl-sn-glycerol is synthesised by the procedure used for the 1,2-product by starting from 2-benzyl-3-allyl-*sn*-glycerol.

1-Stearoyl-2-benzyl-sn-glycerol is prepared from 2-benzyl-3-allyl-*sn*-glycerol.

One of the most common naturally occuring lecithins, 1-stearoyl-2-oleoyl-*sn*-glycero-3-phosphocholine, is made from 1-stearoyl-2-benzyl-*sn*-glycerol *via* phosphorylation with bromoethyl phosphoric acid dichloride (R.Hirt and R.Berchthold, Pharm. Acta. Helv., 1958, 33, 349; H.Eibl and A.Nicksch, Chem. Phys. Lipids, 1978, 22, 1), followed by catalytic debenzylation of the 1-stearoyl-2-benzyl-*sn*-glycero-3-phosphoric acid bromoethyl ester. This can be reacylated with oleic acid anhydride (H.Eibl, Chem. Phys. Lipids, 1980, 26, 239). The resulting diacyl-*sn*-glycero-3-phosphoric acid bromoethyl ester is then converted into 1-stearoyl-2-oleoyl-*sn*-glycero-3-phosphocholine by amination with trimethyl amine.

1,3-Dibromopropan-2-ol is used as the starting material for the synthesis of triglycerides (A.Bhati, R.J.Hamilton and D.A.Steven, Fats and Oils: Chemistry and Technology, Applied Science,London, 1980, 59). This substance is made from glycerol by reaction with bromine in the presence of red phosphorus (G.Braun, Org. Synth., 1967, Coll. Vol. II, 308) to give the product as an oil with b.p. 110-120°/ 20mm.. When 1,3-dibromo-propan-2-ol is reacted with either sodium, potassium or silver stearate the reaction mixture produces

2,3-epoxypropyl stearate, 3-bromo-2-hydroxypropyl stearate, 1-bromomethyl-2-hydroxyethyl stearate, glycerol 1,3-distearate and glycerol 1,2-distearate (A.Bhati *et al.*, J. Chem. Soc. Perkin Trans II, 1983, 1553):

$CH_2Br-CH_2OH-CH_2Br$ —(i)→ $H_2C(-O-)HC-CH_2Br$ —(ii)→ $H_2C(-O-)HC-CH_2OCOR$ + $CH_2OCOR-CHOH-CH_2Br$

$CH_2OCOR-CHOH-CH_2OCOR$ $CH_2OCOR-CHOCOR-CH_2OH$ ←(iii)— + $CH_2OH-CHOCOR-CH_2Br$

Reagents: (i) RCOOM at 70°; (ii) RCOOM/ROOH; (iii) RCOOM/RCOOH.
M = Na, K, Ag. R = $CH_3(CH_2)_{16}$.

The formation 2,3-epoxypropyl stearate indicates that the major reaction is dehydrobromination to provide initially 3-bromo-1,2-epoxypropane which can then undergo reaction with the metal stearate and stearic acid to give the remaining products. This complication may be due to the neighbouring group participation of the hydroxy-function at C-2 in the 1,3-dibromopropan-2-ol, which can lead to the elimination of hydrogen bromide, prior to nucleophilic attack by the carboxylate anion at either the 1- or 3-position.

In order to suppress this, the hydroxyl group can be blocked by substituting its hydrogen with an electron-withdrawing group, e.g. by the formation of 2-bromo-1-(bromomethyl)ethyl palmitate by reaction with palmitoyl chloride in the presence of pyridine. The product is a crystalline substance, m.p. 35.5°.

$$
\begin{array}{l}CH_2Br\\ |\\ CHOCOR'\\ |\\ CH_2Br\end{array} \longrightarrow \begin{array}{l}CH_2Br\\ |\\ CHOCOR'\\ |\\ CH_2OCOR\end{array} + \begin{array}{l}CH_2OCOR\\ |\\ CHOCOR'\\ |\\ CH_2OCOR\end{array}
$$

$$
\begin{array}{l}CH_2\\ ||\\ CHOCOR'\\ |\\ CH_2Br\end{array} + \begin{array}{l}CH_2\\ ||\\ CHOCOR'\\ |\\ CH_2OCOR\end{array}
$$

Reagent: RCOOM at 69°.
M = Na, K, Ag, $(C_2H_5)_4N^+$, $(C_{10}H_{21})_3(CH_3)N^+$, 18-crown-6-K.
R = $CH_3(CH_2)_{16}$; R' = $CH_3(CH_2)_{14}$.

Glycerol 1,3-distearate-2-palmitate, (m.p. 61-61.5°), is synthesised by the reaction of 2-bromo-1-(bromomethyl)ethyl palmitate with *tris*(decyl)methyl ammonium stearate (F.H.Mattson and L.W.Beck, J. Biol. Chem., 1955, 214, 115). Other triglycerides can be synthesised in the same way.

1-Stearoyloxy-2-palmitoyloxy-3-bromopropane is made by the reaction of 2-bromo-1-(bromomethyl)ethyl palmitate with silver stearate.

The synthesis of modified triglycerides is of interest in the study of biologically active compounds such as platelet activating factors, and for the preparation of compounds having technical uses (A.Berg, Dragoco Rep., 1976, 23, 159; A.V.Calogero, Cosmetic Toiletries, 1979, 77, 94; Y.Gama and H.Narazaki, J. Pat., 1979, 106, 425 ; S.Watanbe *et al.*, J. Amer. Oil Chemists' Soc., 1983, 60, 116). There are no known vegetable oils containing 2-alkyl branched triacyl glycerols, but such lipids can be identified in the preen glands of birds (G.A.Garton, Chem. Ind. 1985, 295; P.Nuhn *et al.*, Fette Seifen Anstrichmittel, 1985, 87, 135).

Acylisopropylidene glycerols can be alkylated *via* their enolate anions by alkyl halides of varying length (M.Herslof and S.Gronowitz, Chemica Scripta, 1983, 22, 230; 1985, 25, 257;J. Amer. Oil Chemists' Soc., 1985, 62, 1013). They also react with disulphides (B.M.Trost and T.N.Salzmann, J. Amer. Chem. Soc., 1973, 95, 6840; B.M.Trost, Chem. Rev., 1978, 78, 363) and diselenides. Carboxylation of the enolates is also possible with either carbon dioxide (S.Reiffers *et al.,* Tetrahedron Lett., 1971, 3001) or methyl chloroformate (M.Herslof and S.Gronowitz, Chemica Scripta, 1985, 25, 257).

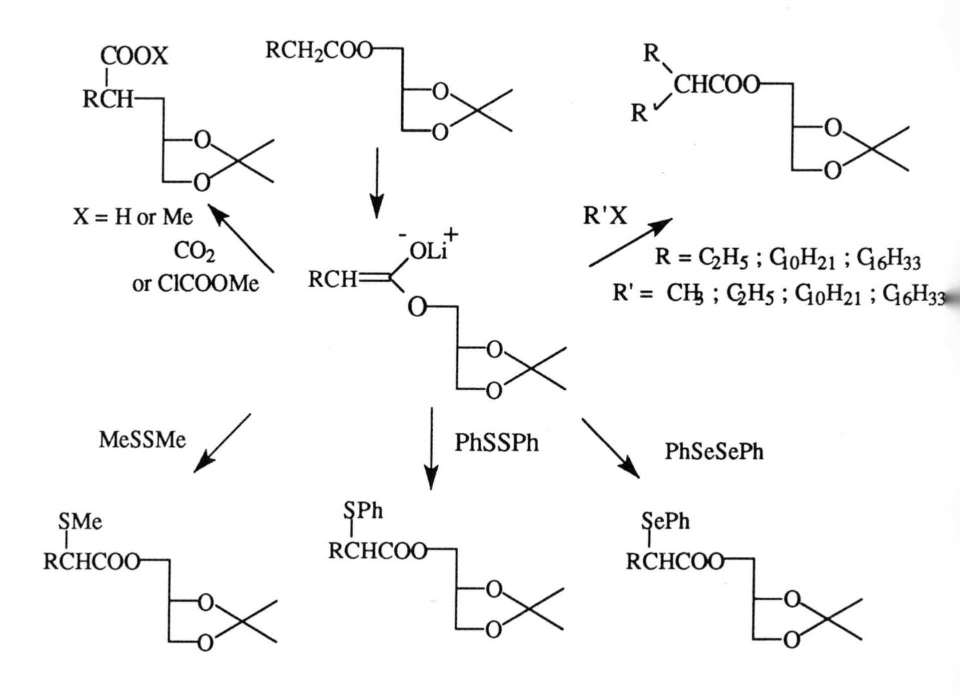

The yields of the sulphur-containing products are affected by the reactivity of the disulphide and by the chain-length of the acylisopropylidene glycerol. Dimethyl disulphide can react with the anions of dodecanoyl and octadecanoylisopropylidene glycerols and yield 2-methylthio-substituted acylisopropylidene glycerols (B.M. Trost and T.N. Salzmann, J. Org. Chem., 1975, 40, 148). Similarly, the enolate anions of acylisopropylidene glycerols undergo reaction with iodomethane in acetone in the presence of potassium carbonate to give 2-carboxy-acylisopropylidene glycerols (G.G.Moore *et al.*, J. Org. Chem., 1979, 44, 2425).

A tridecanoyl glycerol containing one butyl group is also prepared from mono-2-butyldecanoyl glycerol, whilst tridecanoylglycerols with two and three butyl groups are obtained by reaction of monododecanoyl glycerol or glycerol with 2-butyldodecanoyl chloride:

R'(R)CHCOCl

RCH_2COO–CH₂–CH(OH)–CH₂–OH HO–CH₂–CH(OH)–CH₂–OH

RCH_2COO–CH₂–CH(OOCCH(R)R')–CH₂–OOCCH(R')R R'(R)CHCOO–CH₂–CH(OOCCH(R')R)–CH₂–OOCCH(R')R

$R = C_{10}H_{21}$ $R' = C_4H_9$

Triacyl glycerols containing three 2-decyldodecanoyl groups can also be made. However, this method cannot be applied to the synthesis of triacyl glycerols. Triacyl glycerols have a tendency to undergo self condensation; this can be overcome by trapping the enolate anion by use of a reactive electrophile, before the condensation can take place. The most common electrophile is trimethylsilyl chloride which is very reactive at low temperatures and is compatible with lithium diisopropylamide used as the base (E.J.Corey and A.W.Gross, Tetrahedron Lett., 1984, 25, 495). *O*-Trimethylsilyl ketene acetals can be formed by reaction of ester enolates giving stable compounds that can be handled at room temperature, but the presence of acid needs to be avoided (M.W.Rathke and D.F.Sullivan, Synth. Commun., 1973, 3, 67; Tetrahedron Lett., 1973, 15, 1297). These substances are versatile intermediates since derivatives can be obtained by

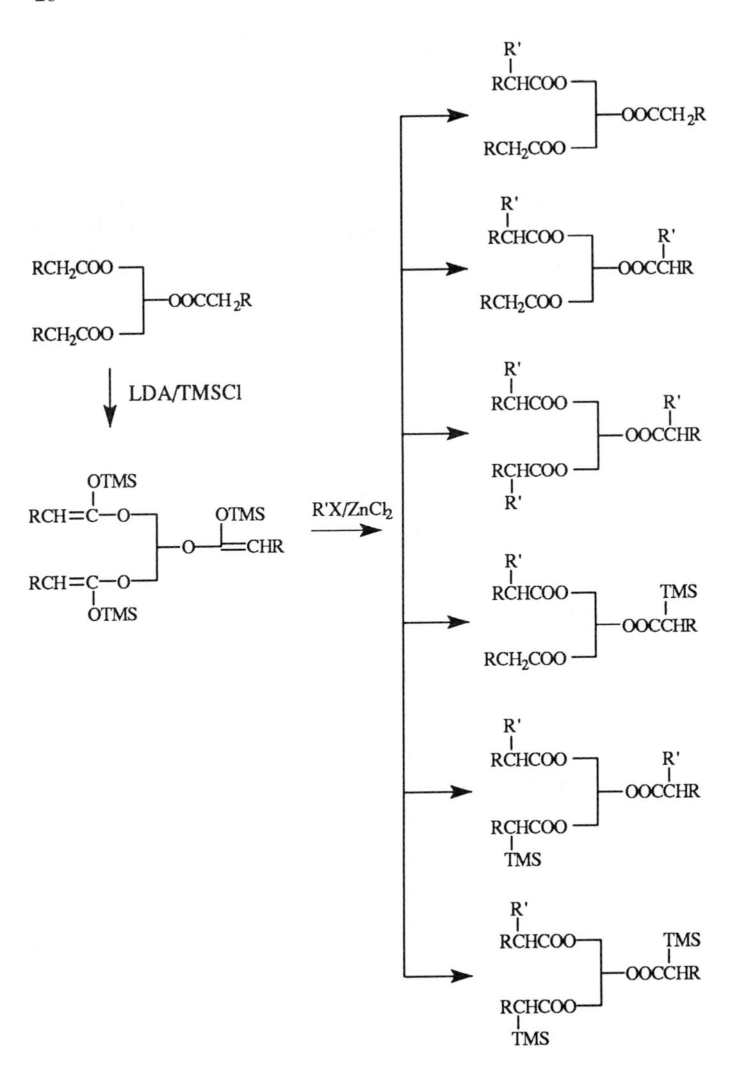
RCH2COO
OOCCH2R
RCH2COO
LDA/TMSCl
OTMS
RCH=C—O
O
OTMS
=CHR
RCH=C—O
OTMS
R'X/ZnCl2
R'
RCHCOO
OOCCH2R
RCH2COO
R'
RCHCOO
R'
OOCCHR
RCH2COO
R'
RCHCOO
R'
OOCCHR
RCHCOO
R'
R'
RCHCOO
TMS
OOCCHR
RCH2COO
R'
RCHCOO
R'
OOCCHR
RCHCOO
TMS
R'
RCHCOO
TMS
OOCCHR
RCHCOO
TMS

Scheme 3

further reaction. It is possible to perform 2-*tert.*-alkylations (M.T.Reetz *et al.*, Chem. Ber., 1983, 116, 3708; C.Lion and J. -E.Dubois, Tetrahedron, 1981, 37, 319), 2-acylations, and aldol additions (Y.Yamamoto *et al.*, Tetrahedron Lett., 1984, 25, 1075; T.H.Chan *et al.*, Tetrahedron Lett., 1979, 42, 4029), Michael additions (T.V.Rajan-Babu, J. Org. Chem., 1984, 49, 2083; M.Miyashita *et al.*, J. Amer. Chem. Soc., 1984, 106, 2149) and oxidations (G.M.Rubottom *et al.*, J. Org. Chem., 1983, 48, 4940). Thus, alkenyl and alkyl groups can be introduced into the 2-position through alkylation with the phenylsulphinyl chloride adduct with alkenes (S.K.Patel and I.Paterson, Tetrahedron Lett., 1983, 24, 1315) or through the alkylation with chloromethyl phenyl sulphide (I.Paterson and I.Fleming, Tetrahedron Lett., 1979, 11, 993).

The addition of triacyl glycerols to cooled solutions of excess lithium diisopropylamide and excess trimethylsilyl chloride in tetrahydrofuran yields silyl ketene acetals containing trimethylsilyloxy groups.

The trimethylsilyl ketene acetals from tributanoyl glycerol and tridodecanoyl glycerol can undergo alkylation reactions with *tert.* butyl chloride or 1-adamantyl bromide in the presence of zinc chloride in methylene chloride (M.T.Reetz and K.Schellnus, Terahedron Lett., 1973, 17, 1455; M.T.Reetz *et al.*, Chem. Ber., 1983, 116, 3708). The reaction of the trimethylsilyl ketene acetal of tributanoyl glycerol with *tert.*-butyl chloride yields three compounds: the tributanoyl glycerol containing one, two or three 2-*tert.*-butyl groups. **(Scheme 3)**

The reactions are complex and in some products trimethylsilyl groups are retained bonded to carbon rather than oxygen. These rearrangements occur during the alkylations and it is possible to desilylate the products by treatment with potassium fluoride.

Archaebacteria are unique life forms that have diverged from other organisms. Among the many features that distinguish archaebacteria from eubacteria and eukaryotes are the unusual structures of the archaebacterial membrane lipids. The hydrocarbon region of these membranes is made from isoprenoid alcohols joined to glycerol through ether linkages instead of the normal straight chain fatty acid glycerol ester structures (M.Kates in "Ether Lipids: Chemistry and Biology"; Ed. F.Synder, Academic Press; New York, 1972, pp 351-398). The basic structural unit is a phytanyl glyceryl diether, 2,3-di-*O*-[(R,R,R)-3',7',11',15'-tetramethylhexadecyl]-*sn*-glycerol

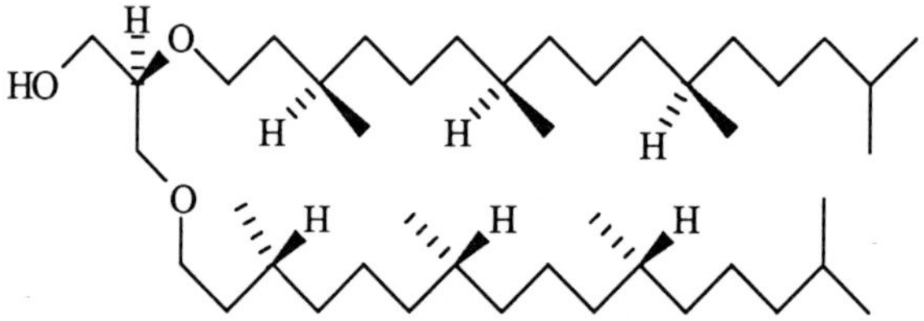

HO H O H H H O H H H

The isoprenoid chain is capable of modification, and the rigidity of these membranes is bought about by a novel 4-4' linkage between the phytanyl units (C.H.Heathcock *et al.*, Science (Washington, D.C.) 1985, 229, 862; M.Kates *et al.*, Biochemistry, 1967, 6, 3329).

The synthesis of 2,3-di-*O*-phytanyl-*sn*-glycerol and the 1,2-isomer is outlined in the following scheme:

HO OH OBn (i) → t-BuPh$_2$SiO OH OBn

(ii) ↓

t-BuPh$_2$SiO Ophytanyl OBn (v) → HO Ophytanyl OBn

(iv,iii) ↓ phytanylO Ophytanyl OH

(iii,iv,v) ↓ HO Ophytanyl Ophytanyl

Reagents: (i) *t*-BuPh$_2$SiCl, imidazole; (ii) NaH, phytanyl triflate; (iii) H_2, Pd/$BaSO_4$; (iv) KH, phytanyl triflate; (v) *n*-Bu_4NF.

3-O-Benzyl-sn-glycerol is made from D-mannitol by cleavage of the 1,2,5,6-di-*O*-isopropylidene derivative, followed by sodium borohydride reduction, benzylation and deprotection, b.p. 140-142° (0.7 mm) and [α] +6.30^{o} (J.-L.Debost *et al.*, J. Org. Chem., 1983, 48, 1381; J.LeCocq and C.E.Ballou, Biochemistry, 1964, 3, 976; B.T.Golding and P.V.Ioannou, synthesis, 1977, 423). This compound on reaction with *tert.*-butyldiphenylsilyl chloride and imidazole in dimethylformamide gives 1-*O*-(*tert.*-butyldiphenylsilyl)-3-*O*-benzyl-*sn*-glycerol as a colourless oil, [α] -1.1^{o}.

Phytanol, (3R,S,7R,11R)-3,7,11,15-tetramethylhexadecan-1-ol, can be converted into (3R,S,7R,11R)-3,7,11,15-tetramethyl hexadecan-1-yl triflate, by reaction with triflic anhydride in pyridine. This triflate reacts with the glycerol derivative in tetrahydrofuran in the presence of sodium hydride to give 1-*O*-(*tert.*-butyldiphenylsilyl)-2-*O*-[(3'R,S,7'R,11'R)-3',7',11', 15'-tetramethylhexadec-1-yl]-3-*O*-benzyl-*sn*-glycerol, as a colourless oil, [α] -4.3°. Hydrogenolysis of this compound with palladium on barium sulphate results in the removal of the benzyl group to give 1-*O*-(*tert.*-butyldiphenylsilyl)-2-[(3'R,S,7'R, 11'R)-3,7,11,15-tetramethylhexadec-1-yl]-*sn*-glycerol as a colourless oil, [α] -9.8°. This derivative can be converted into 1-*O*-(*tert.*-butyldiphenylsilyl-2,3-di-*O*-[3'R,S,7'R,11'R)-3',7', 11',15'-tetramethylhexadec-1-yl]-*sn*-glycerol, a colourless oil, [α] -0.76° by reaction with sodium hydride in tetrahydrofuran. Using tetra-n-butylammonium fluoride in tetrahydrofuran on this product gives 2,3-di-*O*-[(3'R,S,7'R,11'R)-3',7',11',15'-tetra methylhexa- dec-1-yl]-*sn*-glycerol, as a colourless oil, [α] +7.1° The *tert.*-butyldiphenylsilyl protecting group of this derivative is removed by reaction with tetra-*n*-butylammonium fluoride to give 2-*O*-[(3'R,S,7'R,11'R)-3',7',11',15'-tetramethylhexadec-1-yl]-3-*O*-benzyl-*sn*-glycerol, a colourless oil, [α] +8.3°. This compound is converted, on treatment with potassium hydride, into 1,2-di-*O*-[(3'R,S,7'R,11'R)-3',7',11',15'-tetramethylhexadec-1-yl]-3-*O*-benzyl-*sn*-glycerol, as a colourless oil, [α] -0.4°. The benzyl group of this compound can be removed by hydrogenolysis to provide 1,2-di-*O*-[(3'R,S,7'R,11'R)-3',7',11',15'-tetramethyl hexadec-1-yl]-*sn*-glycerol, as a colourless oil, [α] -7.0°. Natural 2,3-di-*O*-(3'R,S,7'R,11'R)-phytanyl-*sn*-glycerol, [α] +7.8° can be isolated from *Methanobacterium thermoautotrophicum* (E.G.Bligh *et al.*, J. Biochem. Physiol., 1959, 37, 911).

Interest in the calcium-lipid-regulated protein kinase (*protein kinase C, PKC*) as a mediator of biological functions has increased and requires the synthesis of specific activators, diacylglycerols, of this enzyme. These glycerols are useful to activate PKC in both homogenate and in living cells. They are also potential tumor promoters.

1,2-Diacyl-*sn*-glycerols are obtained from the corresponding acyl phosphatidylcholines with phospholipases (R.D. Mavis *et al.*, J. Biol. Chem., 1972, 247, 2835). Thus, 1-oleoyl-2-acetyl-*sn*-glycerol with heterologous substitution in the 1 and 2 positions can be achieved by digesting 1-oleoyl-2-acetylglycerophosphorylcholine with *phospholipase C* (M.Kates, "Techniques of Lipidology", North-Holland Publ., Amsterdam, 1972). Deoxy, choro and sulphydryl analogues are also synthesised by this technique (B.R.Ganong and R.M.Bell, Biochemistry, 1984, 23, 4977). These compounds can be tested for their activation of *protein kinase C* in purified preparations from the brain, and this can be correlated with their ability to stimulate luteinising hormone release from pituitary cell cultures (P.M.Conn *et al.,* Biochem. Biophys. Res. Commun., 1985, 126, 532).

Second Supplements to the 2nd Edition of Rodd's Chemistry of Carbon Compounds, Vol. IE/F/G, edited by M. Sainsbury

Chapter 19

TRIHYDRIC ALCOHOLS AND THEIR OXIDATION PRODUCTS (continued)

B.J.COFFIN

Dihydroxycarboxylic and Trihydroxycarboxylic Acids and Related Compounds.

(1) Dihydroxycarboxylic acids

Mevalonic acid (1) and its lactone (2):

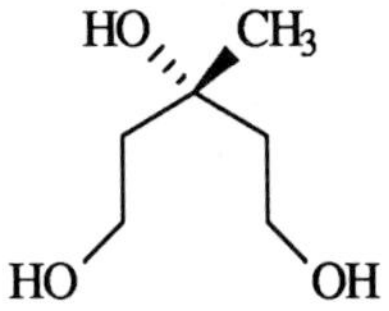

(1)

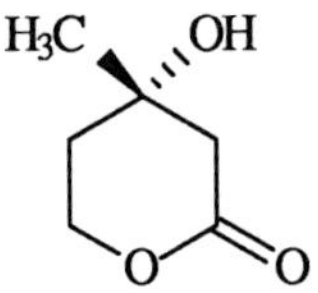

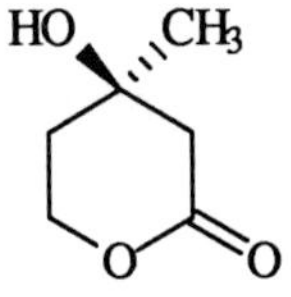

(2)

(S)-(+)-form

(R)-(-)-form

continue to be biologically important substances, and many ways of synthesising them have been studied since they were first identified in extracts of distillers' solubles (L.D.Wright *et al.*, J. Amer. Chem. Soc., 1956, 78, 5273; D.E.Wolf *et al., ibid.*, 1957, 79, 1486). Synthetic work has concentrated on methods for the introduction of isotopic labels into the molecules for use in biosynthetic studies (R.A.Ellison and P.K.Bhatnagar, Synthesis, 1974, 719; J.A.Lawson *et al., ibid.*, 1975, 729; F.-C.Huang *et al.*, J. Amer. Chem. Soc., 1975, 97, 4144; H.Daido *et al.*, Bull. Chem. Soc. Jpn., 1977, 1021; E.Abushanab *et al.*, Tetrahedron Lett., 1978, 31, 3415; M.Fetizon *et al.*, Tetrahedron, 1975, 31, 171).

(i) Preparation of mevalonic acid and mevalonolactone

(1) [3-^{13}C]Mevalonolactone is synthesised by a route, proceeding *via* ethyl [3-^{13}C]acetoacetate (A.Banerji *et al.*, J. Chem. Soc. Perkin Trans. I, 1976, 2221). In turn ethyl [3-^{13}C]acetoacetate is made by the reaction of ethyl hydrogen malonate with magnesium ethoxide, followed by reaction with *N*-[1-^{13}C]acetyl imidazole. A mixture of this acetoacetate with ethylene glycol and concentrated sulphuric acid provides ethyl [3-^{13}C]acetoacetate ethylene acetal which, on reduction with lithium aluminium hydride yields 4-hydroxy[2-^{13}C]butan-2-one ethylene acetal. This, on reaction with acetic anhydride, gives 3-oxo[3-^{13}C]butyl acetate ethylene acetal which is converted to 3-oxo[3-^{13}C]butyl acetate, by reaction with trityl tetrafluoroborate. 3-Oxo[3-^{13}C]butyl acetate reacts with ketene at -20° in the presence of boron trifluoride in ether to give [3-^{13}C]mevalonolactone **(Scheme 1).**

(2) An alternative synthesis of (±)-mevalonolactone begins with a reaction between ethyl acetate and allyl magnesium bromide which affords the tertiary alcohol (3). Ozonolysis and oxidation with hydrogen peroxide yields 3-hydroxy-3-methylpentane-1,5-dioic acid (4). Treatment of the diacid with *N,N'*-dicyclohexylcarbodiimide (DCC) affords the anhydride (5), which upon reduction with sodium borohydride gives mevalonolactone. Should the anhydride be *O*-acetylated, prior to reduction, *O*-acetyloxymevalonolactone (6) is obtained together with the 2,3-dehydrolactone (7) (P.Lewer and J.MacMillan, J. Chem. Soc. Perkin Trans. 1, 1983, 1417; E.Bardshiri *et al., ibid.*, 1984, 1765; A.I.Scott and K.Shishido, J. Chem. Soc. Chem. Commun., 1980, 400).

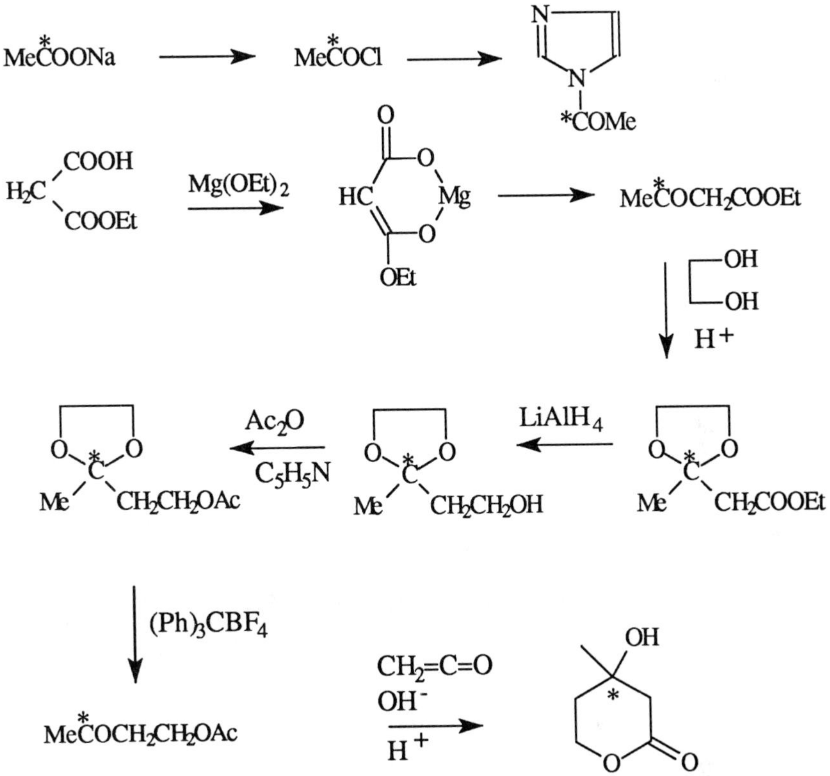

Me*COONa → Me*COCl → N-(*COMe)imidazole

$H_2C(COOH)(COOEt)$ —$Mg(OEt)_2$→ Mg chelate (HC, O, O, Mg, OEt) → $Me\overset{*}{C}OCH_2COOEt$

↓ $HOCH_2CH_2OH$, H^+

Me(*C)(dioxolane)CH_2COOEt —$LiAlH_4$→ Me(*C)(dioxolane)CH_2CH_2OH —Ac_2O, C_5H_5N→ Me(*C)(dioxolane)CH_2CH_2OAc

↓ $(Ph)_3CBF_4$

$Me\overset{*}{C}OCH_2CH_2OAc$ —$CH_2{=}C{=}O$, OH^-, H^+→ lactone (OH, *, O, O)

Scheme 1

CH_3COOEt + MgBr → (3)

O_3, H_2O_2/HOAc → HOOC, COOH, OH (4)

DCC → (5)

$NaBH_4$ → (i) Ac_2O (ii) $NaBH_4$ → OAc (6) + (7)

(3) An asymmetric synthesis of mevalonolactone is achieved using the (S)-morpholino-1,1′-binaphthyl-2,2′-diamine (8) as a chiral auxilary. This compound when reacted with 3-hydroxy-2,4-dimethylglutaric anhydride affords the diastereomeric amides (9). These can be methylated at the carboxylic acid group by diazomethane and then reduced with lithium borohydride to the amido alcohols (10). Hydrolysis and separation of the diastereomers leads to (-)-(R)-mevalonolactone in an overall yield of 44% (Y.V. Ventatachalapathi *et al.*, Biochemistry, 1982, 21, 5502) **(Scheme 2).**

Other anhydrides can replace 3-hydroxy-2,4-dimethylglutaric anhydride in this type of procedure and have been used to synthesise the

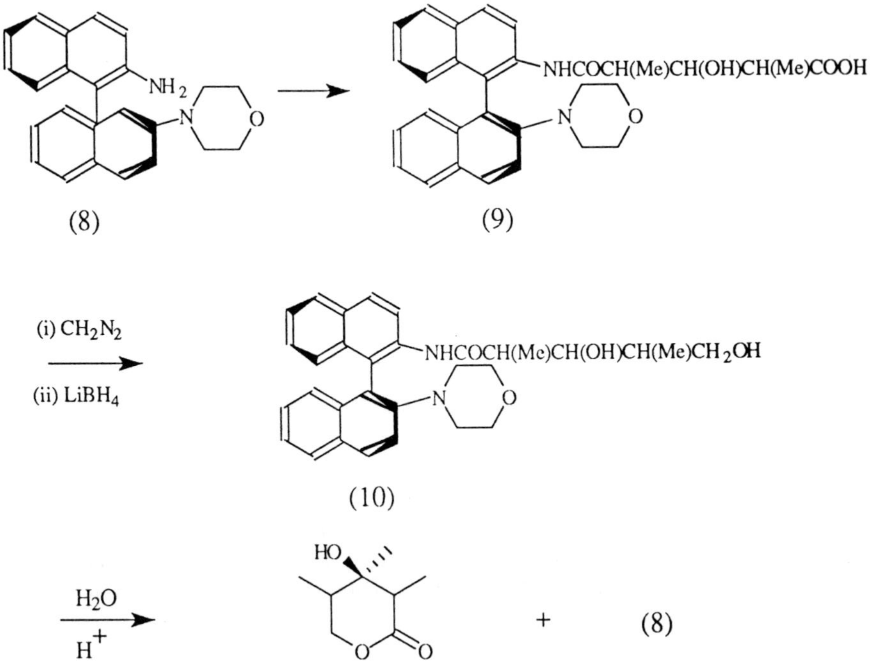

NH_2 N O → NHCOCH(Me)CH(OH)CH(Me)COOH N O

(8) (9)

(i) CH_2N_2 → (ii) $LiBH_4$ NHCOCH(Me)CH(OH)CH(Me)CH_2OH N O

(10)

H_2O → H^+ HO O O + (8)

Scheme 2

deoxymevalolactones (11a) and (11b).

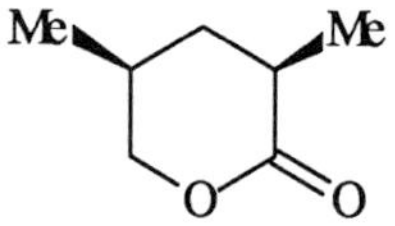

(11a) 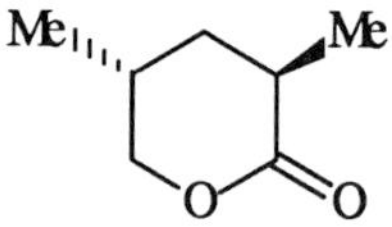　　　　(11b)

(4) Asymmetric syntheses of both (R)-(-)- and (S)-(+)-mevalonolactone in over 98% enantiomeric excess use (R)-(-)-3-hydroxy-3-methyl-4-phenylbutanonitrile (13) as the common starting material. This is synthesised in three steps from (2R,4aS,7R,8aR)-2-acetyl-4a,5,6,7,8,8a-hexahydro-4,4,7-trimethylbenz-1,3-oxathiane (12) (E.L.Eliel and K.Sokai, Tetrahedron Lett., 1981, 22, 2859; K.-Y.Ko *et al.*, Tetrahedron, 1984, 40, 1333; S.V.Frye and E.L.Eliel, J. Org. Chem., 1985, 50, 3402). The acyloxathiane is converted into a carbinol by reaction with benzyl magnesium bromide, and on reaction with *N*-chlorosuccinimide and reduction with sodium borohydride, this provides (R)-(+)-2-methyl-3-phenyl-1,2-propanediol (13a) in an enantiomeric excess of 95%. This is then converted into its tosylate which, on reaction with potassium cyanide, yields the nitrile (13b). On reaction with acetic anhydride containing 4-(dimethylamino)pyridine, followed by reaction with sodium periodate and ruthenium trichloride, this product gives (S)-(-)-3-acetoxy-4-cyano-3-methylbutanoic acid which, on reduction, provides the dihydroxynitrile (14). Hydrolysis with aqueous sodium hydroxide and hydrogen peroxide then gives pure (R)-(-)-mevalonolactone **(Scheme 3).**

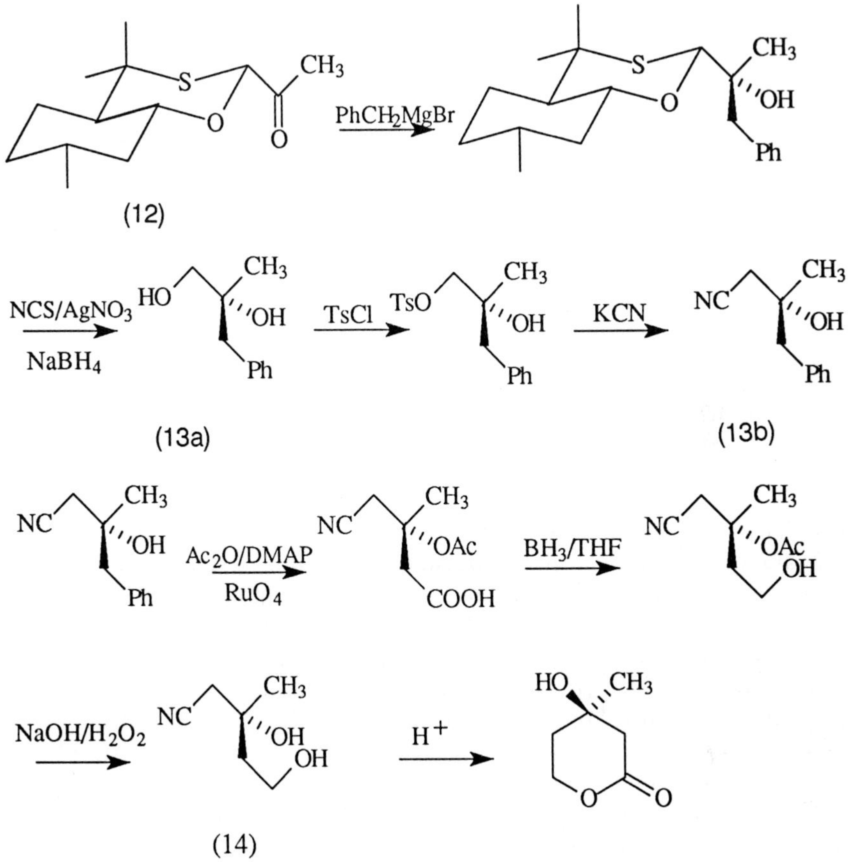

CH_3 S O O (12) $PhCH_2MgBr$ S O CH_3 OH Ph

NCS/$AgNO_3$ $NaBH_4$ HO CH_3 OH Ph (13a) TsCl TsO CH_3 OH Ph KCN NC CH_3 OH Ph (13b)

NC CH_3 OH Ph Ac_2O/DMAP RuO_4 NC CH_3 OAc COOH BH_3/THF NC CH_3 OAc OH

NaOH/H_2O_2 NC CH_3 OH OH (14) H^+ HO CH_3 O O

Scheme 3

(S)-(+)-Mevalonolactone, [α] +21.7° is obtained in 98% enantiomeric excess, by a similar procedure:

(5) The syntheses of both the (R)- and (S)-enantiomers of mevalonolactone are also possible from the α-allyllactone (15) (S.Takano *et al.,* J. Chem. Soc. Chem. Commun., 1984, 84). This compound previously obtained from either (S)-glutamic acid (S.Takano *et al.,* J. Chem. Soc. Chem. Commun., 1980, 616; M.Taniguchi *et al.,* Tetrahedron, 1974, 30, 3547) or from D-mannitol (S.Takano *et al.,* Heterocycles, 1981, 16, 951) can be oxygenated to afford the carbinol (16a), if its enolate is treated with oxydiperoxy(pyridine) (hexamethylphosphoric triamide) molybdenum (MoOPH) (E.Vedejs *et al.,* J. Org. Chem., 1978, 188). Oxygenation of the lithium enolate is confined to attack at the less hindered face of the molecule, and the carbinol may now be treated with lithium aluminium hydride to give the diol (16b). *O*-Tosylation and further reaction with lithium aluminium hydride leads to the methyldiol (17). When this is reacted with aqueous hydrochloric acid the triol (18) is formed, which, with sodium borohydride and cleavage with sodium periodate, affords mainly the R-lactol (19). After purification this compound may be oxidised with pyridinium chlorochromate (PCC) to give (R)-mevalonolactone **(Scheme 4).**

Alternatively, the triol (18) can be firstly cleaved with sodium

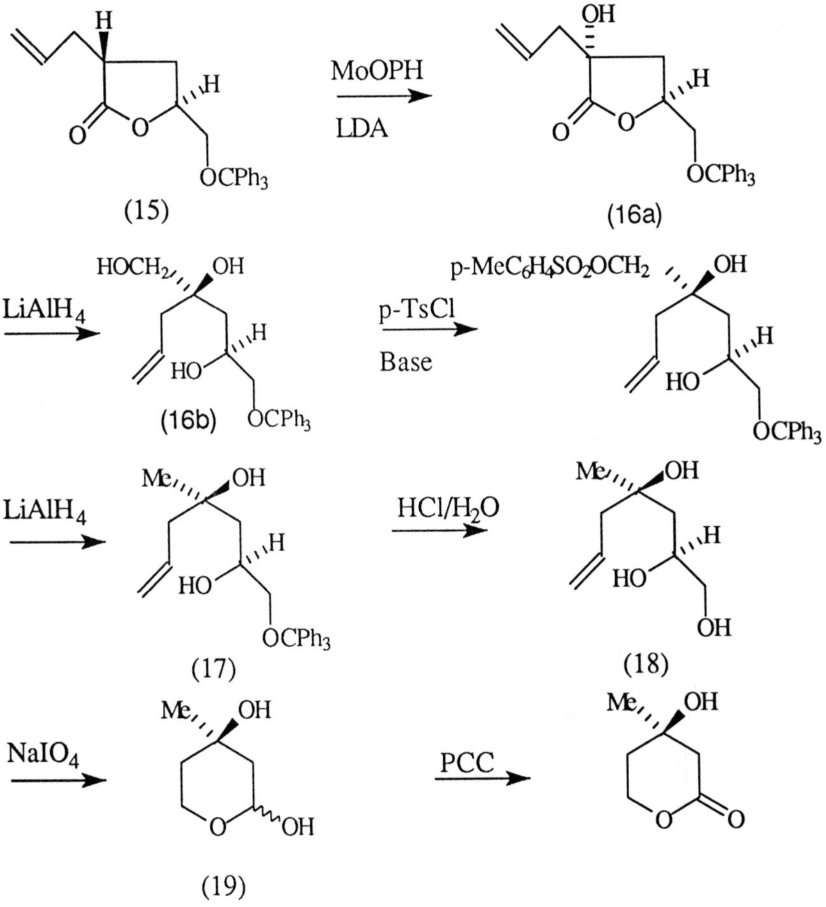

(15) MoOPH / LDA → (16a)

$LiAlH_4$ → (16b)

p-TsCl / Base → p-$MeC_6H_4SO_2OCH_2$

$LiAlH_4$ → (17)

HCl/H_2O → (18)

$NaIO_4$ → (19)

PCC →

Scheme 4

periodate and then reduced with sodium borohydride to give the diol (20). This, on ozonolysis, gives the (S)-lactol (21) as a mixture of epimers which is converted into (S)-mevalonolactone by oxidation with pyridinium chlorochromate.

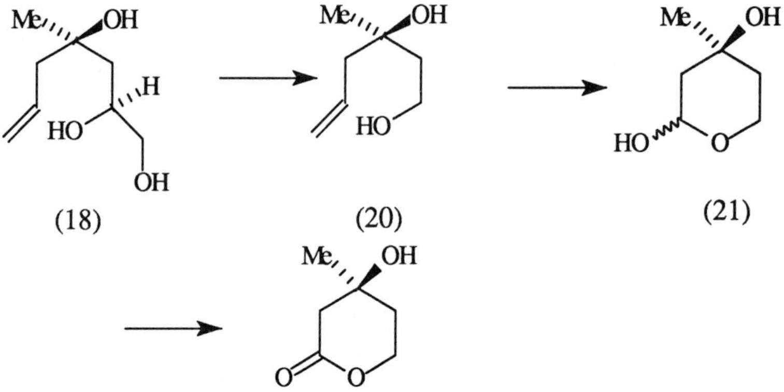

Me OH H HO OH (18) → Me OH HO (20) → Me OH HO O (21) → Me OH O O

(6) 3-Methylpentane-1,3,5-triol (22) can be oxidised by a number of *Gluconobacter spp.* to (S)-mevalonolactone (M.Eberle and D.Arigoni, Helv. Chem. Acta, 1960, 43, 1508; J.W.Cornforth *et al.*, Tetrahedron, 1962, 18, 1351):

HO Me HO OH (22) → HO Me O O

G. scleroideus IAM 1842 exhibits the highest selectivity, whilst *G roseus* provides poorer results.

The reaction pathway for the formation of (S)-mevalonolactone can be interpreted in two ways:

a) *Gluconobacter* oxidise preferentially the pro-S hydroxyethyl group attached to the prochiral centre in 3-methylpentane-1,3,5-triol to give (S)-mevalonolactone,

b) The microorganisms do not distinguish the *pro*-S and the *pro*-R site of 3-methylpentane-1,3,5-triol, resulting in the formation of (±)-mevalonolactone The microorganism then selectively degrades the natural (R)-mevalonolactone leaving the (S)-enantiomer unchanged.

The first mechanism is preferred since prolonged incubation of *Gluconobacter scleroideus* with 3-methylpentane-1,3,5-triol gives (S)-mevalonolactone in 34% yield and optical purity of 66% which is very similar to that obtained for only 3 days incubation. If the second pathway were in operation the optical yield of the mevalonolactone would become higher with longer periods of incubation. In addition, exposure of (±)-mevalonolactone to *G. scleroideus* for 3 days results in the recovery of the racemic mevalonolactone (H.Ohta *et al.*, J. Org. Chem., 1982, 47, 2400). (S)-Mevalonolactone is also obtained by oxidation of 3-methyl pentane-1,3,5-triol by *Flavobacterium oxydans*. In this case oxidation of both *pro*-R and *pro*-S hydroxyethyl groups occurs, followed by selective decomposition of (R)-mevalonolactone. (F.-C.Huang *et al.*, J. Amer. Chem. Soc., 1975, 97, 4144).

(ii) Biosynthesis of isoleucine and valine

It has been reported that the pathways of isoleucine and valine biosynthesis are unusual since the last four steps are catalysed by a common set of enzymes (V.W.Rodwell in 'Metabolic Pathways', ed. D.M.Goldberg, Academic Press, New York and London, 1969, 3rd. Edn, Vol. 1, p.357). These enzymes need to be able to accommodate the structural differences between the intermediates of the two pathways. The enzyme, *2,3-dihydroxyacid hydrolyase, (EC 4.2.1.9)* which catalyses the conversion of the dihydroxycarboxylic acids (23) (R = Me) and (R = Et) into the corresponding 2-oxocarboxylic acids (24) (R = Me) and (R = Et) has a strict requirement for substrates with the 2R-configuration. Both (2R,3R)- and (2R,3S)-2,3-dihydroxy-3-methylpentanoic acid are substrates for the enzyme (F.B.Armstrong *et al.*, Biochem. Biophys. Acta, 1977, 498, 282). From growth studies with mutants of *Salmonella*

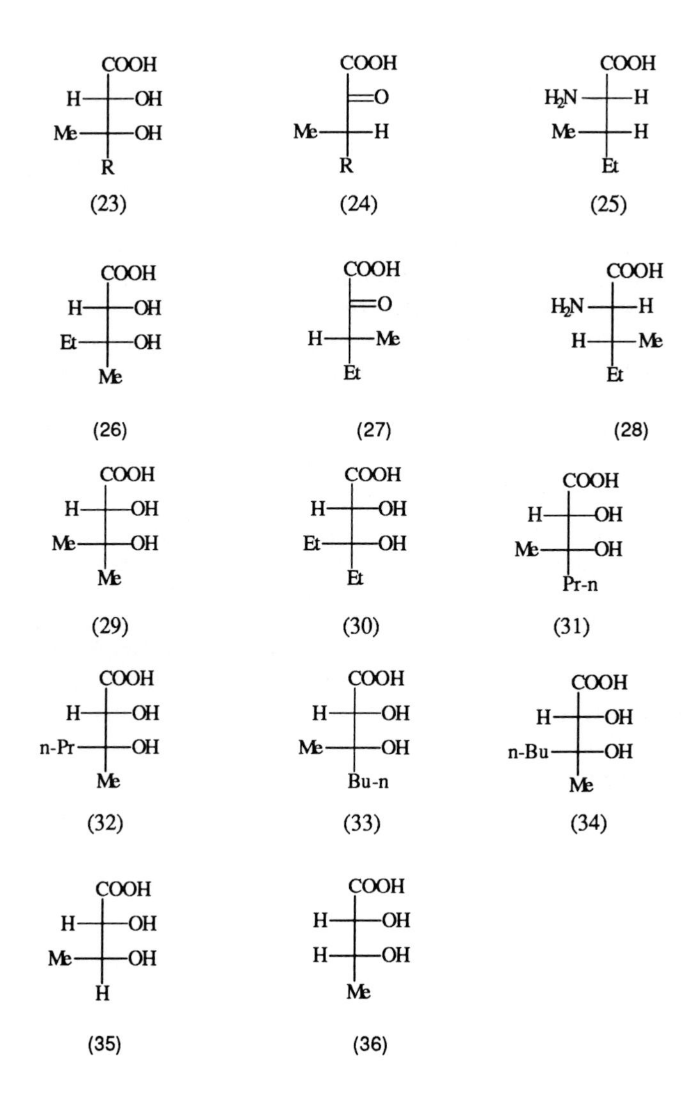
COOH
H
OH
Me
OH
R
(23)
COOH
O
Me
H
R
(24)
COOH
H2N
H
Me
H
Et
(25)
COOH
H
OH
Et
OH
Me
(26)
COOH
O
H
Me
Et
(27)
COOH
H2N
H
H
Me
Et
(28)
COOH
H
OH
Me
OH
Me
(29)
COOH
H
OH
Et
OH
Et
(30)
COOH
H
OH
Me
OH
Pr-n
(31)
COOH
H
OH
n-Pr
OH
Me
(32)
COOH
H
OH
Me
OH
Bu-n
(33)
COOH
H
OH
n-Bu
OH
Me
(34)
COOH
H
OH
Me
OH
H
(35)
COOH
H
OH
H
OH
Me
(36)

typhimurium, the conversion is known to be stereospecific, and, for example (2R,3R)-2,3-dihydroxy-3-methylpentanoic acid (23) (R = Et) is converted into (S)-3-methyl-2-oxopentanoic acid (24) (R = Et) which acts as the keto acid precursor of L-isoleucine (25). (2R,3S)-2,3-Dihydroxy-3-methylpentanoic acid (26) is converted into the (R)-2-oxo-3-methylpentanoic acid (27) which has a configuration at C-3 corresponding to that of L-alloisoleucine (28). Thus, the enzyme acts stereospecifically, but it is not unusually stereoselective with respect to the configuration at C-3 in the substrate. This indicates that there must be the same relative configurations of the C-2 hydrogen atom and the C-3 hydroxy group which are eliminated during dehydration in spite of the absolute configurations of the substrates. The volume of the binding site of the enzyme which can accommodate a methyl group in the natural intermediates of the pathway i.e. (R)-2,3-dihydroxy-3-methylbutanoic acid (29) and (2R,3R)-2,3-dihydroxy-3-methylpentanoic acid also needs to be able to accommodate an ethyl group if the substrate is the abnormal precursor (2R,3S)-2,3-dihydroxy-3-methylpentanoic acid:

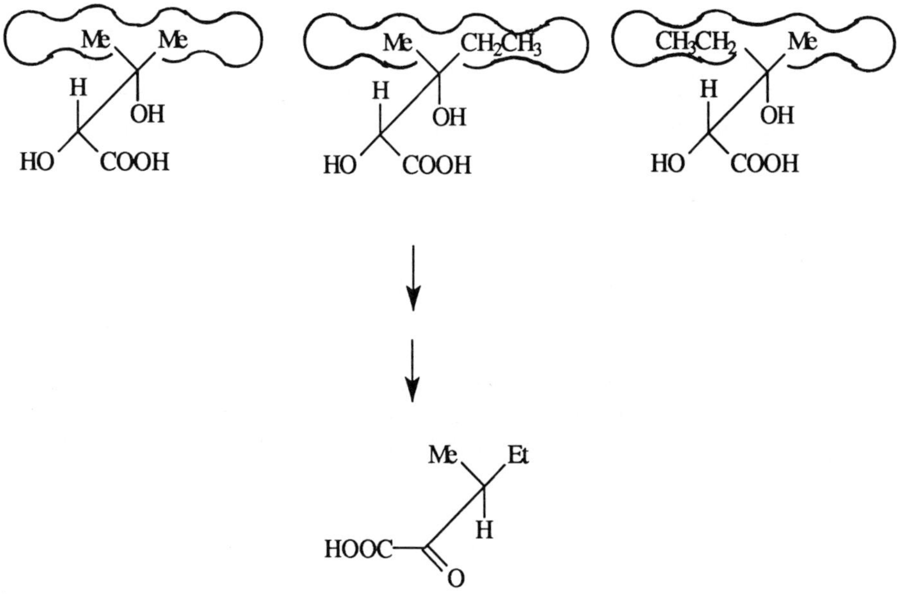

The influence of the size of the C-3 substituent in substrates for the

2,3-dihydroxyacid dehydratase from *Salmonella typhimurium* has been studied using of a group of 2,3-dihydroxycarboxylic acids. These have been synthesised by either *cis*- or *trans*-hydroxylation of the corresponding 2,3-unsaturated carboxylic acids, or by use of the Reformatsky reaction (S.Kobayashi *et al.*, Chem. Lett., 1973, 1097).

(2R,3R)-2,3-Dihydroxy-3-methylpentanoic acid (23) (R = Et) has been synthesised (F.B.Armstrong *et al.,* Biochem. Biophys. Acta, 1977, 498, 282).

(2R,3R)(2S,3S)-2,3-Dihydroxy-3-methylhexanoic acid (31) is prepared from (E)-3-methylhex-2-enoic acid by reaction with hydrogen peroxide in formic acid as an oil, which is purified as its dicyclohexyl ammonium salt, m.p. 175-177^{0}. The free acid is recovered by passage of the salt through an ion exchange resin. (E)-3-Methylhex-2-enoic acid is obtained by heating ethyl 3,3-bis(thiophenyl)butanoate with zinc chloride at 110^{0} (F.B.Armstrong *et al.*, J. Chem. Soc. Perkin Trans. I, 1985, 691).

(2R,3S)(2S,3R)-2,3-Dihydroxy-3-methylhexanoic acid (32) is synthesised by the reaction of osmium tetraoxide with (E)-3-methylhex-2-enoic acid and *N*-methyl-morpholine-*N*-oxide dihydrate. It is an oil, which can be purified as its dicyclohexylammonium salt, m.p. 167-169^{0}.

(2R,3R)(2S,3S)-2,3-Dihydroxy-3-methylheptanoic acid (33) is prepared by the reaction of hydrogen preoxide on (E)-3-methylhept-2-enoic acid. It is an oil; dicyclohexylammonium salt, m.p. 188-191^{0}.

(2R,3S)(2S,3R)-2,3-Dihydroxy-3-methylheptanoic acid (34) is prepared from the same starting material, but using *N*-methylmorpholine-*N*-oxide; dicyclohexylammonium salt, m.p. 152-154^{0}.

2,3-Dihydroxy-3-ethylpentanoic acid (30) is made from ethyl

3-ethylpent-2-enoate, (from the reaction of pentan-3-one with ethyl 2-bromoacetate in the presence of zinc). This is reacted with osmium tetraoxide in the presence of *N*-methylmorpholine *N*-oxide dihydrate, followed by hydrolysis. The oily product is purified as its dicyclohexylammonium salt, m.p. 170-171.5°.

(2R,3S)(2S,3R)-2,3-Dihydroxybutanoic acid (35) is prepared by the reaction of osmium tetraoxide with crotonic acid and *N*-methyl morpholine *N*-oxide dihydrate. It can be characterised as its dicyclohexylammonium salt, m.p. 150.5-151.5°.

(2R,3R)(2S,3S)-2,3-Dihydroxybutanoic acid (36) is obtained from the reaction of crotonic acid with hydrogen peroxide in formic acid. Once again the free acid is an oil; dicyclohexylammonium salt, m.p. 144.5-145.5°.

(2) Amino hydroxy carboxylic acids

This group of compounds continues to be of interest because aminohydroxy acids are related to many naturally occuring substances, and because of their use as starting materials for synthesis.

There are a number of synthetic methods available to obtain a wide range of these compounds, for example the enantioselective decarboxylation of 2-alkyl-2-amino malonates using chiral cobalt(III) complexes (M.J.Jun *et al.,* J. Chem. Soc. Dalton Trans., 1983, 999). 3-Hydroxy-2-amino carboxylic acids are also obtained from aminosilyl ketene acetals using a Lewis acid catalysed addition reaction with aldehydes (G.Guanti *et al.,* Tetrahedron Lett., 1985, 26, 3517).

Threo-3-hydroxy-2-amino acid derivatives, $(PhCH_2)_2NCH_2(CHR^1OH)COOR^2$, are obtained by the reactions of lithium enolates of *N,N*-dibenzylglycine esters with aldehydes, followed by sodium borohydride reduction (G.Guanti *et al.,* Tetrahedron Lett., 1984, 25, 4693).

2-Alkoxy-2-aminocarboxylic acids can be obtained from Schiff bases, $Ph_2C{=}NCH_2COOEt$, by bromination with *N*-bromosuccinimide, followed by bromide displacement with an alkoxide ion, and hydrolysis (M.J.O'Donnell *et al.,* Tetrahedron Lett., 1985, 26, 695).

6-Hydroxy-2-aminopent-4-enoic acids can be synthesised from allylglycines. These in protected forms may be cleaved by ozonolysis to yield the corresponding aldehydes, which in turn may be reacted under Wittig or Horner-Emmons conditions to afford (Z)- or (E)-isomers of the title acids after deprotection. The initial stereochemistry of the allylglycine is unaffected and is thus transferred to the products (Y. Ohfume *et al.*, Tetrahedron Lett., 1986, 27, 6079).

H COOMe
NHBoc
→ →
H COOH
$HOCH_2$
NH_2

Threo-3-hydroxy-2-aminocarboxylic acid derivatives undergo dehydration with Et_2NSF_3.Py reagent involving sidechain hydroxy group participation in the elimination and stereospecific conversion into benzyl-dehydroamino carboxylic acids (L.Somekh and A.Shanzer, J. Org. Chem., 1983, 48, 907). 2-(*N*-Acylamino)-3-hydroxyalkanoic acids are also dehydrated with p-toluenesulphonyl chloride/triethylamine reagent (T.Kolasa, Synthesis, 1983, 539). *N*-Benzyl- or Boc-2-amino-3-hydroxy-alkanoic acids are dehydrated with *N,N'*-dicarbonyldi-imidazole (R.Andruszkiewicz and A.Czerwinski, Synthesis, 1982, 568).

(3) 2-Amino-2-alkenoic acids (didehydroaminocarboxylic acids)

The chemistry of these acids has been well reviewed (U.Schmidt *et al.*, Progress in the Chemistry of Organic Natural Products Vol 37, Springer-Verlag, Wien 1979, pp 251-327; C.H.Stammer in: Chemistry and Biochemistry of Amino Acids, Peptides and Proteins, Vol. 6, John Wright and Sons Ltd., London, 1982, pp33-74; K.Noda *et al.,* in: The Peptides, Vol. 5, Academic Press, New York, 1983, pp285)

Synthesis

Synthesis via Elimination Reactions.

The elimination of water from β-hydroxy-α-amino acid derivatives is particularly valuable method, especially from serine, threonine and phenylserine derivatives (J.V.Edwards *et al.,* Int. J. Pept. Protein Res., 1986 28, 603). The recommended reagents are disuccinimidyl carbonate (H.Ogura *et al.,* Tetrahedron Lett., 1981, 22, 4817; H.Ogura and K.Taleda, Ger. Pat., 3016831, 1980), *N,N*-dicarbonyldiimidazole (R.Andruszkiewicz and A.Czerwinski, Synthesis, 1982, 968), or a base and acetic anhydride (T.Kato *et al.,* Jap. Pat., 60/19074, 1985). This reaction proceeds stereospecifically with (diethylamino)sulphur trifluoride in pyridine as reagent. The *threo* derivatives yield Z-olefins whilst the *erythro* compounds provide E-olefins (L.Somekh and A.Shanzer, J. Org. Chem., 1983, 48,907). Olefins or aziridines are produced using the Mitsunobu reagent and, whichever is formed, depends on the structure of the starting material (K.Nakajima *et al.*, Pept. Chem.,1982, 20, 19). Arylideneserine esters also lose water when treated with *N,N*- dicarbonyliimidazole to give the corresponding didehydroamino acids (G.Wulff and H.Bohnke, Angew. Chem., 1984, 96, 362). Tosyl compounds eliminate *p*-toluene sulphinic acid (T.Kolasa and E.Gross, Int. J. Pept. Protein Res., 1982, 20, 259; Y.Shimohigashi and C.H.Stammer, *ibid.,* 1982, 19, 54).

Aminoacid esters and *N*-acylamino acid esters undergo *N*-chlorination with *tert.*-butyl hypochlorite, followed by loss of hydrogen chloride (H.Schmidt, Chem. Berr., 1975, 108, 2547; *ibid.*, Angew, Chem., 1976, 88, 295; Chem. Ber., 1977, 110, 942). A mixture of the imine and the enamine is obtained from the amino acid ester. The enamine hydrochoride is then obtained by treatment with hydrogen choride in ether:

R_2 COOMe R_1 NH_2

(i)t-BuOOCl
(ii) DBU

R_2 COOMe R_1 NH + R_2 COOMe R_1 NH_2

DBU = diazabicycloundecene

Depending on the solvent and the base used, *N*-chloroamides provide either α-acylamino-α-alkoxycarboxylates or acylimino compounds. Presumably *N*-acylimines are formed initially when sodium alkoxides are the reagents, but these products are then attacked by the alkoxide anions acting as nucleophiles. *N*-Acyldidehydroamino acid derivatives are obtained by treatment of the former with acid or the latter with base (M.Seki *et al.*, Agric. Biol. Chem., 1984, 48, 1251; M.D.Grim *et al.*, J. Org. Chem., 1981, 46, 2671; Y.Shimohigashi and C.H.Stammer, J. Chem. Soc. Perkin Trans. 1, 1983, 803; A.K.Sharma *et al.*, Indian J. Chem. Sect. B 1985, 24, 7; G.F.J.Stoll *et al.*, Liebigs Ann. Chem., 1986, 1968).

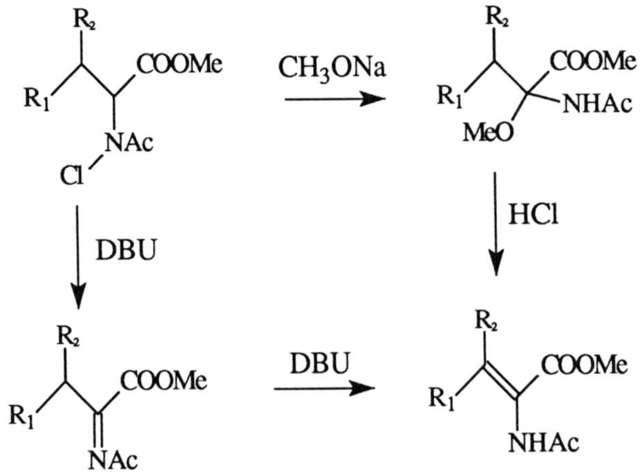

α,β-Dibromocarboxylates react with azide anion to provide α-azidoacrylates (M.Kakimoto *et al.,* Chem. Lett., 1982, 525) which, on electrolytic reduction, give didehydroamino acids in high yield (D.Knittel, Monatsh. Chem., 1984, 115, 1335; *ibid.*, 1985, 116, 1133). Phosphinimes are formed in the presence of phosphites or phosphinites, these then undergo rearrangement to phosphoric amides or phosphinic amides of the corresponding didehydroamino acids respectively (Y.Yonezawa *et al.*, Tetrahedron Lett., 1979, 3851). Acrylates also undergo addition reactions with thiols and alcohols in alkaline solution. The resulting β-phenylthio- and α-alkoxy-α-azidocarboxylates are not isolated but, on reaction with alkoxides, undergo cleavage of the azide group to yield the corresponding β-substituted didehydroamino acids (M.Kakimoto *et al.*, Chem. Lett., 1982, 4, 527).

The rearrangement of α-azidocarboxylates using lithium diisopropyl amide or lithium ethoxide gives a mixture of α-enamino esters and α-imino esters which, on acylation, yield acyldidehydroamino acids (P.A.Manis and M.W.Rathke, J. Org. Chem.,1980, 45, 4952).

α-Azidoacrylates are also converted into acyldidehydroamino acids by reaction with dirhenium heptasulphide in acetic anhydride in the presence of acetic acid saturated with hydrogen chloride. The reaction presumably proceeds *via* a nitrene intermediate from which an imine is formed either by hydride shift, or by an insertion reaction to give an aziridine which undergoes subsequent ring-opening to provide the product (F.Effenberger and T.Beisswenger, Angew. Chem., 1982, 94, 210).

R_2 R_1 COOMe N_3

Re_2S_7/Ac_2O

R_2 R_1 COOMe NHAc

α-Azidocarboxamides and α-azidolactams are also used as sources of didehydroamino acids (*ibid.,* Chem. Ber., 1984, 117, 1513; 1984, 117, 1497; F.Effenberger *et al.*, Ger. Pat., 3508564, 1986). In addition, the photolysis of azidomalonate derivatives yields didehydroamino acids (J.Frank *et al.,* Liebigs Ann. Chem.,1986, 1990).

In the presence of tertiary amines, 3-phenylaziridine-2-carboxylates rearrange to *N*-benzyloxycarbonyldidehydroalanine esters (K.Okawa, Jap. Pat., 57/175150, 1982).

This reaction has been used to prepare peptides containing aziridine carboxylic acids (K.Nakajima *et al.,* Bull. Chem. Soc. Jpn., 1982, 55, 174).

α-Oxo acids form α,α-diaminocarboxylic acid derivatives or enamino acid derivatives when heated with carboxamides. For instance, the addition of acetamide to pyruvic acid yields α-hydroxy-α-acetyl aminopropionic acid which either reacts with a second molecule of acetamide to yield a *bis*(acylamino) compound or undergoes cleavage of water to give an α-(acetylamino)propenoic acid.

$CH_3COCOOH$ + $AcNH_2$ → H_3C, COOH, HO, NHAc → H_3C, COOH, AcHN, NHAc

↓

COOH, NHAc

(U.Schmidt *et al.*, Progress in the Chemistry of Organic Natural Products Vol. 6, John Wright and Sons Ltd., London, 1982, 33; M.Bergmann and K.Grafe, Z. Physiol. Chem., 1930, 187, 183; M.Makowski *et al.*, Justus Liebigs Ann. Chem., 1985, 5, 893; *idem. ibid.*, 1984, 5, 920; C.G.Shin *et al.*, Tetrahedron Lett., 1985, 26, 85; R.Labia and C.Morin, J. Org Chem., 1986, 51, 249 D.W.Graham *et al.*, J. Med. Chem., 1987, 30, 1074).

Didehydroamino acids with aromatic and heterocyclic substituents are obtained from the cleavage of arylmethylene-5(4H)-oxazolones, preformed by the Erlenmeyer method. The azlactones are obtained by melting an aldehyde, acylglycine, acetic anhydride and sodium acetate at approximately 140^0 (C.Halldin and B.Laangstroem, Turun Yliopiston Julk, Sar. D, 1984, 17, 200; Y.Shu *et al.*, Yiyao Gonye, 1985, 16, 116; A.Gaset and J.P.Gorrichon, Synth. Commun., 1982, 12, 71; I.Arenal *et al.*, An. Quim Ser. C. 1981, 77, 56; K.Takahashi *et al.*, J. Labelled Compd. Radiopharm., 1986, 23, 1; C.Cativiela *et al.*, An. Quim. Ser. C 1985, 81, 56; J. Suh *et al.*, J.Org. Chem., 1985, 50, 977; J.Folkers *et al.*, Int. J. Pept. protein Res., 1984, 24, 197; Y.Yuan and C.Yang, Dalian Gongxueyuan Xuebao, 1984, 23, 135; M.A.Hassan *et al.*, Indian J. Chem. Soc. Sect. B, 1983, 22, 637; C.Cativiela *et al.*, Synthesis, 1983, 899). If the aldehyde is unstable 5(4H)-oxazolones are prepared by reaction of acylglycines/chlorocarbonates and triethylamine in the cold, then by condensation of the product with an aldehyde in the presence of diazabicycloundecane (D.J.Phelps and F.C.A.Gaeta, Synthesis, 1982, 234;

V.O.Topuzyan *et al.*, Khim.-Farm. Zh., 1986, 20, 675; A.M.Tikdari *et al.*, Chem. Ind. (London), 1986, 825). The correponding aldehyde imines are also used (P.K.Tripathy and A.K.Mukerjee, Synthesis, 1985, 285, 418; A.K.Sen *et al.*, Indian J. Chem. Sect. B, 1983, 22, 939; A.K.Mukerjee and P.Kumar, Can. J. Chem., 1982, 60, 317). (Z)-Oxazolones are also formed and are cleaved by nucleophiles to yield (Z)-didehydroamino acid derivatives (D.J.Phelps and F.C.A.Gaeta, Synthesis, 1982, 234; A.A.Afifi *et al.*, J. Chem. Soc. Pak., 1986, 8, 297; V.O.Topuzyan *et al.*, Khim.-Farm. Zh., 1986, 20, 675; A.M.Takdari *et al.*, Chem. Ind. (London),1986, 825). These (Z)-oxazolones undergo isomerisation to the (E)-isomers under the influence of phosphoric acid or hydrogen bromide (I.Arenal *et al.*, An. Quim Ser. C, 1981, 77, 56; C.Cativiela *et al.*, Synthesis, 1983, 899). *N*-Acyldidehydroamino acid esters are also prepared from the arylmethyleneoxazolones by acid or base catalysed alcoholysis (D.J.Phelps and and F.C.A.Gaeta, Synthesis, 1982, 234; C.Cativiela *et al.*, Tetrahedron, 1986, 42, 583; S.Solyom *et al.*, Justus Liebigs Ann. Chem., 1987, 153; K.Takahashi *et al.*, J. Labelled Compd. Radiopharm., 1986, 23, 1).

N-Formyldidehydroamino acid esters are prepared *via* oxazolines by the condensation of aldehydes and ketones with isocyanoacetates in aprotic solvents (U.Schollköpf *et al.*, Justus Liebigs Ann. Chem., 1972, 766, 116; U.Schollköpf and R.Meyer, *ibid,* 1981, 1469; J.Rachon and U.Schollköpf, *ibid,* 1981,99; H.H.Wust *et al., ibid,* 1985, 1825):

$R_1R_2C{=}O$ + $NC{-}CH_2{-}COOR_3$

for R_3 = t-C_4H_9, NaCN → oxazoline (O, N; R_1, R_2, $COOR_3$)

for R_3 = C_2H_5, BuLi or NaOH or t-BuOK → $R_1R_2C{=}C(COOR_3)NHCHO$

These mild reaction conditions permit the use of sensitive aromatic and aliphatic carbonyl compounds. α,β-Unsaturated carbonyl compounds undergo condensation in the presence of zinc chloride and copper(1) chloride (Y.Ito *et al.,* Tetrahedron Lett., 1985, 26, 5781).

N-Formyldidehydroamino acid esters readily lose the formyl group on treatment with hydrogen chloride in an alcohol (Tanabe Seiyaku Co. Ltd., Jap. Pat., 57/116032, 1982).

Acyldidehydroamino acid esters are obtained by the acylation of the potassium salt of *N*-formylaminoacrylates, giving *N*-formyl-*N*-acyl didehydroaminoacid esters, which can be *N*-decarbonylated by acid treatment (U.Schollköpf and R.Meyer, Justus Liebigs Ann. Chem., 1981, 99).

The direct condensation of carbonyl compounds with *tert.*-butyl diphenylphosphino(isocyano)acetate gives α-isocyanoacrylates (J.Rachon and U.Schollköpf, Justus Liebigs Ann. Chem., 1981, 99).

The condensation of isocyanoacetates with amide acetals gives rise to the formation of didehydroamino acid derivatives (W.Kantlehner *et al.,* Justus Liebigs Ann. Chem., 1980, 344; C.Herdeis and U.Nagel, Heterocycles, 1983, 20, 2163).

Didehydroamino acid esters are prepared from *N*-acyldialkoxy phosphorylglycine esters (U.Schmidt *et al.*, Angew. Chem. Int. Ed. Engl., 1982, 94, 797; U.Schmidt, A Lieberknecht and J.Wild, Synthesis, 1984, 53). *N*-Benzyloxycarbonyldimethoxyphosphorylglycine esters are readily available from glyoxalic acid and suitably protected phosphorylglycine esters (R.Kober and W.Steglich, Justus Liebigs Ann. Chem., 1983, 599).

(4) Representative amino-hydroxy carboxylic acids and related compounds

2-Carbamoylserine (serine carbamate):

$$H_2N \blacktriangleright \overset{\displaystyle COOH}{\underset{\displaystyle CH_2OCONH_2}{C}} \blacktriangleleft H$$

The (R)-D-form of 2-carbamoylserine is produced by *Streptomyces spp., as* needles, m.p. 226-34°; [α] -19.6°. The (S)-form is also known. It is a glutamine antagonist (G.Hagemann *et al.,* Biochim. Biophys. Acta, 1955, 17,240; C.G.Skinner *et al.,* J. Amer. Chem. Soc., 1956, 78, 2412; Y.Okami *et al.,* J. Antibiot., 1962, 15, 147; Biochem. Prep.,1963, 10, 18; C.G.Skinnner *et al.,* Biochem. Biophys. Res. Commun., 1963 12,68).

Azaserine (USAN, serine diazoacetate, serine O-diazoacetate, O-diazoacetylserine):

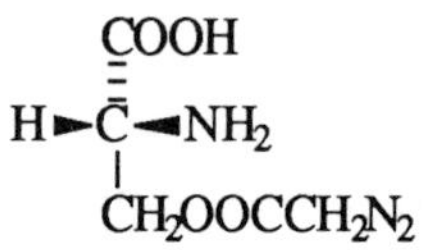

COOH
H–C–NH_2
$CH_2OOCCH_2N_2$

This compound is carcinogenic in rats, and its (S)-L-form is isolated from a culture broth of *Streptomyces spp.*. It is also a glutamine antagonist which inhibits purine biosynthesis, and acts as a antifungal, antibiotic and antitumour agent (B.D.Roebuck *et al.,* Experentia, 1983, 39, 324; N.I.Sax, Dangerous Properties of Industrial Materials, 5th., Ed.,Van Nostrand-Reinhold, 1979, 967). It forms light yellow needles with m.p. 146-62° and [α] -0.5° (T.J.Curphey *et al.,* J. Org. Chem., 1978, 43, 4666).

4-Amino-3-isoxazolidinone (cycloserine, oxamycin, orientomycin):

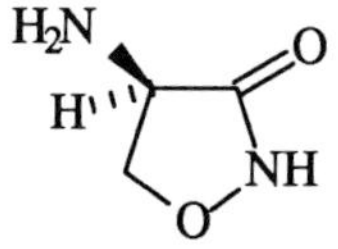

H_2N O
H NH
O

is produced in its (R)-form by *Streptomyces garyphalus, S. orcidaceus, S. lavendulae and S. nagasakiensis.* It exhibits antibiotic activity primarily against mycobacteria (C.H.Stammer *et al.,* J. Amer. Chem. Soc., 1955, 77, 2344; 2345; 2346). It is of limited clinical use due to its toxicity (R.L.Harned *et al.,* Antibiot. Chemother. (Washington D.C.),

1955, 5, 204). Its synthesis has been reported (C.H.Stammer *et al.*, J. Amer. Chem. Soc., 1957, 79, 3236); m.p 155-6°.

3-Amino-2-hydroxypropanoic acid (isoserine):

COOH
H►C◄OH
CH_2NH_2

(R)-form.

has been synthesised (C.L.Kenneth *et al.*, J. Org.Chem., 1962, 27, 438; T.Miyazawa *et al.*, Agric. Biol. Chem., 1976, 40, 1651. (S)-Isoserine is available from L-malic acid monoamide using the Hofmann rearrangement reaction mediated by $(CF_3COO)_2IPh$ in acetic anhydride/pyridine mixture (R.Andruszkiewicz *et al.*, Synthesis, 1983, 31).

3-Amino-2-hydroxy-3-phenylpropanoic acid (3-phenylisoserine):

COOH
H►C◄OH
H►C◄NH_2
Ph

(2R,3R)-form.

can be prepared (E.Fourneau *et al.*, Bull. Soc. Chim. Fr., 1940, 7, 593; K.Harada *et al.*, Bull. Chem. Soc. Jpn., 1974, 47, 2911; E.Kaji *et al.*, *idem*, 1976, 49, 3181). The (2R,3R)-(+)*erythro*-form has m.p. 243-4°; [α] +58.6°; (2S,3R)-(+)*threo*-form, m.p. 256-7°; [α] +14.9°.

Carnitine (2-hydroxy-3-(N,N,N-trimethylammonium)propanoate inner salt):

$$\begin{array}{c} CH_2COO^- \\ HO \blacktriangleright \overset{\equiv}{C} \blacktriangleleft H \\ | \\ CH_2\overset{+}{N}Me_3 \end{array}$$

(R)-form.

m.p. 196-198^0 and [α] -23.5^0, is a constituent of striated muscle, liver and whey. It is used medically as a stimulator of gastric juice secretion. It is synthesised from 3-ketoglutaric acid by conventional methods (A.S.Gopalian and C.J.Sih, Tetrahedron Lett., 1984, 25, 5235; K.Boch *et al.,* Acta Chem. Scand., Ser. B, 1983, 37, 341; K.Tomita *et al.,* Bull. Chem. Soc. Jpn., 1974, 47, 1988). It is also prepared by the aminolysis of $ClCH_2CH(OH)CH_2COOEt$, followed by hydrolysis (Y.Chen *et al.,* Hunan Yixuayuan Xuebeo, 1983, 8, 82).

Cysteine (2-amino-3-mercaptopropanoic acid)

$$\begin{array}{c} COOH \\ H_2N \blacktriangleright \overset{\equiv}{C} \blacktriangleleft H \\ | \\ CH_2SH \end{array}$$

(R)-form.

is found in the hydrolysates of proteins as its (R)-L-form (L.Fowden, Ann. Rev. Biochem., 1964, 33, 173). *S*-3-Ethoxycarbonylcysteine occurs in the seeds of several *Calliandra* species (J.T.Romeo and L.A.Swain, J. Chem. Ecol., 1986, 12, 2089).

Synthetic cysteine has been reported to have m.p. 178^0 (J.Martens *et al.,* Angew. Chem. , Int. Ed. Engl., 1981, 20, 668). It yields a number of derivatives of the (R)-L-form useful in peptide synthesis: *N*-acetyl (a mucolytic agent), m.p. 109-10^0 (N.Suzuki *et al.,* Bull. Chem. Soc. Jpn., 1976, 49, 3155; T.A.Martin *et al.,* J. Med. Chem.,

1968, 11, 625); *N,S*-diacetyl, m.p.179-82°; *S*-methyl (found in human urine and in the kidney bean), m.p. 207-11°; *S*-propyl (occurs as a glutamyl peptide in garlic); *S*-benzyl, m.p. 208-11°; *S*-methoxycarbonyl (Carbocistine, BAN, Mucodyne, Thiodril, Loviscol), m.p. 204-7°.

2-Alkylcysteines are prepared *via* the alkylation of oxazolines (D.Seebach and J.D.Aebi, Tetrahedron Lett., 1983, 24, 3311).

A number of cysteine derivatives, including the four stereoisomers of 3-methyllanthione, are synthesised by conventional routes in which (2S,3S)- or (2R,3R)-3-methylcysteine is reacted with the D- or L-3-chloroalanine (T.Wakamiya *et al.,* Bull. Chem. Soc. Jpn., 1985, 58, 536).

The synthesis of (R)-2-methyl-S-benzylcysteine uses *bis*-lactim ethers derived from dioxopiperazines (U.Groth and U.Schöllkopf, Synthesis, 1983, 37).

S-Substituted-D-cysteines are prepared from 3-chloro-D-alanine by reactions with the appropriate thiols. The reactions are mediated by an enzyme from *Pseudomonas putida* (T.Nagasawa *et al.,* Biochem. Biophys. Res. Commun., 1983, 111, 809).

3,3-Dialkylcysteines are made by the reaction of phosphorus pentasulphide upon *N*-formyldehydroamino acid esters (C.F.Stanfield *et al.,* J. Org. Chem., 1986, 51, 5153).

Threo-(-)-*S*-2-propyloxycarbonyl-L-cysteine is available by alkylation of (S)-(-)-$HSCH_2CHMeCOOMe$ using chiral aziridine-carboxylate esters (R.J.Parry and M.V.Naidu, Tetrahedron Lett., 1983, 24, 1133).

The pyrolysis of cysteine in a sealed tube at 130° yields ammonia and cystine whereas at 180° it yields thiophene, diethyl disulphide, 4-ethyl-2-methyl pyridine, di- and tetramethylthiolanes,1,2-dithiane and *N*-ethyl acetamide (H.Takano *et al.,* Nihon Daigaku Nojuigakubu Gijitsu Kenkyu Hokoku, 1986, 37).

Cysteine can be converted into its thiolsulphonate (D.L.H.Williams, Chem. Soc. Rev., 1985, 14, 171), and also into its *S*-sulphonate (S.Pongor *et al.,* Arch. Biochem. Biophys., 1985, 238, 458). Cysteine readily yields derivatives by reaction with methoxycarbonylsulphenyl chloride. The last compounds yield cysteine-*S*-thiol by reaction with potassium hydrogen sulphide (D.J.Smith and V.Venkatraghavan, Synth. Commun., 1985, 15, 945).

Homocysteine (2-amino-4-mercaptobutanoic acid):

$$\begin{array}{c} COOH \\ H_2N \blacktriangleright \overset{|}{C} \blacktriangleleft H \\ | \\ CH_2CH_2SH \end{array}$$

(S)-form.

is a component of abnormal human urine in its (S)-L-form, and it is an intermediate in the metabolic conversion of methionine into cysteine.

L-Homocysteine is converted into *S*-adenosyl-L-homocysteine using beef liver *S*-adenosyl-L-homocysteine hydrolase (B.Chabannes *et al.,* Prep. Biochem., 1983, 12, 395). L-Homocysteine and 5'-deoxy adenosine react with sodium in liquid ammonia to generate the thiolate anion which then adds to purine to provide *S*-adenosyl-L-homocysteine (A.Holy and I.Rosenberg, Coll. Czech. Chem. Commun., 1985, 50, 1514; P.Serapinowski, Synthesis, 1985, 926).

Homocysteic acid (3-amino-4-carboxybutane sulphonic acid):

$$\begin{array}{c} COOH \\ H \blacktriangleright \overset{|}{C} \blacktriangleleft NH_2 \\ | \\ CH_2CH_2SO_3H \end{array}$$

(R)-form.

is a neurotoxin and its (R)-form is isolated from the red alga, *Palmaria palmata,* m.p. 270° dec. (M.V.Lipcock *et al.,* Phytochemistry, 1979, 18, 1220; S.H.Lipton, J. Agric. Food Chem., 1978, 26, 1406).

2-Methylcysteine (2-amino-3-mercapto-2-methylpropanoic acid):

COOH
H_3C ►C◄ NH_2
CH_2SH

(S)-form.

as its (S)-form is a potential enzyme inhibitor. Its *S*-benzyl, methyl ester has been prepared, b.p.$_{0.1}$ 100-10^{0} (U.Groth *et al.*, Synthesis, 1983, 37; O.W.Griffith *et al.*, J. Biol. Chem., 1983, 258, 1591).

3-Methylcysteine (2-amino-3-mercaptobutanoic acid):

COOH
H_2N ►C◄H
H►C◄SH
CH_3

(2R,3R)-form.

has been synthesised in its various forms. The (2R,3R)-L-*threo*-form has been characterised as its hydrochloride, m.p. 148.5-153^{0} dec. The (2S,3S)-D-*threo*-form, m.p. 199.5-200^{0} is a component of the antibiotics, Nisin and Subtilin. It undergoes oxidation to yield 3,3'-dimethylcystine (T.Wakamiya *et al.*, Bull. Chem. Soc. Jpn., 1982, 55, 3878; 1983, 56, 1559).

Penicillamine (3-mercaptovaline, 2-amino-3-mercapto-3-methyl butanoic acid, 3,3-dimethylcysteine, cuprimine, depamine, distamine):

$$H_2N \blacktriangleright \overset{COOH}{\overset{|}{C}} \blacktriangleleft H$$
$$\quad\quad | $$
$$\quad\quad C(CH_3)_2SH$$

(R)-form.

is a characteristic degradation product of the penicillins. It is used as a chelating agent in the treatment of metal poisoning, Wilson's disease, scleroderma and rheumatoid arthritis (N.I.Sax, Dangerous Properties of Industrial Materials, 5th. Ed., Van Nostrand-Reinhold, 1979, 887).
It undergoes oxidation with hydrogen peroxide to give D-3,3-dimethylcysteic acid (A.Calvo *et al.,* J. Crystallogr. Spectrosc. Res., 1984, 14, 59).

Selenocysteine (3-selenylalanine, 2-amino-3-selenylpropanoic acid), $HSeCH_2CH(NH_2)COOH$, is formed in seleniferous plants, eg. onion, cabbage and *Candida albicans* (J.W.Hamilton, J. Agric. Food Chem., 1975, 23, 1150). It is an unstable compound and its physical properties are not well documented. Selenocysteine has been synthesised by the alkylation of methyl glycinate with bromomethyl selenides (H.J.Reich *et al.,* J. Org. Chem., 1986, 51, 2981).
L-Selenocystine and L-selenohomocysteine are made by the reaction of sodium selenide with *O*-acetylserine and *O*-acetylhomoserine respectively. The processes are catalysed by *O-acetylhomoserine sulphydrylase* (P.Chocat *et al.,* Agric. Biol. Chem., 1985, 49, 1143).
Selenocysteine yields L-selenodjenkolate *via* the reductive selenation of 3-chloro-L-alanine using selenium and sodium borohydride (P.Chocat *et al.,* Anal. Biochem., 1985, 148, 485).

Cystine (3,3'-dithiobis(2-aminopropanoic acid), dithiodialanine):

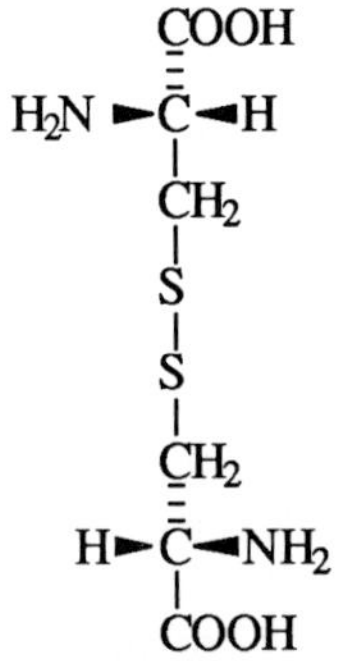

$$\begin{array}{c} COOH \\ H_2N \blacktriangleright C \blacktriangleleft H \\ | \\ CH_2 \\ | \\ S \\ | \\ S \\ | \\ CH_2 \\ H \blacktriangleright C \blacktriangleleft NH_2 \\ | \\ COOH \end{array}$$

(R,R)-form.

is widely distributed as a component of proteins and is isolated as plates or prisms as its (R,R)-L-form, m.p. 258-61°. It is reduced to cysteine, and its *N*-protected derivatives are used in the synthesis of peptides.

L-Cystine undergoes pyrolysis to give a mixture of ammonia, water, hydrogen sulphide and carbon dioxide (Y.Mori *et al.,* Chem. Pharm. Bull., 1985, 33, 916).

Cystine thiosulphinates are reduced by tris(dialkylamino)phosphines to provide unsymmetrical cystines and lanthionines (R.K.Olsen *et al.,* J. Org. Chem., 1985, 50, 4332).

Homoserine (2-amino-4-hydroxybutanoic acid):

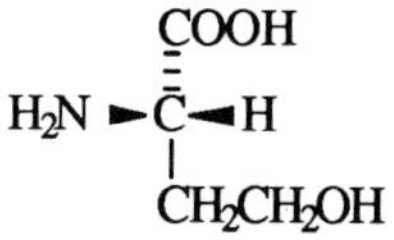

$$H_2N \blacktriangleright \overset{COOH}{\underset{CH_2CH_2OH}{C}} \blacktriangleleft H$$

(S)-form.

occurs in peas, Jack bean seeds *(Canavalia ensiformis).* The (R)-L-form is found in the seedlings of many leguminous plants. Homoserine is an intermediate in the conversion of aspartic acid into homocysteine and α-ketobutyrate in microbia and fungi (J.M.Lawrence, Phytochemistry, 1973, 12, 2207). It is also a structural component of 4'-*O*-diacetylglyceryl-*N*-trimethyl homoserine, which is obtained from the fronds of the fern, *Adiantum capillus veneris,* (N.Sato and M.Furuya, Plant Cell Physiol., 1983, 24, 1113).

The synthesis of homoserine has been reported (D.Guenther *et al.,* Chem. Ber., 1964, 2678 ; A.Kollonitsch *et al.,* J. Amer. Chem. Soc., 1964, 86, 1857; J.P.Grenstein *et al.,* Chemistry of the Amino Acids, 1965, Wiley, New York, 3, 2612). (R)-D-Homoserine has m.p. 203° dec. and [α] -8° in water.

2-Amino-4-(aminooxy)butanoic acid (O-aminohomoserine, canaline):

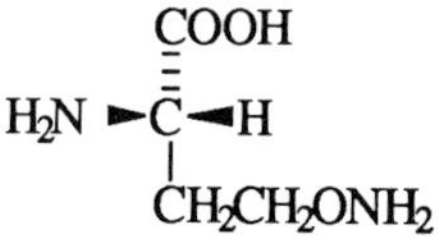

$$H_2N \blacktriangleright \overset{COOH}{\underset{CH_2CH_2ONH_2}{C}} \blacktriangleleft H$$

(S)-form.

is present as its (S)-L-form in *Canavalia ensiformis* and in the unripened seed of *Astragalus sinicus* (J.D.Williamson *et al.,* Life Sci., 1974, 14, 2481). It is a potent inhibitor of pyridoxal phosphate-containing enzymes. Its synthesis has been reported (C.Gilon *et al.,*

Tetrahedron, 1967, 23, 4441; A.J.Ozinskas and G.A.Rosenthal, J. Org. Chem., 1986, 51, 504). It exists as needles with m.p. 214° and [α] -8.31^o.

4-Acetoxy-2-aminobutanoic acid (O-acetylhomoserine, homoserine acetate):

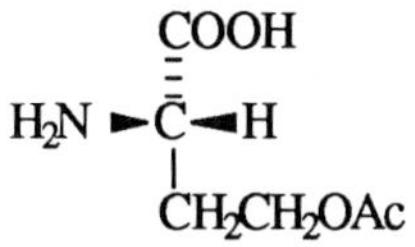

$$H_2N \blacktriangleright \overset{\displaystyle COOH}{\underset{\displaystyle CH_2CH_2OAc}{C}} \blacktriangleleft H$$

(S)-form.

in its (S)-form is found in *Pisum sativum* and has been isolated as crystals with m.p. 200° and [α] +4.5^o (N.Grobbelaar *et al.,* Nature (London), 1958, 182, 1358; H.A.Wald *et al.,* J. Amer. Chem. Soc.,1959, 81, 4367; N.Grobbelaar *et al.,* Phytochemistry, 1969, 8, 553).

Threonine (2-amino-3-hydroxybutanoic acid):

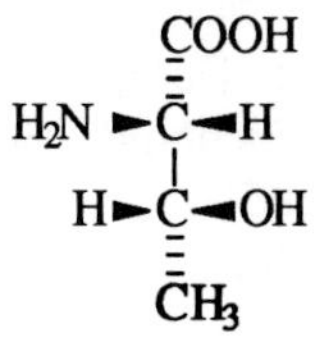

$$\begin{array}{c} COOH \\ H_2N \blacktriangleright C \blacktriangleleft H \\ H \blacktriangleright C \blacktriangleleft OH \\ CH_3 \end{array}$$

(2S,3R)-form.

(2S,3R)-L-Threonine is isolated from a wide variety of protein hydrolysates, m.p. 251-3°. Its *N*-protected derivatives are useful in peptide synthesis.

L-Threonine and its *allo*-isomer are obtained by the alkylation of acetaldehyde by nickel(II) complexes of (N-benzyl-L-propyl)-*o*-aminobenzaldehyde (Yu. N. Belokon *et al.,* Izv. Akad. Nauk. S.S.S.R., Ser. Khim., 1984, 804).

Methionine (2-amino-4-methylthiobutanoic acid, 4-methyl-mercapto-2-aminobutanoic acid):

$$H_2N \blacktriangleright \overset{\overset{\displaystyle COOH}{|}}{\underset{\underset{\displaystyle CH_2CH_2SCH_3}{|}}{C}} \blacktriangleleft H$$

(S)-form.

is obtained by the hydrolysis of many proteins, and its *N*-ethoxycarbonyl derivative is secreted by the healthy crown gall, induced by *Agrobacterium kumefaciens* (E.Messens *et al.,* E.M.B.O.J., 1985, 4, 571).

Methionine is synthesised from *S-tert.*-butylhomocysteine *via* *S*-methylation with iodomethane/silver perchlorate, followed by storage of the sulphonium salt at 40° for two hours (G.Chassainy *et al.,* Tetrahedron Lett., 1985, 26, 623).

The (S)-L-form provides a number of protected derivatives which are useful in peptide synthesis: *N*-formyl, m.p. 98-9°. It is produced by *Clostridium pasteurianum* and has been isolated from bee thorax; *N*-acetyl, m.p. 104°, also isolated from bee thoraxes; methyl ester hydrochloride, m.p. 150° (N.I.Sax, Dangerous Properties of Industrial Materials, 5th. Ed., Van Nostrand-Reinhold, 1979, 801; M.T.Clandinin *et al.,* Phytochemistry, 1974, 13, 585).

L-Methionine is diazotised to give a *S*-nitroso derivative which undergoes a spontaneous (S→N) intramolecular transfer reaction (T.A.Meyer and D.L.H.Williams, J. Chem. Soc. Chem. Commun., 1983, 1067).

DL-Methionine is converted into D-2-aminobutanoic acid by the action of *Methionine lyase* and *D-aminotransferase* (H.Tanaka *et al.,* Agric. Biol. Chem., 1985, 49, 2525).

Methionine reacts with hydroxyl radicals which are generated from Ti(III)/hydrogen peroxide to give an *S*-centred radical cation which initiates oxidative decarboxylation (M.J.Davis *et al.,* J. Chem. Soc. Perkin Trans. 2, 1983, 731).

S-Tert.- butylhomocysteine is made from methionine *via tert.* butylation, followed by equilibration with a nucleophile (G.Chassaing *et al.,* J. Org. Chem., 1983, 48, 1757).

Methionine sulphoxime (S-(3-amino-3-carboxypropyl)-S-methyl sulphoximine):

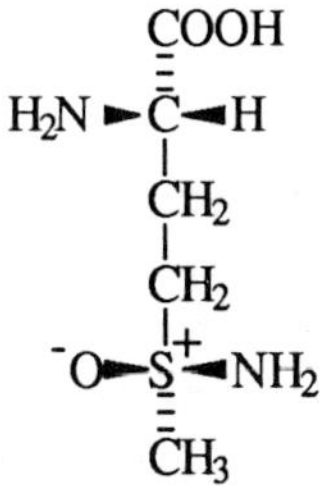

COOH
H_2N ► C ◄ H
CH_2
CH_2
^-O ► S^+ ◄ NH_2
CH_3

$(S)_C,(S)_S$-form.

is a toxic component of *Chestis glabra* (V.L.R.Jeanoda *et al.,* Phytochemistry, 1985, 24, 854). Its $(S)_C(S)_C$-form crystallises as prisms, m.p. 239°, whilst its $(S)_C(R)_C$-form has m.p. 235° (Y.Sugiyama *et al.,* Tetrahedron Lett., 1983, 24, 1471).

2,3-Diaminobutanoic acid:

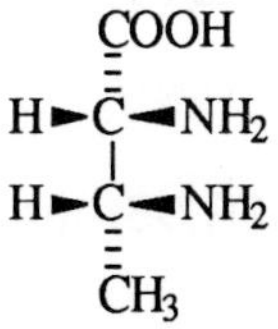

COOH
H ► C ◄ NH_2
H ► C ◄ NH_2
CH_3

(2R,3R)-form.

is known in its various forms (W.K.Hausmann *et al.,* J. Antibiot., 1969,

22, 207; E.Atherton *et al., ibid,* 1972, 25, 539; *Idem,* Hoppe-Seyler's Z. Physiol. Chem., 1973, 354, 689): (2S,3S)-L-*erythro*-form, hydrochloride, m.p. 202-4°; (±)-*erythro-,* hydrochloride, m.p. 178°; (2R,3S)-D-*threo*-, hydrochloride, m.p. 225-6°.

2,4-Diaminobutanoic acid:

$$\begin{array}{c} COOH \\ H \blacktriangleright C \blacktriangleleft NH_2 \\ | \\ CH_2CH_2NH_2 \end{array}$$

(R)-form.

The (R)-D-form of this acid is known as the oxalate, m.p. 205°; dipicrate, m.p. 187°, and *N2,N4*-dibenzoyl, m.p. 149°, derivatives. Its (S)-L-form can be isolated as the hydrochloride, m.p. 225°; its dipicrate, m.p. 187° and its *N4*-acetyl derivative, m.p. 208-11°, which is found in sugar beet and *Euphorbia pulcherrima* (F.Effenberger *et al.,* Angew. Chem. Int. Ed. Engl., 1979, 18, 474; L.Fowden, Phytochemistry, 1972, 11, 2271).

2,4-Diamino-3-methylbutanoic acid (4-aminovaline):

$$\begin{array}{c} COOH \\ H \blacktriangleright C \blacktriangleleft NH_2 \\ | \\ H \blacktriangleright C \blacktriangleleft CH_3 \\ CH_2NH_2 \end{array}$$

(2R,3S)-form.

can be isolated as its (2R,3S)-form from *Rhizobium* bacteria in *Lotus tenius* root nodules (G.J.Shaw *et al.,* Phytochemistry, 1981, 20, 1853).

2-Amino-4-hydroxy-3-methylbutanoic acid (4-hydroxyvaline):

$$
\begin{array}{c}
\mathrm{COOH} \\
| \\
\mathrm{H-C-NH_2} \\
| \\
\mathrm{H-C-CH_3} \\
| \\
\mathrm{H-C-H} \\
| \\
\mathrm{OH}
\end{array}
$$

is isolated from crowngall tumours of *Kalanchoe daigremontiana* m.p. 215°, 228° (J.J.Usher, J. Chem. Res. (S), 1980, 30; E.Galantay *et al.*, J. Org. Chem., 1963, 28, 98).

4-Amino-2-hydroxy-3-methylbutanoic acid

$$
\begin{array}{c}
\mathrm{COOH} \\
| \\
\mathrm{H-C-OH} \\
| \\
\mathrm{H-C-CH_3} \\
| \\
\mathrm{H-C-H} \\
| \\
\mathrm{NH_2}
\end{array}
$$

is an GABA inhibitor. Its various diastereomers can be obtained: (R,R)-form, m.p. 199-201°; (R,S)-form, m.p. 203-6°, and racemic form, m.p. 195-6° (B.Ringdahl *et al.,* Acta Chem. Scand., Ser. B, 1980, 34, 731; L.Brehm *et al., ibid,* 1979, 33, 52; T.Yoneta *et al.,* Bull. Chem. Soc. Jpn., 1978, 51, 3296; H.Sato *et al., ibid*, 1976, 49, 2815).

3-Amino-2-hydroxy-4-phenylbutanoic acid:

COOH
HO►C◄H
H►C◄NH_2
CH_2Ph

(2S,3R)-form.

has been synthesised and characterised as the *N*-benzyloxycarbonyl, m.p. 175-6⁰ (Y.-E.Shih *et al.,* Heterocycles, 1978, 9, 1277; R.Nishizawa *et al.,* J. Med. Chem., 1977, 20, 510).

4-Amino-3-hydroxybutanoic acid:

CH_2COOH
HO►C◄H
CH_2NH_2

(R)-form.

is synthesised from D-arabinose by a route which can be adapted for the synthesis of (S)-carnitine (K.Bock *et al.,* Acta Chem. Scand., Ser. B, 1983, B37, 341).

(S)-4-Amino-3-hydroxybutanoic acid is made by conventional methods from (S)-malic acid *via* its cyclic anhydride (B.Rajashekhar and E.T.Kaiser, J. Org. Chem., 1985, 50, 5480). Its *N*-trimethyl analogue, carnitine, can be made from $Me_2NCH_2COCH_3$ and diethyl carbonate in the presence of sodium hydride, followed by reduction and *N*-methylation (C.A.Hosner and R.N.Comber, J. Org. Chem., 1985, 50, 3627).

The (R)-form is obtained from (2S,4R)-4-hydroxyproline using a decarboxylative electrochemical methylation (P.Renaud and D.Seebach, Synthesis, 1986, 424; Angew. Chem., 1986, 98, 836.

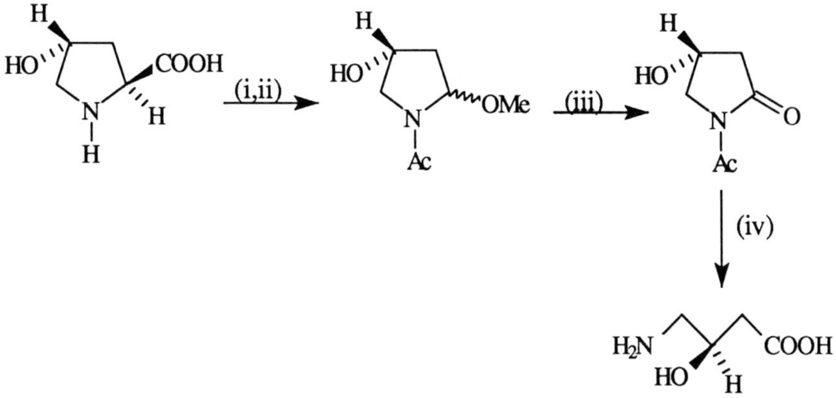

Reagents: (i) Ac_2O; (ii) e, MeOH, carrier electrolyte; (iii) AcO_2H; (iv) 4M HCl.

Ethyl 2-amino-3-oxobutanoate (ethyl 2-aminoacetoacetate, 2-aminoacetoacetic ester), $H_3CCOCH(NH_2)COOEt$, is not known in its free state, but its hydrochloide, m.p. 95^0 dec., can be obtained as yellow hydroscopic needles (T.W.Doyle *et al.,* Can. J. Chem., 1977, 55, 484; H.Idia *et al.,* Synth. Commun., 1973, 225).

(3S,4S)-Statine is known as its *N*-Boc ester, $Me_2CHCH_2CH(NHBoc)CH(OH)CH_2COOMe$. It has been synthesised from *N*-Boc-L-leucinal and (-)-gabaculene (B.Rague *et al.,* Bull. Soc. Chim. Fr., 1983, 230), and also from Boc-L-leucinal *via enantio-* and *erythro* selective aldol condensation with (S)-4-(1-methylethyl)-3-[(methylthio)acetyl]-2,5-oxazolidione (P.W.K.Woo, Tetrahedron Lett., 1985, 26, 2973).

4-Amino-2,3-dihydroxybutanoic acid:

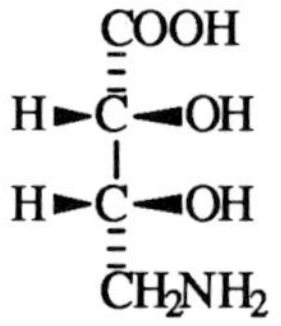

COOH
H–C–OH
H–C–OH
CH_2NH_2

is known as the (2R,3R), m.p. 217-9°, and (2R,3S), m.p. 221-5° (J.A.Musich *et al.*, J. Amer. Chem. Soc., 1978, 100, 4865; N.R.Howe *et al.*, Science, 1975, 189, 386).

4-Amino-2,3-dihydroxy-3-methylbutanoic acid:

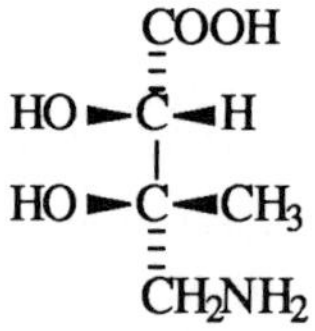

COOH
HO–C–H
HO–C–CH_3
CH_2NH_2

is isolated as its (2S,3S)-form from Carzinophilin; it has m.p. 108-110.5° (P.Garner *et al.*, Tetrahedron Lett., 1985, 26, 3299).

2-Amino-2-hydroxy-1,2,4-butanetricarboxylic acid (2-amino-4-carboxy-4-hydroxyadipic acid:

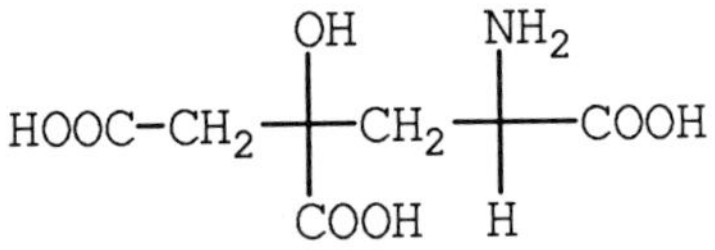

$HOOC{-}CH_2{-}C(OH)(COOH){-}CH_2{-}C(NH_2)(H){-}COOH$

is isolated from *Caylusea abyssinica*, in diastereomeric forms, ^1H n.m.r. data have been reported (O.Olsen *et al.,* Phytochemistry, 1980, 19, 1717).

2-Amino-3-hydroxypentanoic acid (β-Hydroxynorvaline):

COOH
H_2N ►C◄ H
HO ►C◄ H
CH_3

(2S,3S)-form.

as its (2S,3S)-L-*erythro*-form is a component of Cycloheptamycin. It has been synthesised (W.O.Godtfredson *et al.,* Tetrahedron, 1970 26, 4931). The racemic *erythro*-form has m.p. 244^{0} dec., *N*-benzoyl, m.p.180-1^{0}, *N*-benzoyl ethyl ester, m.p. 115.5-116^{0}. The racemic *threo*-form has m.p. 220-1^{0} dec., *N*-benzoyl, m.p. 151-3^{0}.

2-Amino-4-hydroxypentanoic acid (4-hydroxynorvaline):

COOH
H_2N ►C◄ H
CH_2
H ►C◄ OH
CH_3

(2S,3S)-form

as its (2S,4R)-L-*threo*-form is present in the seeds of legumes, notably *Lathyrus odoratus* (sweet pea). It can also be isolated from the carpopores of *Boletus satanas*, and is produced by *Streptomyces griseosporus* (L.Fowden, Nature(London), 1966, 209, 807). It has been synthesised (G.B.Barlow *et al.,* J. Chem. Soc., 1964, 141; P.Matzinger *et al.,* Helv. Chim. Acta, 1972, 55, 1478), as plates, m.p. 194-5^{0} dec.. Its lactone hydrochloride is the antibiotic AL 719Y, which is a weak antiviral agent.

It forms needles, m.p. 198-200^{0}. The (2S,4S)-L-*erythro*-form can be isolated from the carpophores of *B. satanas* as needles, m.p. 192-3^{0}, it produces a lactone hydrochloride, m.p. 166-7^{0} (S.Narayanan *et al.,* J. Antibiot., 1980, 33, 1249).

2-Amino-4-hydroxy-3-methylpentanoic acid (4-hydroxyisoleucine):

COOH
H_2N ►C◄ H
H_3C ►C◄ H
H ►C◄ OH
CH_3

(2S,3R,4R)-form

as its (2R,3R,4R)-form is a minor amino acid constituent of *Trigonella foenumgraecum* seeds, whilst the (2S,3R,4R)-form is the major constituent (L.Fowden *et al.,* Phytochemistry, 1973, 12, 1707).

4-Amino-3-hydroxy-2-methylpentanoic acid:

COOH
H_3C ►C◄ H
HO ►C◄ H
H ►C◄ H
NH_2

is a structural component of bleomycin, and it is made from Boc-D-alanine by treatment with methyllithium, followed by reaction with ethyl acetate and butyllithium (R.M. Di Pardo and M.G. Bock, Tetrahedron Lett., 1983, 24, 4805).

2-Amino-4-chloropentanoic acid:

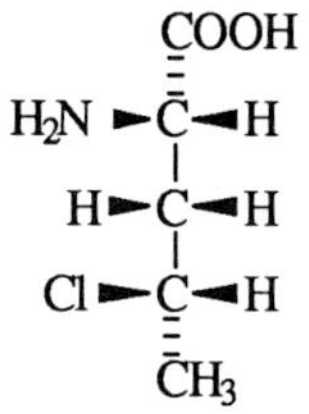

$$
\begin{array}{c}
COOH \\
H_2N \blacktriangleright C \blacktriangleleft H \\
| \\
H \blacktriangleright C \blacktriangleleft H \\
| \\
Cl \blacktriangleright C \blacktriangleleft H \\
CH_3
\end{array}
$$

is isolated from *Streptomyces griseosporus* in its (2S,4S)-form. It is weakly active against gram-negative bacteria and viruses, but its effects are reversed by L-leucine. It is obtained as plates, m.p. 162-5° (S.Narayanan *et al.*, J. Antibiot., 1980, 33,1249).

2-Amino-3-methyl-4-oxopentanoic acid, $H_3CCOCH(CH_3)CH(NH_2)COOH$, can be isolated from *Bacillus cereus 439* fermentations; its synthesis has been reported (D.Perlman *et al.*, Bioorg. Chem., 1977, 6, 263).

Ornithine (2,5-diaminopentanoic acid):

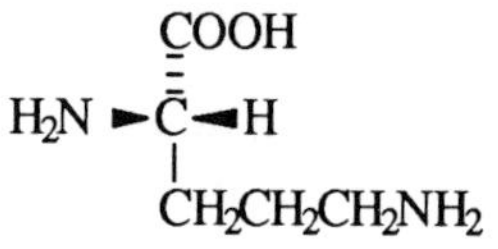

$$
\begin{array}{l}
\quad\quad\;\; COOH \\
H_2N \blacktriangleright C \blacktriangleleft H \\
\quad\quad\;\; | \\
\quad\quad\;\; CH_2CH_2CH_2NH_2
\end{array}
$$

(S)-form

is a constituent of a pepidoglycan from *Coryne bacterium* (K.H.Schleifer *et al.*, Arch. Microbiol., 1983, 134, 243). *N*-Ethoxycarbonyl-L-ornithine is isolated from *Streptococcus lactis*, grown in ornithine supplemented media (J.Thompson *et al.*, J. Bacteriol., 1986, 167, 522).

The synthesis of ornithine (m.p. 140°) has been reported (T.J.Gilbertson *et al.*, J. Amer. Chem. Soc., 1967, 89, 7085; S.H.Hedges *et al.*, Phytochemistry, 1981, 20, 2064). (S)-L-Ornithine provides a number of derivatives, useful in peptide synthesis: *N*-benzyloxycarbonyl, m.p. 254-5°; *N,N-bis*(benzyloxycarbonyl), methyl ester, m.p. 112-4°; *N*-acetyl, m.p. 226-7°.

Lysine (2,6-diaminohexanoic acid):

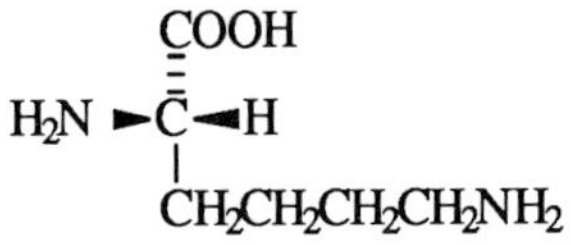

$$H_2N \blacktriangleright \overset{\underset{\equiv}{COOH}}{C} \blacktriangleleft H$$
$$|$$
$$CH_2CH_2CH_2CH_2NH_2$$

is found widely in the (S)-L-form in protein hydrolysates, e.g. casein, egg albumen, fibrin, beet molasses, blood corpuscles and conifer seeds (L.Fowden, Phytochemistry, 1972, 11, 2271). *N*-Methyllysine is a constituent of the flagellins from *Proteus norganic* (B.S.Baker *et al.*, Microbios. Lett., 1983, 23, 7). Lysine is also a constituent of hypusine, (*N'*-(4-amino-2-hydroxybutyl)lysine), which can be isolated from brain tissue (M.H.Park *et al.*, Methods Enzymol., 1983, 94, 458).

Lysine can be isolated as the (S)-form, m.p. 224-5^0 and [α] +14.6^0 (L.Sodek, Phytochemistry, 1976, 15, 1903). It yields numerous derivatives : dihydrochloride, m.p. 193^0 and $[\alpha]_D$ +15.3^0; *N*^{6}-benzoyl, m.p. 235^0 and $[\alpha]_D$ +20.12^0; *N*^{2},*N*^{6}-dibenzoyl (lysuric acid), m.p. 149-50^0; 2-*N*-(2,4-dinitrophenyl), m.p. 260^0; *N*^{6}-methoxycarbonyl, isolated from *Saittaria pygmacea*, m.p. 265-70^0 (H.Matsutani *et al.*, Phytochemistry, 1979, 18, 661; S.Sifniades *et al.*, J. Amer. Chem. Soc., 1976, 98, 3738).

Lysine undergoes oxidative conversion into 2-ketoglutaric acid, a reaction catalysed by *Saccharpine dehydrogenase* (M.S.Simonson and R.E.Eckel, Anal. Biochem., 1985, 147, 230).

L-Lysine is converted into L-pipecolic acid *via* *N*-benzylation of the *N*^{6}-benzylidene derivative, followed by treatment with sodium hypobromite solution and the debenzylation with hydrogen bromide (H.Mihara *et al.*, Mem. Fac. Sci., Kyushu Univ., Ser. C, 1983, 14, 123).

Hydroxylysine undergoes oxidation with periodic acid to yield glutamic acid *via* the semialdehyde, but with sodium periodate in alkaline solutions it is oxidised to pyrroline-5-carboxylic acid (G.Y.Wu and S.Seifter, Anal. Biochem., 1985, 147, 403).

Hypusine is made by the condensation of *N*-benzyloxycarbonyl-L-lysine benzyl ester with the chiral isooxazolidine,

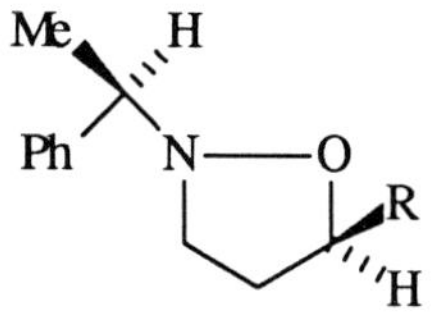

followed by reduction of the resulting imine with sodium borohydride and hydrogenolysis over Pd/C (C.M.Rice and B.Ganem, J. Org. Chem., 1983, 48, 5048).

3,5-Diaminohexanoic acid:

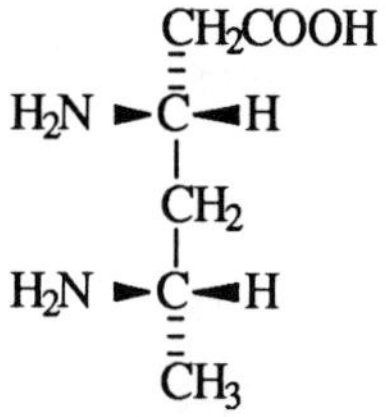

$$\begin{array}{c} CH_2COOH \\ H_2N \blacktriangleright C \blacktriangleleft H \\ CH_2 \\ H_2N \blacktriangleright C \blacktriangleleft H \\ CH_3 \end{array}$$

(3S,5S)-form

as its (3S,5S)-form, is a metabolite from the fermentation of β-lysine by *Clostridium sticklandii.* It is known as the dihydrochloride, m.p. 204-8°, and also: *N*^{3},*N*^{5}-dibenzoyl, m.p. 199-204°. It forms a β-lactam, 4-amino-6-methyl-2-piperidone, m.p. 101-2° (L.Tsai *et al.,* Arch. Biochem. Biophys., 1968, 125, 210; F.Kunz *et al.,* Helv. Chim. Acta, 1978, 61, 1139; J.Retey *et al, idem,* 1978, 61, 2989).

2-Amino-4-methyl-4-hexenoic acid:

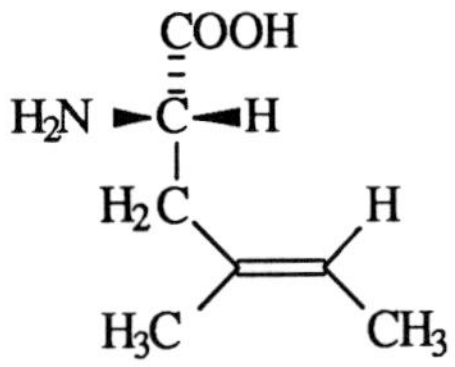

$$\begin{array}{c} COOH \\ H_2N \blacktriangleright \overset{|}{C} \blacktriangleleft H \\ | \\ H_2C \quad\quad H \\ \quad \diagdown \; \diagup \\ \quad C{=}C \\ \quad \diagup \; \diagdown \\ H_3C \quad\quad CH_3 \end{array}$$

(S)-(E)-form

is obtained as its (S)-(E)-L-form from the seeds of *Aesculus californica* with [α] -61°. The (±)-(E)-form has been synthesised, m.p. 219° dec. (J.Edelson *et al.*, J. Amer. Chem. Soc., 1959, 81, 5150; L.Fowden *et al.*, Phytochemistry, 1968, 7, 5150; J.E.Boyle *et al.*, *ibid*, 1971, 10, 2671; E.Gellert *et al.*, *ibid*, 1978, 17, 802).

2-Amino-6-diazo-5-oxohexanoic acid (6-diazo-5-oxonorleucine);

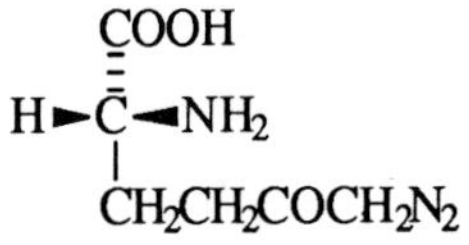

$$\begin{array}{c} COOH \\ H \blacktriangleright \overset{|}{C} \blacktriangleleft NH_2 \\ | \\ CH_2CH_2COCH_2N_2 \end{array}$$

(R)-form

is produced from *Streptomyces ambofaciens* as a yellow crystalline solid, m.p. 142-55°. It is unstable in aqueous solution, and very unstable in both acid and alkaline solution (H.W.Dion *et al.*, J. Amer. Chem. Soc., 1956, 78, 3075; D.A.Cooney *et al.*, Biochem. Pharmacol., 1976, 25, 1859). Its racemic form is a tumour inhibitor, m.p. 145-55° (H.A. de Wald *et al.*, J. Amer. Chem. Soc., 1958, 80, 3941; G.Pettit *et al.*, *J. Org. Chem.*, 1983, 48, 741).

Second Supplements to the 2nd Edition of Rodd's Chemistry of Carbon Compounds, Vol. IE/F/G, edited by M. Sainsbury

Chapter 20

TRIHYDRIC ALCOHOLS AND THEIR OXIDATION PRODUCTS (continued)

B.J.COFFIN

Hydroxy- and amino- dicarboxylic acids.

Aminopropandioic acid (aminomalonic acid), $H_2NCH(COOH)_2$ can be synthesised to give a hydrated crystalline product , m.p. 109°, which readily undergoes decarboxylation (N.H.Khan *et al.,* Indian J. Chem., Sect. B, 1977, 15, 573). It provides a number of derivatives: hydrochloride, m.p. 162° dec.; dimethyl ester hydrochloride, m.p. 159°; diamide, m.p. 167°; *N*-methyl, m.p. 137-42° dec. (E.Hardegger *et al.,* Helv. Chim. Acta, 1956, 39, 980; J.Hess, Chem. Ber., 1972, 105, 441).

Diethyl acetamidomalonate can be used as the starting material for a number of α-amino acids and related compounds. These include, for example, 2,6-dihalotyrosins (R.A.Pascal and Y.C.J.Chan, J. Org. Chem., 1985, 50, 408) and β-methyleneglutamic acid (N.A.Sasaki *et al.,* Int. J. Pept. Protein Res., 1986, 227, 360). The half nitrile of this reagent can be reacted with various alkyl halides and used to synthesise betaines. It has also been used to obtain alanine derivatives (M.Smulkoski *et al.,* Pol. J. Chem., 1982, 56, 699). (For related work see C.Petermann and J.L.Fauchere, Helv. Chim. Acta, 1983, 66, 1513).

Acylamino malonates are also substrates for asymmetric syntheses of β–amino acids (Y.H.Paik and P.Dowd, J. Org. Chem., 1986 51, 2910; R.Miller and K.Ulsson, Acta Chem. Scand., Ser. B, 1985, B39, 717; L.S.Payne and J.Bager, Synth. Commun., 1985, 15, 1277; C.R.Creveling and K.K.Kirk, Biochem. Biophys. Res. Commun., 1985, 130, 1123; D.S.Kemp and T.P.Curran, J. Org. Chem., 1986, 51, 2377; N.A.Sasaki *et al.,* Int. J. Pept. Protein Res., 1986, 27, 360; O.Tiba and C.G.Overberger,

Polm. Sci. Technol. (plenum), 1985, 31, 419).

Optically pure (L)-vinylglycine has been synthesised from (L)-glutamate esters of *N*-hydroxy-2-selenopyridine (D.H.R. Barton *et al.*, Tetrahedron, 1985, 41, 4347).

Diethylbenzamidomalonate can be reacted with trimethylsilylpropargyl bromide to give *N*-benzoylpropargylglycine which, on decarboxylation followed by reaction with *N*-iodosuccinimide, gives (Z)-2-benzamido-4-hydroxy-5-iodopent-4-enoic acid. This reaction with phenyl and benzyl substituted benzamidomalonate esters can be used to prepare 2-phenyl and 2-benzyl analogues of unsaturated α-amino acids (M.J. Sofia *et al.*, J. Org. Chem., 1983, 48, 3318).

Cyanoselenylbenzylacetamido malonate esters can used similarly to provide m- and p-methylselenylphenylalanines (C.A.Loeschorn *et al.*, Tetrahedron Lett., 1984, 25, 3387).

2-Amino-2-methylpropanedioic acid (1-aminoisosuccinic acid, isoasparaginic acid, isoaspartic acid) :

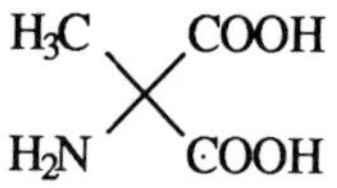

H_3C COOH
H_2N COOH

can be obtained as prisms which explode at about 250°. It provides a diethyl ester as its *N*-acetyl derivative, m.p. 90°, and a diamide, m.p. 200-1° dec. (J.W.Thanassi *et al.*, Biochemistry, 1962, 1, 975; R.G.Asperger *et al.*, Inorg. Chem., 1967, 6, 796; J.W.Thanassi *et al.*, J. Org. Chem., 1971, 36, 3019; R.C.Job, J. Chem. Soc. Chem. Commun., 1977, 258; R.S.Hosame *et al.*, J. Amer. Chem. Soc., 1982, 104, 235).

Aspartic acid (aminobutanedioic acid, aminosuccinic acid):

$$\begin{array}{c} COOH \\ H_2N \blacktriangleright \overset{\vdots}{C} \blacktriangleleft H \\ | \\ CH_2COOH \end{array}$$

can be isolated as its (S)-L-form from sugar cane and sugar beet molasses and from a number of proteins and peptides. It has m.p. 269-71^{0} and [α] +5.05^{0}. *N*-Acetyl-L-aspartic acid is an essential heat stable factor for the conversion of lignoceric acid into cerebronic acid and glutamic acid (H.Shigematsu *et al.,* J. Neurochem., 1983, 40, 814).

L-Aspartic acid can be synthesised from ammonium fumarate by use of immobilised cells of *Alcaligenes metalcaligenes* (V.Vojtisek *et al.,* Biotechnol. Bioeng., 1986, 28, 1072).

Aspartic acid can be made by converting laevulinic acid into its 3-bromo-derivative the oxime of which can be subjected to the Beckmann rearrangement (U.R.Joshi and P.A.Linaya, Indian J. Chem. Sect. B, 1982, 21, 1122).

The alkylation of a dialkyl malonate with *N*-benzyloxycarbonyl-L-alanyl-2-chloro-glycine methyl ester followed by hydrolysis provdes L-aspartic acid (S.Shiono and K.Harada, Bull. Chem. Soc. Jpn., 1985, 58, 1061).

DL-β-Carboxyaspartic acid can be made *via* the alkylation of the Schiff base, ZN=CHCOOMe, with the anion of di-*tert.* butyl malonate (D.H.Rice and M.K.Dhaon, Tetrahedron Lett., 1983, 24, 1671).

β-Carboxy-D-aspartic acid diester can be prepared *via* the lithio-bislactim:

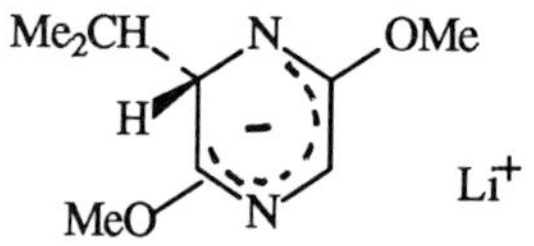

derived from L-valylglycinedioxopiperazine, which undergoes chlorination with Cl_3CCCl_3 followed by reaction with a malonic ester (U.Schollköpf *et al.,* Angew. Chem., 1985, 97, 1065).

Aspartic acid-1-^{13}C can be made using the Strecker synthesis with $EtOOCCH_2CHO$ potassium cyanide-^{13}C (U.Fotader and D.Cowburn, J. Labelled Compd. Radiopharm., 1983, 20, 1003).

L-Aspartic acid is the starting material in a process which allows the formation of both 4-aminobutyrolactones and 2-phenyl-1,3-oxazolines in chiral forms (M.C.Fusee and J.E.Weber, Appl. Environ. Microbiol., 1984, 4, 694; G.J.McGarvey *et al.,* Tetrahedron Lett., 1983, 24, 2733) **(Scheme 1).**

The useful chiral synthon, (S)-3-hydrobutyrolactone, can be made from L-aspartic acid *via* the elaboration of (S)-malic acid which can be formed from L-aspartic acid by its reaction with nitrous acid (S.Henrot *et al.,* Synth. Commun., 1986, 16, 153).

N-Acylaspartate α-esters undergo oxidative β-decarboxylation with sodium hypochlorite to the corresponding *N*-acyldehydroalanines through *N*-chlorination, followed by dehydrochlorination (M.Seki *et al.,* Agric. Biol. Chem., 1984, 48, 1251).

Aspartic semi-aldehyde (protected) reacts with L-proline ethyl ester to give the proline analogue of nicotinamide (J.Faust *et al.,* Tetrahedron, 1983, 39, 1593).

Ethyl *N*-methoxycarbonyl-L-aspartate reacts with benzene under Friedel-Crafts conditions to give phenacylglycine derivatives (T.Moriya *et al.,* Chem. Express, 1986, 1, 157).

(S)-L-Aspartic acid, m.p. 251°, provides a number of derivatives which are useful in the synthesis of peptides: 4-methyl ester, m.p. 181°; dimethyl ester hydrochloride, m.p. 116-7°; diamide, m.p. 131°; *N*-

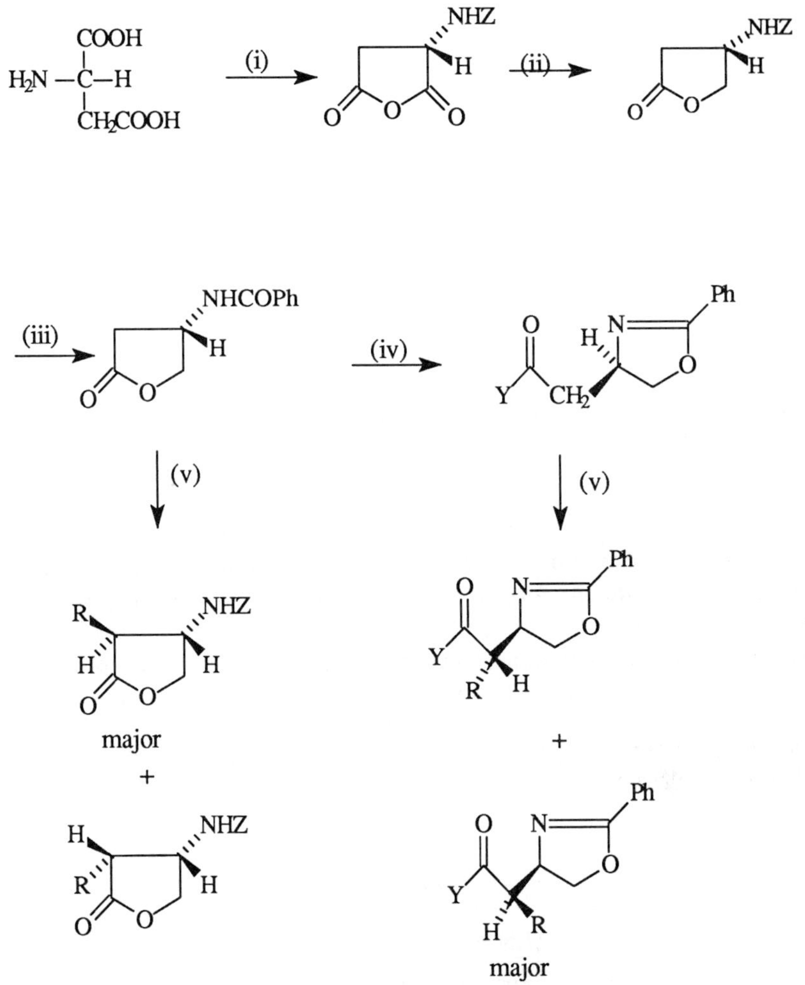

Reagents: (i) ZCl then Ac_2O; (ii)$NaBH_4$; (iii)HBr-AcOH then PhCOCl-py; (iv) EtO^- or Me_2NH; (v) $LiPr_2N$, RX.

Scheme 1

benzoyl, m.p. 171-3°; *N*-2,4-dinitrophenyl, m.p. 186-7° (K.Harada *et al.,* Bull. Chem. Soc. Jpn., 1983, 56, 653; S.Fujiwara *et al., ibid,* 1964, 37, 344).

2-Amino-3-methylbutanedioic acid (3-methylaspartic acid, 2-amino-3-methylsuccinic acid):

```
       COOH
  H ►─ C ─◄ NH2
       |
H3C ►─ C ─◄ H
       COOH
```

(2R,3R)-form

is produced by *Clostridium tetanomorphum* and is a component of the antibiotics, Aspartocin, Amphomycin and Glutamycin, as its (2S,3S)-L-*threo*-form, [α] -12.4°. It has been synthesised in the racemic form and characterised as the di-ethyl ester hydrochloride, m.p. 118° (K.Mori *et al.,* Tetrahedron Lett., 1981, 1127).

2-Amino-3-hydroxybutanedioic acid (3-hydroxyaspartic acid, 2-amino-3-hydroxysuccinic acid):

```
       COOH
H2N ►─ C ─◄ H
       |
 HO ►─ C ─◄ H
       COOH
```

(2S,3R)-form

can be isolated as its (2S,3R)-L-*erythro*-form from *Astragalus sinicus* (T.Ishiyama *et al.,* J. Antibiotic., 1975, 28, 821), whilst the (2S,3S)-L-*threo*-form, m.p. 210^{0} dec. and [α] +46.0^{0}, is a component of the antibiotic derived from *Streptomyces spp.* and *Arthrinium phaeospermum* (M.Teintze *et al.,* Biochemistry, 1981, 20, 6446). *Erythro*-2-amino-3-hydroxybutanedioic acid, m.p. 229^{0}, is a component of bovine protein C, which acts as an anticoagulant vitamin K-dependant plasma protein (T.Drakenberg *et al.* Proc. Navl. Acad. Sci., U.S.A., 1983, 80, 1802). These various forms can be obtained as crystalline solids (A.Singerman *et al.,* Tetrahedron Lett., 1968, 4733; H.Okai *et al.,* Bull. Chem. Soc. Jpn., 1969, 42, 3550). *Erythro*-2-amino-3-hydroxy-L-butanedioic acid can be synthesised *via* an intermediate formed from the epoxide of (+)-diethyl tartarate by ring opening with azide anion, and selective reduction of the resulting azido-ester (G.Zimmerman *et al.,* Liebigs Ann. Chem., 1985, 2165):

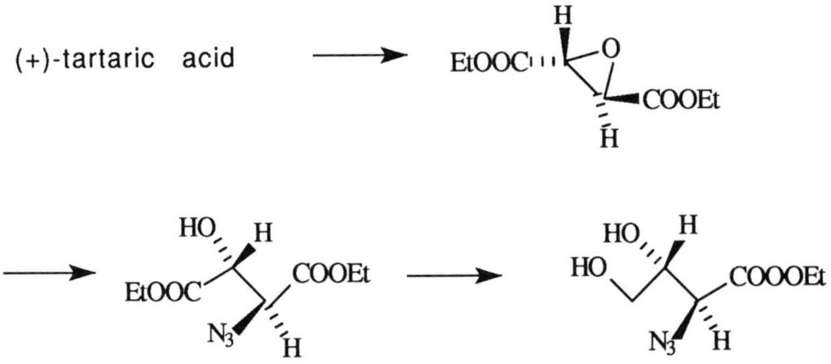

(+)-tartaric acid ⟶ EtOOC, H, O, COOEt, H

⟶ HO, H, EtOOC, COOEt, N_3, H ⟶ HO, H, HO, COOOEt, N_3, H

N-Boc-L-*erythro*-3-benzyloxyaspartic acid can be made from (R,R)-tartaric acid *via* partial debenzylation and other conventional steps (T.G.Hansson and J.O.Kihlberg, J. Org. Chem., 1986, 51, 4490).

2-Amino-1,1,2-ethanetricarboxylic acid (3-carboxyaspartic acid), $(HOOC)_2CHCH(NH_2)COOH$, is a constituent of ribosomal proteins of *E. coli*

and its synthesis has been reported (E.B.Henson *et al.,* Tetrahedron, 1981, 37, 2561; M.R.Christ, Diss. Abstr. Int. B, 1982, 42, 4420; N.E.Dixon *et al.,* J. Amer. Chem. Soc., 1982, 104, 6716).

2-Amino-3-fluorobutanedioic acid (3-fluoroaspartic acid):

$$\begin{array}{c} COOH \\ H \blacktriangleright \overset{|}{C} \blacktriangleleft NH_2 \\ F \blacktriangleright \overset{|}{C} \blacktriangleleft H \\ COOH \end{array}$$

(2RS,3RS)-form

is a cytotoxic agent, m.p. 175^{0} dec.. This property is also shown by its monoamide, m.p. 177^{0} dec. (A.M.Stern *et al.,* J. Med. Chem., 1982, 25, 544).

Asparagine (2-aminosuccinamic acid):

$$\begin{array}{c} COOH \\ H_2N \blacktriangleright \overset{|}{C} \blacktriangleleft H \\ CH_2CONH_2 \end{array}$$

(S)-form

is widely distributed in the plant kingdom. It can be isolated from asparagus, beetroot and beans (J.P.Greenstein *et al.,* Chemistry of the Amino Acids, 1961, Wiley, New York, 2, 1257; 3, 1856). It can be synthesised in the form of rhombic crystals, m.p. 234-5^{0} (rapid heating). It provides a range of derivatives: *N*-acetyl, m.p. 203-4^{0}, *N*-benzoyl, m.p. 189^{0}; *N*-(2,4-dinitrophenyl), m.p. 180-2^{0}; *N*-methyl (isolated from *Corallocarpus epigaeuss*);*N*-ethyl (isolated from *Ecballium elaterium*) (D.O.Gray *et al.,* Nature (London), 1961, 189, 401).

L-Asparagine can be used to prepare pipecolic acid derivatives (B.D.Christie and H.Rapoport, J. Org. Chem., 1985, 50, 1239).

α-Asparagine:

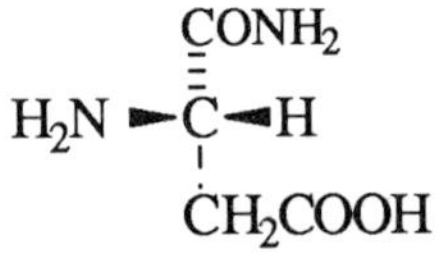

$$\begin{array}{c} CONH_2 \\ H_2N \blacktriangleright C \blacktriangleleft H \\ CH_2COOH \end{array}$$

(S)-form.

can be isolated as the (S)-L-form, [α] +14.8^0 (*N*-benzyloxycarbonyl derivative, m.p. 1030 and [α] -12.8^0) (W.Tittelbach-Helmrich, Chem. Ber., 1965, 98, 2051; R.Straka *et al.*, Collect. Czech. Chem. Commun., 1977, 42, 560).

Glutamic acid (2-aminopentanedioic acid, 2-aminoglutaric acid):

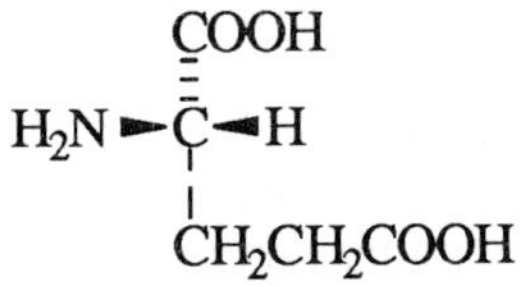

$$\begin{array}{c} COOH \\ H_2N \blacktriangleright C \blacktriangleleft H \\ CH_2CH_2COOH \end{array}$$

(S)-form

can be isolated as its (S)-L-form from the acid hydrolysis of many proteins. It forms rhombic crystals, m.p. 224-5^0 and [α] +17.0^0.

Syntheses of both the racemic and the D-form have also been described (J.P.Greenstein *et al*., Chemistry of the Amino Acids, 1961, Wiley, New York, Chap. 25, 1929; Org. Synth., Coll. Vol., 1, 286; N.I.Sax, Dangerous Properties of Industrial Materials, 5th. Ed., Van Nostrand-Reinhold, 1979, 705). A number of derivatives can be made from (S)-L-glutamic acid: hydrochloride, m.p. 202° dec., *N*-benzoyl, dimethyl ester, m.p. 216°; monoethyl ester, m.p. 194°; *N*-acetyl, m.p. 199° (the last compound acts as an intermediate in ornithine biosynthesis).

Threo- and *erythro*-4-chloro- and 4-bromo-glutamic acids can be made from the corresponding hydroxyglutamic acid by reaction with *N*-methoxycarbonylphthalimide and phosphorus pentahalide(I.M.Kocheva *et al.,* Zh. Org. Khim., 1983, 19, 283).

L-Homoglutamic acid can be synthesised from *N*-acetyl-L-lysine ethyl ester using *tert*-butyl hypochlorite, for a chlorination step, followed by dehydrochlorination and hydrolysis (M.Kondo *et al.,* Bull. Chem. Soc. Jpn., 1985, 58, 1171).

Glutamic acid-3-^{13}C can be made from $EtOOCCH_2CH_2CHO$ by reaction with potassium cyanide-^{13}C, or from $PrOOCCH_2CH_2Br$ *via* the acetamidomalonate route (U.Fotader and D,Cowburn, J. Labelled Compd. Radiopharm., 1983, 20, 1003).

The protected forms of dehydroglutamic acid can be made from the corresponding 2-ketoglutaric acids by condensation with benzyl carbonate (C.Shin *et al.,* Tetrahedron Lett., 1985, 26, 85), or by the addition of malonic esters to nitriles, catalysed by stannous chloride (F.Scarvo and P.Helquist, *ibid,* 1985, 26, 2603).

L-Glutamic acid can be converted into (S)-$H_2NCH_2CH(OH)CH_2CH_2COOH$ *via* (S)-5-carboxybutyrolactone (C.Herdeis, Synthesis, 1986, 232).

δ-*N*-Hydroxy-L-ornithine derivatives, such as the *N*-acetyl analogue, can be prepared from L-glutamic acid (R.K.Olsen *et al.,* J. Org. Chem., 1984, 49, 3527; B.H.Lee and M.J.Miller, Tetrahedron Lett., 1984, 25, 927) **(Scheme 2).**

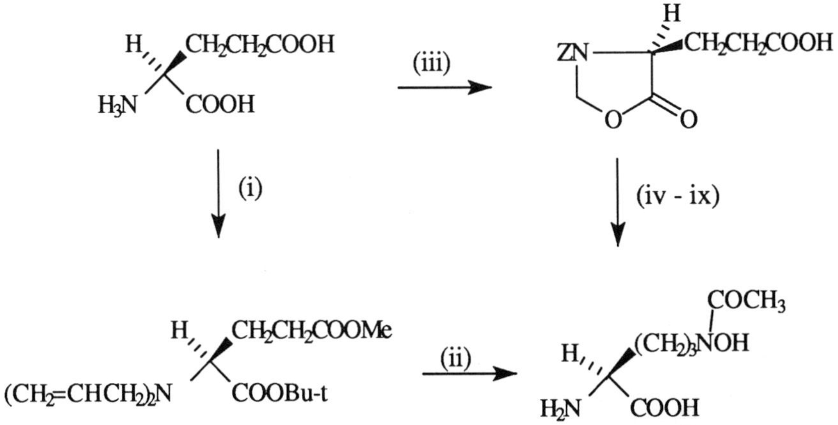

H, CH_2CH_2COOH H_3N COOH —(iii)→ ZN—H CH_2CH_2COOH O O

(i) (iv - ix)

H, CH_2CH_2COOMe $(CH_2{=}CHCH_2)_2N$ COOBu-t —(ii)→ $COCH_3$ H, $(CH_2)_3NOH$ H_2N COOH

Reagents: (i) esterify, $CH_2{=}CHCH_2Br$; (ii) $(Ph_3)_3RhCl$, CF_3COOH, hydrogenolysis; (iii) ZCl, HCHO/TsOH; (iv) $SOCl_2$; (v) Bu_3SnH or Li(OBu-*t*)$_3$AlH; (vi) $PhCH_2ONH_2$; (vii) $NaBH_3CN$: (viii) OH^-; (ix) HBr.

Scheme 2

Glutamic acid can be converted into its hydroxyglutamic acid derivatives (*threo*-4-hydroxy-D-glutamic acid, starting with *N*-phthaloyl-4-bromo-D-glutamic acid dimethyl ester; whilst *threo*-3-hydroxy-L-glutamic acid can be made from L-serine, starting with a stereoselective addition of a serinal derivative onto 1-methoxy-3-(trimethylsiloxy)-1,3-butadiene (P.Garner, Tetrahedron Lett., 1984, 25, 5855) **(Scheme 3).**

L-Glutamic acid-4-semialdehyde can be prepared from glutamic acid *via* *N*-acetyl-L-asparagine methyl ester. The aldehyde can also be converted by a series of further steps into L-tryptophan with negligible loss of chirality at the α-position (F.Masumi *et al.*, Chem. Pharm. Bull., 1982, *30*, 3831) **(Scheme 4).**

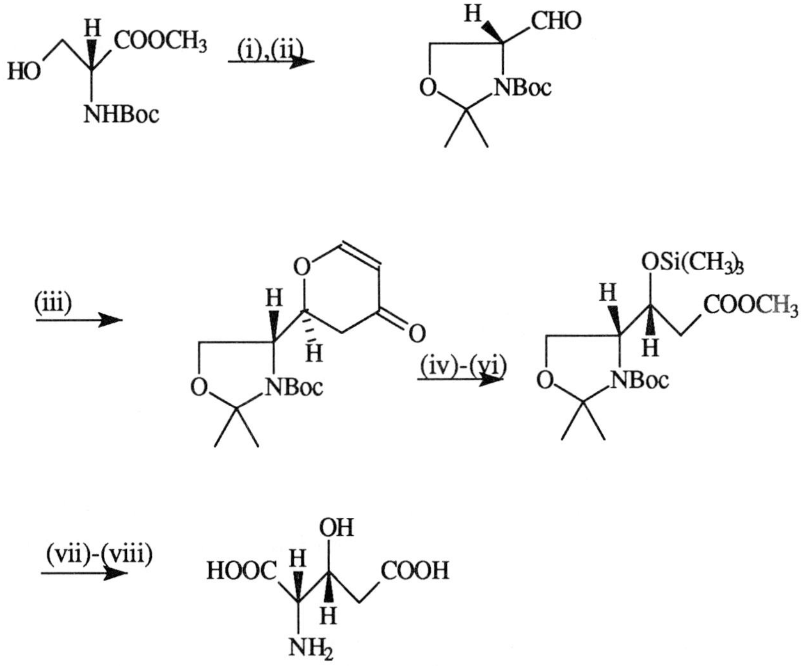

Reagents: (i) dimethoxypropane,TsOH; (ii) di-isobutylaluminium hydride, -78^0; (iii) $MeOCH{=}CHC(OSiMe_3){=}CH_2$; (iv) $NaIO_4$-RuO_2; (v) Et_2NSiMe_3; (vi) MeOH,H^+; (vii) $KMnO_4$; (viii) HCl/H_2O.

Scheme 3

COOH
|
$CHNH_2$ ⟶ H $CH_2CH_2CONH_2$ AcNHCCOOCH$_3$ —(i)⟶ H CH_2CH_2CN AcNHCCOOCH$_3$
|
CH_2CH_2COOH

—(ii)⟶ H, $COOCH_3$, HO, N, H, $COCH_3$ —(iii)⟶ H CH_2CH_2CH=NNHPh AcNHCCOOCH$_3$

—(iv)⟶ COOH, H, NH_2, N, H

Reagents: (i) TsCl/pyridine; (ii) H_2/Raney Ni; (iii) $PhNHNH_2$; (iv) refluxing 0.1M HCl.

Scheme 4

4-Aminobutanoic acid can be obtained enzymatically from L-glutamic acid (R.Januseviciute *et al.,* Khim. Prir. Soedin., 1983, 246).

L-Glutamic acid serves as starting material for (S)-4-amino-4,5-dihydrofuran-2-carboxylic acid, a potent 4-aminobutanolic acid transaminase inhibitor (J.P.Bunkhart *et al.,* Tetrahedron Lett., 1984, 25, 5267). L-Glutamic acid also yields L-proline, using a barley mutant.

The Group IA and IIA metallic salts of L-glutamic acid, when heated, undergo cyclodehydration and dehydration to 5-carboxy-2-pyrrolidone, followed by further dehydration to pyrrole (Z.B.Bakasova *et al.,* Izv. Akad. Nauk. Kirg. SSSR., 1982, 37).

4-Carboxyglutamic acid reacts readily with aldehydes by an intramolecular Mannich reaction to give 5-substituted-4,4-dicarboxy prolines which can in turn undergo monodecarboxylation at 100° (R.Capasso *et al.,* Can. J. Chem., 1983, 61, 2657).

N-Benzyloxycarbonyl-2-methyl-L-glutamic acid ester can be converted into *N*-benzyloxycarbonyl-L-proline methyl ester by a diborane reduction (B.Y.Chung *et al.,* Bull.Korean Chem. Soc., 1985, 6, 177). *N*-Benzyloxycarbonyl-L-glutamic acid affords *cis*-4-β–carboxy-ethylprolines by intramolecular cyclisation (T.C.Ho *et al.,* J. Org. Chem., 1986, 51, 2405).

Z-(*N*-Benzoyl)-2,3-dehydroglutamic acid can be converted into carnosadine (1-amino-2-guanidino-methylcyclopropane-1-carboxylic acid (T.Wakamiya *et al.,* Tetrahedron Lett., 1986, 27, 2143) **(Scheme 5)**.

2-Amino-2-methylpentanedioic acid (2-methylglutamic acid), $HOOCCH_2CH_2C(NH_2)(CH_3)COOH$, can be made in the D-form from the product of the reaction of methyl acrylate with an alanine-based isonitrile (Y.Yamamoto *et al.,* Agric. Biol. Chem., 1985, 49, 1761). Its racemic form has m.p. 169°.

2-Amino-3-methylpentanedioic acid (3-methylglutamic acid) can be obtained as its (±)-*erythro*-form, m.p. 172.5-173.5°, and its (±)-*threo*-form, m.p. 185-6° (A.B.Mauger, J. Org. Chem, 1981, 46, 1032).

2-Amino-4-methylglutamic acid (4-methylglutamic acid) $HOOCCH(CH_3)CH_2CH(NH_2)COOH$, can be synthesised as its (2S,4R)-*erythro*- and its (2S,4S)-*threo* form from (RS)-4-methyl-2-oxo glutaric acid by glutamate dehyrogenase catalysed reductive amination (A.Righini-Tapie and R.Azerad, J. Appl. Biochem., 1984, 6, 361).

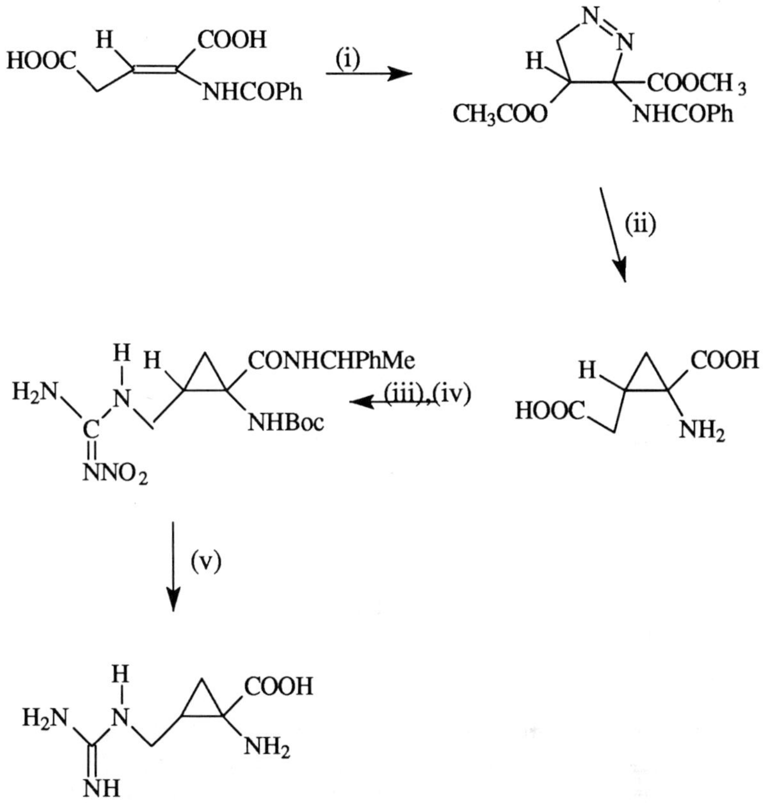

Reagents: (i) CH_2N_2,MeOH; (ii) hυ; 6M HCl; (iii) MeOH,H^+; Boc_2O; NH_3; Br_2/NaOH; (iv) ZCl; (R)-(+)-PhCHMeNH_2; DCCl; H_2/Pt; 3,5-dimethyl-1-nitroguanyl-pyrazole; (v) H_2/Pd; 6M HCl.

Scheme 5

3-Aminopentanedioic acid (3-aminoglutamic acid), $HOOCCH_2CH(NH_2)CH_2COOH$, can be prepared as a crystalline solid, mp. 295°. The *N*-methyl derivative, m.p. 177-9°, is present in the prokaryotic algal symbiont, *Prochloron didemnii* (R.E.Summons, Phytochemistry, 1981, 20, 1125).

2-Amino-4-hydroxy-4-methylpentanedioic acid (4-hydroxy-4-methyl glutamic acid):

COOH
H_2N ►C◄ H
CH_2
H_3C ►C◄ OH
COOH

(2S,4S)-form

is widespread in plants such as *Reseda luteola, Adiantum pedatum, Pandanus veitchii,* and can be isolated in the (2S,4S)-L-*erythro*-form, m.p. 164° dec. (L.K.Meier *et al.,* Phytochemistry, 1979, 18, 1173, 1505; B.Bjerg *et al.,* Acta Chem. Scand., Ser. B, 1983, 37, 321).

2-Amino-3-hydroxy-4-methylpentanedioic acid (3-hydroxy-4-methyl glutamic acid):

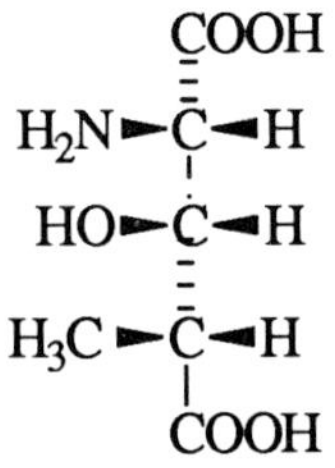

COOH
H_2N–C–H
HO–C–H
H_3C–C–H
COOH

(2S,3S,4R)-form

The (2S,3R,4R)-form, m.p. 193-4°, is a constituent of the seeds of *Gymnocladus spp.* (E.A.Bell *et al.,* Phytochemistry, 1978, 17, 1127).

Glutamine:

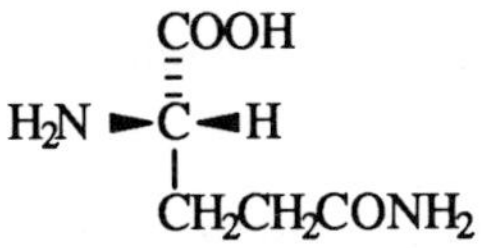

COOH
H_2N–C–H
$CH_2CH_2CONH_2$

(S)-form

is widely distributed in plants, e.g. beetroot. It has been synthesised and forms solid needles, m.p. 184-5° and [α] +8.0° (P.Lindberg *et al.,* J. Chem. Soc. Perkin Trans. I, 1977, 684; T.Kaneko *et al.,* Synthetic Production and Utilisation of Amino Acids, 1974, Halstead Press, 109). (±)-Glutamine-2,5-^{15}N can be prepared from 2-ketoglutaric acid by treatment with $^{15}N_2H_4$, followed by hydrogenation of the resulting cyclic hydrazone (H.Engelmann *et al.,* Zfl. Mitt., 1983, 77, 93).

Its derivatives are useful in the synthesis of peptides: *N*-(2,4-dinitro phenyl), m.p. 189-91°; p-nitrobenzyl ester hydrobromide, m.p. 175°; *N*[4]-(4-hydroxyphenyl), isolated from the fungus, *Agaricus horttensis,* $[\alpha]_D$ +30°.

Isoglutamine:

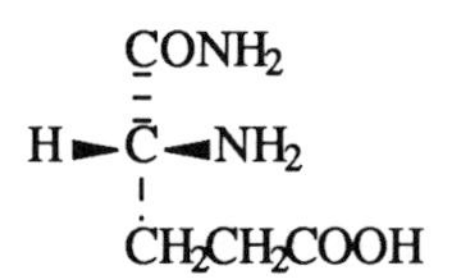

in its (R)-D-form is present in the peptides of the cell walls of *Streptococcus aureus, Myobacterium rosens* and *Streptococcus pyogenes.* Isoglutamine has $[\alpha]_D$ -20.6°; principal derivatives include the hydrobromide, m.p. 215-6° and the *N*-benzyloxycarbonyl, m.p. 175-6°.

The (S)-form is a metabolite of Thalidomide, and can be obtained as a crystalline solid, m.p. 181°, and $[\alpha]_D$ +20.5°. It yields an hydrochloride, m.p. 172-3°, (R.Straka *et al.,* Coll. Czech. Chem. Commun., 1977, 42, 560; I.B.Kristensen *et al.,* Acta Chem. Scand., 1973, 27, 3123).

2-Amino-4,5-dihydroxyhexanedioic acid (2-amino-4,5-dihydroxy adipic acid), $HOOCCH(OH)CH(OH)CH_2CH(NH_2)COOH$, can be isolated from human urine, m.p. 207-9°, and $[a]_D$ +4.4°. Its stereochemistry is not known (S.Yuasa, Biochim. Biophys. Acta, 1978, 540, 93).

2-Aminooctanedioic acid, (2-aminosuberic acid),
$HOOC(CH_2)_5CH(NH_2)COOH$, can be synthesised as its derivative, $BzOCO(CH_2)_5CH(NHAc)COOEt$, *via* the alkylation of the ethyl ester of benzylidene glycine (D.H.Rich *et al.,* J. Org. Chem., 1983, 48, 741).

2-Amino-4-methylheptanedioic acid (2-amino-4-methylpimelic acid),

$HOOCCH_2CH_2CH(CH_3)CH_2CH(NH_2)COOH$, is a constituent of *Lactarius quietus*, and can be isolated as a solid, m.p. 189-91°, and $[\alpha]_D$ -6.3° (S.-I.Hatanaka *et al.,* Phytochemistry, 1975, 14, 1559; H.Oogishi, C.A. 1976, 85,17400).

2,6-Diaminoheptanedioic acid (2,6-diaminopimelic acid) is most commonly encountered as the *meso*-form; chiral isomers are constituents of the polymeric peptidoglycans present in the cell walls of algae and all bacteria except halobacteria. The (2R,6R) form is obtained as a solid, m.p. 309-10°.

The (2RS,6SR)-*meso*-form can be obtained as needles, m.p. >313-5°, and gives a *N,N*-dibenzoyl derivative, m.p. 194-5° (S.E.Hull *et al.,* Acta Crystallogr., Sect. B, 1977, 33, 3832; A.Arendt *et al.,* Rocz. Chem., 1974, 48, 883).

2,6-Disubstituted aminopimelic acids are prepared by intermolecular ene reactions between ethyl acrylate and alkenyl glyoxylates (K.Agouridas and J.M.Girodeau, Tetrahedron Lett., 1985, 26, 3115).

2,6-Diamino-2-hydroxymethylheptanedioic acid (2,6-diamino-6-hydroxymethylpimelic acid:

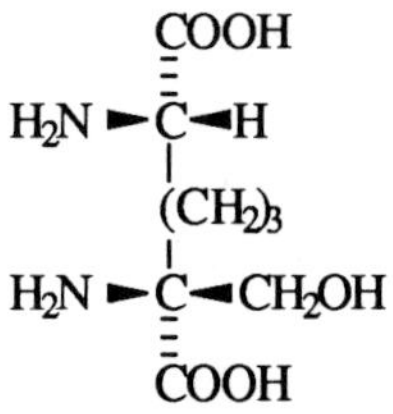

COOH
H_2N ► C ◄ H
$(CH_2)_3$
H_2N ► C ◄ CH_2OH
COOH

(2S,6S)-form

is a constituent of the antibiotic, PA 3534J, and can be isolated from *Mucromonospora chalcea* as an amorphous powder, $[\alpha]_D$ +8.1° (J.Shoji *et al.,* J. Antibiot., 1981, 34, 374).

4-Amino-3-hydroxy-6-methylheptanoic acid (Statine):

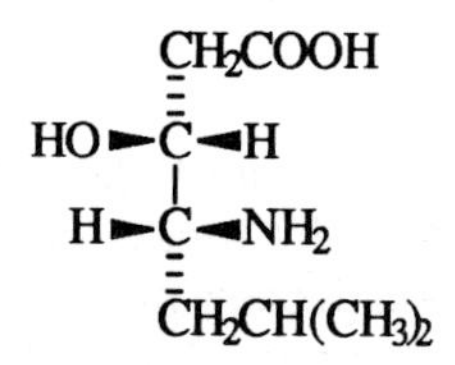

is formed from the hydrolysis of Pepstatin A. The m.p. of the (3R,S,4RS) form is 211.5-213.5°, whereas the (3R,4S) has m.p. 202.5-203.5°(B.Raque et al., Bull. Soc.Chim. Fr. Part II, 1983, 230; K.E.Rittle et al., J.Org.Chem., 1982, 47, 3016; W.S.Liu, J.Med. Chem., 1979, 22, 577; T.Katsuki et al., Bull. Chem. Soc. Jpn., 1976, 49 3287).

Second Supplements to the 2nd Edition of Rodd's Chemistry of Carbon Compounds, Vol. IE/F/G, edited by M. Sainsbury

Chapter 21

PHOSPHOLIPIDS

STEPHEN A. BOWLES

Phospholipids[1] are an important class of biomolecules which are currently the subject of considerable biological and chemical attention. Phospholipids and phospholipid analogues have found use as probes for membrane structure and function and for lipid-protein interactions, elucidation of enzyme mechanisms, and in the preparation of liposomes and vesicles with precise characteristics. Therapeutical applications have included the use of these compounds as drug-carriers, controlled-release agents and drugs *per se*. Morover, a number of enzymes have a specific requirement for phospholipids. The effective study of these processes demands phospholipids of unequivocal structure, and as a consequence a myriad of new synthetic techniques have evolved. This review surveys the

---

[1] Abbreviations: Bn, benzyl; Boc, *t*-Butyloxycarbonyl; DCC, *N,N'*-dicyclohexylcarbodiimide; DMAP, *N,N'*-dimethylaminopyridine; DME, 1,2-dimethoxyethane; DMF, *N,N'*-dimethylformamide; DMSO, dimethyl sulphoxide; NBS, *N*-bromosuccinimide; PAF, platelet-activating factor; PLA_2, phospholipase A_2; PPY, 4-pyrrolidinopyridine; QSAR, quantitative structure-activity relationships; TBDMS, *t*-butyldimetylsilyl; TFA, trifluoro acetic acid; THF, tetrahydrofuran; TMS, trimethylsilyl; TPS, 2,4,6-triisopropylbenzenesulphonyl; Tr, trityl.

analogues published in the literature between 1980-1990. It does not include glycolipids or inositol-phospholipids, which have been reviewed elsewhere (J. Gigg and R. Gigg, *Top. Curr. Chem.*, 1990, **154**, 77; I. M. Morrison, *Carbohydrate Chem.*, 1983, **14**, 345; C. M. Sturgeon, *Carbohydrate Chem.*, 1986, **15**, 618; *Carbohydrate Chem.*, 1983, **14**, 375; *Carbohydrate Chem.*, 1982, **13**, 572; R. Gigg, *Lab. Lipid Gen. Chem., Natl. Inst. Med. Res.*, 1980, **26**, 287).

General References

Chemistry, preparation and analysis of phospholipids: M. Ranny and J. List, *Acta Univ. Carol. Med.*, 1986, **32**, 5; J.- J. Godfroid *et al., Pharmacol. Res. Commun.*, 1986, **18**, 1; I. Lindh, *Chem. Commun., Univ. Stockholm*, 1988, 35; *Phospholipids*, 1982, J. N. Hawthorne and G. B. Ansell (Eds.), Elsevier Biomedical Press, Amsterdam; *Topics in Lipid Research: From Structural Elucidation to Biological Function*, 1986, R. A. Klein and B. Schmitz (Eds.), Royal Society of Chemistry, London; A. R. N. Eberendu, B. J. Venables and K. E. Daugherty, *LC Mag.*, 1985, **3**, 424; *Ether Lipids - Biochemical and Biomedical Aspects*, 1983, H. K. Mangold and F. Paltauf (Eds.), Academic Press, New York; *Phospholipids and Cellular Regulation*, 1985, J. F. Kuo (Ed.), CRC Press, Boca Raton, Florida; W. W. Christie, *HPLC and Lipids: A Practical Guide*, 1987, Pergamon Press, Oxford; R. H. McCluer, M. D. Ullman and F. B. Jungalawa, *Adv. Chromatogr.*, 1986, **25**, 309; W. W. Christie, *Lipid Analysis* (second edition), 1982, Pergamon Press, Oxford.

Phospholipid Liposomes as Drug Carriers: Preparation and Properties: H. Stricker in *Phospholipids: Biochem. Pharm. Anal. Consid., (Proc. Int. Colloq. Lecithin)*, 1990, I. Hanin and G. Pepeu (Eds.), Plenum Press, New York; H. Schrier, *Pharm. Unserer Zeit.*, 1982, **11**, 97; *Liposomes in Biological Systems*, G. Gregoriadis and A. C. Allison (Eds.), 1980, John Wiley and Sons, Chichester; G. Cevc and D. Marsh, *Phospholipid Bilayers:*

Physical Principles and Models, 1987, John Wiley and Sons, New York.

Nuclear magnetic resonance spectroscopy, particularly of the ^{31}P nucleus, is a powerful tool in phospholipid chemistry (*^{31}P NMR*, 1984, D. A. Gorenstein (Ed.), Academic Press, Orlando, Florida). References to some applications of this technique in the major fields are given below.

Structure and quantification: T. E. Merchant and T. Glonek, *J. Lipid Res.*, 1990, **31**, 479; S. Bradamante *et al.*, *Anal. Biochem.*, 1990, **185**, 299; D. B. Fenske *et al.*, *Biochemistry*, 1990, **29**, 3973; J.-L. Eisele *et al.*, *Chem. Phys. Lipids*, 1990, **55**, 351; T. E. Merchant and T. Glonek, *J. Lipid Res.*, 1990, **31**, 479; L. C. Stewart and M. Kates, *Chem. Phys. Lipids*, 1989, **50**, 23; P. Meneses, P. Para and T. Glonek, *J. Lipid Res.*, 1989, **30**, 458; A. Tokumura *et al.*, *J. Lipid Res.*, 1989, **30**, 219; G. Ferrante *et al.*, *J. Lipid Res.*, 1989, **30**, 1601; N. Sotirhos, *Dev. Food Sci.*, 1988, **17**, 443; J. R. Cavanaugh and P. E. Pfeffner, *Chem. Phys. Lipids*, 1988, **49**, 87; P. Meneses and T. Glonek, *J. Lipid Res.*, 1988, **29**, 679; N. Sotirhos, B. Hersloef and L. Kenne, *J. Lipid Res.*, 1986, **27**, 386; L. C. P. Mollevanger *et al.*, *Eur. J. Biochem.*, 1986, **156**, 383.

Phospholipid profile in tissues, both *in vitro* and *in vivo*, and phospholipid metabolism: J. S. Cohen, *Mayo Clin. Proc.*, 1988, **63**, 1199; A. Tommaso *et al.*, *Brain Res.*, 1990, **526**, 108; J. W. Pettegrew, G. Withers and K. Panchalingam in *NMR: Principles and Applications to Biomedical Research*, J. W. Pettegrew (Ed.), 1989, Springer-Verlag, New York, 204; E. J. Murphy *et al.*, *Magn. Reson. Med.*, 1989, **12**, 282; M. Lowry *et al.*, *Biochem. Soc. Trans.*, 1989, **17**, 1053; A. Miccheli *et al.*, *Neurosci. Res. Commun.*, 1989, **4**, 33; P. Meneses, P. F. Para and T. Glonek, *J. Lipid Res.*, 1989, **30**, 458; I. L. Kwee and T. Nakada, *Magn. Reson. Med.*, 1988, **6**, 296; T. Nakada, I. L. Kwee and G. A. Rao, *Biochem. Arch.*, 1988, **4**, 35; P. F. Daly *et al.*, *J. Biol. Chem.*, 1987, **262**, 14875; O. Miatto *et al.*, *Can. J. Neurol. Sci.*, 1986, **13**, 535; S. Cerdan *et al.*, *Magn. Reson. Med.*, 1986, **3**, 432.

Membrane structure and dynamics: C. P. S. Tilcock, P. R. Cullis and S. M. Gruner, *Chem. Phys. Lipids*, 1986, **40**, 47; A. C. McLaughlin *et al.*, *Biophys. J.*, 1982, **37**, 49; W. M. Loffredo, R. T. Jiang and M.- D. Tsai, *Biochemistry*, 1990, **29**, 10912; J. L. Eisele, J. M. Neumann and C. Chachaty, *Chem. Phys. Lipids*, 1990, **55**, 351; L. C. Stewart *et al.*, *Chem. Phys. Lipids*, 1990, **54**, 115; M. M. Basti and L. A. LaPlanche, *Chem. Phys. Lipids*, 1990, **54**, 99; E. J. Defourc, J. M. Bonmatin and J. Dufourcq, *Biochemie*, 1989, **71**, 117; A. Lopez, M. P. Rols and J. Teissie, *Biochemistry*, 1988, **27**, 1222; H. E. Sarvis *et al.*, *Biochemistry*, 1988, **27**, 4625; D. Uhrikova and P. Balgavy, *Chem. Phys. Lipids*, 1988, **47**, 69; E. J. Dufourc, I. C. P. Smith and J. Dufourcq, *Biochemistry*, 1986, **25**, 6448; T. Bayerl *et al.*, *Biochim. Biophys. Acta*, 1986, **858**, 285; L. R. C. Barclay, B. J. Balcom and B. J. Forest, *J. Am. Chem. Soc.*, 1986, **108**, 761; K. Arnold *et al.*, *Stud. Biophys.*, 1985, **107**, 65; R. A. Burns Jr., J. M. Friedman and M. F. Roberts, *Biochemsitry*, 1981, **20**, 5945; J. R. Silvius *et al.*, *Biochemistry*, 1985, **24**, 5388; H. Hauser *et al.*, *Biochemistry*, 1980, **19**, 366; R. A. Burnes Jr. and M. F. Roberts, *Biochemistry*, 1980, **19**, 3100.

1. Synthesis of Phospholipids

A review by Eibl (H. Eibl, *Chem. Phys. Lipids.*, 1980, **26**, 405) comprehensively documents the early methods of synthesising glycerophospholipids. Although many of these methods remain in use, an abundance of new techniques are available that allow for the synthesis of a plethora of phospholipids and phospholipid analogues. In many instances the number of protecting groups required is minimal, and the degree of stereochemical purity achieved is exceptionally high.
It is often considered that the most important stage in the synthesis of phospholipids is the phosphorylation step leading to the formation of phosphoester bonds. This is the subject of a recent review by Russian workers (A. E. Stepanov and V. I. Shvets, *Chem. Phys. Lipids.*, 1986, **41**, 1). The sheer diversity of

modern methodology and starting materials makes this classification a convenient one.

(a) Halophosphate Reagents

The formation of phosphoester bonds is most usually effected by reaction of a glycerol derivative with an appropriate phosphorylating agent in the presence of an organic base (N. N. Karpyshev *et al., Bioorgan. Khim.,* 1980, **6**, 1214). In the synthesis of phosphatidic acids, diphenyl chlorophosphate is the most widely used reagent. Subsequent deprotection is achieved by catalytic hydrogenolysis (H. Brachwitz *et al., Chem. Phys. Lipids.,* 1982, **31**, 33). In one synthetic route (R. Bittman, A. F. Rosenthal and L. A. Vargas, *Chem. Phys. Lipids,* 1984, **34**, 201) dimethyl chlorophosphate was used since the methyl groups are readily cleaved under mild conditions by trimethylsilyl bromide. Dichlorophosphate reagents are occasionally used in the synthesis of phosphatidic acids (Q. Q. Dang *et al., Lipids,* 1982, **17**, 798), but are routinely employed in the formation of phosphodiester bonds. The use of phenyl dichlorophosphate is widespread, but this reagent may give bis-(diacylglycero)phosphoric acids as by-products (Q. Q. Dang *et al., Lipids,* 1982, **17**, 798). Methyl dichlorophosphate has been used for the preparation of racemic mixed-chain phosphatidylcholines (C. J. Lacey and L. M. Loew, *J. Org. Chem.,* 1983, **48**, 5214). This synthetic route includes an unusual step (Scheme 1), whereby a triester (**1**), obtained by sequential phosphorylation, isomerises spontaneously on standing to the allyl choline phosphate (**2**) (C. J. Lacey and L. M. Loew, *Tetrahedron Lett.,* 1980, 2017).

Scheme 1

MeO–P(=O)(O-allyl)–O–CH₂CH₂–NMe₂ ⟶ ⁻O–P(=O)(O-allyl)–O–CH₂CH₂–N⁺Me₃

(**1**) (**2**)

This synthesis is also noteworthy in that the phospholipid backbone is constructed by elaboration of this allyl group, rather than from manipulation of a glycerol derivative.

One of the phosphodiester bonds may be present in the reagent, leaving only one to be formed by reaction with the hydroxyl of a desired glycerol derivative. Haloethyl dihalophosphates are used extensively for the convenient preparation of phospholipids (H. Brachwitz *et al.*, *Chem. Phys. Lipids*, 1982, **31**, 33; R. Berchtold, *Chem. Phys. Lipids*, 1982, **31**, 33; W. J. Hansen *et al.*, *Lipids*, 1982, **17**, 453; H. Brachwitz, P. Langent and J. Schlidt, *Chem. Phys. Lipids*, 1984, **34**, 355; S. K. Bhatia and J. Hajdu, *Tetrahedron Lett.*, 1987, 1729; Y.- F. Wang and C.- H. Wong, *J. Chin. Chem. Soc.*, 1989, **36**, 463). Reaction of these reagents with a lipophilic alcohol, such as a disubstituted glycerol, is followed by hydrolysis, and nucleophilic displacement of the alkyl halide with a selected amine. This method is equally successful for large scale preparations (R. Berchtold, *Chem. Phys. Lipids.*, 1982, **30**, 389), and is compatible with unsaturated glycerol substituents (B. Hupfer, H. Ringsdorf and H. Schupp, *Chem. Phys. Lipids*, 1983, **33**, 355).

An alternative to this procedure has been developed in the synthesis of dialkyl phospholipids containing a carboxyl group bonded to the 1-alkyl chain (R. Berchtold, *Chem. Phys. Lipids*, 1981, **28**, 55). In this case, dichlorophosphoric acid ethylphthalimide was reacted with the glycerol derivative in the presence of base. Hydrolysis was followed by an exchange reaction with hydrazine in, what is essentially, a Gabriel synthesis.

Phosphorus oxychloride is another bifunctional reagent which has been frequently exploited for the synthesis of phosphomonoester (H. Eibl *et al.*, *Methods Enzymol.*, 1983, **98**, 623; H. Brachwitz *et al.*, *Chem. Phys. Lipids*, 1982, **31**, 33; A. Hermetter, F. Paltauf and H. Hauser, *Chem. Phys. Lipids*, 1982, **30**, 35; R. Bittman and N. M. Witzke, *Chem. Phys. Lipids*, 1989, **50**, 99) and phosphodiester bonds (H. Eibl, *Chem. Phys. Lipids*, 1980, **26**, 239; J. W. Cox and L. W. Horrocks, *J. Lipid Res.*, 1981, **22**, 496; H. S. Hendrickson *et al.*, *J. Lipid Res.*, 1983, **24**, 1532), and is particularly attractive when unsaturated fatty acid residues are present. However, the high reactivity of this reagent can often result in undesired side products - a feature which may restrict its utility.

This procedure was employed in the synthesis of phospholipids which are asymmetric at phosphorus (K. Bruzik and M.- D. Tsai, *J. Am. Chem. Soc.*, 1984, **106**, 747).

Scheme 2

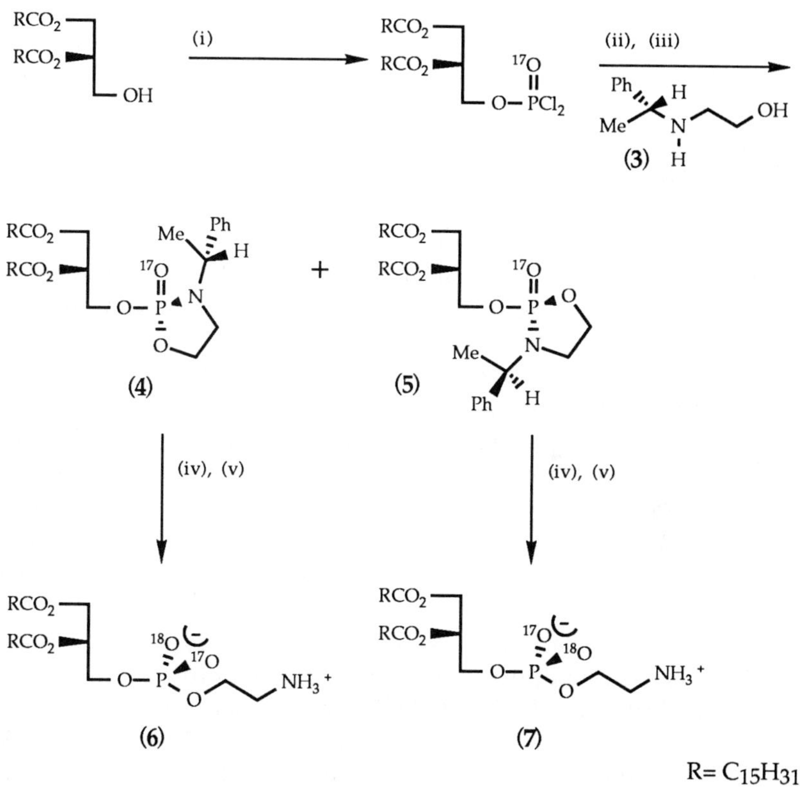

RCO2 / RCO2 / OH —(i)→ RCO2 / RCO2 / O–$P^{17}OCl_2$ —(ii), (iii)→ (3)

(4) + (5)

(iv), (v) ↓ (iv), (v) ↓

(6) (7)

R= $C_{15}H_{31}$

Reagents: (i) $P^{17}OCl_3$, Et_3N, $ClCH{=}CCl_2$, -20°C; (ii) (3), Et_3N, $ClCH{=}CCl_2$, -20°C; (iii) chromatography; (iv) $H_2{}^{18}O$, TFA, DME; (v) H_2, Pd/C, EtOH.

Condensation of ^{17}O labelled phosphorus oxychloride (Scheme 2) with 1,2-dipalmitoyl-*sn*-glycerol and (R)-2-[*N*-(1-

phenylethyl)amino]ethanol (**3**) gave the diastereomeric cyclic oxazaphospholidines, (**4**) and (**5**), which were separated by chromatography. Hydrolysis of the individual isomers with ^{18}O labelled water and trifluoroacetic acid, followed by deprotection gave the (R_p)- and (S_p)- [^{17}O,^{18}O]-1,2-dipalmitoyl-*sn*-glycero-3-phosphatidylethanolamines (**6**) and (**7**). These were used in mechanistic studies to determine the stereochemistry of phospholipase D-catalysed trans-phosphatidylation.

Abdelmageed *et al.*, used phosphorus oxychloride in improved syntheses of phosphatidylcholines, phosphatidylethanolamines and some ^{2}H- and ^{13}C-labelled analogues (O. H. Abdelmageed *et al., Chem. Phys. Lipids*, 1989, **50**, 163). This work included significant improvements to known hydrogenation and quaternisation steps.

A novel approach to building the head groups of synthetic phospholipids is based on the construction of a neutral phosphotriester from phosphorus oxychloride. The phosphate group bears three neutral alkyl groups: one is a glycerol derivative; the second a methyl group which is removed cleanly at a later stage; the third determines the identity of the phospholipid (P. Woolley and H. Eibl, *Chem. Phys. Lipids*, 1988, **47**, 55). This procedure is lengthier than many routes, but has been demonstrated by the preparation of a wide range of enantiomerically pure examples, and its scope extended to include the preparation of lysophospholipids with less than 1% racemisation (H. Eibl and P. Woolley, *Chem. Phys. Lipids*, 1988, **47**, 63).

The use of silver salts for the synthesis of phosphoester bonds has largely been superseded, although examples can still be found (B. Hupfer, H. Ringsdorf and H. Schupp, *Chem. Phys. Lipids*, 1983, **33**, 355; S. Ali and R. Bittman, *J. Org. Chem.*, 1988, **53**, 5547). Silver dibenzylphosphate was used in the preparation of ^{14}C acyl labelled phosphatidic acid as a precursor to labelled phospholipids (J. L. Danan and L. Pichat, *J. Labelled Cmpds. Radiopharm.*, 1981, **18**, 1245), and the silver salt of *N*-tritylaminoethylphosphoric acid to synthesise labelled phosphatidylethanolamines.

The efficiency of phosphorylation can often be increased by the use of cyclic chlorophosphates (V. V. Chupin *et al.*, *Zh. Organ. Khim.*, 1980, **16**, 31; D. A. Predvoditelev *et al.*, *Bioorg. Khim.*, 1980, **6**, 1087; P. N. Guivisdalsky and R. Bittman, *J. Org. Chem.*, 1989, **54**, 4643).
Hajdu and co-workers have synthesised a wide range of phospholipids and phospholipid analogues from novel chiral sources (N. S. Chandrakumar and J. Hajdu, *J. Org. Chem.*, 1983, **48**, 1197). The acetonide of L-glyceric acid was used in a novel stereospecific synthesis of antitumour ether phospholipids (S. K. Bhatia and J. Hajdu *Tetrahedron Lett.*, 1987, 271), in which phosphorylation was achieved with a cyclic chlorophosphate. These workers also reported that 2-chloro-2-oxo-1,3,2-dioxaphospholane gave higher yields and fewer by-products than 2-bromoethyl phosphorodichloridate (S. K. Bhatia and J. Hajdu, *Synthesis*, 1989, 16), and that efficiency could be enhanced by freeze-drying the substrate alcohol from benzene, prior to reaction.
An important discovery was made by Gadek, who found that a slight excess of trimethylsilyltrifluoromethane sulphonate accelerates the nucleophilic opening of cyclic phosphates with tertiary amines, at a rate proportional to the nucleophilicity of the amine (T. R. Gadek, *Tetrahedron Lett.*, 1989, 915). This procedure was applied to the synthesis of ether phospholipids, and is reported to give the products rapidly, in a high state of purity, and isolable without the need for chromatography.
A cyclic reagent was also employed in the synthesis of azathia-phospholipids containing polyhetero side chains (J. M. Ziedler, W. Zimmermann and H. J. Roth, *Chem. Phys. Lipids*, 1990, **55**, 155).

(c) The Cyclic Enediol Phosphoryl (CEP) Method

This is a specific case of the use of cyclic reagents in a double phosphorylation strategy, which has been used extensively in the synthesis of phospholipids, glycolipids and phospholipid conjugates. The application of this method to the preparation of phosphodiesters in a variety of examples has been reviewed (F. Ramirez and J. F. Marecek, *Synthesis*, 1985, 499). The approach (Scheme 3) entails reaction of an alcohol (R^1OH) with the phosphorylating reagent (CEP-X) (**8**) - where X is a good leaving group - in the presence of base, to give a cyclic triester (**9**). This reaction proceeds in high yield, with almost total ring preservation.

Scheme 3

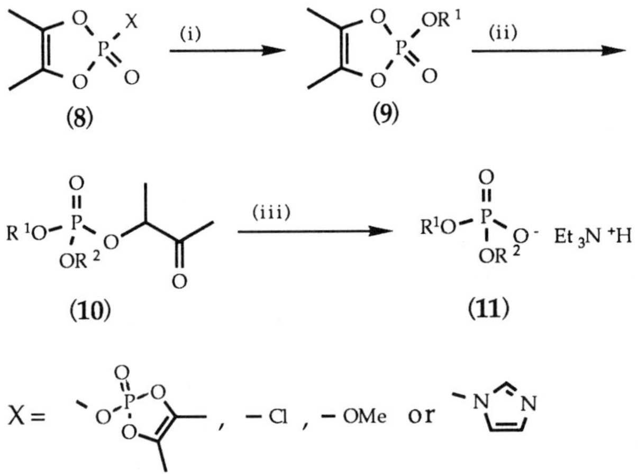

(i) (ii)

(**8**) (**9**)

(iii)

$R^1O-P(=O)(OR^2)-O-$ ... (**10**) → $R^1O-P(=O)(OR^2)O^-$ Et_3N^+H (**11**)

X = , – Cl , – OMe or

Reagents: (i) R^1OH, base; (ii) R^2OH, Et_3N or Imidazole: (iii) Et_3N, H_2O, MeCN or pyridine.

The intermediate (**9**) is then available for phosphorylation with a second alcohol, R^2OH, in the presence of triethylamine or imidazole. These particular bases increase the rate of reaction and allow for the use of a wider variety of aprotic solvents. This is important since the rate of phosphorylation is solvent-dependent, and is noticably slower in more polar aprotic solvents. These amines also enhance the rate of ring opening, preventing transesterification. The final stage is hydrolysis of the dialkyl phosphoacetoin (**10**) with two equivalents of triethylamine in either acetonitrile-water, or pyridine-water, to obtain the triethylammonium salt (**11**). The CEP method is exemplified in the synthesis of radiolabelled phospholipids (J. L. Danan and L. Pichat, *J. Labelled Compd. Radiopharm.*, 1980, **17**, 223), and short chain phosphatidylcholines (P. Kanda and M. A. Wells, *J. Lipid Res.*, 1981, **22**, 879).

(d) Activating Reagents

A highly efficient route to phosphodiesters is available by activation of a phosphomonoester, *in situ*, with a suitable reagent such as trichloroacetonitrile (R. Bittman, A. F. Rosenthal and L. A. Vargan, *Chem. Phys. Lipids*, 1984, **34**, 201), aryl sulphonyl halides (B. Hupfer *et al., Chem. Phys. Lipids*, 1983, **33**, 355) or carbodiimides (Q. Q. Dang and W. Stoffel, *Chem. Phys. Lipids*, 1983, **33**, 33), followed by reaction with an alcoholic component. Hindered aryl sulphonates generally give the best yields, and since there are effectively no by-products formed, facile isolation of the products is possible. This method was employed in the preparation of a nitroxide spin-labelled phosphatidylethanolamine (P. Kertscher *et al., Pharmazie*, 1980, **35**, 10). Similarly, fluorescent labelled phosphatidylethanolamines and phosphatidylcholines have been prepared by reaction of phosphatidic acid with *N*-(β-hydroxymethyl)carbazole in the presence of 2,4,6-triisopropylbenzenesulphonyl chloride (TPS) (J. Lakowicz and D. Hogen, *Chem. Phys. Lipids*, 1980, **26**, 1). Diacyl- and alkylacylglycerophosphoserines (A. Hermetter, F. Paltauf and H. Hauser, *Chem. Phys. Lipids*, 1982, **30**, 32), and ^{14}C acyl-labelled phospholipids (J. L. Danan and L. Pichat, *J. Labelled Cmpds. Radiopharm.*, 1981, **18**, 1245) have also been synthesised with the same activating agent. Roux *et al.*, used TPS successfully to prepare gram quantities of deuterated phosphatidylcholines (M. Roux *et al., Chem. Phys. Lipids*, 1983, **33**, 41).

(e) *Oligonucleotide Methodology*

A number of beneficial synthetic methods have been introduced recently, which employ techniques originally developed for the preparation of oligonucleotides. Lindh and Stawinski have applied hydrogen-phosphonate chemistry (I. Lindh and J. Stawinski, *J. Org. Chem.*, 1989, **54**, 1338). This facile and efficient phosphorylation procedure confers a number of advantageous features upon the synthetic strategy. The most significant benefit is that no phosphate protecting group is necessary - a factor which contributes to the low yields in other routes. The key intermediate (**12**) was prepared from 1,2-dipalmitoyl-*sn*-glycerol, with either a phosphorus trichloride-imidazole (P. J. Garegg *et al.*, *Chemica Scr.*, 1986, **26**, 59), or salicylchlorophosphite (J. E. Marugg *et al.*, *Tetrahedron Lett.*, 1986, 2661) phosphitylating agents (Scheme 4).

Scheme 4

$H_{31}C_{15}COO$, $OCOC_{15}H_{31}$, OH —(i)→ (**12**) —(ii)→ (**16**) —(iii), (iv)→ (**17**)

(**12**): $H_{31}C_{15}COO$, $OCOC_{15}H_{31}$, O–P(=O)(H)–OH

(**16**): $H_{31}C_{15}COO$, $OCOC_{15}H_{31}$, O–P(=O)(H)–O–CH2CH2–NHBoc

(**17**): $H_{31}C_{15}COO$, $OCOC_{15}H_{31}$, O–P(=X)(OH)–O–CH2CH2–NH_2

Reagents: (i) PCl_3-imidazole, $PhCH_3$, Et_3N; pyridine-H_2O; (ii) $HOCH_2CH_2NHBoc$, pyridine, (**13**) or (**14**) or (**15**); (iii) 2% I_2, pyridine-H_2O; (iv) TFA, $HClO_4$, CH_2Cl_2.

This glycerohydrogen-phosphonate was then condensed with a suitably protected hydroxylic component, in the presence of a coupling agent (**13**), (**14**), or (**15**).

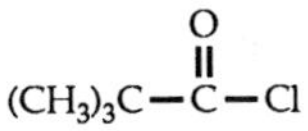

$(CH_3)_3C-C(=O)-Cl$ **(13)** 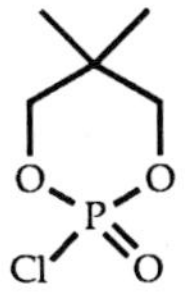 **(14)** 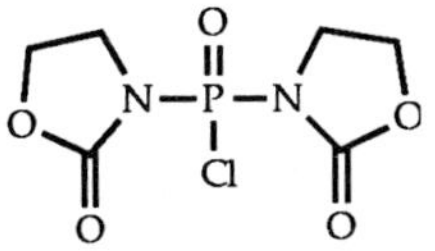 **(15)**

Simultaneous quench and oxidation of the glycerohydrogen-phosphonate diester (**16**) with iodine in aqueous pyridine, followed by deprotection of the nitrogen, if necessary, yielded the phospholipid (**17**, X=O) in approximately 80% overall yield from (**12**). This methodology was applied to the preparation of phospholipids bearing choline, ethanolamine and serine polar head groups, and for analogues with modified phosphorus centres (**17**, X=S, Se).

Martin *et al.*, have developed a novel protocol for the efficient preparation of a variety of phospholipids and their derivatives (S. F. Martin and J. A. Joey, *Tetrahedron Lett.*, 1988, 3631). This uses a highly effective phosphite coupling procedure (Scheme 5) in which phenyl dichlorophosphite (**18**) is reacted sequentially with the lipophilic and hydrophilic alcohols, then oxidised with hydrogen peroxide to form a phosphotriester (**19**). This may then be readily transformed into the corresponding phosphodiester (**20**) by reductive cleavage of the protecting group.

Scheme 5

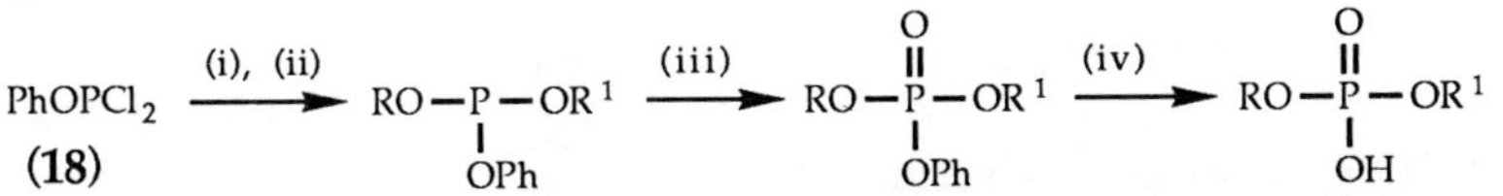

$$\underset{\textbf{(18)}}{PhOPCl_2} \xrightarrow{\text{(i), (ii)}} RO-P(OPh)-OR^1 \xrightarrow{\text{(iii)}} \underset{\textbf{(19)}}{RO-P(=O)(OPh)-OR^1} \xrightarrow{\text{(iv)}} \underset{\textbf{(20)}}{RO-P(=O)(OH)-OR^1}$$

Reagents: (i) ROH, iPr_2NEt, THF, -78°C; (ii) R'OH, THF, -78°C; (iii) H_2O_2, CH_2Cl_2, 0°C; (iv) H_2, Pd-PtO_2, AcOH.

In such a way, phospholipids were constructed with serine and ethanolamine head groups. Sequential acylation of 3-*sn*-benzyl glycerol was achieved efficiently without 1,2-acyl migration. The coupling protocol, followed by oxidation and hydrogenolysis gave the 1,2-diacyl-*sn*-phosphatidylserine analogues in high overall yields. An important aspect of this chemistry is that it overcomes the problems associated with the synthesis of biologically active phospholipids containing unsaturated *sn*-2-acyl chains. A slight modification gave a more general procedure, which allowed construction of a linoleoyl phospholipid, *via* a lysophsopholipid intermediate.

Another new technique uses a phosphite-amide-triester strategy (P. Lemmen, K. M. Buchweitz and R. Stumpf, *Chem. Phys. Lipids*, 1990, **53**, 65). Phosphorylative coupling by means of the highly reactive P(III)-amide chlorides has been adapted to make use of the 1,1,1-trichloro-2-methylpropyl (TCB) residue as a protecting group for the phosphodiester. An amino alcohol, N-protected by a 1,1,1-trichloro-2-methyl-2-propyloxycarbonyl (TCBOC) group is reacted with the phosphorylating agent (**21**), in the presence of triethylamine (Scheme 6), and the intermediate phosphorus diester amide coupled with a protected glycerol derivative by activation with a weak acid. Mild oxidation of the resulting triester (**22**) with iodine in THF-water was followed by deprotection of the glycerol hydroxyl groups, and acylation under standard conditions. The TCB and TCBOC protecting groups are cleaved simultaneously by a mild, reductive fragmentation to yield the ethanolamine phospholipid (**23**). Alternatively, TCBOC protected serine esters may be used in place of the protected amino alcohol for the synthesis of phosphatidylserines.

Scheme 6

OTCB
$Me_2N-P-Cl$
(21)

(i), (ii)

OTCB
O—P—O NHTCBOC
(22)

(iii)

OTCB
O—P—O NHTCBOC
O

(iv), (v)

RO
RO
OTCB
O—P—O NHTCBOC
O

(vi)

RO
RO
O^-
O—P—O N^+H_3
O
(23)

$R = C_{13}H_{27}, C_{15}H_{31}$

Reagents: (i) N-Boc-ethanolamine, Et_3N, CH_2Cl_2; (ii) 1, 2-isopropylidene-*sn*-glycerol, dimethylaniline hydrochloride, CH_2Cl_2; (iii) I_2, pyridine, THF-H_2O; (iv) $HClO_4$, CH_2Cl_2; (v) RCOCl, pyridine, DMAP; (vi) Li[Co(I)phthalocyanine], MeOH.

A similar method was reported by Stec and co-workers (K. S. Bruzik, G. Salamonczyk and W. J. Stec, *J. Org. Chem.*, 1986, **51**, 2368), who employed chloro-(*N*,*N*-diisopropylamino)-methoxyphosphine in a condensation, first with a glycerol derivative, then a protected amino alcohol. This was followed by oxidation and deprotection steps. The same reagent has been employed successfully in more recent syntheses (G. Lin *et al.*, *J. Biol. Chem.*, 1988, **263**, 13208; W. Yuan, K. Fearon and M. H. Gelb, *J. Org. Chem.*, 1989, **54**, 906).

(f) Glycidol Derivatives

With the recent advances in asymmetric epoxidation techniques, several groups have recognised the possibility of total synthesis of phospholipids from optically active glycidol derivatives. This has been assisted by related work (M. Caron and K. B. Sharpless, *J. Org. Chem.*, 1985, **50**, 1557) on titanium mediated regioselective epoxide ring-opening with a variety of nucleophiles, including carboxylate anions.

Johnson and co-workers have developed a flexible general synthesis which allows for variation of each structural component of the final phospholipid (C. E. Burgos, D. E. Ayer and R. A. Johnson, *J. Org. Chem.*, 1987, **52**, 4973). Asymmetric epoxidation of allyl alcohol (Scheme 7) gave (S)-glycidol, with a minimum of 86% enantiomeric excess (ee). This enantiomer is required for the natural (R)-configuration of the final phospholipid. Titanium isopropoxide assisted opening with stearic acid, followed by a single recrystallisation, yielded the diol (**24**), which was silylated cleanly to give monosilyl ether (**25**). High field NMR analysis of a Mosher ester derivative of (**25**) showed an ee of 93%, demonstrating no significant loss of optical activity during epoxide opening. Acylation of the secondary alcohol was accomplished with a range of fatty acid chlorides in pyridine, allowing facile access to mixed-chain phospholipids. Extensive acyl migration is known to accompany standard methods of desilylation, however *N*-bromosuccinimide in THF-DMSO-water effects this transformation with only a trace of acyl migration. The synthesis from (**26**) was completed using standard chemistry

Scheme 7

(24)

(25)

(26)

R, R' = $C_{17}H_{35}$

Reagents: (i) $Ti(OiPr)_4$, (+)-diisopropyl tartrate, cumene hydroperoxide, CH_2Cl_2 -10°C ; (ii) $Ti(OiPr)_4$, RCOOH, Et_2O; (iii) TBDMSCl, imidazole, THF ; (iv) R'COCl, pyridine; (v) NBS, DMSO/H_2O/THF; (vi) $Cl_2P(O)OCH_2CH_2Br$; hydrolysis; (vii) Me_3N.

Bittman and co-workers have utilised the commercially available homochiral glycidyl tosylate in a similar manner, in a short synthesis of symmetric-chain glycerophospholipids (S. Ali and R. Bitmann *J. Org. Chem.*, 1988, **53**, 5547)
Key steps are the facile boron trifluoride etherate-assisted diacylation with fatty acid anhydride and conversion of the tosylate into a phosphocholine group, with retention of configuration at C-2 throughout (P. G. Guivisdalsky, *Diss. Abstr. Int. B.*, 1990, **50**, 5067). This approach has been extended to include the preparation of both enantiomers of a cytotoxic unnatural ether-linked phospholipid (P. N. Guivisdalsky and R. Bittman, *J. Org. Chem.*, 1989, **54**, 4637). These unnatural ether-linked phospholipids show potent cytotoxicity, where they accumulate in the membranes of tumour cells, and are

thought either to alter tumour cell invasion, or to interfere with cellular phospholipid metabolism. The same methodology was used to prepare platelet-activating factor analogues (P. N Guivisdalsky and R. Bittman, *J. Org. Chem.*, 1989, **54**, 4693). These approaches by Bittman and co-workers are noteworthy in that no protecting groups are required. However, an alternative three step procedure to a pivotal intermediate from (R)-(+)- and (S)-(-)-oxiranemethanol-*t*-butyldiphenylsilyl ethers was also explored by this group (P. N. Guivisdalsky and R. Bittman, *J. Org. Chem.*, 1989, **54**, 4637).

(g) Semi-Synthetic Methods

The preparation of many mixed-chain glycerophospholipids are, in reality, only semi-synthetic in nature. A universally employed technique involves specific hydrolysis at the *sn*-2 position with PLA_2, followed by reacylation of the resulting 2-lysophospholipid with the desired carboxylic acid derivative. Early methods of acylation suffered from a number of disadvantages, not least of which were unreliable yields, and acyl migration with consequent racemisation (D. R. Kodali, *Chem. Phys. Lipids*, 1990, **52**, 163). These problems prompted numerous studies devoted to increasing the efficiency of the acylation procedure.

(i) Methods of Acylation

Acylation of lysophospholipids can be achieved with fatty acid anhydrides or chlorides, in the presence of either alkali or ammonium salts of these acids (J. Lakowicz and D. Hogen, *Chem. Phys. Lipids*, 1980, **26**, 1), or perchloric acid (H. Eibl, *Chem. Phys. Lipids*, 1980, **26**, 239). The cadmium chloride complex of the glycerol derivative may also be acylated with these reagents (P. D. Brimble and P. G. Choy, *Prep. Biochem.*, 1982, **11**, 525). Another method which is frequently employed, uses DMAP with the fatty acid derivative (W. Stoffel *et al.*,

Hoppe-Seyler's Z. Physiol. Chem., 1982, **363**, 1; D. S. Johnston *et al., Biochim. Biophys. Acta*, 1980, **602**, 57). This method is compatible with unsaturated and radiolabelled functionality (B. Perly, E. J. Dufourc and H. C. Jarrell, *J. Labelled Compds. Radiopharm.*, 1984, **21**, 1), and photosensitive groups (R. Radhakrishnan *et al., Methods Enzymol.*, 1981, **72**, 408). DMAP was also used to prepare phospholipid analogues containing a cyclopropyl ring (E. J. Dufourc, I. C. P. Smith and H. C. Jarrell, *Chem. Phys. Lipids*, 1983, **33**, 153) which are found in some bacteria and protozoa.

A similar acylation mediator is 4-pyrrolidinopyridine (PPY), which was utilised in a partial synthesis of isomerically pure mixed-chain phosphatidylcholines. Typically these were obtained in good yields, and in a high state of purity (J. T. Mason, A. V. Brocolli and C. Huang, *Anal. Biochem.*, 1981, **113**, 96). This catalyst was used for the esterification of lysolecithins with a fatty acid anhydride containing a masked hydroxyl group (Y. Isaacson *et al., Chem. Phys. Lipids*, 1990, **52**, 217), which was subsequently revealed. Hydroxylated fatty acids are formed enzymatically through the lypoxygenase pathways, and may then be esterified into membrane phospholipids in certain cell types. This has the effect of introducing a hydrophilic hydroxyl group into the hydrophobic portion of the phospholipid, with a consequent alteration of the biophysical character of the membrane. The compounds prepared by these workers are useful in modelling these changes. This method of acylation was also applied successfully to the synthesis of a hydroxyeicosatetraenoyl derivative.

A recent survey of the effects of solvents and catalysts on the acylation of phospholipids (D. Mangroo and G. E. Gerber, *Chem. Phys. Lipids*, 1988, **48**, 99) concluded that: 1) the rate of acylation is inversely proportional to the polarity of the solvent, with chloroform optimal; 2) activation of anhydrides with 4-pyrrolidinopyridine is twice as effective as DMAP; and 3) the phosphocholine phosphate group interferes with the acylation procedure by a process which can be reversed by use of a two hundred-fold excess of PPY. The authors suggested that a mixed anhydride is formed with the phosphate, which is reversed at high concentrations of catalyst, to produce the reactive acylating agent.

Acylation of lysophospholipids has sometimes been carried out by 2-thiopyridyl esters of fatty acids with silver perchlorate catalysis (A. W. Nicholas *et al., Lipids,* 1983, **18**, 434). Catalysis by trifluoroacetic anhydride has been used successfully for chains of eight or fewer carbon atoms, in yields of between 70 and 79% (P. Kanda and M. A. Wells, *J. Lipid Res.,* 1981, **22**, 877). Acylation with imidazolides of fatty acids offers another efficient method of esterification (B. Schmitz *et al., Z. Physiol. Chem.,* 1980, **361**, 1338), and is generally carried out in the presence of a catalytic amount of methylsulphinylmethide ion (A. Hermetter H. Stütz and F. Paltauf, *Chem. Phys. Lipids,* 1983, **32**, 145; A. Hermetter, F. Paltauf and H. Hauser, *Chem. Phys. Lipids,* 1982, **30**, 35; D. J. Vaughan and N. Z. Stanacev, *Can. J. Biochem.,*1980, **58**, 143). The substrate must be suitably protected, or may be reacted as the cadmium (II) chloride complex (A. Hermetter and F. Paltauf, *Chem. Phys. Lipids,* 1981, **28**, 111). Imidazolides have also been formed effectively from unsaturated fatty acids (B. Schmitz and H. Egge, *Chem. Phys. Lipids,* 1984, **34**, 139).

Hermetter and co-workers have developed an efficient protocol for the preparation of symmetric- or mixed-acid phosphatidylethanolamines (A. Hermetter *et al., Chem. Phys. Lipids,* 1982, **32**, 145; A. Hermetter *et al., Chem. Phys. Lipids,* 1987, **43**, 69), in which phospholipids isolated from any natural source are initially protected as the *N*-trityl phosphatidylethanolamine. This novel intermediate can then be hydrolysed specifically at the *sn*-1 position with boron trifluoride-methanol, at the *sn*-2 position with PLA_2, or non-specifically under mild basic conditions. Reacylation of the resulting lysophospholipid with acyl imidazolides in the presence of methylsulphinylmethide anion, followed by detritylation using trifluoroacetic acid furnishes the phosphatidylethanolamine derivative. This process is applicable to the large scale preparation of phospholipids.

Enzymic acylation of lysophospholipids was accomplished with fatty acyl-Co As (H. Okuyama and M. Inoue, *Methods*

Enzymol., 1982, **86**, 370) for the preparation of radiolabelled phosphatidylcholines and phosphatidylethanolamines containing ^{3}H and ^{14}C labelled polyunsaturated chains.
Whilst these recent advances in coupling protocol have overcome the problems of acyl migration and racemisation associated with earlier methods, they invariably employ an excess of the acylating agent, which may be undesirable in certain instances, for example where a valuable radioactive, spin-labelled or photoactive acid derivative is used.
Eibl has reported a high yielding method for the semi-synthesis of labelled phospholipids from 1,2-isopropylidene-*sn*-glycerol (H. Eibl *et al., Methods Enzymol.*, 1983, **98**, 623), in which the key sequences are the phospholipase A_2-mediated hydrolysis of phosphatidic acid derivatives, and subsequent re-esterification of the resulting lysophosphatidic acid with a labelled fatty acid, using *N,N'*-dicyclohexyl carbodiimide (DCC) and DMAP. The use of DCC as a coupling agent avoids the need for prior activation of the carboxylic acid, and allows the esterification of acids which contain sensitive functionality. Unfortunately, the presence of the reactive phosphodiester group precludes the use of this highly efficient coupling procedure, so that the reactions are generally carried out on the phosphomonoester derivatives.
Schreiber and co-workers (J. M. Delfino, S. L. Schreiber and F. M. Richards, *Tetrahedron Lett.*, 1987, 2327) have overcome this drawback by using a phosphotriester (**27**). Compound (**27**) is obtained from myristoyl-*N*-Boc-lysophosphatidylethanolamine and diazomethane (Scheme 8), and is available for esterification by the DCC coupling procedure using a limited amount of fatty acid. In a range of examples, yields for the esterification step varied from 79-84% based on the carboxylic acid. This partial synthesis was completed with two high-yielding deprotection steps, using sodium iodide, and then triflouroacetic acid.

Scheme 8

HO— $OCOC_{13}H_{27}$ / O–P(=O)(OH)–O–CH_2CH_2–NHBoc —(i)→ HO— $OCOC_{13}H_{27}$ / O–P(=O)(OMe)–O–CH_2CH_2–NHBoc **(27)** —(ii)→

RCOO— $OCOC_{13}H_{27}$ / O–P(=O)(OMe)–O–CH_2CH_2–NHBoc —(iii), (iv)→ RCOO— $OCOC_{13}H_{27}$ / O–P(=O)(O^-)–O–CH_2CH_2–N^+H_3

R = $(CH_2)_3CH_3$, $(CH_2)_{12}CH_3$, $(CH_2)_{12}CO_2H$, Z-$(CH_2)_7CH{=}CH(CH_2)_7CH_3$, Z,Z-$(CH_2)_7CH{=}CHCH_2CH{=}CH(CH_2)_4CH_3$

Reagents: (i) CH_2N_2, Et_2O; (ii) 0.8 eq. RCO_2H, DCC, DMAP, CH_2Cl_2; (iii) NaI, 2-butanone, 85°C; (iv) TFA, CH_2Cl_2.

(ii) Transesterification

The polar head groups of phospholipids may be interconverted *via* transesterification (C. J. Dekker *et al., Chem. Phys. Lipids,* 1983, **33**, 93; J. G. Moldkovsky *et al., Bioorgan Khim.,* 1980, **6**, 594; *Bioorgan Khim.,* 1980, **6**, 144) or hydrolysis with phospholipase D (H. Eibl and S. Kovatchev, *Methods Enzymol.,* 1981, **72**, 632; J. P. 02 02 381, 1990). The stereochemistry of phospholipase D-catalysed trans-phosphatidylation has been investigated (K. Bruzik and M.- D. Tsai, *J. Am. Chem. Soc.,* 1982, **104**, 863). Isolated yields are generally lower than those obtained using chemical methods, but Farren *et al.,* have prepared gram quantities of labelled phosphatidylethanolamines, phosphatidylserines, and phosphatidic acids from phosphatidylcholines (S. B. Farren, E. Sommerman and P. R. Cullis, *Chem. Phys. Lipids,* 1984, **34**, 279).

2. Phospholipid Analogues

Phospholipid analogues are playing an increasingly important role in a number of fields: their utility as drug carriers (D. Walde *et al., Chem. Phys. Lipids,* 1990, **53**, 265) and slow-release agents has been demonstrated (J. P. 63 96 200, 1988), and they may also act as therapeutic entities in their own right (see sections 3 and 4). Compounds have also been proposed as devices for solar energy conservation (S. L. Regen *et al., J. Am. Chem. Soc.,* 1982, **104**, 791; J. H. Fendler, *Acc. Chem. Res.,* 1980, **13**, 7), but by far the greatest use of phospholipid analogues is within the field of membrane research; and it is for this purpose that the majority of variants have been prepared.

(a) Analogues Containing Sulphur

Tsai and co-workers have prepared 1,2-dipalmitoyl-*sn*-glycero-3-thiophosphocholine (DPPsC) (**28**) as a mixture of diastereoisomers using modifications of literature methods (Scheme 9). The starting material is 1,2-dipalmitin, which is in turn obtained in seven steps from D-mannitol (K. Bruzik, R.- T. Jiang and M.- D. Tsai *J. Am. Chem. Soc.,* 1982, **104**, 4682).

Scheme 9

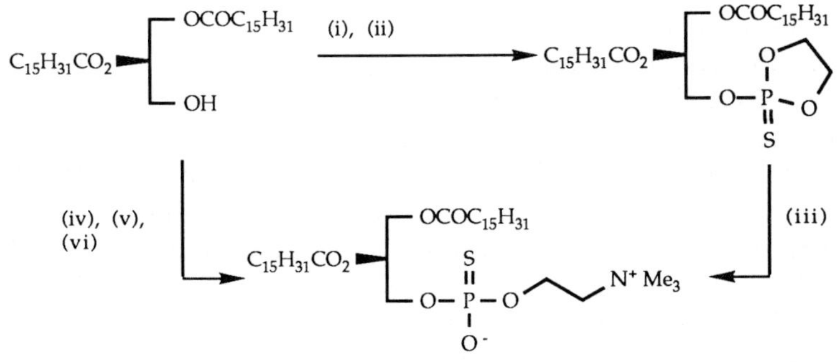

(28)

Reagents: (i) 2-chloro-2-oxo-1,3,2-dioxaphospholane; (ii) S_8; (iii) Me_3N; (iv) $PSCl_3$; (v) choline tosylate; (vi) H_2O.

These compounds are potentially useful in probing the mechanism of enzyme reactions involving phospholipids, since phospholipase A_2 was shown to hydrolyse one isomer of (**28**) specifically, whereas the other isomer was hydrolysed specifically by phospholipase C. This propensity also allowed the individual isomers of DPPsC to be separated and characterised. DPPsC (**28**) had been prepared previously without resolution (I. Vasilenko, B. De Kruijiff and A. J. Verkleij, *Biochem. Biophys. Acta*, 1982, **685**, 144) for application in ^{31}P NMR studies of lipid polymorphism, and the synthesis and stereospecific hydrolysis of the diastereoisomers by phospholipases A_2 and C has also been reported (G. A. Orr, C. F. Brewer and G. Heney, *Biochemistry*, 1982, **21**, 3202). In this case the products were not isolated. An alternative method of synthesis of DPPsC has been developed recently (I. Lindh and J. Stawinski, *J. Org. Chem.*, 1989, **54**, 1338).

The preceding examples illustrate specific applications of phospholipids which possess a chiral label at phosphorus. In general, the use in enzyme reactions of chiral phosphothioates or phosphate groups labelled asymmetrically with oxygen isotopes, may yield useful information on the steric course of nucleophilic displacement at this phosphorus atom. In addition, phosphothioates can be employed with advantage, to determine the stereochemical constraint of substrate binding, and the nature of the enzyme active site. This approach has been adopted for the study of phosphatidylserine synthase (C. R. H. Raetz *et al.*, *Biochemistry*, 1987, **26**, 4022), phospholipase A_2 (T. Rosario-Jansen, *Phosphorus Sulfur*, 1987, **30**, 601; T.- C. Tsai *et al.*, *Biochemistry*, 1985, **24**, 3180), phospholipase C (G. Lin, F. C. Bennett and M.- D. Tsai, *Biochemistry*, 1990, **29**, 2747; G. Lin and M.- D. Tsai, *J. Am. Chem. Soc.*, 1989, **111**, 3099), phospholipase D (R.- T. Jiang, Y.- J. Shyy and M.- D. Tsai, *Biochemistry*, 1984, **23**, 1661; K. Bruzik and M.- D. Tsai, *Biochemistry*, 1984, **23**, 1656), and lecithin-cholesterol acyltransferase (W. M. Loffredo and M.- D. Tsai, *Bioorg. Chem.*, 1990, **18**, 78).

Other phospholipids which are chiral at phosphorus have been reported by Tsai's group (W. M. Loffredo and M.- D. Tsai,

Bioorg. Chem., 1990, **18**, 78; K. Bruzik and M.- D. Tsai, *J. Am. Chem. Soc.*, 1984, **106**, 747; K. S. Bruzik, R.- T. Jiang and M.- D. Tai, *Phosphorus Sulfur*, 1983, **18**, 369; K. Bruzik, R.- T. Jiang and M.- D. Tsai, *Biochemistry*, 1983, **22**, 2478).

A novel synthesis of thiophospholipids has been reported, in which thiosuccinimides and thiophthalimides react spontaneously with trimethyl- or tris(trimethylsilyl)-phosphites in a Michaelis-Arbuzov fashion (Scheme 10), to yield thiophosphoric acid esters (C. E. Muller and H. J. Roth, *Tetrahedron Lett.*, 1990, 501). After deprotection, the phosphatidic acid analogues were precipitated as their cyclohexylamine salts (**29**). The thioimides described are easy to prepare and are highly reactive, so that this method provides an efficient means of preparing thiophospholipids with a C-S-P linkage.

Scheme 10

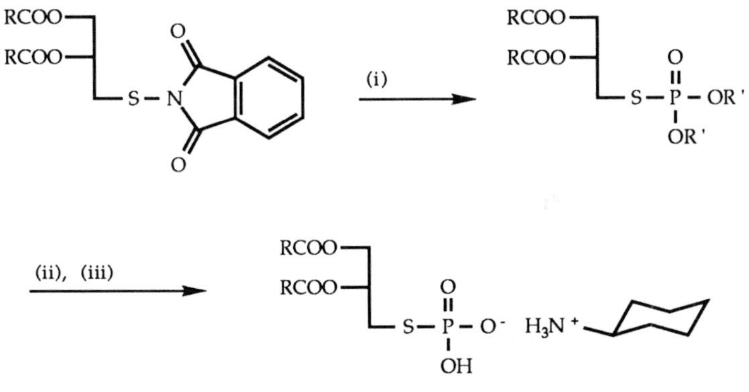

RCOO, RCOO, S – N, O, O, (i), RCOO, RCOO, S – P – OR', O, OR', (ii), (iii), RCOO, RCOO, S – P – O⁻ H_3N^+, O, OH

(29)

R = $C_{13}H_{27}$, $C_{17}H_{35}$
R' = Me, Me_3Si

Reagents: (i) $P(OR')_3$, $PhCH_3$; (ii) Deprotection; (iii) cyclohexylamine.

Sulphur has also been incorporated into the side chains of phospholipids, with thioester analogues finding use as substrates for the assay of phospholipases, and for critical micellar concentration studies (J. W. Cox and L. A. Horrocks, *J. Lipid Res.*, 1981, **22**, 496; G. L. Kucera *et al.*, *J. Biol. Chem.*, 1988, **263**, 12964; A. J. Aarsman *et al.*, *Chem. Phys. Lipids*, 1985, **36**, 229; H. S. Hendrickson and E. A. Dennis, *J. Biol. Chem.*, 1984, **259**, 5734; J. Hajdu and J. M. Sturtevant, *Chem. Phys. Lipids*, 1990, **55**, 323). These were originally prepared in a racemic form from 1,2-dimercaptopropanol, but a chiral synthesis has since been completed (H. S. Hendrickson, E. K. Hendrickson and R. H. Dybvig, *J. Lipid Res.*, 1983, **24**, 1532) using 1-trityl-*sn*-glycerol.

Hajdu and co-workers have reported several stereospecific syntheses of phospholipids containing side chain thioester and thioether linkages (S. K. Bhatia and J. Hajdu, *Synthesis*, 1989, 16; S. K. Bhatia and J. Hajdu, *Tetrahedron Lett.*, 1987, 3767; *Tetrahedron Lett.*, 1988, 31). These compounds were designed for use as an enzyme probe (S. K. Bhatia and J. Hajdu, *Tetrahedron Lett.*, 1987, 1729), and as antitumour agents (S. K. Bhatia and J. Hajdu, *J. Org. Chem.*, 1988, **53**, 5034).

Familiar chemistry was employed in a versatile asymmetric synthesis of sulphur analogues of glycerophospholipids (H. S. Hendrickson and E. K. Hendrickson, *Chem. Phys. Lipids*, 1990, **53**, 115). Both enantiomers of tritylglycidol are available by Sharpless epoxidation, followed by tritylation. The trityl protecting group was chosen since it can be removed selectively in the presence of acyl esters. The flexibility of this approach means that a range of thioester and thioether phospholipids can be synthesised from a common intermediate, simply by varying the choice of nucleophile.

Azathia analogues which contain polyhetero side chains have been synthesised from 1-acetylthio-3-hydroxy-2-propanamide hydrochloride (J. M. Ziedler, W. Zimmermann and H. J. Roth, *Chem. Phys. Lipids.*, 1990, **55**, 155). Reaction with an alkyl halide and concommitant acyl migration, allowed formation of the thioether- and acetamido- linkages simultaneously.

Novel 1-β-D-arabinofuranosylcytosine conjugates of thioether phospholipids have been prepared as potential antitumour drugs (I. C. Hong *et al.*, *Cancer Drug Delivery*, 1986, **3**, 101), and analogues of a naturally occurring phospholipid bearing a thioester linkage at the *sn*-1 position, also exhibit antitumour properties (W. E. Berdel *et al., Cancer Res.*, 1983, **43**, 5538). Five sulphur-containing phospholipid analogues were prepared, which exhibited neoplastic cell growth inhibitory properties (S. Morris-Natschke *et al., J. Med. Chem.*, 1986, **29**, 2114).

(b) Backbone Modifications

Bittman and Witzke have prepared a conformationally restricted phospholipid from methallyl alcohol (R. Bittman and N. M. Witzke, *Chem. Phys. Lipids*, 1989, **50**, 99), in which bond rotation in the glycerol backbone is severely impeded by the methyl group at the 2-position. This analogue is useful for studying the influence of conformational changes at the lipid-water interface of lipid bilayers. Similar conformation studies have been carried out by other workers (H. Hauser *et al., Biochemistry*, 1980, **19**, 366; H. Hauser, *Biochim. Biophys. Acta*, 1981, **646**, 203; R. Ghosh and J. Seeling, *Biochim. Biophys. Acta*, 1982, **691**, 151). An alternative approach to conformational restriction is to constrain the backbone as a ring structure. Using this idea the syntheses of cyclopentanoid analogues of glycerophospholipids were reported (A. J. Hancock, *Methods Enzymol.*, 1981, **72**, 640). In addition, seven geometrical isomers of dipalmitoyl cyclopentanophosphoric acid were prepared to study the effect of backbone geometry on thermotropic properties (T. Y. Ahmad *et al., Chem. Phys. Lipids*, 1990, **55**, 231).

It has long been established that in addition to glycerophospholipids, similar derivatives of C2, C3, and C4 diols are found in plant and animal organs. These appear to be generated at the most intensive stages of growth, and act as bioregulators. Recent work describing the synthesis of novel mimics of such compounds has been reported (S. D. Stamatov *et al., Chem. Phys. Lipids*, 1988, **46**, 199). This paper describes the preparation of 1-*O*-acylethane-2-*O*-amidothionphosphonates, and their R,S-α-tocopherol derivatives (**30**) as a new class of phospholipids.

OCOR
R'
O–P
NEt 2
S

(30)

R = stearoyl, oleoyl, linoleoyl
R' = NEt_2, (R,S)-α-tocopherolyl

McGuigan and co-workers have reported the application of phosphoramidite methodology to the synthesis of novel phospholipids related to ethylenediamine (M. S. Anson and C. McGuigan, *J. Chem. Soc., Perkin I*, 1989, 715). In a three stage procedure (Scheme 11), low temperature reaction of 2-chloro-1,3-dimethyl-1,3-diazaphosphacyclopentane (**31**) with an alcohol, followed by oxidation with dinitrogen tetroxide, gave the cyclic phosphate phosphite triester (**32**), which was then hydrolysed. The method allowed rapid access into a series of phosphoryl-*N*-methylethanolamine analogues (**33**).

Scheme 11

Me N P–Cl N Me (i), (ii) → Me N P O OR N Me (iii) → Me NH_2^+ O N–P–OR Me O^-

(31) **(32)** **(33)**

R = C_6H_{13}, C_8H_{17}, $C_{18}H_{37}$, Δ^9-$C_{18}H_{35}$

Reagents: (i) ROH, Et_3N, CH_2Cl_2, -60°C; (ii) N_2O_4, CH_2Cl_2, -60°C; (iii) H_2O, THF, H^+.

A new class of ethanolamino-phospholipids have been identified (D. G. I. Kingston *et al.*, *Tetrahedron Lett.*, 1989,

6665). These polyunsaturated plasmalogens were shown to be biological precursors of a potent faecal mutagen, fecapentaene-12.

(c) Phosphonolipids

A useful review of phosphonolipids has been published (T. Hori and Y. Nozawa in *Phospholipids*, 1982, J. N. Hawthorne and G. B. Ansell (Eds.), Elsevier Biomedical Press, Amsterdam, 95).

Engel and co-workers prepared a series of chiral phosphono analogues of phosphatidic acid and phosphatidylcholine (P. W. Schwartz, B. E. Tropp and R. Engel, *Chem. Phys. Lipids*, 1988, **48**, 1) in which an olefinic linkage was incorporated into the backbone (Scheme 12). These analogues necessarily possess a greater structural rigidity than simple phosphonic acid analogues of phospholipids. Replacement of the esteric phosphate oxygen with an isosteric methylene group results in a compound in which there is negligable variation in spacial relationships between the remaining functionalities, but a total resistance to phosphorus elimination by enzymes. However, this results in a marked reduction in solubility. To counter this, an hydroxyl group was introduced adjacent to the phosphonyl centre (P. W. Schwartz, B. E. Tropp and R. Engel, *Chem. Phys. Lipids*, 1988, **49**, 131). All compounds are available from the common phosphonodiester intermediate (**34**), which was obtained by a Wittig reaction of tetraisopropyl methylenebisphosphonate and isopropylidene-*sn*-glycerol. The stereochemistry is entirely *trans*, and is maintained throughout. Reduction gave the saturated phosphono analogue (**35**), and hydroboration gave the α-hydroxyphosphono analogue (**36**) of phosphatidic acid. Deprotection, followed by reaction with choline gave the analogue of phosphatidylcholine (**37**).

Scheme 12

(i) → (ii), (iii) →

CHO

$O{=}P(OiPr)_2$

OCOR RCOO $O{=}P(OiPr)_2$ (34) (iv), (v) → OCOR RCOO PO_3H_2 (35)

(vi), (v) → OCOR RCOO OH PO_3H_2 (36)

(v), (vii) ↓ OCOR RCOO $O{=}P{-}O$ N^+Me_3 O^- (37)

$R = C_9H_{19}, C_{15}H_{31}$

Reagents: (i) tetraisopropylmethlenebisphosphonate, BuLi, hexane; (ii) Dowex 50(H^+); (iii) RCOCl, $CHCl_3$, pyridine; (iv) H_2, Pd-C; (v) Me_3SiBr, H_2O; (vi) H_3BSMe_2; H_2O_2, NaOH: (vii) choline tosylate, pyridine, Cl_3CN.

Some phosphono analogues of a bioactive phospholipid, platelet-activating factor, have been prepared from a protected glycerol derivative (H. Disselnkoetter *et al.*, *Arch. Pharm. (Weinheim, Ger.)*. 1985, **318**, 695). Moschidis *et al.*, have also prepared phosphono analogues of PAF in which the oxygen atoms were replaced by methylene groups (M. C. Moschidis, C.

A. Demopoulos and L. Kritikou, *Chem. Phys. Lipids*, 1983, **33**, 87; M. C. Moschidis and C. A. Demopoulos, *Chem. Phys. Lipids*, 1985, **37**, 45).

Direct esterification of alkylethanediols with 2-bromoethyl phosphonic acid in the presence of TPS-pyridine, followed by standard amination protocol, has been used for the synthesis of 1-*O*-alkylethylene glycol phosphonic acid analogues of lecithin and cephalin (M. C. Moschidis, *Chem. Phys. Lipids*, 1987, **43**, 39).

The phosphono analogue of cardiolipin has also been prepared (M. C. Moschidis, *Chem. Phys. Lipids*, 1988, **46**, 253).

(d) Vesicles

Phospholipid bilayer vesicles have received growing attention as models for biological membranes. Eibl's excellent article (H. Eibl, *Angew. Chem., Int. Ed. Engl.*, 1984, **23**, 257) details the four levels of membrane research: 1) synthesis; 2) analogues for the study of physical properties of phospholipid-water systems; 3) reconstitution and reactivation experiments; 4) selective modification of natural cell membranes. Regan *et al.*, introduced the concept of polymerised vesicles (S. L. Regan, B. Czech and A. J. Singh, *J. Am. Chem. Soc.*, 1980, **102**, 6638) as a new class of materials which retain the properties of promoting the separation of charged photoproducts, as well as the entrapment and slow release of drugs, but which exhibit greater stability than liposomes (vesicles formed from naturally occurring phospholipids). These workers have reported the synthesis and properties of polymerised vesicles from phosphatidylcholine analogues (S. L. Regan *et al.*, *J. Am. Chem. Soc.*, 1982, **104**, 791). A vast number of phospholipid analogues bearing polymerisable groups have since been prepared for this purpose

(L. Gros, H. Ringsdorf and H. Schupp, *Angew. Chem., Int. Ed. Engl.*, 1981, **20**, 305; R. Radhakrishnan *et al.*, *Methods Enzymol.*, 1981, **72**, 408; E. Hasegawa *et al.*, *Makromol. Chem. Rapid. Commun.*, 1984, **5**, 779; E. Tsuchida *et al.*, *Chem. Lett.*, 1985,

969; E. Hasegawa *et al., Polym Bull,* 1985, **14**, 31; E. Hasegawa *et al., Synth,* 1987, 60). B. Ostermeyer *Chem. Phys. Lipids,* 1986, **41**, 265; A. Furakawa *Makromol Chem,* 1986, **187**, 311; T. Kunitake *Am. N. Y. Acad. Sci.,* 1986, **471**, (*Int. Symp. Bioorg. Chem.,* 1985), 70; P. Yager *et al., Chem. Phys. Lipids,* 1988, **46**, 171). Rhodes *et al.*, reported the preparation of a tubule-forming phospholipid (D. G. Rhodes *et al., Chem. Phys. Lipids,* 1988, **49**, 39), and Granzer *et al.*, have disclosed the synthesis and aggregation properties of chiral phospholipids with a pantoic acid backbone (U. H. Granzer *et al., Liebigs Ann. Chem.,* 1989, 59) from pantolactone.

(e) Membrane Probes

The topology of the interior of phospholipid bilayers may be investigated by physical methods such as fluorescence, NMR, EPR and calorimetry, or by the use of chemical probes. $\Delta^{5,7,9}$-Cholestatrien-3β-ol has been coupled to phosphatidic acid by the CEP method as a membrane probe (F. Ramirez *et al., Phosphorus Sulfur,* 1983, **17**, 67). Phospholiposterols have been synthesised for the examination of their membrane-forming properties, with a view to shedding light on the unspecified phospholipid-cholesterol complexes which are known to exist in some membranes (M. K. Jain *et al., Biochim. Biophys. Acta,* 1980, **600**, 678). The interactions of these complexes in hydrated bilayers have also been reviewed (J. B. Finean, *Chem. Phys. Lipids,* 1990, **54**, 147). Lebeau *et al.*, synthesised similar compounds which were linked to the steroidal hormone derivatives dexamethasone and cortisol, in which the chain length and degree of unsaturation were varied (L. Lebeau, C. Mioskowski and P. Oudet, *Chem. Phys. Lipids,* 1988, **46**, 57). These compounds showed a high affinity for the glucocorticoid receptor. Amidothiophospholipid derivatives of cholecalciferol have also been prepared (S. D. Stamatov and S. Gronowitz, *Lipids,* 1990, **25**, 149).
Schrieber and co-workers devised a general strategy for the synthesis of membrane-spanning bipolar phospholipid probes

(J. M. Delfino *et al .*, *Tetrahedron Lett.*, 1987, 2323). A similar transmembrane probe has been designed, which contains a 4,4'-dihydroxybenzophenone linkage (Y. L. Diyizou *et al.*, *Tetrahedron Lett.*, 1987, 5743).

Photolabelled phospholipids have been used to provide useful information concerning the interaction of membrane components. Photolabile precursors are used, which form reactive carbenes and nitrenes on illumination, and cross-link with neighbouring molecules with broad specificity (W. Stoffel, K.- P. Salm and M. Muller *Hoppe Seyler's Z. Phisiol. Chem.*, 1982, **363**, 1). The [4-(trifluoromethyl)diazirinyl]phenyl group is often incorporated into the phospholipid fatty acid chain (J. Brunner *et al.*, *Biochemistry* , 1983, **22**, 3812; Y. Takagaki *et al.*, *J. Biol. Chem.*, 1983, **258**, 9128). Similarly a 2-diazirinophenoxy group has been used as a carbene precursor (Y. Takagaki *et al.*, *J. Biol. Chem.*, 1983, **258**, 9136). Brunner and Richards have synthesised phospholipid analogues in which the fatty acid chain contains a disulphide bridge and either a carbene or nitrene precursor (J. Brunner and F. M. Richards *J. Biol. Chem.*, 1980, **225**, 3319), and the synthesis of phospholipids containing the 2-diazocyclopentadienylcarbonyl group has also been reported (M. O. Karyukhina *et al.*, *Bioorg. Khim.*, 1988, **14**, 1256). Arylazido phospholipids labelled with ^{14}C or ^{3}H have been effectively exchanged with boundary lipids and successfully cross-linked upon exposure to ultraviolet light (R. Bisson *et al.*, *Biochem. J.*, 1981, **193**, 757).

Fluorescent anisotropy is one of the most widely used techniques for studying membrane fluidity. This is just one of a host of measurable parameters which can be obtained from photoluminescent probes. There are two commonly encountered luminescences: fluorescence and phosphorescence, with the latter most extensively employed. The relative merits of each have been reviewed (N. J. Turro, M. Grätzel and A. M. Braun, *Angew. Chem., Int. Ed. Engl.*, 1980, **19**, 675).

Beck *et al.*, used the methodology developed by Schreiber and co-workers to prepare a fluorescent membrane probe (A. Beck *et al.*, *Chem. Phys. Lipids*, 1990, **55**, 13). This method was used to probe phospholipid-PAF interactions (T. Thursen *et al.*,

Chem. Phys. Lipids, 1990, **53**, 129). An azobenzene derivative has been reported for the study of lipid transfer between liposomes (E. Hantz *et al., Chem. Phys. Lipids,* 1989, **51**, 83).

(f) Electron transport

The electron transport systems in biological membranes are responsible for respiratory and photosynthetic energy transduction. In many cases this involves membrane-bound quinones. Little is known of the mechanisms of these processes (G. von Jagow and W. D. Engel, *Angew. Chem., Int. Edn. Engl.,* 1980, **19**, 659; E. L. Ulrich *et al., J. Biol. Chem.,* 1985, **24**, 2501), but development of model systems may aid in the elucidation process. Such a system was described by Struck *et al.,* who appended two fluorophores to the free amino group of phosphatidylethanolamine (D. K. Struck, D. Hoekstra and R. E. Pagano, *Biochemistry,* 1981, **20**, 4093). Leidner *et al.,* described the synthesis and properties of a plasmalogen quinone (C. R. Leidner *et al., Tetrahedron Lett.,* 1990, 189), which was used for the preparation of quinone-functionalised liposomes in a simple chemical model for redox and transport reactions. Tabushi *et al.,* have reported an artificial flavophospholipid which was shown to facilitate transmembrane electron movement (I. Tabushi and I. Hamachi, *Tetrahedron Lett.,* 1986, 5401). This electron transport could be controlled by a thermally-induced phase transition (I. Tabushi, I. Hamachi and Y. Kobuke, *Tetrahedron Lett.,* 1987, 5899).

(g) Miscellaneous Analogues

Phospholipid analogues are growing in importance as medicinal substances in their own right: in addition to PAF agonists and antagonists (section 3) and PLA_2 inhibitors (section 4), phospholipids have found use as antineoplastic agents (R. Zeisig, D. Arndt and H. Brachwitz, *Pharmazie,* 1990, **45**, 809; J. P. 59 163 389, 1984; Eur. P. 262 876, 1988 ; Y. Honma *et al.,*

Cancer Res., 1981, **41**, 3211; U.Kiyoshi *et al., Chem. Pharm. Bull.*, 1989, **37**, 1249; W. E. Berdel, *Lipids*, 1987, **22**, 970), immunostimulants (J. P. 01 258 691, 1989), and pharmaceutical carriers (J. P. 01 47 793, 1989). A radioiodinated analogue of a naturally occuring alkyl lysophospholipid was synthesised for evaluation as a potential tumour-localising imaging agent (K. L. Meyer, A. W. Schwendner and R. E. Counsell, *J. Med. Chem.*, 1989, **32**, 2142), and a series of fluoro- and chloro-containing analogues of alkyl lysophospholipids have been prepared as potential antimetabolites of natural phospholipids (H. Brachwitz *et al., Chem. Phys. Lipids*, 1982, **31**, 33-check; H. Brachwitz, P. Langent and J. Schildt, *Chem. Phys. Lipids*, 1984, **34**, 355). Some of these compounds have been found to exhibit a strong inhibitory effect on the proliferation of certain carcinoma cells *in vitro*.

Phosphatidyl choline derivatives of 1-*O*-(2'-methoxy-hexadecyl)glycerol have been made (G. A. M. Staellberg, *Acta. Chem. Scand.*, 1990, **44**, 368), and the thermotropic phase behavior of analogues bearing a branched fatty acid at C-1 were studied (G. Brezesinski *et al., Chem. Phys. Lipids*, 1987, **43**, 257). Amido phospholipids were prepared in both racemic and enantiomeric forms by competitive inhibition of lipolytic enzymes (D. Ruud, N. Dekker and G. H. De Haas, *Biochim. Biophys. Acta*, 1990, **1043**, 67).

In a novel medical application, phospholipids containing imidazole have been prepared as ligands for stable ribosome-enclosed iron porphyrin complexes. These play an important role in the storage and transport of oxygen (J. P. 63 297 392, 1988). Novel compounds were also synthesised for use in studies on phosphorylcholine-binding immunoglobulins (T. F. Spande, *J. Org. Chem.*, 1980, **45**, 3081).

Djerassi's group reported the synthesis of phospholipids containing unusual fatty acids isolated from sponges (E. Ayanoglou *et al., Chem. Phys. Lipids*, 1988, **47**, 165). Over fifty such acids have been isolated, and synthetic phospholipids have been prepared which incorporate a selection of these acids, and some close structural analogues. In addition, the interactions of these compounds with conventional phospholipids and cholesterol in model membranes was studied. Similar compounds embracing the 1,2-di-(6Z,9Z)-6,9-hexacosadienoyl group have also been prepared (H. Li *et al.*, *Chem. Phys. Lipids*, 1988, **48**, 109).

3. Platelet-Activating Factor (PAF[1])

Platelet-Activating Factor (**38**) is a biologically active, ether-linked phospholipid, identified as 1-*O*-hexadecyl/octadecyl-2-acetyl-*sn*-glycero-3-phosphorylcholine. It is synthesised and secreted directly from the membranes of a variety of cells involved in inflammatory responses, and elicits a range of potent and specific effects. PAF has been implicated in a wide spectrum of disease states, including asthma, pulmonary dysfunction, endotoxic shock, and allergic skin diseases.

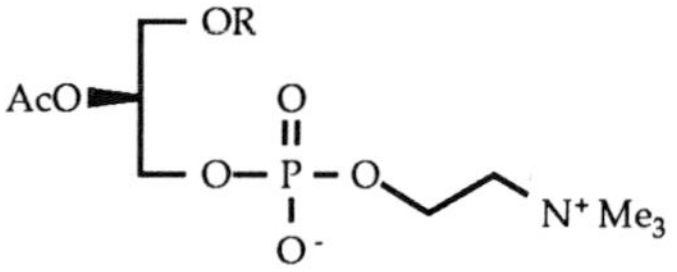

OR
AcO
O—P—O
N^+Me_3
O^-

(38)

R = $C_{16}H_{33}$ or $C_{18}H_{37}$

A prodigious amount of literature has been published on PAF as a result of the widespread interest shown in this phospholipid since its discovery. Much of the biological activity is summarised in a number of authoritative reviews (F. Snyder, *Ann. Rep. Med. Chem.*, 1982, **17**, 243; M. C. Venuti, *Ann. Rep. Med. Chem.*, 1985, **20**, 193; D. A. Handley, *Drugs of the Future*, 1988, **13**, 137; D. J. Hanahan, *Ann. Rev. Biochem.*, 1986, **55**, 483; P. Braquet *et al.*, *Pharmacol. Rev.*, 1987, **39**, 97; R. N. Pinchard, J. C. Ludwig and L. M. Manns in *Inflammation: Basic Principles and Clinical Correlates.*, 1988, J. I. Gallin and I.

---

[1] The terms AGEPC (acetyl glyceryl ether phosphocholine) and PAF acether are also used.

M. Goldstein (Eds.), Raven Press Ltd., New York, 139; F. Snyder, *Proc. Soc. Exp. Biol. Med.*, 1989, **190**, 125; F. Snyder, *Med. Res. Rev.*, 1985, **5**, 107; *Platelet Activating Factor and Related Lipid Mediators*, 1987, F. Snyder (Ed.), Plenum Press, New York; D. Hosford *et al.*, *Phytotherapy Res.*, 1988, **2**, 1; F. H. Valone, *West. J. Med.*, 1989, **150**, 334; D. J. Hanahan *et al.*, *Prog. Lipid. Res.*, 1987, **26**, 1; S. M. Prescott, G. A. Zimmerman and T. M. McIntyre, *J. Biol. Chem.*, 1990, **265**, 17381; *Frontiers in Pharmacology and Theraputics: Platelet-Activating Factor and Human Diseases*, 1989, P. J. Barnes, C. P. Page and P. M. Henson (Eds.), Blackwell Scientific Publications, Oxford; *New Horizons in Platelet-Activating Factor Research*, 1987, C. M. Winslow and M. L. Lee (Eds.), John Wiley and Sons, Chichester).

(a) Syntheses of PAF

Since its discovery there have been numerous syntheses and partial syntheses of PAF and its analogues. Some employ the conventional techniques illustrated in Section 1, but the prominence of PAF in the literature has acted as a stimulus for the development of new methodology, much of which is of general applicablity to the synthesis of phospholipids. The syntheses of PAF have been compiled without commentary in two articles (*Synform*, 1990, **8**, 313; 1988, **6**, 115).

(i) Syntheses from D-*Mannitol*

A number of syntheses have employed D-mannitol as their starting material, however, the stereochemistry of this compound is such that routes are more readily structured to give *enantio*-PAF. An inversion of the *sn*-2 centre is therefore required to obtain PAF, which invariably involves protecting group manipulations.
The first total synthesis of PAF was accomplished by Godfroid *et al.*, (*FEBS Lett.*, 1980, **116**, 161), and was important in providing

conclusive proof of the structure which had been proposed for PAF. These workers employed standard phospholipid chemistry (Scheme 13), but the synthesis was flawed by partial epimerisation suffered during the preparation of an alkyl glycerol intermediate (**39**). Nevertheless, (**39**) was selectively protected at the primary hydroxyl, then acetylated to give (**40**). Conventional phosphorylation conditions yielded a compound which was identical to PAF isolated from natural sources.

Scheme 13

$OC_{18}H_{37}$
HO
OH
(i), (ii)
(39)

$OC_{18}H_{37}$
AcO
OTr
(iii), (iv), (v)
(40)

$OC_{18}H_{37}$
AcO
O
O– P – O
OH
Br
(vi), (vii)

PAF
(38)

Reagents: (i) TrCl, pyridine; (ii) AcCl, pyridine; (iii) H_2, Pd-C; (iv) $Cl_2P(O)OCH_2CH_2Br$; (v) hydrolysis; (vi) Me_3N; (vii) Ag_2CO_3.

This group and others have subsequently reported a number of similar methods, all utilising D-mannitol as a source of chirality. These routes involve either manipulation of D-mannitol to obtain (**39**), or alternatively employ a protected form of glyceraldehyde, obtained from D-mannitol, as the starting point for novel chemistry (F. Heymans *et al.*, *C. R. Acad. Sc. Paris, Ser. II*, 1981, 293, **49**; F. Heymans *et al.*, *Biochim. Biophys. Acta*, 1981, **666**, 230; M. C. Borrel *et al.*, *Agents Actions*, 1982, **12**, 709; G. Hirth and R. Barner, *Helv. Chim. Acta*, 1982, **65**, 1059).

The synthesis of enantiomeric PAF has also been completed (C. A. A. van Boeckel *et al.*, *Synthesis*, 1982, 399), and this synthesis is unusual in being one of the few routes in which acetylation was not the ultimate step. In this procedure

(Scheme 14), acetylation was achieved early in the synthesis, and phosphorylation of an advanced intermediate (**41**) was accomplished with a hydroxybenztriazole-activated phosphate (**42**). The unisolated intermediate (**43**) was treated with choline tosylate in the presence of 1-methyl imidazole, then deprotected with *N,N,N',N'*-tetramethylguanidinium-*syn*-4-nitrobenzald-oximate (**44**).

Scheme 14

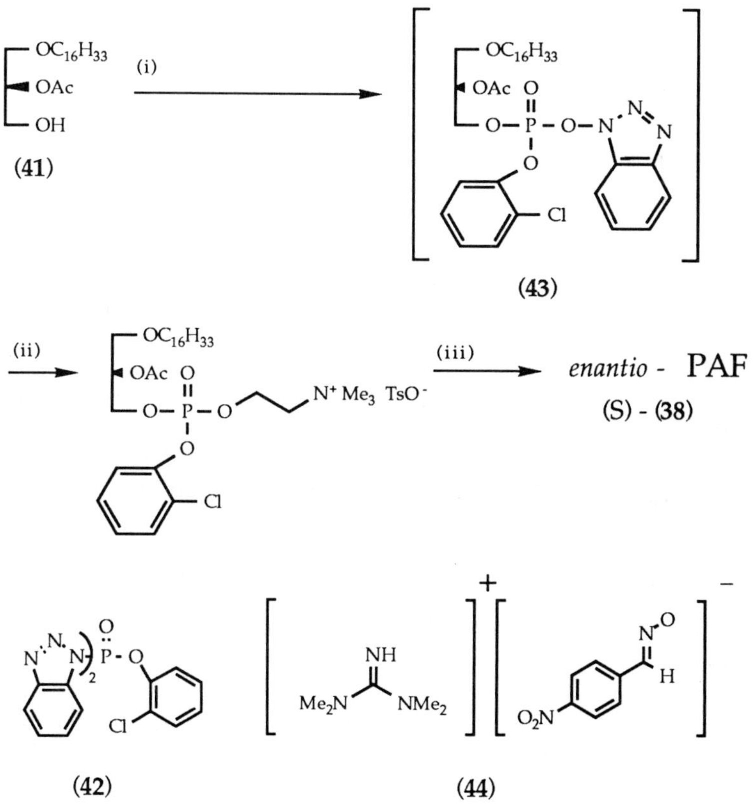

$OC_{16}H_{33}$ OAc OH (**41**) (i) → [$OC_{16}H_{33}$ OAc O–P(=O)(O-2-chlorophenyl)–O–N (benzotriazolyl)] (**43**)

(ii) → $OC_{16}H_{33}$ OAc O–P(=O)(O-2-chlorophenyl)–O–CH_2CH_2–N^+Me_3 TsO^- (iii) → *enantio* - PAF (S) - (**38**)

(**42**) (benzotriazolyl-N)$_2$P(=O)–O-2-chlorophenyl

(**44**) $[(Me_2N)_2C{=}NH]^+$ [4-O_2N-C_6H_4-CH=N-O]$^-$

Reagents: (i) (**42**), THF; (ii) choline tosylate, pyridine, 1-methylimidazole, MeCN; (iii) (**44**), THF.

Several of the routes to PAF cited above, and indeed many classical syntheses of phospholipids in general, commence with 1,2-isopropylidene-*sn*-glycerol, which is obtained in several steps from D-mannitol. However, its utility has been questioned (H. Eibl, *Chem. Phys. Lipids*, 1981, **28**, 1) due to problems of racemisation encountered on storage, and its notoriously unreliable optical purity. Eibl has reported a synthesis of PAF which avoids this problem (H. Eibl, *Angew. Chem., Int. Edn. Engl.*, 1984, **23**, 257), by using 3,4-isopropylidene-D-mannitol. This intermediate is stable and easily prepared, and has also been used as a general intermediate in the syntheses of a number of other important bioactive phospholipids.

A recent synthesis of PAF from D-mannitol was reported (N. Nakamura *et al.*, *Tetrahedron Lett.*, 1990, 699), which uses novel Mitsunobu chemistry (Scheme 15). These workers followed a literature procedure to obtain a protected alkylglycerol derivative (**45**) from 1,2-isopropylidene-*sn*-glycerol. They found that if worked up at pH 7.2, and treated with aqueous sodium hydrogen carbonate solution prior to distillation, the specific rotation of 1,2-isopropylidene-*sn*-glycerol could be preserved for months. Compound (**45**) was then treated with dimethyl azodicarboxylate and triphenylphosphine, in a Mitsunobu reaction, with 3-hydroxyisoxazole as the acidic component. The intermediate (**46**) was then deprotected to give (R)-(**47**). Reductive cleavage of the isoxazole group, and acetylation gave PAF. Alternatively, a second inversion reaction on (**46**), and a similar phosphorylation sequence gave (S)-(**47**). Both intermediates (**47**) were found to exhibit potent PAF agonistic activity.

Scheme 15

D - MANNITOL

(45)

(vi), (vii)

PAF

(38)

(R)- (47)

(ii), (iii) (iv), (v)

(45) (i)

(46)

(vi), (ii)

(ii), (iii) (iv), (v)

(S)- (47)

Reagents: (i) 3-hydroxyisoxazole, $MeOCON{=}NCO_2Me$, PPh_3; (ii) TsOH; (iii) $Cl_2P(O)OCH_2CH_2Br$, Et_3N; (iv) H_2O; (v) Me_3N, Ag_2CO_3; (vi) H_2, Pd-C; (vii) Ac_2O, Et_3N.

Racemic syntheses of PAF from differentially protected glycerol derivatives have also been reported. Tin chemistry (M. H.

Marx and R. A. Wiley, *Tetrahedron Lett.*, 1985, 1379), and conventional chemistry have been used (D. A. Bass, L. C. McPhail and J. D. Schmitt, *J. Biol. Chem.*, 1988, **263**, 19610), and synthetic procedures may include a resolution step in order to obtain homochiral PAF (H. P. Kertscher, *Pharmazie*, 1983, **38**, 421; H. P. Kerscher and G. Ostermann, *Pharmazie*, 1986, **41**, 596; J. R. Surles *et al.*, *J. Med. Chem.*, 1985, **28**, 73). Key intermediates have also been prepared from D-mannitol (U. Peters, W. Bankova and P. Welzel, *Tetrahedron*, 1987, **43**, 3803; S. Takano, A. Masashi and K. Ogasawara, *Chem. Pharm. Bull.*, 1984, **32**, 791).

(ii) Other Natural Sources of Chirality

Ohno and co-workers have accomplished several complimentary variations of an enantioselective synthesis of both enantiomers of PAF, together with a number of its analogues, from D- and L-tartaric acids (K. Fujita *et al.*, *Tetrahedron Lett.*, 1982, 3507; M. Ohno *et al.*, *Chem. Pharm. Bull.*, 1985, **33**, 572)
Tsuri and Kamata described a formal synthesis in which a key intermediate was prepared from (S)-(-)-malic acid (T. Tsuri and S. Kamata, *Tetrahedron Lett.*, 1985, 5195), and Berti *et al.*, described a preparation in which ascorbic acid was converted, in several steps, to 2,3-isopropylidene-*sn*-glycerol. This was then transformed into PAF by a multi-step process (F. Berti *et al.*, *Pharm. Res. Commun.*, 1986, **18**, 557).

(iii) Other Syntheses of PAF

Wang and Wong have devised an enantioselective synthesis of PAF (Y.- F. Wang and C.- H. Wong, *J. Chin. Chem. Soc.*, 1989, **36**, 463), in which the chiral precursor (**48**), was prepared *via* lipase catalysed enantioselective acetylation of 2-benzyl glycerol. This differentially protected glycerol derivative was subsequently converted to PAF using a literature procedure.

Formal syntheses utilising enzyme catalysed asymmetric hydrolyses have also been reported (H. Sueme *et al.*, *Chem. Pharm. Bull.*, 1986, **34**, 3440; H. Sueme *et al.*, *Chem. Pharm. Bull.*, 1987, **35**, 3112).

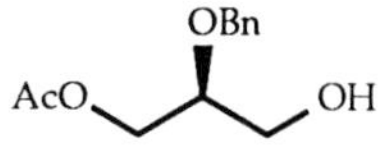

OBn
AcO OH
HO $OC_{18}H_{37}$ SMe_2 BzO CO_2H BzO COO

(48) 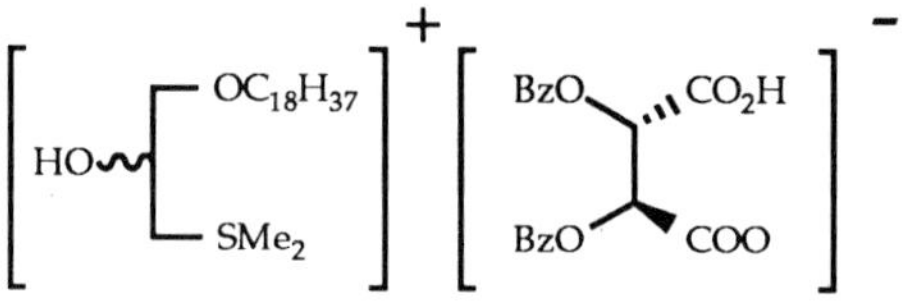 **(49)**

Julia and co-workers completed a short synthesis of PAF from epichlorohydrin (B. Cimetiere, L. Jacob and M. Julia, *Tetrahedron Lett.*, 1986, 6329), which involved the resolution of a racemic intermediate (**49**). Eight recrystallisations of the salt were followed by treatment with base to give (S)-alkyl glycidol with 95% ee. Opening of this epoxide with phosphorylcholine was followed by an acetylation step to furnish homochiral PAF. Bittman's group have extended methodology previously applied to other bioactive ether lipids (section 1), to produce an efficient stereocontrolled route to both enantiomers of PAF (P. N. Guivisdalsky and R. Bittman, *J. Org. Chem.*, 1989, **54**, 4643). This employed a boron trifluoride etherate-catalysed regio- and stereospecific opening of either (R)- or (S)-glycidyl arenesulphonate with 1-hexadecanol (Scheme 16). This synthesis has two key features: an exceptionally mild benzylation step, in which competitive epoxide formation by displacement of the tosyl group from (**50**) is avoided, and the *in situ* conversion of the resulting benzyl tosylate (**51**) to benzyl glycidol (**52**).

Scheme 16

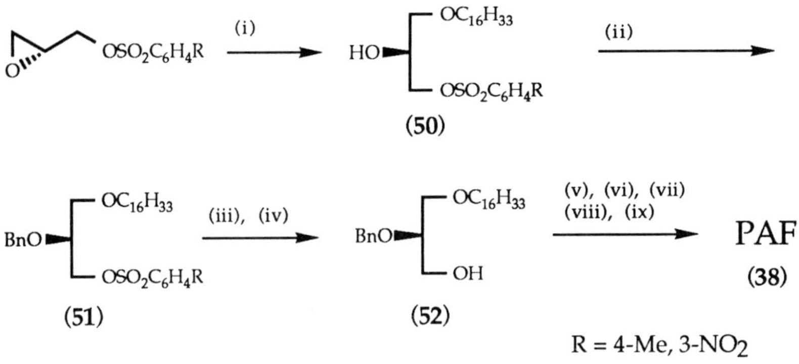

Reagents: (i) $C_{16}H_{33}OH$, $BF_3.Et_2O$, CH_2Cl_2; (ii) BnOH, $(F_3CSO_2)_2O$, 2,6-di-*t*-butyl-4-methyl pyridine, CH_2Cl_2, -78°C; (iii) CsOAc, DMF-DMSO; (iv) $LiAlH_4$, Et_2O, 0°C; (v) $POCl_3$, Et_3N; (vi) choline tosylate, pyridine; (vii) H_2O; (viii) H_2, $Pd(OH)_2$-C, MeOH, H_2O; (ix) Ac_2O, DMAP, $CHCl_3$.

This is achieved by displacement of the tosyl group with caesium acetate, followed by lithium aluminium hydride reduction. The synthesis was completed using conventional phosphorylation chemistry.
Russian workers prepared PAF from 3-octadecyl-2-myristoyl-*sn*-glycero-1-phosphocholine by inversion of the phospholipid configuration with reagents coupled to an ion-exchange resin (K. Y. Gordeev, G. A. Serebrennikova and R. P. Evstigneeva, *Bioorg. Khim.*, 1987, **13**, 234).
A formal synthesis (H. Sueme, A. Akashi and K. Sakai, *Chem. Pharm. Bull.*, 1985, **33**, 1055) in which chirality is introduced *via* an asymmetric reduction with lithium aluminium hydride-binaphthol complex, has also been reported in the literature.

(b) PAF Analogues

With the intense pharmaceutical interest in PAF has come the introduction of systematic structural variants of the PAF molecule, prepared to establish the structural requirements for activity, to obtain theraputically selective agonists, and to discover PAF receptor antagonists. A vast number of naturally occuring and synthetic PAF agonists and antagonists are now known. These are reviewed in detail elsewhere (D. Horsford and P. Braquet, *Prog. Med. Chem.*, 1990, **27**, 325; M. C. Venuti in *Comprehensive Medicinal Chemistry*, 1990, **3**, J. C. Emmett (Ed.), Pergamon Press, Oxford, 715; D. A. Handley, *Med. Res. Rev.*, 1990, **10**, 351; K. Cooper and M. J. Parry, *Ann. Rep. Med. Chem.*, 1989, **24**, 81). Only the phospholipid analogues of PAF are discussed here.

(i) PAF Receptor Agonists

(1) Variations at C-1

The effects on platelet aggregation and bronchoconstriction with respect to alkyl chain length have been studied (D. J. Hanahan *et al., Biochem. Biophys. Res. Commun.*, 1981, **99**, 183; J.- J. Godfroid and P. Braquet, *Trends Pharmacol. Sci.*, 1986, **7**, 368). QSAR analysis of linear chain homologues from C1 to C20 showed a close correlation between agonistic activity and substituent hydrophobicity (J.- J. Godfroid *et al., J. Med. Chem.*, 1987, **30**, 792). Insertion of a phenyl spacer gave a general decrease in agonistic activity, with the *m*- more active than the *p*-substituted compounds. The *o*-substituted analogue was inactive (A. Wissner *et al., J. Med. Chem.*, 1984, **27**, 1174; R. C. Anderson, B. E. Reitter and C. M. Winslow, *Chem. Phys. Lipids*, 1986, **39**, 73). Agonistic activity for unsaturated chains is marginally increased by the presence of up to two double bonds (J. R. Surles *et al., J. Med. Chem.*, 1985, **28**, 73). Terminal fluorination of chains of various length decreases activity (K. Shigenobu *et al., J. Pharmacobiodyn.*, 1985, **8**, 128; K. Fujita *et al., Chem. Pharm. Bull.*, 1987, **35**, 647), as does multiple oxygen substitutions (A. Wissner, C. A. Kohler and B. M. Goldstein, *J. Med. Chem.*, 1986, **29**, 1315).

The ether linkage appears to be essential for activity (P. Braquet *et al., Pharmacol. Rev.*, 1987, **39**, 97), since replacement with

sulphur (I. Hillmar, T. Muramatsu and N. Zöllner, *Hoppe Seyler's Z. Physiol. Chem.*, 1984, 365, 33; J. M. Mencia-Huerta *at al.*, J. *Immunol.*, 1982, 29, 804), or methylene (C. Broquet *et al.*, *Eur. J. Med. Chem.*, 1984, **19**, 229; N. Nakamura *et al.*, *Chem. Pharm. Bull.*, 1984, **32**, 2452; K. Y. Gordeev, G. A. Serebrennikova and R. P. Evstigneeva, *Bioorg. Khim.*, 1987, **13**, 234) results in a significant loss of activity. One synthesis of an 1-octadecylthio analogue of PAF (B. Garrigues *et al.*, *Phosphorus Sulfur*, 1984, **21**, 171), was envisioned as a general means of synthesising phospholipids, since a TBDMS deprotection step, effected with boron trifluoride etherate, occurs without transacylation.
Hirth *et al.*, have prepared an olefinic analogue of PAF, its enantiomer, and positional isomers (G. Hirth *et al.*, *Helv. Chim. Acta*, 1983, **66**, 1210).

(2) Variations at C-2

The stimulus for many changes at this position arises from the discovery of a specific enzyme in the plasma which causes rapid degradation of PAF (M. C. Blank *et al.*, *J. Biol. Chem.*, 1981, **256**, 175).
Epimerisation of the natural R configuration of PAF leads to a significant attenuation of agonistic activtiy (D. J. Hanahan and R. Kumar, *Prog., Lipid Res.*, 1987, **26**, 1), however, quite diverse changes at C-2 are possible without a notable reduction in potency, as long as the R configuration is retained. Compounds with relatively small substituents show PAF-like actions, as do 2-desoxy-2-alkyl analogues (R. L. Wykle *et al.*, *FEBS Lett.*, 1982, **141**, 29). Long-chain ester analogues, and both saturated and unsaturated ether phospholipids, were prepared by nucleophilic epoxide opening of glycidyl arenesulphonates (P. N. Guivisdalsky and R. Bittman, *J. Org. Chem.*, 1989, **54**, 4643). The 2-*N*-methoxycarbamyl substituted compound is a biologically potent, non-metabolisable PAF agonist (T. G. Tessner, J. T. O'Flaherty and R. L. Wykle, *J. Biol. Chem.*, 1989, **264**, 4794). The propionyl ester (P. Hadvary and H. R. Baumgartner, *Thrombosis Res.*, 1983, **30**, 143), thioester (S. K. Bhatia and J.

Hajdu *et al.*, *Tetrahedron Lett.*, 1987, 1721; K. Y. Gordeev, G. A. Serebrennikova and R. P. Evstigneeva, *Bioorg. Khim.*, 1987, **13**, 234) and both enantiomers of a novel 2-*O*-(3-isoxazolyl) analogue (N. Nakamura *et al.*, *Tetrahedron Lett.*, 1990, 699), show potent PAF agonism.
Arachidonyl-PAF has been found to be an important precursor in the biosynthesis of both PAF and arachidonic acid, and thus of prostaglandins leukotrienes and other inflammatory mediators (N. Yasuhito *et al.*, *Prog. Lipid Res.*, 1989, **28**, 205).
Of the isosteric replacements prepared, the trihaloester (D. M. Humphrey *et al.*, *Lab. Invest.*, 1982, **46**, 422), ketone (C. Broquet *et al.*, *Eur. J. Med. Chem.*, 1984, **19**, 229), and amide (N. S. Chandrakumar and J. Hajdu, *Tetrahedron Lett.*, 1982, 1043; N. S. Chandrakumar and J. Hajdu, *J. Org. Chem.*, 1983, **48**, 1197) substituents are generally less active, as are the ether (J. T. O'Flaherty *et al.*, *J. Immunol.*, 1981, **103**, 70), and halo-substituted analogues (G. Ostermann, H. Brachwitz and U. Pill, *Biomed. Biochem. Acta.*, 1984, **43**, 349).
2-Azido-2-deoxy-1-*O*-hexadecyl-*sn*-glycero-3-phosphorylcholine was prepared from D-mannitol (M. M. Ponpipom and R. L. Bugianesi, *Chem. Phys. Lipids*, 1984, **35**, 29). In this synthesis, nucleophilic displacement of a 2-methanesulphonate group by azide ion occurred with inversion of configuration, and without racemisation. This azide was reduced to the amine derivative- a versatile intermediate for the preparation of amide analogues of PAF. The synthesis of a 2-fluoro derivative was also described in this article.

(3) Variations at C-3

The distance between phosphoryl and the polar head groups has been examined. A range of analogues have been prepared in which the number of methylene groups were varied, and in which phenyl spacer or chain branching were incorporated (A. Wissner *et al.*, *J. Med. Chem.*, 1986, **29**, 328). A correlation was made between increasing chain length and a gradual decrease in hypotensive and platelet aggregating responses. Kertscher and co-workers described the structure-activity relationship between

synthetic analogues of PAF, including compounds in which the trimethylammonium is replaced by alkylamino, and heterocyclic systems (G. Ostermann *et al., Biochim. Biophys. Acta,* 1988, **47**, 236), and they emphasise the requirement for a positive charge for expression of platelet stimulating activity.
Systematic removal of methyl substituents from the quaternary nitrogen (K. Satouchi *et al., J. Biol. Chem.,* 1981, **256**, 4425) gives analogues in which activity decreases as the methyl groups are replaced. These compounds were prepared primarily by phospholipase D modification of the corresponding phosphatidylcholine. A number of variants contain cyclic systems in place of the quaternary nitrogen, and in some cases are more potent than PAF itself (Y. Kasuga, Y. Masuda and K. Shigenobu, *Can. J. Physiol. Pharmacol.,* 1986, **131**, 179; F. Berti *et al., Pharmacol. Rev. Commun.,* 1986, **18**, 557). For instance, an analogue bearing a pyrrolidino head group shows ten times the platelet aggregation inducing activity of PAF itself (E. Coeffier *et al., Eur. J. Pharmacol.,* 1986, **18**, 557).
In order to probe the structural requirements of the PAF phosphate group and its possible involvement with a receptor, Tsai and co-workers have prepared phosphorothioate analogues of PAF, which are chiral at the phosphorus atom (T. Rosario-Jansen *et al., Biochemistry,* 1988, **27**, 4619).
Replacement of the phosphate with a phosphonate group has been achieved by replacement of either the glycerol or choline oxygen with a methylene (N. Nakamura *et al., Chem. Pharm. Bull.,* 1984, **32**, 2452), or by deletion of one of the oxygen atoms (H. Disselnkoetter *et al., Arch. Pharm. (Weinheim, Ger.),* 1985, **318**, 695; A. Wissner *et al., J. Med. Chem.,* 1986, **29**, 328). Moschidis and co-workers have also reported the chemical synthesis of phosphono analogues of PAF (M. C. Moschidis, C. A. Demopoulos and L. Kritikou, *Chem. Phys. Lipids,* 1983, **33**, 87), and 2-amido-PAF (M. C. Moschidis and C. A. Demopoulos, *Chem. Phys. Lipids,* 1985, **37**, 45).

(4) Modifications to the Glycerol Backbone

Positional and stereoisomers of PAF have been prepared (G. Hirth and R. Barner, *Helv. Chim. Acta.*, 1982, **65**, 1059; G. Hirth *et al.*, *Helv. Chim. Acta.*, 1983, **66**, 1210). The natural (R)-PAF is three thousand times more active than its antipode (M. Tence *et al.*, *Biochim. Biophys. Acta*, 1983, **755**, 526), a trend which is also reflected by the enantiomers of a number of PAF analogues. The effect of conformational restriction has been studied by substitution of methyl groups at the three possible positions on the glycerol backbone (M. Ohno *et al.*, *J. Med. Chem.*, 1986, **29**, 1814). These substitutions were made possible by employing the tartaric acid synthetic strategy (section 3(a)), so that C-1 and C-3 methyl groups were introduced diastereo- and enantioselectively. All were weak agonists, of which the 1-(S)-methyl-PAF is the most active. C-2 Substituted methyl analogues have been reported (E. N. Lewis *et al.*, *Biochim. Biophys. Acta*, 1986, **861**, 44; R. Bittman *et al.*, *J. Lipid Res.*, 1987, **28**, 733), and four *O*-carba-PAF analogues were prepared (N. Nakamura *et al.*, *Chem. Pharm. Bull.*, 1984, **32**, 2452), all of which show reduced potency.
Homologation of the glycerol backbone by insertion of an additional methylene group (A. Wissner *et al.*, *J. Med. Chem.*, 1985, **28**, 1181), and inclusion of an additional acetoxymethylene unit (R. C. Anderson and R. C. Nabringer, *Tetrahedron Lett.*, 1983, 2741) gave compounds with significantly reduced agonistic activity.
Conformationally restricted butyrolactone and tetrahydrofuran derivatives have been prepared which retain the natural stereochemistry, but these show no agonistic activity (M. C. Phillips and R. Bonjouklain, *Carbohydr. Res.*, 1986, **146**, 89).

(ii) PAF Receptor Antagonists

The most successful PAF analogue antagonists incorporate several simultaneous changes in basic structure, so these will be classified into non-constrained and constrained compounds for convenience.

(1) Non-Constrained Analogues

These are all open-chain analogues, and include the first specifically-designed, structurally related PAF antagonist. This was Takeda's CV-3988 (**53**, R^1=$CONHC_{18}H_{37}$, R^2=OMe, X=O, Y=thiazolium) (Z.- I. Terashita *et al.*, *Life Sci.*, 1983, **32**, 1975)

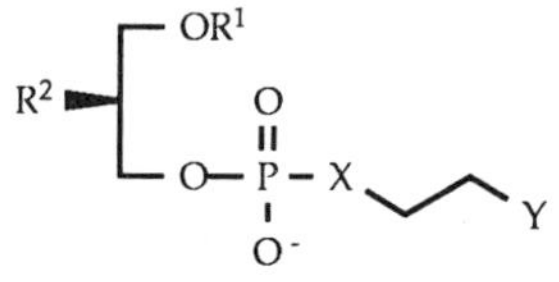

(53)

The fundamental phospholipid structure is modified with an octadecyl carbamate at C-1, a methyl ether at C-2, and a thiazolium ethyl phosphate at C-3, and this compound provided the starting point for a multitude of PAF-related antagonists. Takeda subsequently reported structurally analagous compounds containing an ester group at C-2, which showed antitumour activity, and a complete loss of PAF antagonistic action (E. P. 209 239, 1989). However, compounds which possess C-1 carbamate, C-2 ether (U.S. P. 4 582 824, 1986), C-1 ether, C-2 cyclic imide (U.S. P. 4 650 791, 1987), and C-1 carbamate, C-2 ester (U.S. P. 4 675 430, 1987) functionalities have been reported, and are also effective as PAF antagonists.
The amidophosphonate (**53**, R^1=$C_{18}H_{37}$, R^2=CH_2NHAc, X=CH_2, Y=NMe_3^+), shows antagonistic activity, although it has been suggested that the mechanism of action of such compounds is by inhibition of phospholipase A_2 or phospholipase C (M. Steiner *et al.*, *Biochim. Biophys. Res. Commun.*, 1985, **133**, 851).
Hanahan's group have tested compounds in which the phosphate and polar head groups are separated by chains of various lengths (A. Tokumura, M. Homma and D. J.

Hanahan, *J. Biol. Chem.*, 1985, **260**, 12710). The hexanolamino analogues (**53**, R^1=$C_{16}H_{33}$ or $C_{18}H_{37}$, R^2=OAc, X=$(CH_2)_4$, Y=NMe_3^+) also show inhibition of phospholipid turnover and secretion (M. Rouis *et al.*, *Biochim. Biophys. Res. Commun.*, 1988, **156**, 1293; A. Tokumura, H. Homma and D. J. Hanahan, *J. Biol. Chem.*, 1985, **260**, 12710).
A number of antagonists retain the glycerol backbone, but contain either alkyl, ketone or carboxylate isosteres for the phosphate ester (G. Grue-Sorensen *et al.*, *J. Med. Chem.*, 1988, **31**, 1174; B. Wichrowski *et al.*, *J. Med. Chem.*, 1988, **31**, 410).

(2) Constrained Analogues

Both enantiomers of a novel spirocyclic tetrahydrofuranyl-PAF antagonist SRI-63-072 (**54**), from Sandoz, exhibit similar receptor binding characteristics (W. Hsueh *et al.*, *Eur. J. Pharmacol.*, 1986, **123**, 79). Another, less active, Sandoz compound is derived from a constrained PAF backbone, and thiamine phosphate, which is itself a modest PAF antagonist *in vivo* (M. L. Lee *et al.*, *Prostaglandins*, 1985, **30**, 690).
The Sankyo tetrahydropyranyl derivative (**55**) shows greater activity (U.S. P. 4 619 917, 1986). A number of lactone and cyclic ether analogues of PAF, including (**55**), were synthesised, and their biological activity evaluated (H. Miyazaki *et al.*, *Chem. Pharm. Bull.*, 1989, **37**, 2379). In addition, a number of variations to these compounds were made (H. Miyazaki *et al.*, *Chem. Pharm. Bull.*, 1989, **37**, 2391), and it was shown that positional and stereochemical changes to this structure decrease potency, but that replacement of the carbamate oxygen with sulphur resulted in no significant effect on activity

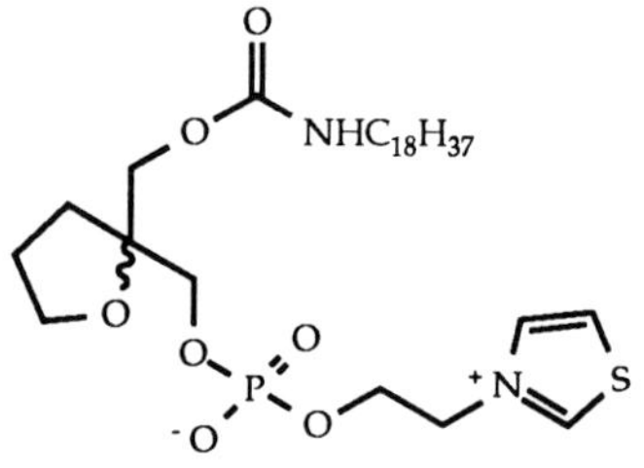

(54)

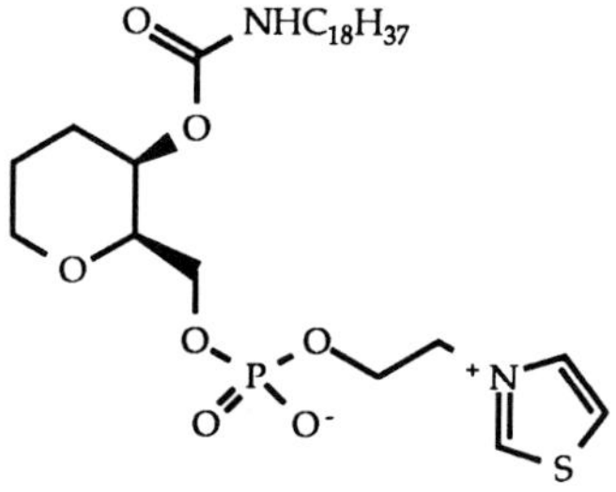

(55)

A series of antagonists have been reported, such as (**56**), which are also based on the PAF structure (E. P. Appl. 178 261, 1986). These are built around the constraint of a 2,5-disubstituted tetrahydrofuran ring (D. A. Handley, J. C. Tomesch and R. N. Saunders, *Eur. J. Pharmacol.*, 1987, **141**, 409). Modifications have been made to these compounds to enable the potential sites of metabolism to be identified (D. A. Handley *et al.*, *Adv. Prostagland. Thrombox. Leukotr. Res.*, 1989, **19**, 367).

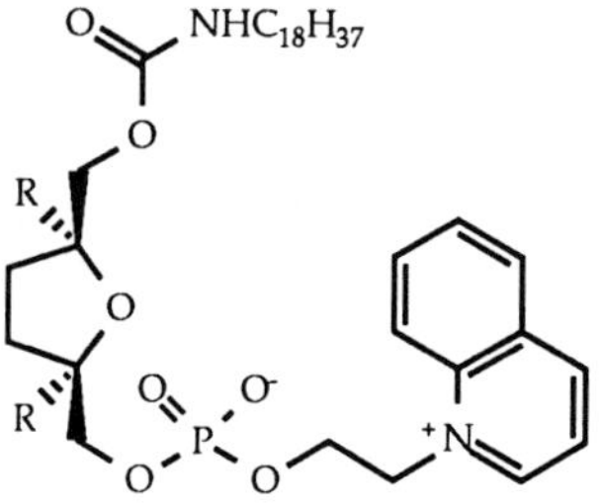

(56)

R = Me, H

A series of cyclic phosphates also show inhibition of PAF-induced platelet aggregation (P. Hadvary and T. Weller, *Helv. Chim. Acta*, 1986, **69**, 1862).

(iii) Miscellaneous PAF Analogues

2-Thio-PAF is an isosteric thioacetyl analogue of PAF which shows potent antihypertensive properties (S. K. Bhatia and J. Hajdu, *Tetrahedron Lett.*, 1987, 1729). The route to this compound, has been refined and augmented to allow differential functionalisation of each position, so that the scheme offers a flexible, general method of synthesis of thio-analogues of phospholipids (S. K. Bhatia and J. Hajdu, *Synthesis*, 1989, 16).

The same workers employed also employed L-glyceric acid as a novel chiral source, for the preparation of a range of phospholipid analogues (S. K. Bhatia and J. Hajdu, *Tetrahedron Lett.*, 1987, 271; *J. Org. Chem.*, 1988, **53**, 5034). The synthesis requires only a single protecting group, since introduction of the *sn*-2 alkyl substituent, by silver fluoroborate-catalysed alkylation, leaves the neighbouring carbomethoxy group unaffected. The resulting intermediate can then be used for the introduction of either an ether or thioether linkage at the *sn*-1 site. Cyclic phosphate methodology is used to introduce the polar head group.

The 2-methyl ether analogue of PAF shows potent antitumour and antimetastatic activity (M. H. Runge, *J. Natl. Cancer Inst.*, 1980, **64**, 1301), as do analogues in which the number of methylene units in the polar head group are increased (U.S. P. 4 710 579, 1987).

The 1-thioether-2-acetamido analogue of PAF has been prepared from cysteine (B. Garrigues, G. Bettrtrand and J.- P. Maffrand, *Synthesis*, 1984, 870), and Hajdu has published the synthesis of some amide derivatives (N. S. Chandrakumar and J. Hajdu, *J. Org. Chem.*, 1983, **48**, 1197), by utilising the asymmetric α-carbon of L-serine to obtain the correct chiral configuration. Modification of this route allows for the introduction of substituents at the *sn*-2 position after the phospholipid skeleton has been assembled. The resulting amides show antihypertensive properties, and are cytotoxic agents. Salient features of this method are the protection of the nucleophilic amino alcohol as an oxazoline, and the use of cyclic

chlorophosphate chemistry to introduce the phosphocholine group after amide formation.
In order to understand the underlying mechanisms of the PAF response and to determine its concentration in biological fluids, specific antibodies are required. The molecular weight of PAF is too low to induce antibodies, but if conjugated to proteins, the resulting haptens can cause antibody production. The syntheses of two potential haptens have been reported recently. These incorporate potential protein binding groups on the *sn*-1 alkyl chain, since the other positions of the PAF molecule were identified as being essential for optimum activity. Prashad *et al.*, chose a carboxylic acid as the active group. The synthesis was designed to incorporate a double bond in the chain to allow for tritium labelling, and thus facililitate characterisation of the conjugate (M. Prashad *et al., Chem. Phys. Lipids*, 1990, **53**, 121). An alternative method (C.- J. Wang and H. H. Tai, *Chem. Phys. Lipids*, 1990, **55**, 265) described the preparation of an aldehydic derivative of PAF by ozonolysis of a terminal double bond. This compound was not isolated, but conjugated directly to the thyroglobulin *via* reductive amination, and used to immunise rabbits for the production of specific antibodies.

4. Phospholipase A_2 Inhibitors

Membrane phospholipids contain three types of ester bonds that are susceptible to hydrolysis by enzymes, known collectively as phospholipases (scheme 17). This subject has been reviewed extensively (H. Van Den Bosch in *Phospholipids*, J. N. Hawthorne and G. B. Ansell (Eds.), Elsevier Biomedical Press, 1982, 313; E. G. Lapetina, *Ann. Rep. Med. Chem.*, 1984, **19**, 213; C. Little, *Biochem. Soc. Trans.*, 1989, **17**, 271). External phosphodiester bonds can be cleaved by phospholipase D (PLD) which is found mainly in higher plants. Phospholipase C (PLC) is responsible for the hydrolysis of internal phosphodiester bonds, which in mammalian tissue is invariably restricted to inositol-phospholipids.

Scheme 17

PLB
$R^2C(=O)-O$, $O-C(=O)R^1$, H, $O-P(=O)(O^-)-O-X$
PLA_1 PLA_2
PLC PLD

Recent interest has focused on inhibitors of PLC, which has been associated with inflammatory responses and cell division. The subject is beyond the scope of this review, but the following articles are relevant (H. M. Verheij, *Physiol. Biochem. Pharmacol.*, 1981, **91**, 91; S. D. Shukla, *Life Sci.*, 1982, **30**, 1323; E. A. Dennis, *The Enzymes*, 1983, **XVI**, 307; M. J. Berridje, *J. Biochem.*, 1984, **220**, 345; L. Hokin, *Ann. Rev. Biochem.*, 1985, **54**, 205; D. J. R. Massey and P. Wyss, *Helv. Chim. Acta*, 1990, **73**, 1037; M. Whitman and L. Cantley, *Biochem. Biophys. Acta*, 1988, **948**, 327).

Hydrolysis of the fatty acyl ester bond is catalysed by A-type phospholipases to give lyso-phospholipids, which may be further hydrolysed at the remaining side chain by B-type phospholipases. More specifically, acyl ester bonds at the *sn*-1 position are cleaved by phospholipase A_1 (PLA_1), and at the *sn*-2 position by phospholipase A_2 (PLA_2). It is the latter enzyme which is responsible for the deacylation-reacylation processes vital for the general repair and maintenance of cell membranes. PLA_2 also participates in a number of regulatory processes, including platelet aggregation, cardiac contraction and excitation, prostaglandin biosynthesis, and aldosterone-dependent sodium transport. During inflammation PLA_2 activity is heightened, and its action on membrane phosphoglycerides causes the formation of lyso-phospholipids, with the consequent liberation of fatty acids, predominantly arachidonic acid. These products are then available for the conversion to pro-inflammatory mediators, the eicosanoids (prostanoids, leukotrienes, lipoxins, hydroxy fatty acids and thromboxanes) in the arachidonic acid cascade (G. A. Higgs, E. A. Higgs and S. Moncado in *Comprehensive Medicinal Chemistry*, 1990, **2**, J. C. Emmett (Ed.), Pergamon Press, Oxford, 147). In some cases the lysophospholipid fragment is also a pro-inflammatory mediator, since PAF may be synthesised by acylation of lyso-PAF. The mode of action of PLA_2 has been reviewed recently (M. Waite in *Phospholipases, Vol 5, Handbook of Lipid Res.*, 1987, Plenum Press, New York, NY, 1). Elevated levels of PLA_2 activity have been associated with a number of disease states (J. Chang *et al., Biochem. Pharm.*, 1987, **36**, 2429), and so the preparation of specific inhibitors of PLA_2 is therefore seen as a novel approach to the design of anti-inflammatory agents (E. G. Lapetina, *Trends Pharm. Sci.*, 1982. 115; E. A. Dennis. *Drug Dev. Res.*, 1987, **10**, 205).

(a) Synthesis of PLA_2 Inhibitors

Many agents have been found which exhibit PLA_2 inhibitory properties. These include a variety of naturally occurring substances such as proteins, plant extracts, and some chemicals, as well as synthetic substrate mimics. The gamut of PLA_2 inhibitors has been comprehensively documented in two excellent reviews (H. Van Der Bosch in *Comprehensive Medicinal Chemistry*, 1990, **2**, J. C. Emmett (Ed.), Pergamon Press, Oxford, 515; D. Mobilo and L. A. Marshall, *Ann. Rev. Med. Chem.*, 1989, **24**, 157). Only the phospholipid-related substrate analogues are reviewed here.

Hajdu and co-workers developed an expeditious synthesis of an amido-PLA_2 inhibitor from D-serine (N. S. Chadrakumar and J. Hajdu, *J. Org. Chem.*, 1982, **47**, 2144). The design of this compound was based on early findings that replacement of the *sn*-2 ester group with an amide linkage in short-chain lecithin analogues, completely abolishes catalytic hydrolysis by the enzyme. The key element in this route (Scheme 18) is the use of an oxazoline residue which is multi-functional: it serves to protect the amino and hydroxyl groups, prevents loss of chirality during the reduction sequence, and acts as the precursor for the ester group in the final product. Formation of the oxazoline (**57**) was readily accomplished by condensation of methyl-D-serine with an imidate. Reduction of the ester, followed by phosphorylation using cyclic chlorophosphate chemistry, gave the intermediate (**58**). The oxazoline group was hydrolysed by the addition of one equivalent of dilute hydrochloric acid, and then *N*-acylation with either an acid chloride or anhydride in the presence of DMAP, gave the final product (**59**). It was only at this stage that chromatography was required.

Scheme 18

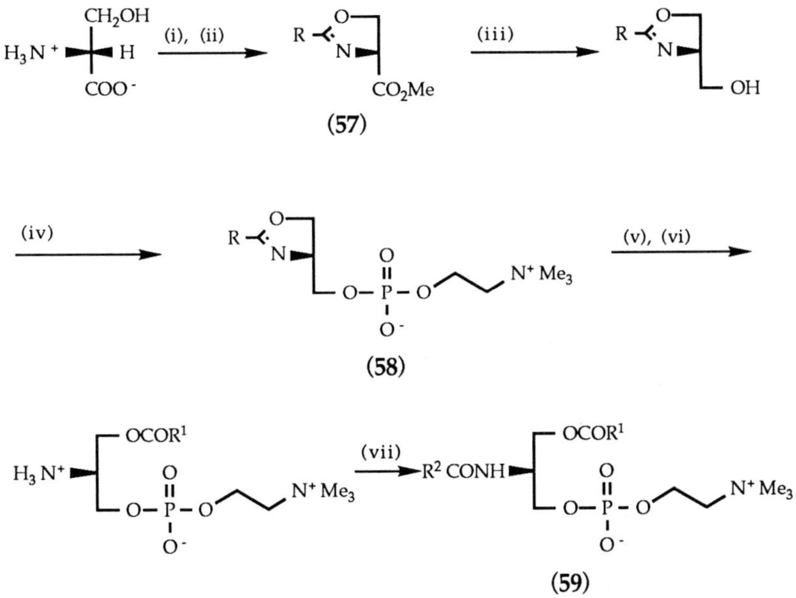

$R^1, R^2 = (CH_2)_{16}CH_3$

Reagents: (i) HCl, MeOH; (ii) Ethyl octadecanimidate hydrochloride, Et_3N, CH_2Cl_2; (iii) $LiAlH_4$, Et_2O, 0°C; (iv) 2-chloro-2-oxo-1,3,2-dioxaphospholane, C_6H_6; (v) Me_3N, MeCN; (vi) HCl; (vii) $(RCO)_2O$, DMAP, $CHCl_3$.

This synthetic strategy was preferred over an earlier route to similar compounds (N. S. Chandrakumar and J. Hajdu, *Tetrahedron Lett.*, 1981, 2949), in which *N*-acylation was carried out early in the scheme, and tetrahydropyranyl-protection of the alcohol was employed prior to phosphorylation. Compound (**59**) has also been prepared through a sequence in which chirality was introduced by lipase-catalysed enantioselective transformations (Y.- F. Wang and C.- H. Wong, *J. Chin. Chem. Soc.*, 1989, **36**, 463).

Compound (**59**) shows potent PLA_2 inhibition, but is rapidly hydrolysed by PLA_1. However, replacement of the ester with an ether linkage at C-1 (**60**, $R^1 = C_{18}H_{37}$, $R^2 = H, CH_3, CF_3, C_{15}H_{31}$) (N. S. Chandrakumar and J. Hajdu, *J. Org. Chem.*, 1983, **48**, 1197), circumvents the problem of PLA_1 hydrolysis.

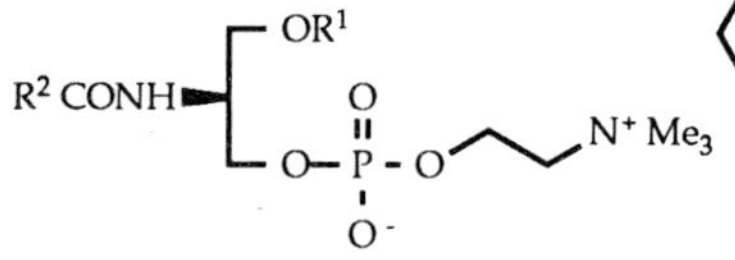

(60) 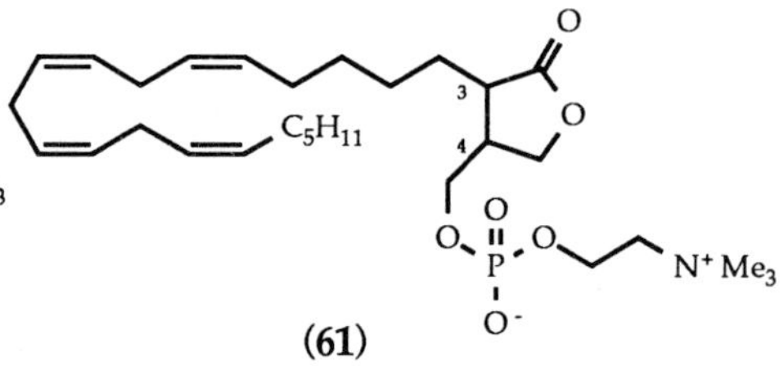 (61)

The analogous compound (**60**, $R^1 = C_{17}H_{35}$, $R^2 = C_{17}H_{35}$) is also an effective PLA_2 inhibitor (F. F. Davidson, J. Hajdu and E. A. Dennis, *Biochem. Biophys. Res. Commun.*, 1986, **137**, 587). The *sn*-2 amide group in these compounds is capable of a strong interaction with the PLA_2 active-site electrophile through hydrogen bonding, and the replacement of the acyloxy- with an alkyloxy- group at the *sn*-1 position enhances hydrophobic interaction with the enzyme binding site. This gives a reversible inhibitior of PLA_2.

The active site domain of PLA_2 has been identified by experiment (H. M. Verheij *et al.*, *Biochemistry*, 1980, **19**, 743), and the active site electrophile suggested to be Ca^{2+} (D. Dinur, E. R. Kantrowitz and J. Hajdu, *Biochem. Biophys. Res. Commun.*, 1981, **100**, 785). More recently the X-ray crystallographic structure of an inhibitor covalently bound to

PLA_2 has appeared in the literature (B. W. Dijkstra *et al.*, *J. Mol. Biol.*, 1988, **200**, 181), and this formed the basis of a molecular recognition study by Campbell *et al.*, (*J. Chem. Soc., Chem. Commun.*, 1988, 1560) in which molecular modelling techniques were used to generate an image of the PLA_2 active site-phospholipid substrate bound structure. The model was tested with a known PLA_2 inhibitor, and then utilised to investigate various conformations-both *cis/trans* isomers and keto/enol tautomers-of the novel constrained phospholipid analogue (**61**). Synthetic methodology was then developed to enable this prediction to be tested (M. M. Campbell *et al.*, *Tetrahedron*, 1989, **45**, 4551).

Molecular modelling of PLA_2 inhibitors has also been carried out by other workers (W. C. Ripka, W. J. Sipio and J. M. Blaney, *Heterocyclic Chem.*, 1987, **IX**, 95).

To mimic the tetrahedral intermediate formed during phospholipase-catalysed lipolysis, Gelb designed a transition state analogue inhibitor (**66**) (M. H. Gelb, *J. Am. Chem. Soc.*, 1986, **108**, 3146). This is an interesting synthesis (Scheme 19), in which the dioxolane (**62**) was lithiated, reacted with a difluoroester, then treated with acid to give a lactol (**63**). This could be selectively acylated and then reduced with sodium borohydride to the diol (**64**). Selective phosphorylation of the more nucleophilic, non-fluorinated hydroxy group, and oxidation of the other hydroxyl was followed by deprotection to give the phosphatidic acid (**65**). This was converted to a phosphodiester (**66**) by esterification and deprotection. The resulting compound proved to be a tight-binding inhibitor of PLA_2.

Scheme 19

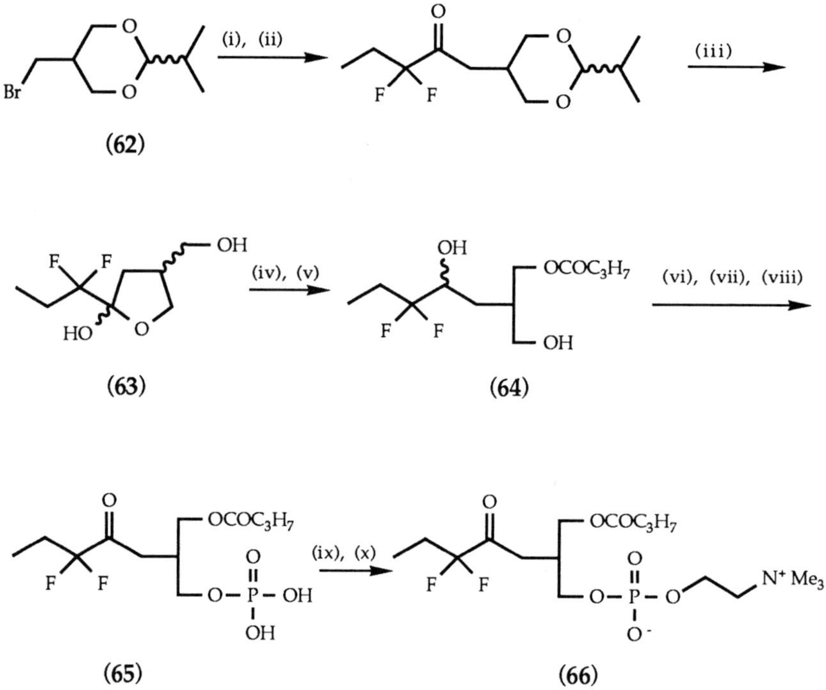

Reagents: (i) tBuLi, Et_2O, -78°C; (ii) $CH_3CH_2CF_2CO_2CH_3$, -78°C; (iii) HCl; (iv) C_3H_7COCl, pyridine; (vii) CrO_3.(pyridine)$_2$, CH_2Cl_2; (viii) Pt_2O, H_2, MeOH; (ix) $TrNHCH_2CH_2OH$, TPSCl, $CHCl_3$; (x) TFA, CH_2Cl_2, 0°C.

The initial success of this concept lead to new synthetic routes for a series of phospholipid analogues containing fluoro-ketone, ketone, and alcohol bioisosteres, of the C-2 ester group of the glycerol backbone (W. Yuan, R. J. Berman and M. H. Gelb, *J. Am. Chem. Soc.*, 1987, **109**, 8071). Transition state analogues were also prepared which incorporated two alkyl chains of varying lengths, and compounds which are analogues of C-2 diols. The best of these inhibitors proved to be the single chain fluoro ketone (**67**).

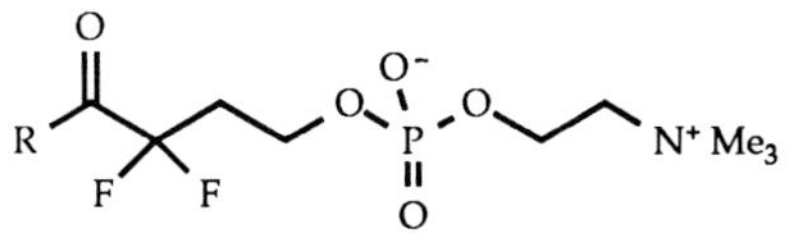

(67)

$R = C_{16}H_{33}$

However, potent inhibition was observed only with those compounds which were significantly hydrated in the micellar phase. This suggested that compounds containing an enforced tetrahedral group would also be potent inhibitors of PLA_2. The phospholipid analogues (**68**) and (**69**), which incorporate a phosphonate group in place of the ester at C-2, were subsequently prepared, and also shown to be tight-binding inhibitors of PLA_2 (W. Yuan and M. H. Gelb, *J. Am. Chem. Soc.*, 1988, **110**, 2665).

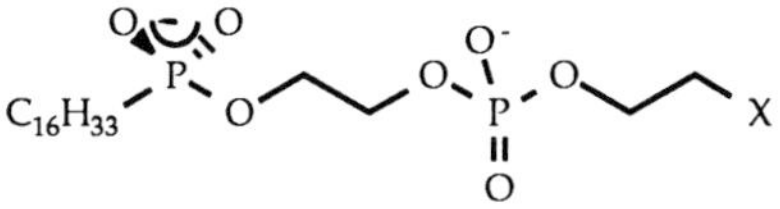

(68)

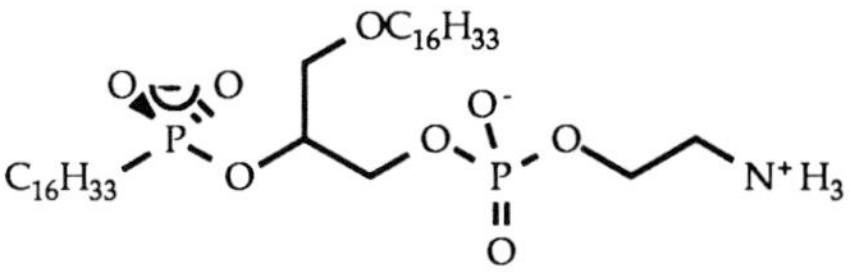

(69)

$X = NH_3^+, NMe_3^+$

Gelb and co-workers have further extended this concept to the synthesis of thiophosphonate analogues (W. Yuan, K. Fearon and M. H. Gelb, *J. Org. Chem.*, 1989, **54**, 906). Here one of the non-bridging oxygens is replaced by a sulphur atom. Both stereoisomers were prepared, in a sequence (Scheme 20) in which the alkyl glycidol (**70**) was selectively monomethoxytritylated, and then reacted with *N,N*-diisopropylmethylphosphonamidic chloride to give the phosphoramidite (**71**). Treatment of (**71**) with hydrogen sulphide in the presence of tetrazole gave (**72**), which, after reaction with a terminal olefin under radical conditions, yielded the diastereomeric thiophosphonates (**73**). After detritylation the alcohols (**74**) and (**75**) could be separated by chromatography.

Scheme 20

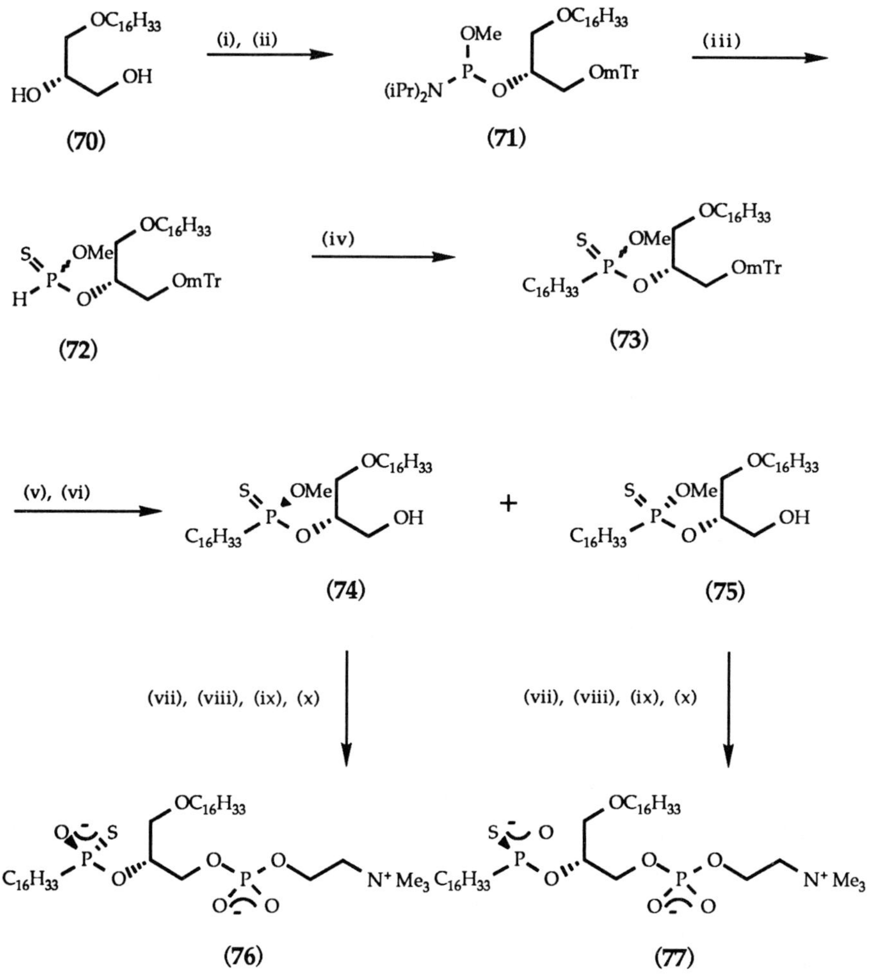

(70)

(i), (ii)

(71)

(iii)

(72)

(iv)

(73)

(v), (vi)

(74) + (75)

(vii), (viii), (ix), (x)

(76)

(77)

mTr = monomethoxytrityl

Reagents: (i) m-TrCl; (ii) $(iPr)_2NP(OMe)Cl$, Et_3N, CH_2Cl_2; (iii) H_2S, tetrazole, CH_2Cl_2; (iv) $C_{14}H_{29}CH{=}CH_2$, AIBN; (v) BF_3, MeOH; (vi) chromatography; (vii) $POCl_3$, Et3N, CH_2Cl_2; (viii) $HOCH_2CH_2NH_2$, Et_3N, CH_2Cl_2; (ix) HCl, iPrOH, $CHCl_3$; (x) Me_3N, $PhCH_3$.

Compounds (**74**) and (**75**) were individually subjected to a standard phosphorylation procedure, and this was followed by demethylation with trimethylamine in toluene, which gave the phospholipid analogues (**76**) and (**77**). These workers also described the preparation of the phosphonates (**78**).

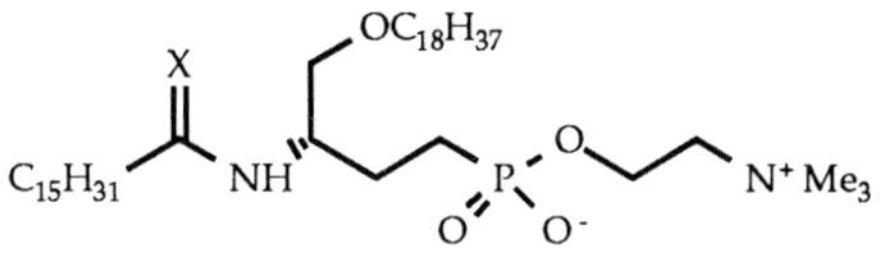

$C_{15}H_{31}$ NH $OC_{18}H_{37}$ X P O O O⁻ N^+Me_3

(**78**)

X = O, S

A series of *sn*-1-thioalkylphospholipids (J. Hajdu and J. M. Sturtevant, *Chem. Phys. Lipids*, 1990, **55**, 323) were reported to show a ten-fold increase in potency over their *O*-alkyl counterparts (L. Yu *et al., J. Biol. Chem.*, 1990, **265**, 2657). High resolution X-ray crystal structures of one of these phosphonate inhibitors complexed to two types of PLA_2 (S. P. White *et al., Science*, 1990, **250**, 1560; D. L. Scott *et al., Science*, 1990, **250**, 1563) have enabled this group to propose both a binding mechanism and the chemical basis of catalysis of PLA_2 action (D. L. Scott *et al., Science*, 1990, **250**, 1541), which was simultaneously, and independently corroborated by the work of Thunnissen *et al.*, (M. G. M. Thunnissen *et al., Nature*, 1990, **347**, 689).
It is well established that most PLA_2 enzymes hydrolyse aggregated phospholipid substrates much more readily than monomeric substrates, and that the activity of PLA_2 is greatly enhanced once the substrate is present above its critical micellar concentration. Numerous models have been proposed to account for this phenomenon (see for example H. M. Verheij, A. J. Slotboom and G. H. Haas, *Rev. Physiol. Biochem. Pharmacol.*, 1981, **91**, 91; J. J. Volwerk and G. H. Haas, *Lipid Protein Interactions*, 1982, **1**, 69). Several workers have initiated studies to test the 'substrate conformation model,' which suggests that this activation is due to a change in substrate conformation in the micellar phase. Accordingly, conformationally restricted phospholipid analogues such as (**79**)

have been synthesised, in which the glycerol backbone has been replaced by a cyclopentane triol structure (P. N. Barlow *et al.*, *Chem. Phys. Lipids*, 1988, **46**, 157; *J. Biol. Chem.*, 1988, **263**, 12954; G. Lin *et al.*, *J. Biol. Chem.*, 1988, **263**, 13208).

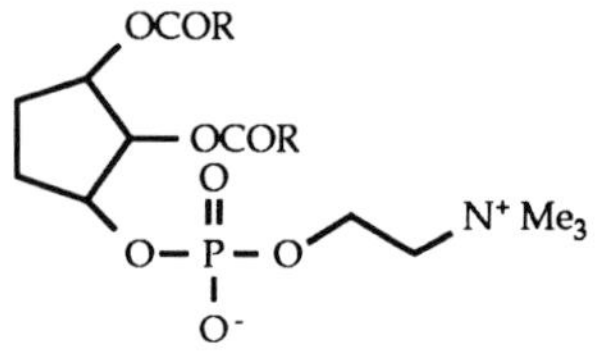

OCOR

OCOR

O

O–P–O

O⁻

N^+Me_3

(79)

$R = C_3H_7, C_5H_{11}, C_7H_{15}$

The conclusions from these studies have provided strong evidence to support the 'substrate conformation model.' In addition, the (+)-enantiomer of all *cis*-(**79**) is a strong PLA_2 inhibitor.

Dennis and co-workers (A. Plückthun *et al.*, *Biochemistry*, 1985, **24**, 4201) have prepared a series of short-chain phospholipids which are resistant to hydrolysis, and they have used these to demonstrate the requirement for an interface in PLA_2 activation.

A PLA_2 inhibitor based on monolysocardiolipin has been reported recently (U.S. P. 4 959 357, 1990), and another patent describes a series of alkoxy-, hydroxy- and benzyloxy- substituted phospholipid analogues, which also show PLA_2 inhibitory properties (E. P. 300 397, 1989). Typical of the compounds is the phospholipid analogue (**81**), which was prepared by a short synthesis (scheme 21) in which Grignard addition of vinyl magnesium bromide to a suitable aldehyde, followed by protection of the resulting hydroxyl group, gave the benzyloxyalkene (**80**). This was followed by hydroboration with an oxidative work up, phosphoesterification, and aminolytic ring-cleavage using standard techniques.

Scheme 21

Reagents: (i) $C_{17}H_{35}CHO$, THF; (ii) BnBr, NaH, DMF; (iii) hydroboration (9-BBN); [O]; (iv) 2-chloro-2-oxo-1,3,2-dioxaphospholane; (v) Me_3N.

Magolda and co-workers have reported an efficient general synthetic method to alkyl-, thioalkyl-, and alkylglycoether-phospholipid-like compounds as PLA_2 inhibitors (R. L. Magolda and P. R. Johnson, *Tetrahedron Lett.*, 1985, 1167), and some analogues of PAF which show PLA_2 inhibition, antihypertensive and antiinflammatory effects (U.S. P. 4 702 864, 1987).

A major problem which pervades this area of research, is that direct comparisons between PLA_2 inhibitors are difficult, since compounds are tested against different forms of PLA_2 which have been isolated from different sources, and the analytical protocol employed may vary enormously between research groups.

The author gratefully acknowledges the help of Mr Neil Gow, and Drs. Neil Mathews, Andy Miller and Mark Whittaker in the preparation of this manuscript.

Second Supplements to the 2nd Edition of Rodd's Chemistry of Carbon Compounds, Vol. IE/F/G, edited by M. Sainsbury

Chapter 22

POLYHYDRIC ALCOHOLS AND THEIR OXIDATION PRODUCTS

P. D. JENKINS

1. Introduction

The following chapter is directly based on the thorough treatment provided in the First and Second Editions of the Chemistry of the Carbon Compounds and in the First Supplement to the main work. The contents of each section are approximately parallel to those of the earlier editions. However, the emphasis of this chapter is to select specific examples which outline the growth areas of the subject matter. A larger proportion of this chapter is devoted to asymmetric synthesis, reflecting the rapid development of enantioselective syntheses in the last twenty years.

2. Preparation and Synthesis of Alditols

(a) Reduction

Alditols are readily prepared by the reduction of the appropriate aldoses or ketoses. Many methods have been developed to achieve these transformations. Most methods have already been covered in the First Edition, however the use of high pressure hydrogenation over metal catalysts has received continued interest over recent years. Continuous and semi-continuous methods have also been developed for industrial applications (A. G. Ivchenko, Y. S. Khakimov and M. F. Abidova, Uzb. Khim. Zh., 1990, 66). The list of metal and metal alloy catalysts is impressive and includes nickel, nickel-molybdenum, copper, aluminium-nickel, aluminium-nickel-molybdenum-magnesium, ruthenium-carbon, nickel-aluminium-titanium and platium-carbon. The rate and selectivity of the hydrogenation of xylose to xylitol at 100-110°C in the presence of either an aluminium-nickel catalyst or a copper catalyst can be significantly increased by the addition of metal promoters such as cadmium, molybdenum, cerium, lanthanum, manganese or chromium (M. P. Karimkulova, Y. S. Khakimov and M. F. Abidova, Khim. Prir. Soedin., 1989, 426).

The catalytic hydrogenation of glucose to sorbitol, over a nickel-titanium catalyst is found to show first order kinetics in both substrate and hydrogen. The reaction is faster in 2-propanol/water mixture than in water (N. F. Shumatara *et al.*, Izv. Akad. Nauk Naz SSSR, Ser. Khim., 1984, 27).

Catalytic transfer hydrogenation of glucose-fructose syrups can be achieved using platinum-carbon or rhodium-carbon catalysis in alkaline solution at 25°C. The reaction yields equal amounts of glucoronic acid, mannitol and sorbitol (A. J. Van Hengstum, Starch/ Staerke, 1984, 36, 317).

The selectivity of the formation of D-mannitol from D-fructose by hydrogenation over a copper-silica catalyst is substantially higher than that when using other catalysts such as nickel, copper, ruthenium, rhodium, osmium, iridium or platinum in aqueous media at 60-80°C and hydrogen at 20-75 atmospheres. D-Fructose is preferentially hydrogenated *via* its furanose forms by attack of a copper hydride-like species at the anomeric carbon atom. This results in an inversion of configuration. The proposed mechanism explains both the enhanced yield of D-mannitol from boric esters of D-fructose and the diastereoselectivity exhibited during the hydrogenation of other ketoses (M. Makee *et al.*, Carbohydr. Res., 1985, 138, 225).

Hydrolysis of glycosides and reduction of the resulting sugars to afford alditols can be achieved with sodium cyanoborohydride in 2M trifluoroacetic acid. Although acid-labile sugars are not degraded under these conditions, several anhydroalditols are formed along with the required alditols (P. J. Gregg *et al.*, Carbohydr. Res., 1988, 176, 145).

(b) Electrolysis

Paired electrolyses of aldoses, in a divided flow cell, produce xylitol in the cathode and glucoronic acid in the anode compartments respectively by simultaneous anodic oxidation and cathodic reduction (T. C. Chou *et al.*, J. Chin. Chem. Soc., 1987, 34, 141).

(c) Enzymatic

Many procedures are available for the enzymatic reduction of sugars to alditols. Enzymatic processes for sorbitol and gluconate production have been reviewed (P. L. Rogers and U. H. Chun, Aust. J. Biotechnol., 1987, 1, 51). Yeasts are generally used for these transformations (C. Gong *et al.*, Biotechnol. Bioeng., 1983, 25, 85 ; V. Vongsuvanlert and Y. Tani, J. Ferment. Bioeng., 1989, 67, 356). Recently isolated yeasts have been shown to produce very high yields of polyols (K. Wako *et al.*, Hakko Kogaku Kaishi, 1988, 66, 209).

(d) Syntheses

Synthetic methods for the preparation of monosaccharides are relevant to this section as alditols can be prepared from them by reduction (e.g. the synthesis of all four possible pentoses, A. W. M. Lee *et al.*, J. Am. Chem. Soc., 1982, 104, 3515).

A mixture of linear, deoxy- and branched chain alditols are formed in the self-condensation of formaldehyde in the presence of calcium oxide (N. G. Medvedeva *et al.*, Zh. Prikl. Khim., 1983, 56, 2708).

Xylitol and other sugar-like substances are formed by ultraviolet irradiation of formaldehyde absorbed on zeolites (F. Seel *et al.*, Z. Naturforsch, B: Anorg. Chem. Org. Chem., 1981, 36B, 1451).

A method has been developed where acyclic carbohydrates can be obtained by a reaction of formaldehyde with syngas in the presence of chlorotris(triphenylphosphine)rhodium(I) carbonyl and a tertiary amine (T. Okano *et al.*, Chem. Lett. 1986, 1731). The view that carbohydrates can be thought of as syngas polymers has reciprocally led to a new approach

5 Steps

HO OH OH

(i) p-NO_2-$C_6H_4CO_3H$

(ii) Ac_2O , py

(1)

AcO OAc OAc O + AcO OAc OAc O

(i) $^nBu_4NOAc$, Ac_2O , 112 °C

(ii) Chromatography

(2) (3)

OAc OAc AcO OAc OAc

(4)

to the synthesis of alditols, deoxyalditols and glycosylalditols. Unprotected C_n aldoses are decarbonylated by chlorotris(triphenylphosphine)rhodium(I) to give the corresponding C_{n-1} alditols (M. A. Andrews, G. L. Gould and S. A. Klaeren, J. Org. Chem., 1989, 54, 5257).

O CHO

(5)

(i) $Ph_3P{=}CHCHO$

(ii) DIBAH

O OH

Ti ($^iPrO)_4$, (-)-Diethyltartrate, CH_2Cl_2, - 20°C

Ti ($^iPrO)_4$, (+)-Diethyltartrate, CH_2Cl_2, - 20°C

O OH O

(6)

O OH O

(7)

NaOH, H_2O, Me_3COH, 70 °C

NaOH, H_2O, Me_3COH, 70 °C

O OH OH OH

(8)

O OH OH OH

(9)

A synthetic approach to the pentitols has been described, utilising the stereoselective epoxidation of divinyl methanol (D. Holland and J. F. Stoddart, Carbohydr. Res., 1982, 100, 207). However, the diastereoselectivities characterising the epoxidations and the regioselectivities which govern the epoxide ring openings are not sufficiently high to constitute an attractive, general synthesis. Xylitol, a commercially important carbohydrate sweetener (see A. Baer, Food Sci. Techol., 1986, 17, 185) can be synthesised *via* the *cis*-trihydroxypentene (1), available in five steps from cyclopentadiene (D. Holland and J. F. Stoddart, J. Chem. Soc., Perkin Trans I, 1983, 1553). Epoxidation of (1) with *p*-nitroperbenzoic acid, followed by acetylation, leads to a 30:70 mixture of the diastereomers (2) and (3), which can be separated by chromatography. However, treating the mixture with tetrabutylammonium acetate in acetic anhydride at 112°C, followed by chromatography gives a 21% isolated yield of xylitol pentaacetate (4).

The development of the Sharpless asymmetric epoxidation reaction has led to a new approach to sugars. This approach has been demonstrated by the synthesis of tetrols,

(10) $\xrightarrow[VO(acac)_2]{^tBuOOH}$ (12)

Hacac = pentane-2,4-dione

(11) $\xrightarrow[VO(acac)_2]{^tBuOOH}$ (13)

$\xrightarrow[\text{(ii) L-Selectride}]{\text{(i) Swern}}$ (14)

pentitols, and hexitols (T. Katsuki *et al.*, J. Org. Chem., 1982, 47, 1373). The method involves the titanium catalysed asymmetric epoxidation of allylic alcohols and selective opening of the resultant 2,3-epoxyalcohols. The process is illustrated by the synthesis of the monoacetonides of D-arabinitol (9) and ribitol (8) from D-glyceraldehyde acetonide (5).

A similar approach has been developed for the synthesis of deoxyalditols (P. Ma *et al.*, J. Org. Chem., 1982, 47, 1378).

Epoxides of the type (6) and (7) can be regioselectively ring opened by azide ion. Subsequent reduction affords 2-amino-2-deoxypentitol derivatives. The epoxides also undergo ring opening, with varying selectivities, when reacted with sulphur and selenium nucleophiles (C. H. Behrens and K. B. Sharpless, Aldrichimica Acta, 1983, 16, 67).

The reaction of 1-trimethylsilylvinyl copper compounds and D-glyceraldehyde acetonide (5) proceeds in a highly stereoselective manner to afford the *syn* or *anti* addition products, (10) and (11), respectively, depending on the reaction conditions employed (M. Kusakabe and F. Sato, J. Chem. Soc. Chem. Commun., 1986, 989). Asymmetric epoxidation using *tert*-butyl hydroperoxide and a vanadium catalyst leads to the epoxides (12) and (13), whilst Swern oxidation of (13) to the ketone, followed by stereoselective reduction, affords the *anti*-epoxyalcohol (14). The epoxides (12), (13) and (14) can be converted into peracylated D-lyxitol (15), ribitol, and xylitol (4) respectively by the method of Masamune and Sharpless (C. H. Behrens *et al.*, J. Org. Chem., 1985, 50, 5687).

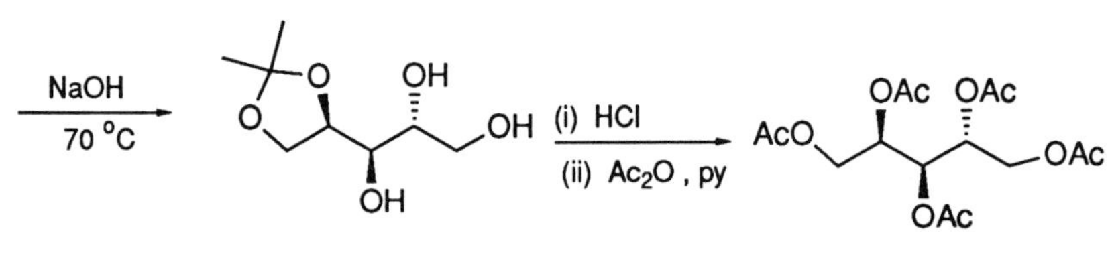

^{t}BuOK , Bu_4NF

NaOH, 70 °C

(i) HCl (ii) Ac_2O , py

(15)

A highly stereoselective protocol for the chain elongation of alkoxy aldehydes into 1,2-polyalkoxy higher homologues has been described (A. Dondoni *et al.*, J. Org. Chem., 1989, 54, 693). D-Glyceraldehyde acetonide (5), L-threose acetonide and dialdogalactopyranose diacetonide have been converted into higher homologues bearing nine, seven and ten carbon atoms respectively. The methodology consists of the iterative repetition of a linear one-carbon chain extension reaction that involves the *anti* diastereoselective addition of 2-trimethylsilylthiazole (16) to a chiral alkoxy aldehyde and the unmasking of the formyl group from the thiazole ring in the resulting adduct.

(5)

(i) A , (ii) B
Repeat 5 Times

A: (i) (16) , $^{n}Bu_4NF$; (ii) NaH , $^{n}Bu_4NI$, BnBr.
B: (i) MeI , MeCN ; (ii) $NaBH_4$; (iii) $HgCl_2$, MeCN / H_2O.

(16)

Benzyl 2-deoxy-2-*C*-methylpentanopyranosides ring open *via* attack at the anomeric centre by organoaluminium reagents (Me_2AlR) to give chiral, partially protected, branched 1,2,3,5-tetrol derivatives. For example, when the β-L-ribo derivative (17) is treated with four molar equivalents of trimethylaluminium, a 10:1 diastereomeric mixture of the chain extended tetrol derivatives (18) and (19) is isolated. Initially, a regioselective opening of the epoxide occurs to yield the 2-deoxy-2-*C*-methylpentanopyranoside which then undergoes pyranosidic ring-opening (T. Inghardt and T. Frejd, J. Org. Chem., 1989, 54, 5539).

(17) (18) 10 : 1 (19)

Pentitols and their esters have been prepared in the presence of a platinum or rhodium complex, *via* intramolecular hydrosilylation of 2-alkoxy-3-(1,3-dioxolan-4-yl)-3-silyloxy-1-propene derivatives (Y. Ito and K. Tamao, J. P. 01,230,536, 1989).

A resurgence of interest in the synthesis of higher-carbon sugars has followed the discovery of important antibiotics such as the amino-octodioses (S. O'Connor *et al.*, J. Org. Chem., 1976, 41, 2087), and the undecose hikosamine (20), which is isolated from the nucleoside hikimycin (J. A. Secrist III and K. D. Barnes, J. Org. Chem., 1980, 45, 4526).

Of the methods used to ascend the aldose series, the Fischer-Kiliani cyanohydrin synthesis has been the most extensively applied to the synthesis of higher-carbon sugars (J. S. Brimacombe and J. M. Webber, 'The Carbohydrates: Chemistry and Biochemistry', eds. W. Pigman and D. Horton, Academic Press, New York and London, 1972, Vol. 1A, p479). In recent times, however, the cyanohydrin procedure has been augmented by other methods. These permit extension of the sugar chain by two or more carbon atoms in a single step, but seldom in a wholly predictable way (H. Paulsen *et al.*, Liebigs Ann. Chem., 1981, 2009).

The development of the titanium catalysed asymmetric epoxidation reaction and the catalytic osmylation reaction for the asymmetric hydroxylation of unsaturated sugars gives a more versatile approach to higher-carbon sugars. For example, stereoselective syntheses of D-*erythro*-D-*galacto*-octitol, L-*erythro*-D-*galacto*-octitol and L-*threo*-D-*galacto*-octitol can be achieved from a single precursor, (E)-6,7-dideoxy-1,2:3,4-di-*O* - isopropylidene - α - D-*galacto*-oct-6-enopyranose (21), by way of either catalytic osmylation, or titanium catalysed asymmetric epoxidation (J. S. Brimacombe, A. K. M. S. Kabir and F. Bennett, J. Chem. Soc. Perkin Trans I, 1986, 1677).

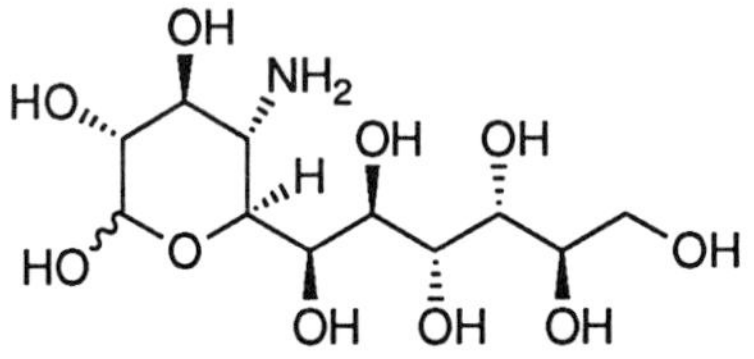

(20)

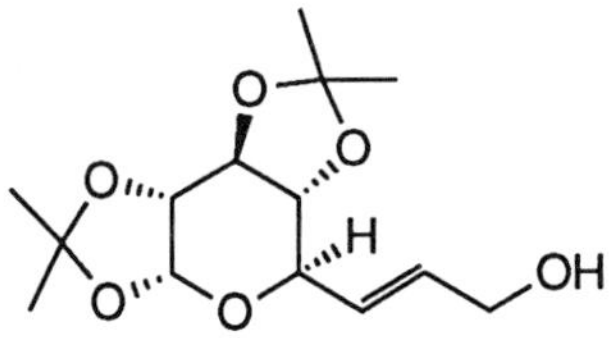

(21)

The stereochemistry of the new higher-carbon sugars, obtained by catalytic osmylation of unsaturated precursors, can be predicted by the application of Kishi's empirical rule (J. K. Cha, W. J. Christ and Y. Kishi, Tetrahedron, 1984, 40, 2247). The procedure has been used in the preparation of octitols (J. C. Barnes *et al.*, J. Chem. Soc., Perkin Trans. I, 1988, 3391), and nonitols (J. S. Brimacombe and A. K. M. S. Kabir, Carbohydr. Res., 1986, 158, 81). The catalytic osmylation reaction and the Sharpless asymmetric epoxidation reaction have both been utilised for the synthesis of new decitols (J. S. Brimacombe, R. Hanna and

A. K. M. S. Kabir, J. Chem. Soc., Perkin Trans. I, 1987, 2421). In an illustration of the methodology, L-*threo*-D-*gluco*-octitol (23) and L-*threo*-D-*manno*-octitol were synthesised by catalytic osmylation of methyl (E)-2,3,4-tri-*O*-benzyl-6,7-dideoxy-α-D-*gluco*-oct-6-enopyranoside (22) and methyl (E)-2,3,4-tri-*O*-benzyl-6,7-dideoxy-α-D-*manno*-oct-6-eno-pyranoside respectively. (This chemistry has been reviewed by J. S. Brimacombe in 'Studies in Natural Products Chemistry', ed. Atta-Ur Rahman, Elsvier Science Publishers, Amsterdam, 1989, Vol. 2).

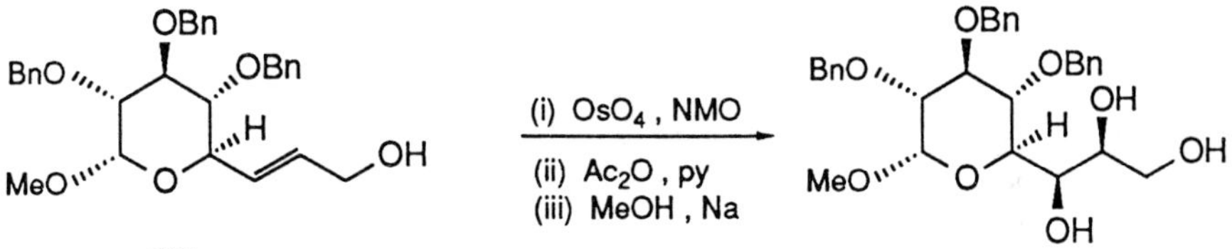
OBn
BnO
OBn
H
OH
MeO
O
(i) OsO4 , NMO
(ii) Ac2O , py
(iii) MeOH , Na
OBn
BnO
OBn
OH
H
OH
MeO
O
OH

(22)

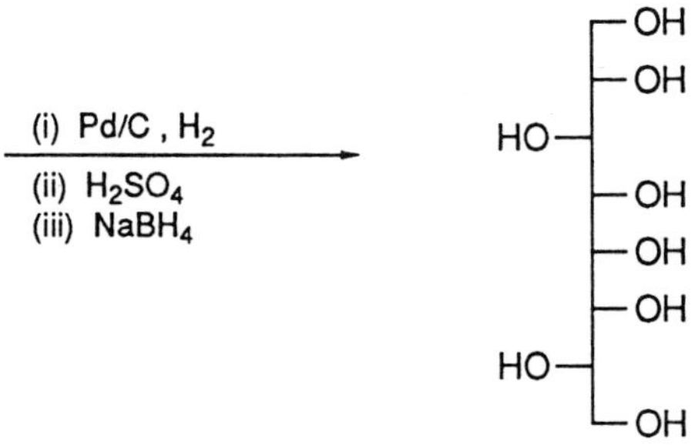
(i) Pd/C , H2
(ii) H2SO4
(iii) NaBH4
OH
OH
HO
OH
OH
OH
HO
OH

(23)

Calditol (24) is a representative of a new class of branched-chain nonitols. It was found to be part of complex, macrocyclic tetraether lipid (25), isolated from the membrane of thermoacidophilic archaebacteria of the Calderilla group. The configuration of C-4 to C-7 in calditol has yet to be established. Two diastereomers of calditol (24) and one diastereomer of 4-deoxycalditol have been prepared *via* the (-)-lactone (26), which in turn was synthesised from D-glyceraldehyde and (+)-(1R,4R)-7-oxabicyclo[2.2.1]hept-5-en-2-one (S. Jeganathan and P. Vogel, Tetrahedron Lett., 1990, 31, 1717).

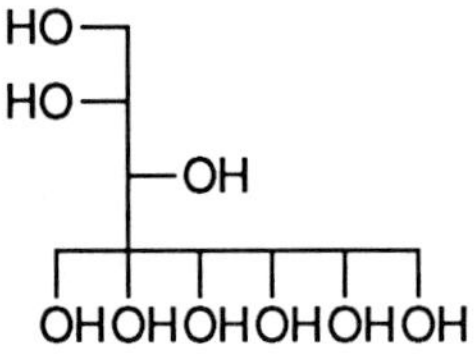

(24)

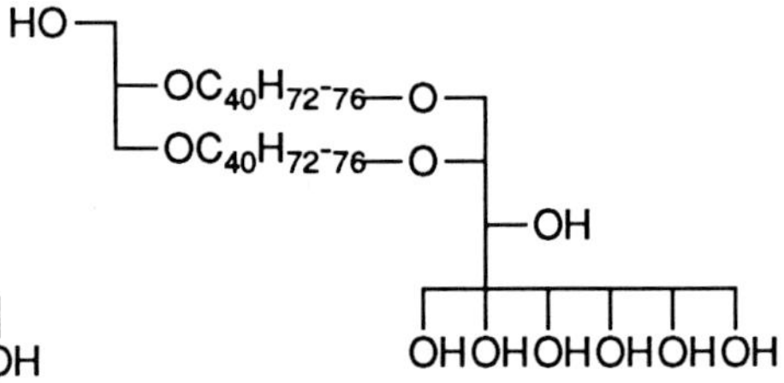

(25)

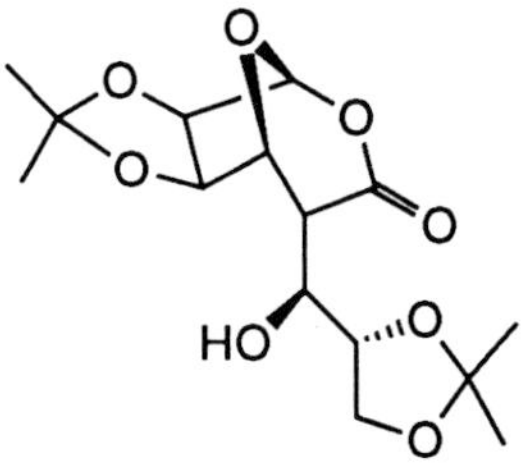

(26)

3. Conformation, chelation and chromatography

(a) Conformation and stereochemistry

Crystal structure determinations have led to the announcement of the general rule that alditols exist as planar zig-zag carbon chains, unless such a conformation leads to parallel 1,3-interactions between C-O bonds. In this case, bond rotation through 120° takes place to produce a non-linear, sickle shaped array. Energy calculations have been described for some alditols and a planar zig-zag conformation is established as the most favoured. However, the energy difference between the bent chain and straight chain conformations is less in the case of D-glucitol than those for either D-mannitol or galactitol. It has been suggested that, when the energy difference between straight and bent chain conformations is small, the lattice energy and the solvent energy play dominant roles in determining the most favoured conformations (M. Ranganathan and V.S.R. Rao, Curr. Sci., 1981, 50, 933).

Routine ^{1}H NMR spectra of alditols with free hydroxy groups are usually uninformative, but at high fields (e.g. 400 or 620 MHz), a degree of analysis is often possible. Indeed, the advent of more powerful NMR spectrometers now allows chemical shift data and coupling constants to be obtained by computer simulation. The acquired shifts and coupling constants can then be used for accurate conformational analysis (G. E. Hawkes and D. Lewis, J. Chem. Soc., Perkin Trans. II, 1984, 2073; D. Lewis, *ibid*, 1986, 467).

A recent 620 MHz ^{1}H NMR study, combining both experiment and computational simulation, was performed on deuterium-exchanged alditols for solutions in deuterium oxide, [$^{2}H_{5}$]pyridine and [$^{2}H_{6}$] acetone. Significant differences between the ^{3}J coupling constants were found for the spectra determined in aqueous and non-aqueous solvents, but only minor differences were noted for the spectra run in deuterated pyridine and acetone. From an application of the Karplus equation to the ^{3}J values, the authors conclude that it is unlikely that single unique conformations of the alditols exist in solution. Instead, the results are interpreted in terms of different populations of rotamers that are rapidly equilibrating between dihedral angles of 60 and 180° (F. Franks, R. L. Kay and J. Dadok, J. Chem. Soc., Faraday Trans. I, 1988, 84, 2595 and references therein).

Later work has shown that certain alditols crystallise with the oxygen atom of the hydroxymethyl group extending the chain. This is also the main conformer of the group when the alditols are dissolved in water. Arabinitol crystallises with the C-O bond of the C-1 hydroxymethyl group *trans* to the adjacent C-O bond, and this is the geometry adopted when alditols are dissolved in very polar aprotic solvents (e.g. hexamethylphosphorous triamide, or dimethyl sulphoxide) (F. B. Gallwey *et al.*, J. Chem. Soc., Perkin Trans. II, 1990, 1979).

A chemical procedure has been developed to distinguish between *syn* and *anti*-1,2-diol functionalities in complex, acyclic polyhydroxylated compounds. Borate complexes of the acyclic polyhydroxylated compound are formed, then acetylated and decomposed to diols, which are separated and analysed by ^{1}H NMR spectroscopy. The basis for differentiation relies on the property of borate ion to complex preferentially with *syn*-1,2-diols rather than with *anti*-1,2-diols, or terminal glycols in acyclic systems (R. E. Moore, J. R. Barchi Jr. and G. Bartolini, J. Org. Chem., 1985, 50, 374). A non-degradative method of unambiguously assigning the stereochemistry of 1,3-diols can be achieved by inspection of the ^{13}C NMR spectrum of the corresponding acetonide (S. D. Rychnovsky and D. J. Skalitzky, Tetrahedron Lett., 1990, 31, 945).

Other methods, apart from high-field NMR studies have also been used to assign the stereochemistry of alditols. Computational molecular-dynamics simulation of the conformation of polyols has attracted recent interest. Although simulation does not provide much new information, it was confirmed that the conformations of alditols in aqueous solution are similar to those observed in the crystalline state. It was also suggested that the conformations of alditols in solution are dictated by the characteristics of the solvent (J. P.

Grigera, J. Chem. Soc., Faraday Trans. I, 1988, 84, 2603)

An important general procedure has recently been developed wherein the relative and absolute stereochemistry of 1,2,3-triols, 1,2,3,4-tetrols, and 1,2,3,4,5-pentols can be assigned by circular dichroism (C.D.), after a simple two-step derivatisation with exciton-coupling chromophores. Thus, selective 9-anthroylation of primary hydroxyl groups followed by per-*p*-methoxycinnamoylation of secondary hydroxyls affords bichromophoric derivatives, the C.D. of which are characteristic and predictable for certain stereochemical patterns (W. T. Wiesler and K. Nakanishi, J. Am. Chem. Soc. 1989, 111, 9205).

General rules for the preferred conformations of glycosylalditols can be outlined by the comparative assessment of the conformational features of known glycosylalditols, determined by X-ray analysis, in combination with Jeffrey's alditol rules and the exo-anomeric effect (F. W. Lichtentaler and H. J. Lindner, Liebigs Ann. Chem., 1981, 2372).

Alditols display almost zero optical rotation in water, independent of conformation, but they have significant and differing rotations in non-aqueous media such as methanol, dimethyl sulphoxide, pyridine, and hexamethylphosphorous triamide (D. Lewis, J. Chem. Soc., Perkin Trans. II, 1991, 197).

(b) Complexes and chelation

Metal-sugar complexes play an important role in biological systems and have found useful applications in chemical synthesis (see Section 4a (ii)), NMR spectroscopy, paper electrophoresis, and chromatography. (A review of metal-carbohydrate complexes has appeared; S. J. Angyal, Chem. Soc. Rev., 1980, 9, 415).

Polyhydric alcohols complex and chelate with many metal ions e.g. Na^+, Mg^{2+}, Al^{3+}, Ca^{2+}, Cr^{3+}, Fe^{3+}, Ni^{2+}, Cu^{2+}, Zn^{2+}, Sr^{2+}, Cd^{2+}, Ba^{2+}, Pt^{2+}, Mn^{2+}.

Copper (II) complexes of polyhydric alcohols have been used for thin-layer ligand exchange chromatography and allows excellent resolution of mixtures of carbohydrates not easily achieved by other methods (J. Briggs *et al.*, Carbohydr. Res., 1981, 97, 181). Standard solutions of a Sn(II)-sorbitol complex can be used in volumetric analyses of potassium dichromate, potassium ferricyanide, and inorganic or organic peroxides (J. Zima *et al.*, J. Microchem., 1981, 26, 506).

The tetrahydroxyborate ion, $B(OH_4)^-$, complexes with almost universal ease with polyols and carbohydrates, and the use of polyhydroxy compounds as a means of increasing the strength of boric acid for titration purposes has been known for many years. Borate formation has also been used to characterise carbohydrates. The development of ^1H, ^{13}C, and ^{11}B NMR has allowed more specific identification of the reaction sites in polyols/carbohydrates and enables respective stoichiometries to be calculated (J. G. Dawber *et al.*, J.

Chem. Soc., Faraday Trans. I, 1988, 84, 41). The exchange rate between boric acid, its monoesters and diesters, at room temperature, is slow on the ^{11}B NMR time scale, thus facilitating this experimental technique for the analysis of borate esters of sugar acids and alditols (M. Van Duin *et al.*, Tetrahedron, 1985, 41, 3411).

Aqueous alkaline solutions of borates and polyhydroxy carboxylates (e.g. D-glucarate) possess good cation sequestering properties. The divalent cations Mg^{2+}, Ca^{2+}, Ni^{2+}, Sr^{2+}, Cd^{2+}, and Ba^{2+} are coordinated by the borate diesters but the univalent cations K^+, Na^+, and Ag^+ do not show preferential coordination. Cations that ionise the α-hydroxyl functions of D-glucarate (e.g. Cu^{3+}, Zn^{2+}, Pr^{3+}, and Pb^{2+}) and/or compete with borate ions for the diol functions (e.g. Al3+, and Fe3+) are more strongly coordinated by free D-glucarate than by its borate ester. These cation sequestering complexes have useful applications in the galvanic, glass, cement, and pharmaceutical industries. Other oxyacid anions such as aluminate, silicate, germanate, arsenite, selenite, stannate, antimonate, tellurate, and periodate, are also capable of forming oxyacid anion esters (or complexes) with diol functions. Synergic Ca^{2+} sequestration has been reported for aluminate-D-*gluco*heptonate mixtures (M. Van Duin *et al.*, J. Chem. Soc., Dalton Trans. 1987, 2051 and references therein).

Platinum (II) complexes of 1,2-diamino-1,2-dideoxy-D-glucitol have been prepared and found to exhibit marginal antitumour activity (M. Noji *et al.*, Chem. Pharm. Bull., 1986, 34, 2321).

Manganese (II) polyol complexes catalyse the aerobic oxidation of several peroxide-sensitive compounds at high pH, hence giving a relatively inexpensive alternative to hydrogen peroxide for the bleaching of pulp (P. K. Lim *et al.*, Ind. Eng. Chem. Fundam., 1984, 23, 29).The stoichiometries of the metal-carbohydrate complexes vary and are dependent on the metal and the pH of the solution. Phenyl boronic acid forms 1:1 chelates with alditols in aqueous acid solution (E. Huttunen, Finn. Chem. Lett., 1982, 38). Iron (III) complexes with sorbitol and glucoronic acid were thought to be polymeric in both the solid state and in aqueous solutions (M. Tonkovic *et al.*, Inorg. Chim. Acta, 1983, 80, 251). However, the ESR spectra of the Fe(III) complexes indicate dimers or oligomeric complex structures to be present (L. Nagy *et al.*, Inorg. Chim. Acta, 1986, 124, 55).

A polarimetric study of the tungsten (VI)-sorbitol system shows the formation of three stable complexes. At higher pH, the preferred species is monomeric, while the other two are dinuclear complexes of different stoichiometries (A. Cervilla *et al.*, Transition Met. Chem., 1983, 8, 21). Alditols form dinuclear tungstate complexes according to the overall equilibrium:

$$2\,WO_4^{2-} + 2\,H^+ + L \longrightarrow [W_2O_7L]^{2-} + H_2O \qquad \{L{=}Alditol\}$$

When the reaction is fast and complete (e.g. with xylitol or D-glucitol), it allows the acidimeric titration of tungstate. Alditols of the *threo*-configuration react with tungstate faster than those of *erythro*-configuration, but the stability order is *erythro*> *threo* (J. F. Verchere *et al.*, Analyst, 1990, 115, 637).

Alditols form four types of binuclear molybdate complex with ammonium molybdate. The alditols serve as tetradentate donors with four vicinal hydroxyl groups. In two types of alditol-molybdate complex, the alditol chain remains in the shape close to the preferred zig-zag arrangement. In the other two types of complex, the alditol carbon chain is forced into a sickle arrangement (M. Matulova, V. Bilik, and J. Alfoldi, Chem. Pap., 1989, 43, 403). Dinuclear molybdate complexes of alditols can be studied in aqueous acidic solution by ^{95}Mo and ^{13}C NMR (S. Chapelle, J. F. Verchere, and J. P. Savage, Polyhedron, 1990, 9, 1225).

Copper(II) ions generally form weak complexes with polyols. At pH >5, however, strong complex formation is observed. This is attributed to the formation of binuclear ions (S. J. Angyal, Carbohydr. Res., 1990, 200, 181). Telluric acid and hexitols form 1:1 complexes at low pH, but in strongly alkaline media 1:3 polyol:tellurates are present. The 1:2 chelates are thought to behave as transient intermediates (J. Mbabazi, Polyhedron, 1985, 4, 75). Lanthanide (III) compounds such as $Gd(dpm)_3$ and $Eu(fod)_3$ (dpm=2,2,6,6-tetramethyl-heptane-3,5-dionate; fod=6,6,7,7,8,8,8-heptafluoro-2,2-dimethyl-3,5-octanedionate) show a high preferency for adduct formation with alditols. These reagents exhibit a higher selectivity for bidentate coordination with an -O-C-C-OH group in an *erythro*-conformation rather than in a *threo*-conformation (J. A. Peters, W. M. M. J. Bovee, and A. P. G. Kieboom, Tetrahedron, 1984, 40, 2885).

Alditols which have a *threo-threo* sequence of hydroxyl groups complex more strongly with lanthanum rather than calcium cations. Consequently, sugars and alditols have differing mobilities on ion-exchange resin thin-layer plates in the calcium and lanthanum forms and this property can be used to separate them (S. J. Angyal and J. A. Mills, Aust. J. Chem., 1985, 38, 1279).

Although a Cr^{3+}-glucitol complex is stable, glucitol will not complex with Zn^{2+} or Ni^{2+} ions alone. However, the addition of dextrin produces stable complexes which are useful dietary supplements (I. H. Siddiqui *et al.*, Proc. Pak. Acad. Sci., 1987, 24, 17).

In recent years, EXAFS (Extended X-ray Absorption Fine Structure) spectroscopy has become very important in determining the structure of metal complexes of sugar-type ligands (L. N. H. Ohtaki, T. Yamaguchi, and M. Nomura, Inorg. Chim. Acta, 1989, 159, 201).

(c) Chromatography

The development of high-performance liquid chromatography (HPLC) as an analytical tool provides a rapid method for the separation and simultaneous determination of derivatised or non-derivatised low-molecular weight carbohydrates and polyols. The use of various solid phases have been developed, including silica (modified with tetraethylenepentamine) or ion exchange columns (J. G. Baust *et al.*, J. Chromatogr., 1983, 261, 65). A HPLC analysis technique using sulphonated polystyrene-divinylbenzene resin, in the calcium(II) form, as the stationary phase has been reported to effect excellent separations of aldoses and alditols (M. Makee *et al.*, Int. Sugar J., 1985, 87, 63). Neutral alditols and aminoalditols, derived from sugars in complex carbohydrates, can be completely separated by HPLC using a column of Shodex SP-100 (operating in the gel-permeation plus ligand-exchange mode with Pb^{2+} as the cation.). By reducing the sugars with sodium borotritide, NaB_3H_4 , the alditols and aminoalditols can be labelled, allowing a rapid method of monosaccharide composition analysis for glycoproteins and glycolipids (M. Takeuchi *et al.*, J. Chromatogr., 1987, 400, 207). A review on carbohydrate separations using zeolite molecular sieves has appeared (J. D. Sherman and C. C. Chao, Stud. Surf. Sci. Catal., 1986, 28, 1025).

The gas-chromatographic (g.c.) separation of alditols can be achieved and a variation of the solid phase or the use of different derivatives of the alditols is recommended to optimise the results. For example, several alditols derivatised by methoxylation and trimethylsilylation can be separated on glass capillary (SP-2250) or fused silica (SP-2100) columns (O. Pelletier and S. Cadieux, J. Chromatogr. 1982, 231, 225). Improvements in the capillary g.c. analysis of C_3-C_6 alditols has been reported wherein the alditol is initially converted into its *O*-benzyloxime derivative followed by trifluoroacetylation with *N*-methylbis(trifluoroacetamide) (M. A. Andrews, Carbohydr. Res., 1989, 194, 1). Monosaccharides can be resolved as their trimethylsilyl ethers of α-methylbenzyl-aminoalditols by gas-liquid chromatography (g.l.c.) on a carbowax 20M column (R. Oshima *et al.*, J. Chromatogr.,1983, 259, 159). Alternatively, the phenyldimethylsilyl derivatives of alditols can be used to effect their separation (C. A. White *et al.*, J. Chromatogr. 1983, 264, 99). Aldonic acids can be converted to *N*-propylaldomide acetates and alditols into their corresponding acetates to achieve resolution and simultaneous detection by g.l.c. (J. Lehrfeld, Anal. Chem., 1984, 56, 1803).

Capillary g.c. on chiral stationary phases has become increasingly more common in the resolution of chiral compounds and in the determination of enantiomeric purity. Recently, suitable columns with thermostable phases have become commercially available and new methods of derivatisation (such as the formation of carbamates and amides) have enabled the separation of a large number of substances (W. A. Koenig, S. Lutz, and G. Wenz,

Angew. Chem., Int. Edn. Eng., 1988, 27, 979 and references therein). Modified cyclodextrins provide highly enantioselective stationary phases for g.c.. Especially high separation factors are obtained in the case of trifluoroacetylated polyols and 1,5-anhydroalditols (W. A. Koenig *et al.*, Carbohydr. Res., 1988, 183, 11). Enantiomers of arabinitol and mannitol are separated, as their trifluoroacetyl derivatives, by g.c. on fused silica columns coated with XE-60-(L)-valine-(R)-α-phenylethylamide or its diastereomer (W.A. Koenig and I. Benecke, J. Chromatogr., 1983, 269, 19).

4. Reactions and derivatives of the alditols

(a) Use of alditols in asymmetric synthesis

(i) Chiral synthetic intermediates (chirons)

The synthesis of many natural products starting from carbohydrates has been covered by S. Hanessian in 'Total Synthesis of Natural Products. The Chiron Approach.', Pergamon Press, Oxford, 1983.

Alditols and their derivatives have found increasing use in asymmetric synthesis. Generally, the alditol is degraded to utilise the stereochemistry of a section of the original molecule e.g. D-mannitol is a precursor of (R)- or (S)-isoserine (A. Dureault, I. Tranchepain, and J. C. Depezay, Synthesis, 1987, 49). Indeed, D-mannitol is a unique, inexpensive chiral compound and it is the alditol most commonly used in constructions of this type. The recent development of the practical preparation of versatile chiral building blocks (chirons) from D-mannitol has been reviewed (S. Takano and K. Ogasawa, Yuki Gosei Kagaku Kyokashi, 1987, 45,1157).

Enantiomerically pure α-hydroxy aldehydes of (R) or (S)-configuration, and various side chains, can be obtained from D-mannitol. The strategy is based upon the nucleophilic opening of diastereomeric diepoxides (27) and (28). D-mannitol has a two-fold axis of symmetry and if this symmetry is preserved during chemical transformations then C-2 and C-5 will have identical absolute configurations and oxidative cleavage at the C-3-C-4 bond will lead to two identical molecules. Thus, starting from D-mannitol, the "D-mannitol diepoxide" (27) can be prepared with retention of configuration (*via* selective tosylation of the primary hydroxyl groups and displacement). Regiospecific, nucleophilic ring opening (by a general nucleophile, Nu^-) can introduce various groups at C-1 and C-6. Protection of the secondary hydroxyls, hydrolysis and oxidative cleavage of the glycol affords two molecules of the protected (R)-α-hydroxy aldehyde (29). Starting from D-mannitol, but with inversion of configuration at C-2 and C-5, the "L-iditol diepoxide" (28) is obtained (*via* selective benzoylation of the primary hydroxyl groups; tosylation of the hydroxyl groups at C-2 and C-5; debenzoylation and displacement). Following the same reaction sequence, the protected (S)-α-hydroxy aldehyde (29) becomes available (Y. Le Merrer *et al.*, J. Org. Chem., 1989, 54, 2409). A similar approach to that used above has been developed in the synthesis of methyl (9R)-hydroxy-5Z,7E,11Z,14Z-eicosatetraenoate (9-HETE, 30) (M. Saniere *et al.*, Tetrahedron, 1989, 45, 7317).

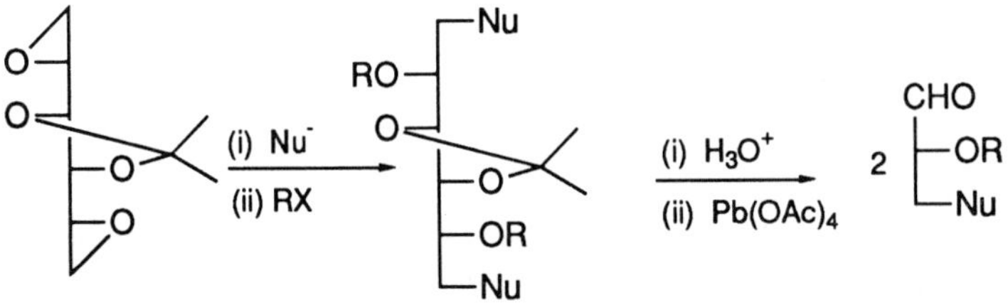

(27) (R) - (29)

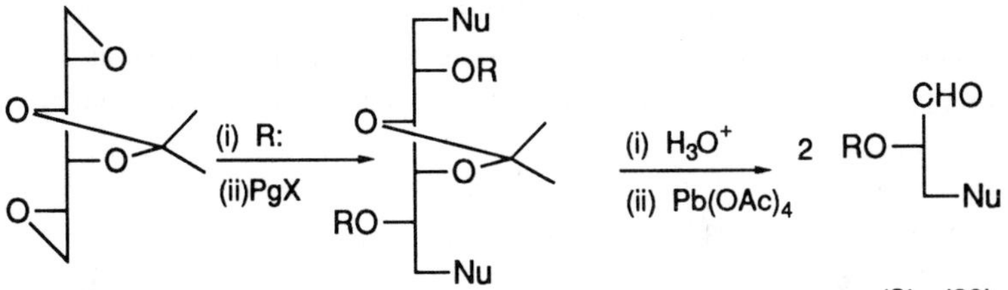

(28) (S) - (29)

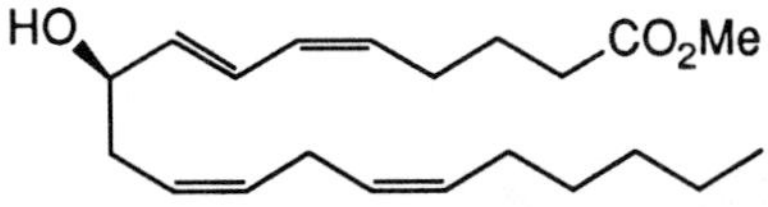

(30)

(R,R)-1,2-Divinylglycol (32) (1,5-hexadien-3,4-diol) is a versatile intermediate in natural product syntheses and can be efficiently prepared from D-mannitol *via* the tetramesylate (31) (A. V. Rama Rao *et al.*, Tetrahedron Lett., 1987, 28, 2183).

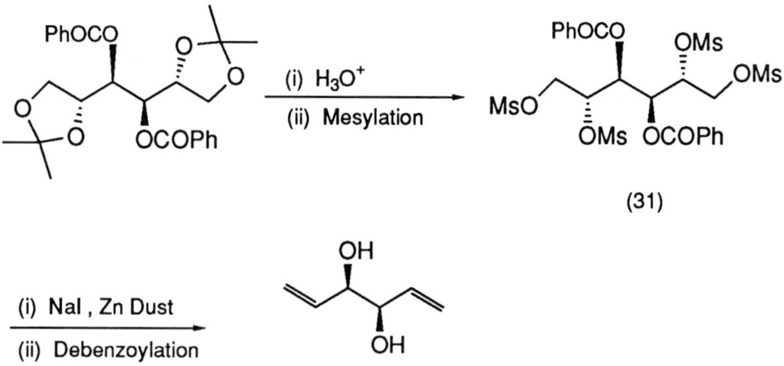

(i) H_3O^+
(ii) Mesylation

(31)

(i) NaI , Zn Dust
(ii) Debenzoylation

(32)

An interesting application of arabinitol as a chiron in the synthesis of natural products has been developed. The important feature of this work is that terminus differentiation can be achieved in a two-directionally homologated chain. Thus, the symmetrically homologated hydroxydiester (33), obtained in eight steps from arabinitol, undergoes a group selective lactonisation in the presence of pyridinium *p*-toluenesulphonate (PPTS) to yield a 6:1 mixture of lactones, with (34) as the dominant isomer. This approach has been entitled a 'two-directional chain synthesis strategy' and has been utilised in the synthesis of the novel immunosuppressant, FK-506 (S. L. Schreiber, T. Sammakia and D. E. Uehling, J. Org. Chem., 1989, 54, 15).

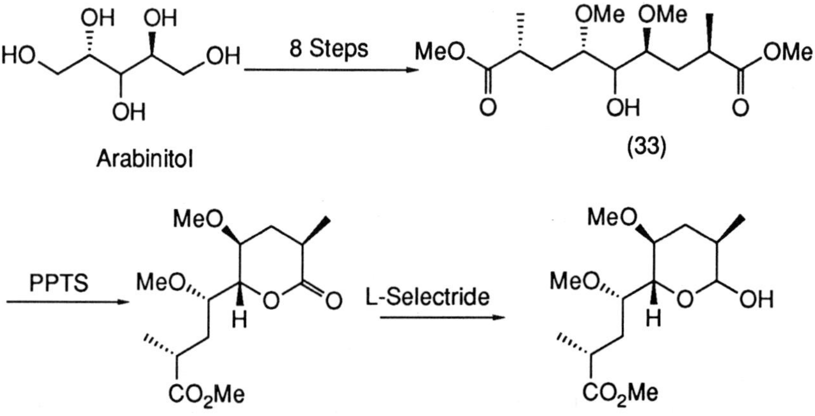

Arabinitol

8 Steps

(33)

PPTS

L-Selectride

(34)

(ii) Reagents and catalysts

Various metal complexes of alditol derivatives are known to be catalysts for the enantio-selective hydrogenation of prochiral substrates. For example, rhodium complexes of the diphosphine 1,4:3,6-dianhydro-2,5-dideoxy-2,5-bis(diphenylphosphino)-L-iditol (35) are

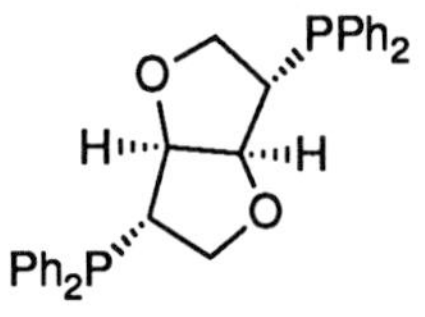

(35)

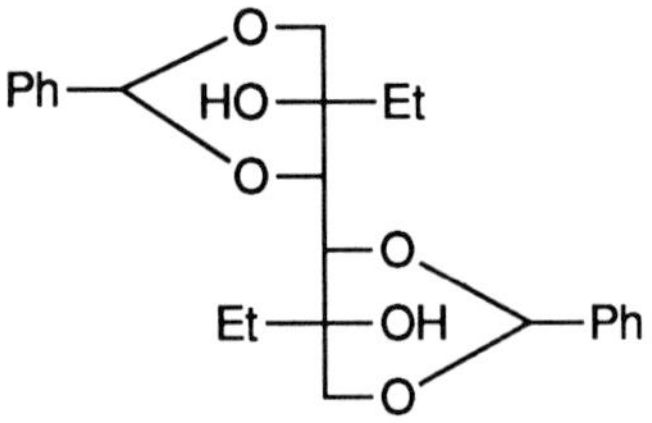

(36)

homogeneous catalysts for the hydrogenation of dehydroamino acids. (S)-Amino acids are obtained in 21-58% optical yields (J. Bakos *et al.*, J. Organomet. Chem., 1983, 253, 249). Acetophenone can be reduced in an asymmetric fashion by complexes formed from

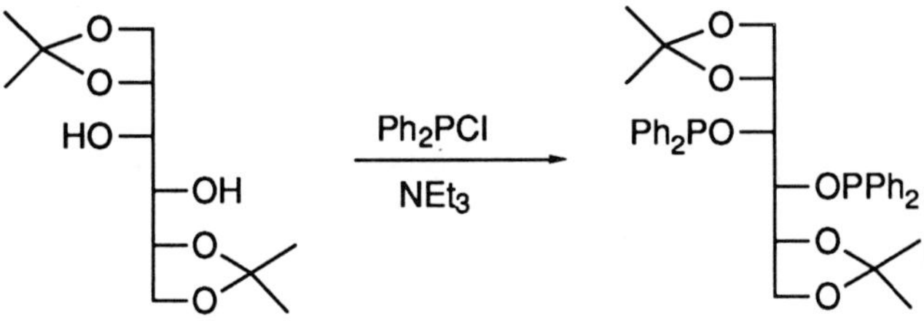

Ph_2PCl, NEt_3

(37)

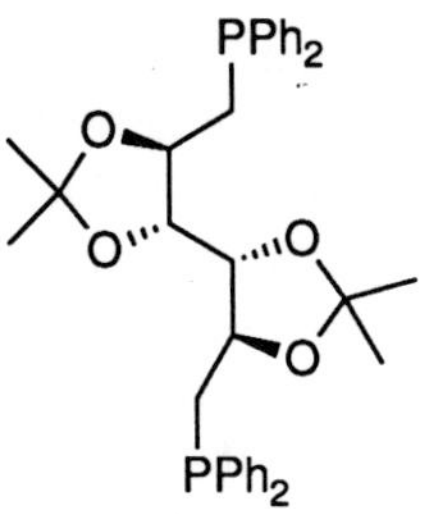

(38)

1,2:5,6-diisopropylidenemannitol-lithium aluminium hydride and an alkanol. In all cases studied, (R)-(+)-phenylmethylcarbinol is produced in excess and the optical yields increase with the increasing size of the alkyl group in the alkanol used (M. Song *et al.*, Beijing Shifan Daxue Xuebo, Ziran Kexueban, 1986, 44). Similarly, hydroaluminate complexes of symmetrical, chiral branched-chain alditol derivatives, e.g. (36), have been reported to influence the stereochemical outcome of the reduction of a number of ketones (N. Baggett *et al.*, Carbohydr. Res., 1987, 162, 153). Hydrogenation of prochiral olefins using a rhodium catalyst with a chiral phosphinate ligand (37) (derived from D-mannitol) have been carried out to obtain chiral substances in optical yields of 8-78% (M. Yamashita *et al.*, Bull. Chem. Soc. Jpn., 1989, 62, 942).

The L-iditol derived biphosphine (38) has been used as a ligand for the rhodium catalysed hydroformylation of styrene (J. M. Brown and S. J. Cook, Tetrahedron, 1986, 42, 5105).

Acetals of D-mannitol have been used as alkoxyligands of titanium to induce poor to modest enantioselectivity in various reactions such as: the nucleophilic transfer of alkyl and phenyl groups to aldehydes (D. Seebach *et al.*, Helv. Chim. Acta, 1987, 70, 954); the ring opening of cyclohexene oxide by trimethylsilylazide to afford *trans*-2-azidocyclohexanol (M. Emziane, K. I. Sutowardoyo and D. Sinou, J. Organomet. Chem., 1988, 346, C7); and the glycolate-ene reaction (K. Mikami, J. Am. Chem. Soc., 1990, 112, 3949). Ti (IV) complexes of alditol derivatives have been used as the chiral catalyst in the Sharpless asymmetric epoxidation reaction but were found to give poorer enantioselectivities than when using R,R-diisopropyltartrate as the catalyst (P. G. Potvin, P. C. C. Kwong, and M. A. Brook, J. Chem. Soc., Chem. Commun., 1988,773 and references therein). For a discussion on the solution structures of chiral Ti (IV) alkoxides, see P. G. Potvin *et al.*, Can. J. Chem., 1989, 67, 1523).

1,2:5,6-Di-*O*-isopropylidene-D-mannitol and its 3,4-di-trimethylsilyl derivative have been used as chiral promoters in the Lewis acid catalysed Diels-Alder reaction between cyclopentadiene and 3-crotonyl-4,4-dimethyl-1,3-oxazolidin-2-one (39) to afford a cycloadduct with a diastereomeric excess of between 92-96% (C. Chapuis and J. Jurczak, Helv. Chim. Acta,1987, 70, 436).

Homotopic 18-crown-6 ethers derived from D-glucitol and D-mannitol form stable 1:1 complexes with chiral ammonium salts. In some cases, these ethers differentiate between enantiomers allowing resolution of the salts (M. I. Struchkova *et al.*, Izv. Akad. Nauk SSSR, Ser. Khim., 1989, 2492). However, it has been reported that coronands derived from D-glucitol display poor, or no, recognition towards (R)- or (S)-α-phenylethylammonium hexafluorophosphates (E. El'Perina *et al.*, Heterocycles, 1989, 28, 805). Chiral macrocyclic ethers, derived from alditols, e.g. the 18-crown-6 ether (40), produced from 1,2:5,6-di-*O*-isopropylidene-D-mannitol, can be immobilised on silica and used for the efficient chromatographic separation of racemic amino acids (J. P. Joly and B. Gross, Tetrahedron Lett., 1989, 30, 4231 and references therein).

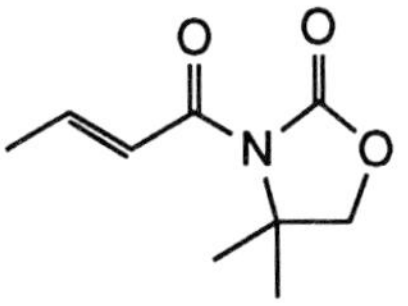

(39)

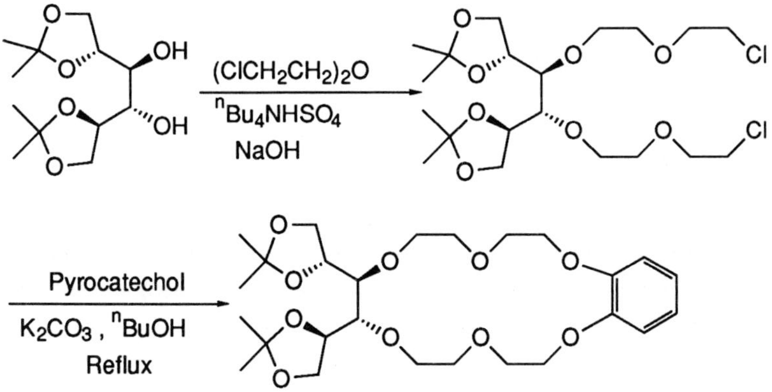

$(ClCH_2CH_2)_2O$
$^nBu_4NHSO_4$
NaOH

Pyrocatechol
K_2CO_3, ^nBuOH
Reflux

(40)

(b) Oxidation and oxidation products

Alditols are the reduction products of monosaccharides, hence both types of compounds share common oxidation products. The oxidation of alditols and monosaccharides and the chemistry of the corresponding products was fully covered by Ferris in the First Supplement, Vol. 1FG, chapter 22 and chapter 23 section 1(i). The chemistry of the monosaccharides, although mentioned throughout the sections of this chapter, will be covered more thoroughly in chapters 23a, 23b and 23c.

A novel synthesis of natural 3-deoxy-D-*manno*octulosonic acid (KDO, 41) was achieved by treatment of 1,2-anhydro-3,4:5,6-di-*O*-isopropylidene-D-mannitol with the anion of diethyl malonate, followed by oxidative decarbonylation and acid hydrolysis of the resulting 2-acetoxy-3-deoxy-5,6:7,8-di-*O*-isopropylidene-D-*manno*-2-octenic acid γ-lactone (Y. A. Zhdanov *et al.*, Bioorg. Khim., 1983, 9, 104). The eight-carbon sugar (41) is a component of lipopolysaccharides isolated from cell walls of gram-negative bacteria. Several other routes have also been published towards the synthesis of KDO (P. A. M. Klein *et al.*, Tetrahedron Lett., 1989, 30, 5477 and references therein).

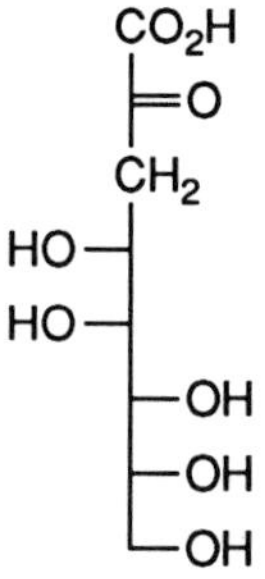

(41)

Oxidation of 1,3:4,6-di-*O*-benzylidene-D-mannitol (42) with ruthenium tetraoxide yields the corresponding diketone (43), after dehydration. The carbonyl groups of this compound are not equivalent and show differing reactivities towards a variety of nucleophiles (N. Baggett and P. Stribblehill, Carbohydr. Res., 1981, 96, 41).

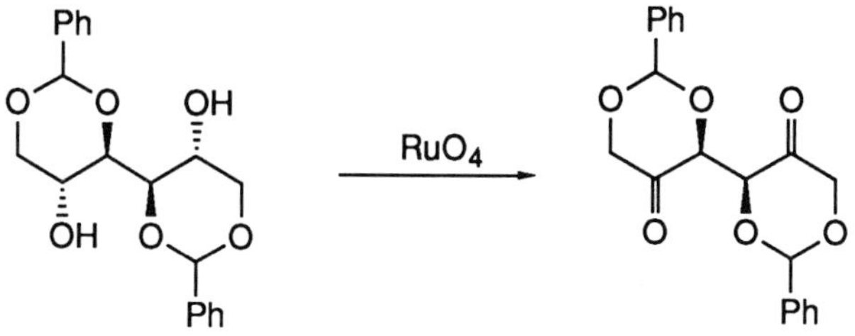

RuO_4

(42) (43)

A new selective oxidation procedure has been developed for the preparation of hexoses. Sequential trimethylsilylation, Collins' oxidation, and deprotection of L-fucitol (44) leads to L-fucose (45) in 87% yield (H. Kristen *et al.*, J. Carbohydr. Chem., 1988, 7, 277).

Oxidation of pentitols and hexitols with bromine in the presence of calcium carbonate affords 2- and 3-uloses, as well as 2,5-hexodiuloses. Derivatives of these products can be separated by capillary gas chromatography (P. Decker and H. Schweer, Carbohydr. Res., 1982, 107, 1; H. Schweer, *ibid*, 1983, 16, 139).

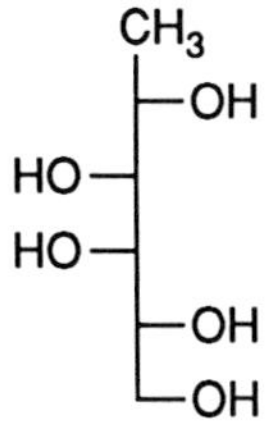

(44) 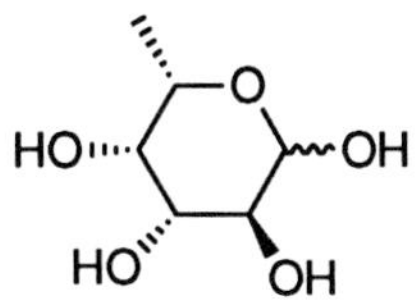 (45)

Polyhydric alcohols are oxidised by a number of metal compounds. The kinetics of these types of oxidations are usually first order with respect to both the oxidant and the substrate. Typical examples of the metal oxidants are manganese (III) pyrophosphate (A. G. Fadnis and S. K. Kulshthrestha, React. Kinet. Catal. Lett., 1982, 19, 267; A. G. Fadnis *et al.*, Carbohydr. Res., 1983, 112, 137); cerium (IV) (A. G. Fadnis *et al.*, Rev. Roum. Chim., 1983, 28, 99); $[Fe(CN)_6]^{3-}$ in the presence of ruthenium tetraoxide (M. P. Singh and S. Kothan, Ind. J. Chem., Sect. A, 1983, 22A, 697); vanadium (V) (A. G. Fadnis and S. K. Kulshthrestha, Cienc. Cult., 1985, 37, 781). The rate of oxidation of mannitol and sorbitol by vanadium (V) ion in sulphuric acid is found to be significantly increased by the addition of copper(II) ions (V. Devi *et al.*, Acta Cienc. India Chem., 1986, 12, 72). Galactitol can be oxidised to galactonic acid by an acidic solution of *N*-bromosuccinimide in the presence of ruthenium trichloride, as a homogeneous catalyst, and mercuric acetate (J. P. Sharma *et al.*, Tetrahedron, 1986, 42, 2739).

Numerous dehydrogenases have been used for the enzymatic oxidation of alditols and their oxidation products (R. C. Mortlake, J. Bacteriol., 1985, 162, 845; J. K. Churry *et al.*, Yakkhai Hoechi, 1988, 32, 386). Enzymes can produce unusual oxidation products, not formed by normal chemical methods. For example, the unprecedented oxidation of methyl D-galactopyranoside (46) to methyl D-galactopyranosiduronic acid (48), *via* aldehyde (47), is achieved, in high yield by application of the enzyme galactose oxidase (S. Matsumura *et al.*, Chem. Lett., 1988, 1747). Galactose oxidase is an enzyme which catalyses the oxidation of the C-6 hydroxymethyl function of D-galactose in the presence of molecular oxygen to an aldehyde group. Unfortunately, the enzyme is known to be relatively non-specific and the exact oxidation products are not always clear. Consequently, the enzyme has been used mainly as an analytical tool for determining the presence of D-galactose and D-galactopyranosides. However, D-galactose oxidase has recently been used in alternative syntheses of unusual sugars, particularly L-sugars. (Unnatural sugars are important because of their potential use as safe, noncaloric sweeteners and as precursors for many natural products.). D-Galactose oxidase converts D-(+)-threitol and xylitol into D-(+)-threose and

L-(-)-xylose respectively. Galactitol and L-(+)-glucitol prove to be poor substrates for the enzyme, but the addition of catalytic amounts of ferricyanide ion increases the reaction rate. Although no side reactions are observed for each reaction, product inhibition occurs and only 10-50% of the alditol is oxidised. To minimise this problem, a column reactor can be employed to enable recycling of the unreacted alditol (R. L. Root, J. R. Durrwachter, and C. H. Wong, J. Am. Chem. Soc., 1985, 107, 2997).

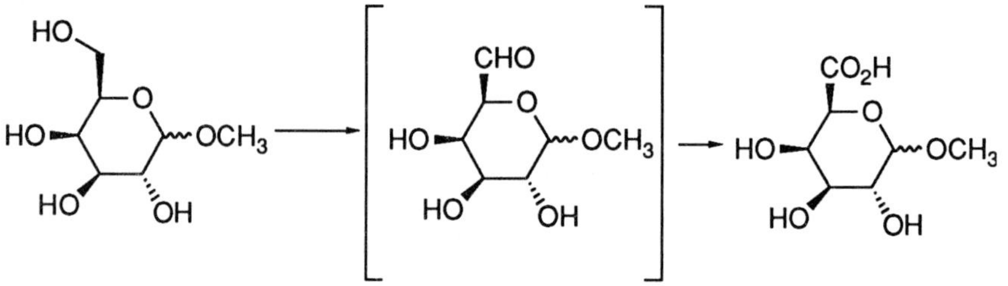

(46) (47) (48)

Alditols and their oxidation products undergo electrochemical oxidations at various metal anodes (e.g. Ni, Pt, or Au electrodes). Recent electrochemical literature concerning the oxidation of small organic molecules as potential fuels for fuel cells has been reviewed (R. Parsons and T. Van der Noot, J. Electroanal. Chem., 1988, 257, 9). The electrooxidation of diacetone-L-sorbose to diacetone-2-keto-L-gulonic acid occurs efficiently at a nickel hydroxide electrode (I. A. Avrutskya *et al.*, Sov. Electrochem., 1975, 11, 1176).

Alternative reagents and procedures for the oxidative cleavage of 1,2-diols have recently been developed. These include a catalytic system consisting of triphenylbismuth, *N*-bromosuccinimide, potassium carbonate, acetonitrile and water (D. H. R. Barton *et al.*, Tetrahedron, 1986, 42, 5627) and *N*-(*tert*-butyl)-*N'*,*N'*,*N''*,*N''*-tetramethylguanidinium-3-iodoxybenzoate in the presence of trichloroacetic acid (D. H. R. Barton *et al.*, Tetrahedron Lett., 1982, 23, 957). Sodium metaperiodate/wet silica gel in the presence of dichloromethane is an efficient reagent for the oxidative cleavage of 1,2-diols, and is a good alternative to sodium metaperiodate (which has low solubility in organic solvents) and quaternary ammonium periodates (which are relatively expensive). The method is especially useful when the product aldehydes easily form hydrates (M. Daumas *et al.*, Synthesis, 1989, 64).

(c) Cyclic acetals

The formation and migration of carbohydrate cyclic acetals has been thoroughly reviewed (D. M. Clode, Chem. Rev., 1979, 79, 491).

Condensation of alditols with aldehydes, or ketones, affords a wide range of cyclic acetals containing 5-, 6- or 7-membered rings. The reaction is usually carried out in the presence of acidic catalysis but other catalysts, such as iodine have been used (K. P. R. Kartha, Tetrahedron Lett. 1986, 27, 3415). 1,2:5,6-Di-*O*-isopropylidene-D-glucitol can be obtained, under neutral conditions, by the reaction of D-glucitol with 2,2-dimethoxypropane in 1,2-dimethoxyethane (G. J. F. Chittenden, Carbohydr. Res., 1982, 108, 81).

The formation of alditol acetals are generally complicated reactions, with the various products dependent on the nature of the carbonyl compound, the stereochemistry of the alditol, and the reaction conditions employed. For example, the reaction of D-glucitol with acetone, catalysed by zinc chloride, yields eleven products (E. Tomari, J. Kuszmann, and G. Howarth, J. Chromatogr., 1984, 286, 381). All of the above isomers can be separated, as their acetylated derivatives, by capillary g.c.. Correlations are found between the methylene unit (M. U.) values and the structures of mono-, di-, and tri-*O*-isopropylidene-D-glucitol derivatives. A relationship is also established between the chromatographic retention indices of the isomers with their preferred conformations.

Many methods have been developed for the formation of specific cyclic acetals of alditols.These compounds have found increasing use as synthetic intermediates and reagents (see section 4a). They have also found uses in polymer synthesis (G. Wulff *et al.*, Macromolecules, 1990, 23, 100). Highly selective acetalisaton of glucitol can be achieved using inverse phase transfer catalysis (G. Fleche, Starch/ Staerke, 1990, 42, 31).

The 2-haloethylidene acetals of alditols can be prepared by transacetalisation of haloacetaldehyde diethylacetals with alditols, e.g. 1,1-diethoxy-2-chloroethane reacts with D-glucitol in the presence of hydrogen chloride to afford the triacetal (49). Monoacetals can be obtained if shorter reaction times are employed (S. Yanai, Carbohydr. Res., 1983, 113, 336).

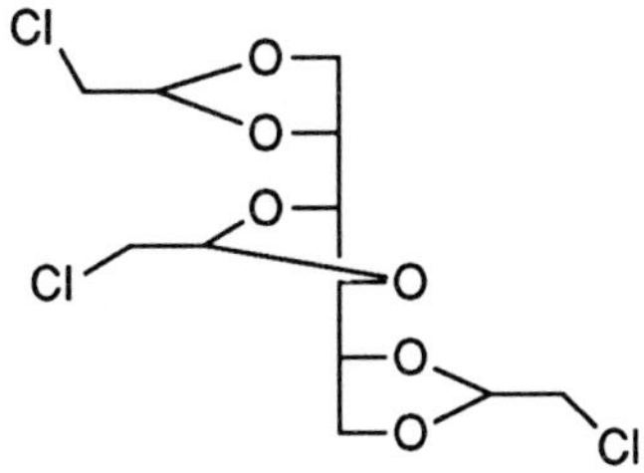

(49)

The 1,3-dioxolane ring of benzylidene acetals of 1,5-dianhydroalditols, e.g. (50), can be selectively cleaved by butyl lithium, with the 1,3-dioxane ring remaining intact. The reaction can be interpreted as proceeding *via* abstraction of the quasi-axial hydrogen atom at position 3. Subsequent expulsion of the elements of benzaldehyde leaves an enolate anion which is quenched to afford the vicinal deoxy ketone (51) (D. Horton and W. Weckerle, Carbohydr. Res., 1988, 174, 305).

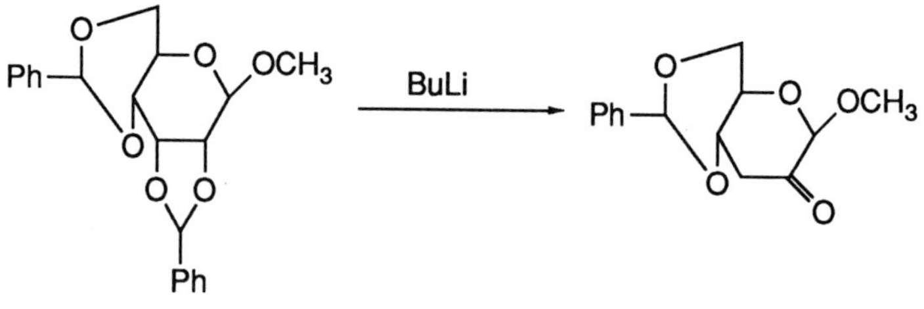

BuLi

(50)

(51)

Mannitol-1,6-dibenzoate (52) is readily available and, in the presence of sulphuric acid and cupric sulphate, reacts with acetone to form the symmetrical monodioxolane (53). This compound cannot be reacted further, even under forcing conditions. However, a reaction of (52) with an excess of 2,2-dimethoxypropane, catalysed by toluene-*p*-sulphonic acid, proceeds cleanly to give a mixture of bisdioxolanes (54) and (55). These compounds can be separated by chromatography and furthur manipulation of these products *via* reactions of the respective dimesylates with lithium diphenylphosphide, yields the corresponding

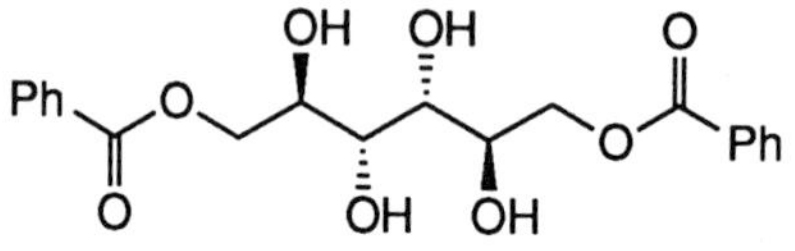

(52)

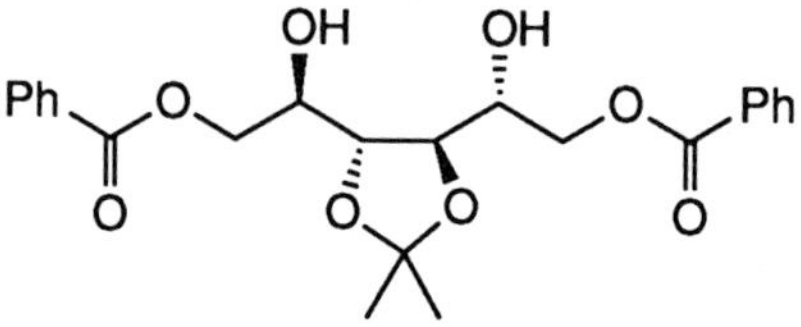

(53)

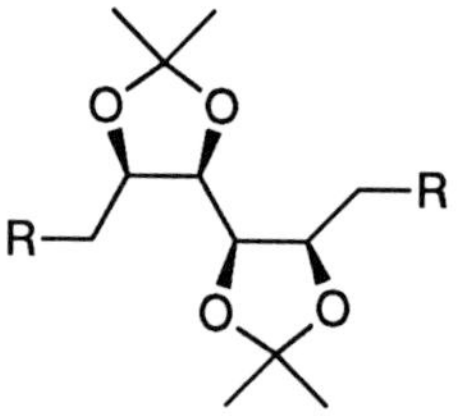

(54) R = OCOPh
(56) R = PPh_2

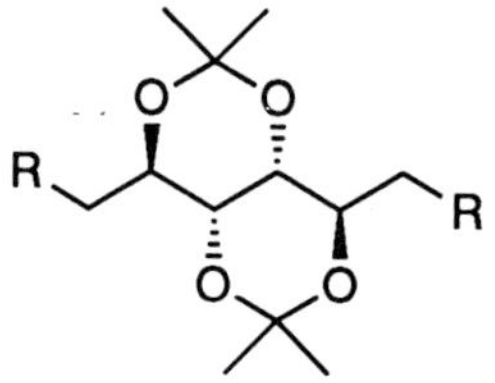

(55) R = OCOPh
(57) R = PPh_2

phosphines (56) and (57). Rhodium-chelate complexes of these products have been formed and used as catalysts for the asymmetric hydrogenation of various enamides (J. M. Brown, F. M. Dayrit, and D. Sinou, J. Chem. Soc., Perkin Trans. II, 1987, 91). (See also section 4(a) (ii).).

Galactitol, which has an *erythro* arrangement of hydroxyl groups at C-3-C4, forms a mixture of 1,2:4,5- and 2,3:4,5-di-*O*-isopropylidene derivatives under the usual acidic conditions used for acetal formation. However, treatment of galactitol with 2,2-dimethoxypropane in dimethyl sulphoxide yields 1,2:3,4:5,6-tri-*O*-isopropylidenegal-actitol. This unexpected product contains a *cis*-dioxolane ring, not predicted by the Bar-ker-Bourne rules. Dimethyl sulphoxide is thought to influence the course of the reaction through solvation of the reactants (G. J. F. Chittenden, Tetrahedron Lett., 1981, 22, 4529).

(d) Esters

Selective esterification at primary hydroxyl groups of alditols can sometimes be achieved. {e.g. by reaction with benzoyl chloride, mesyl chloride, and more recently with allyl chloroformate (P. Boullanger *et al.*, J. Carbohydr. Chem., 1986, 5, 541).}. However, the regioselective acylation of hydroxyl groups in sugars, even discrimation between primary and secondary hydroxyls, usually involves multistep procedures (A. H. Haines, Adv. Carbohydrate Chem. Biochem., 1981, 39, 13). While there is no *general* basis for the positionally-specific acylation of secondary hydroxyl groups, some specific examples are known, e.g. 1,5-anhydro-4,6-*O*-benzylidene-D-galactitol (58) can be selectively esterified, in modest yields, with either benzoyl chloride or tosyl chloride to afford the derivatives (59) and (60) respectively.

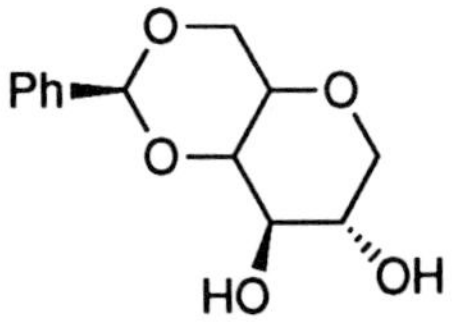

(58)

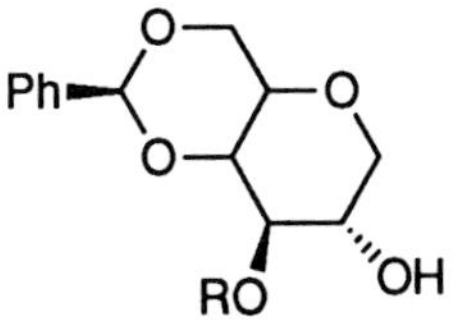

(59) R = COPh (67%)
(60) R = Ts (50%)

A complementary, regioselective acylation of secondary hydroxyl groups has recently been achieved by the use of various lipases. Classical chemical methods for the selective acylation of 1,4-anhydro-5-*O*-hexadecyl-D-arabinitol (61) is unsatisfactory. However, treat-ment of (61) with different lipases and trichloroethyl butyrate, in benzene, leads to excellent regioselective acylation, affording either *O*-acylated product (62) or (63), depending on the lipase used (F. Nicotra *et al.*, Tetrahedron Lett., 1989, 30, 1703).

$C_{16}H_{33}O$

HO OH

(61)

Rhizopus japonicus, Trichloroethyl butyrate

Humicula langinosa, Trichloroethyl butyrate

$C_{16}H_{33}O$

RO OH

2 : 3 , 3 : 97%

(63)

R = -CO^{n}Pr

$C_{16}H_{33}O$

HO OR

2 : 3 , 86 : 14%

(62)

In a general sense, several unrelated lipases (e.g. those from porcine, pancreas, bacteria, yeast, or fungi) have been shown to catalyse transesterification reactions between trichloroethyl butyrate and monosaccharides with protected C-6 hydroxyl groups. The lipases exhibit a remarkable regioselectivity, discriminating between the four available secondary hydroxyl groups in C-6 protected glucose, mannose, and galactose (M. Therisod and A. M. Klibanov, J. Am. Chem. Soc., 1987, 109, 3977). Enzymes can also be used to selectively produce mono- or di-linoleic esters of sorbitol, giving compounds which are useful as antitumour agents, immunostimulants, antibacterials, and plant growth inhibitors (I. Morita and Y. Chikasawa, J.P. 01 31,745, 1989).

Similar sorbitol derived fatty esters can be prepared by transesterification of sorbitol with methyl esters of palmitic, oleic and linoleic acids, using an alkaline catalyst. The fatty acid esters, when heated with propylene oxide at 130-140°C, give propoxylated surfactants (A. A. Samy *et al.*, Kolor Ert., 1989, 31, 42).

Treatment of 1,2:5,6-di-*O*-isopropylidene-D-mannitol (63a) with *N*-dichloromethylene-*N*,*N*-dimethylammonium chloride (Viehe's salt) and triethylamine in dichloromethane affords the cyclic iminium salt (64). Hydrolysis of (64) affords the carbonate (65), whereas treatment of (64) with sodium hydrogen sulphide gives the corresponding thiocarbonate (66) (C. Copeland and R. V. Stick, Aust. J. Chem., 1982, 35,1709). Thermolysis of cyclic iminium salts of the type (64) leads to chlorocarbamates (C. Copeland and R. V. Stick, *ibid*,, 1982, 35, 2257).

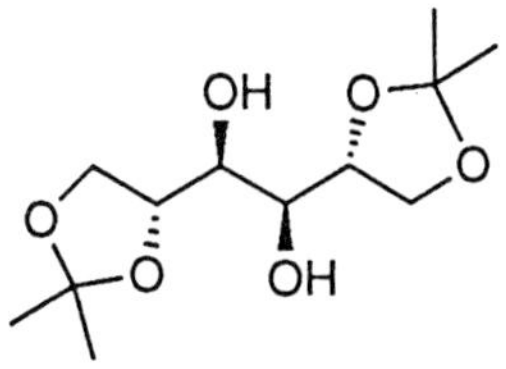

(63a)

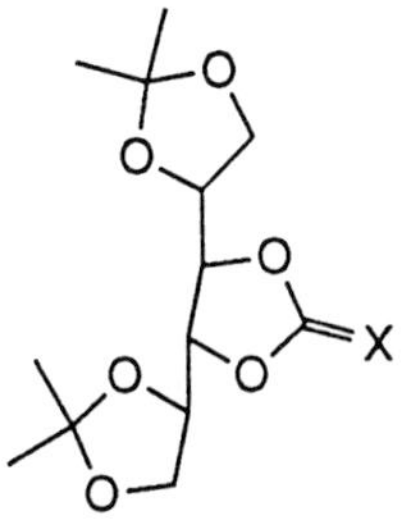

(64) X = $^{+}NMe_2$ ^{-}Cl
(65) X = O
(66) X = S

A method has been developed for the efficient carboalkoxylation of polyols using the appropriate fluoroformates as reagents, dimethylformamide or dimethyl sulphoxide as solvent, and potassium fluoride or triethylamine as the acid scavenger (V. A. Dong and R. A. Olofson, J. Org. Chem., 1990, 55, 1851).

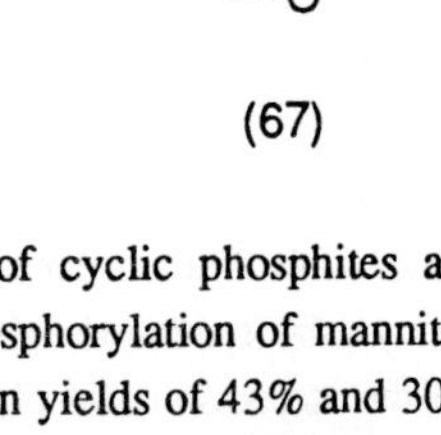

(67)

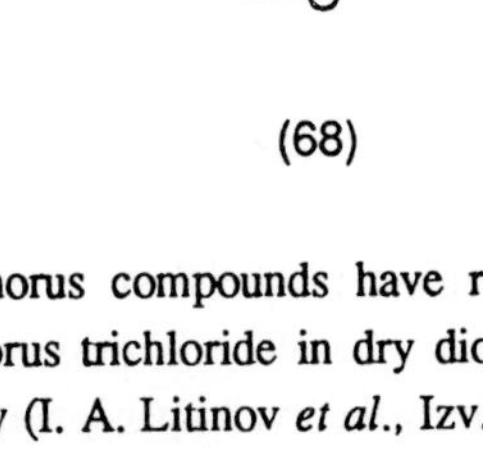

(68)

A number of cyclic phosphites and thiophosphorus compounds have recently been described. Phosphorylation of mannitol by phosphorus trichloride in dry dioxane affords (67) and (68) in yields of 43% and 30% respectively (I. A. Litinov *et al.*, Izv. Akad. Nauk SSSR, Ser. Khim., 1983, 2068). The cyclic phosphites (69), (70) and (71) are obtained in yields of 38-46% by esterification of ribitol, xylitol, and L-arabinitol with triethylphosphite respectively. Alternatively, transesterification of (72), (73) and (74) with ethanol/diethyl ether containing a catalytic amount of pyridine affords (69), (70) and (71) in yields of 36-40% respectively (N. A. Naakarova *et al.*, Izv. Akad. Nauk SSSR, Ser. Khim., 1988, 847).

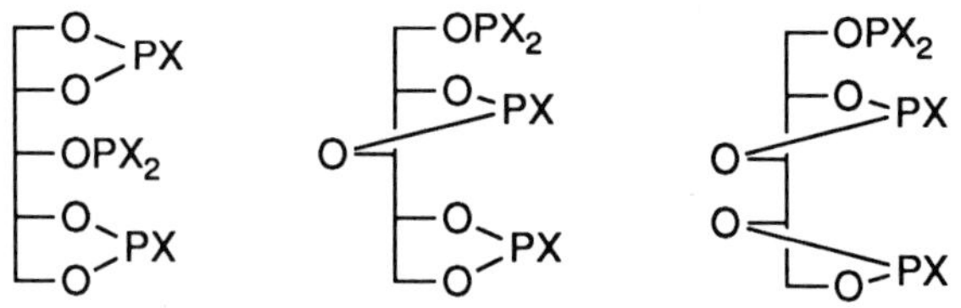

(69) X = OEt (70) X = OEt (71) X = OEt
(72) X = Cl (73) X = Cl (74) X = Cl

Chlorination of the phosphorylated mannitol derivative (67) by chlorine in chloroform gives the chlorophosphate (75) which can be amidated by piperidine to give diastereomeric piperidides, e.g. (76) (L. I. Gurarii *et al.*, *ibid*, 1989, 426).

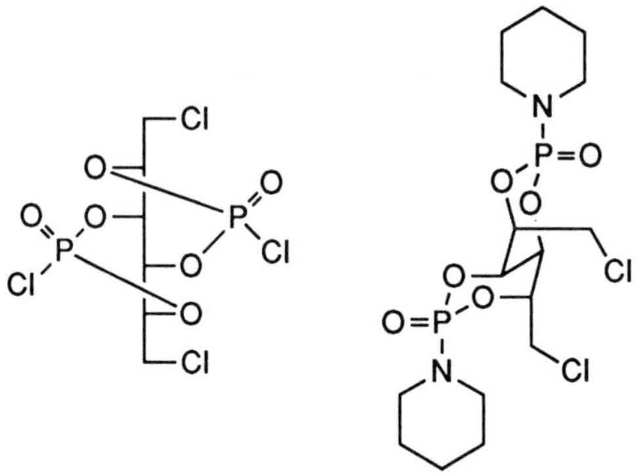

(75) (76)

Similarly, treating 1,6-dichloro-1,6-dideoxy-D-mannitol with three molar equivalents of tris(diethylamino)phosphine in dioxane at 80°C, followed by oxidation with dinitrogen tetraoxide affords the cyclic phosphoramide (77) and the stereoisomeric phosphoramide (78) in respective yields of 40% and 26% (L. I. Gurarii *et al.*, Zh. Obsch. Khim., 1988, 58, 699).

Phosphorylation of D-galactitol by phosphorus trichloride in dry dioxane at 55-60°C

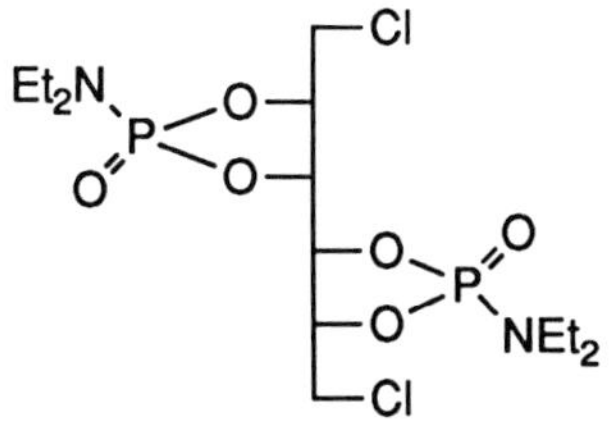

(77) 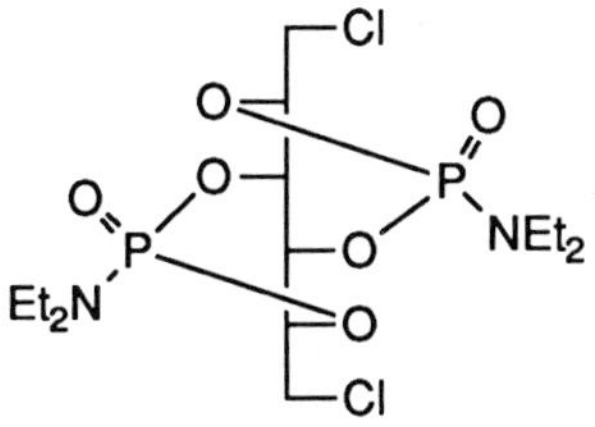 (78)

affords the tetracyclic compound (79), which when treated furthur with diethylamine and sulphur gives the aminothiophosphate (80). A variation of this approach yields a mixture of the diastereomeric compounds (81) and (82) *via* phosphorylation of galactitol with tris-(diethylamino)phosphine, followed by treatment with sulphur (L. I. Gurarii *et al.*, Izv. Akad. Nauk SSSR, Ser. Khim., 1983, 424; 896).

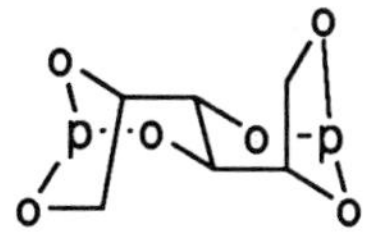

(79) 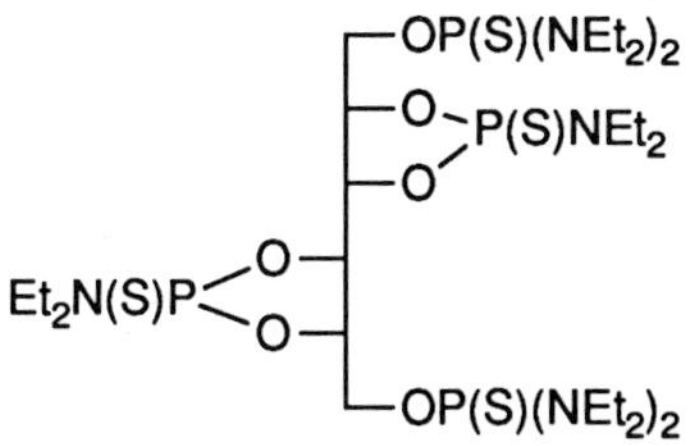 (80)

The tricyclic D-sorbitol derivative (84) was obtained in 23% yield by phosphorylation of 5-*O*-acetyl-6-iodo-1,3:2,4-diethylidene-D-sorbitol (83) with trimethylphosphite. The reaction proceeds *via* initial attack by the phosphine at the C-5 acetoxy group, followed by

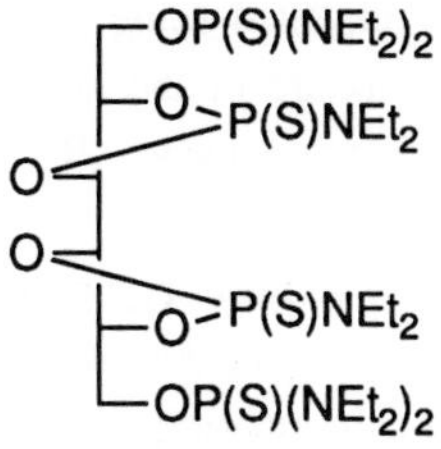

(81) 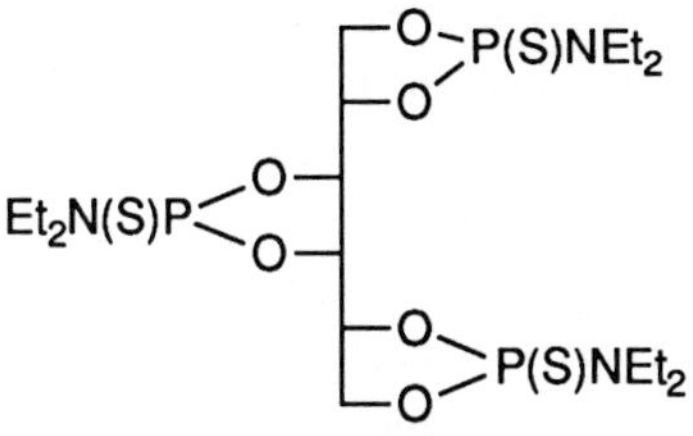 (82)

intramolecular displacement of iodide ion (Y. Zhdanov *et al.*, Zh. Obsch Khim., 1982, 52, 1655).

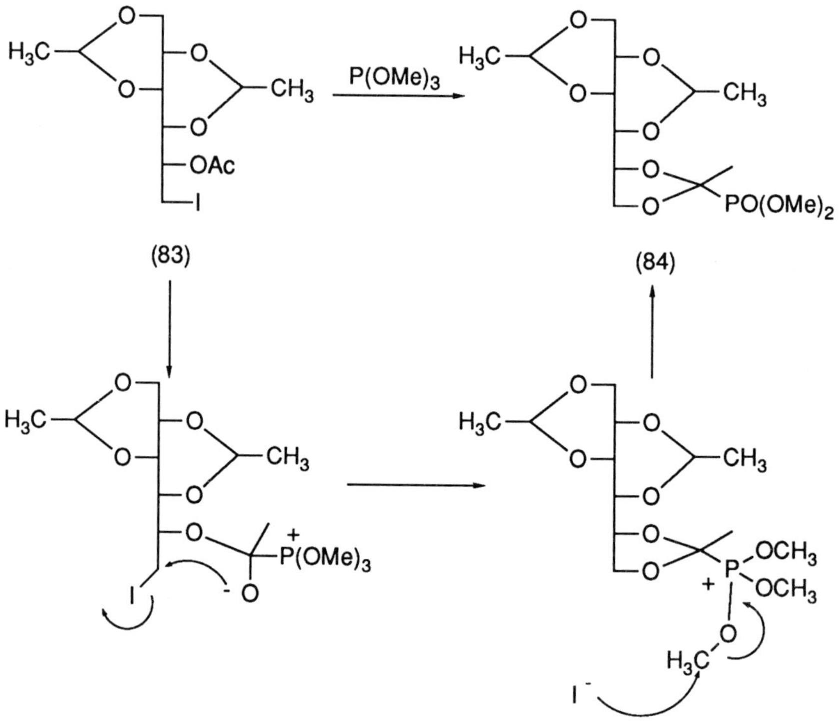

(83)

(84)

(e) Ethers

Generally, water soluble compounds (e.g. carbohydrates) can not be alkylated under phase transfer conditions due to the lack of ion pair formation in the organic phase. However, partially tetrahydropyranylated intermediates can easily be alkylated using phase transfer catalysis. Typical conditions for the alkylation employ a mixture of the alkyl bromide, tetrabutylammonium bromide, sodium hydroxide, water and toluene (R. Nouguier, Tetrahedron Lett., 1982, 23, 3505).

Monoperfluoroalkylated ethers of alditols can be prepared by addition of a perfluoroalkyl iodide to the double bond of a protected alditol allyl ether in a one-step addition-elimination procedure. Monoesters are obtained specifically at position-5 by treating 1,2:3,4-di-*O*-isopropylidenexylitol with perfluoroalkylated acid chlorides in pyridine. These products exhibit remarkable properties as new surfactants for biomedical uses (L. Zarif *et al.*, J. Med.

Chem., 1990, 33, 1262). A new method for the formation of allyl ethers of alditols, in one step under neutral conditions, involves using allyl ethylcarbonate in the presence of Pd(0) (R. Lakhmiri *et al.*, Tetrahedron Lett., 1989, 30, 4669).

5. Anhydroalditols

Anhydrosugars are generally formed after extended heating of polyols in acid solutions (Y. Halpern, R. Riffer, and A. Broido, J. Org. Chem., 1973, 38, 204), or by elimination reactions, e.g. from tosylates (P. Koll, G. John, and J. Kopf, Liebigs Ann. Chem., 1982, 626; P. Koll and G. Papert, *ibid,* 1986, 1568).

Eliminations, thermal or otherwise, to give dianhydroaldohexoses are rare and usually proceed in very low yields. In contrast, dianhydromannitol (isomannide) and dianhydrosorbitol (isosorbide) are readily available by acid catalysed thermal dehydration. A recent development provides 1,4:3,6-dianhydromannopyranose (86), in 75-80% yield, a compound virtually inaccessible by other means. The method involves flash vacuum pyrolysis of isomannide dinitrate (85), forming (86) *via* a radical mechanism.

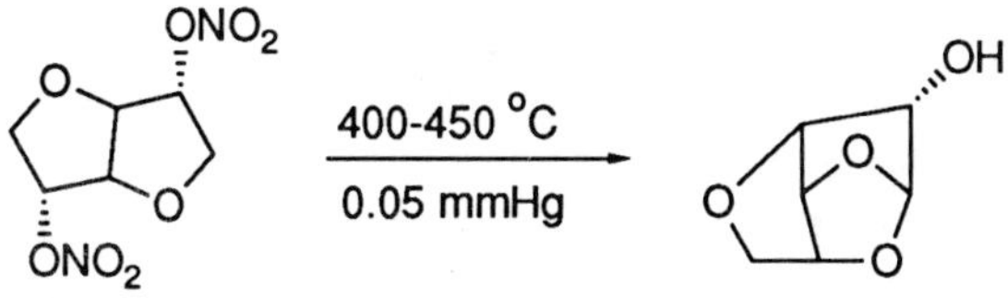

Catalytic oxidation of isosorbide with oxygen and a platinum on carbon catalyst yields (1S,4S,5R)-4-hydroxy-2,6-dioxabicyclo[3.3.0]octan-8-one (87) (F. Jacquet *et al.*, J. Chem. Technol. Biotechnol., 1990, 48, 493).

A study of the dehydration of pentitols in aqueous sulphuric acid shows that four isomeric

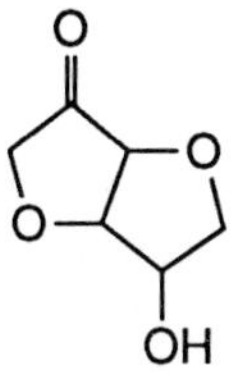

(87)

1,4-anhydropentitols are obtained from D-arabinitol, but only two each from xylitol and ribitol. The number of products can only be explained by assuming inversion of configuration at C-2 and C-4 during 1,4- or 2,5-cyclisation reactions. No product is observed which involves inversion of configuration at C-3 (A. Wisniewski *et al.*, Carbohydr. Res., 1981, 97, 229). The dehydration of pentitols in acetic acid containing an acidic catalyst parallels that in aqueous sulphuric acid. Acetylated alditols undergo similar processes *via* intermediates having free hydroxyl groups. The configurational inversion of 1,4- or 1,5-anhydroalditols is now attributed to the formation of intermediate acyloxonium ions (A. Wisniewski, *ibid*, 1983, 114, 11).

Contrary to the usual behaviour of alditols, 2-amino-2-deoxy-D-galactitol, 2-amino-2-deoxy-D-glucitol, D-mannitol, and 2-acetamido-2-deoxy-D-glucitol do not form anhydrides when treated with methanolic hydrogen chloride at 85°C. This is a significant discovery, with relevance in the quantification of sugar residues by derivatisation and subsequent gas chromatography (G. Gerwig *et al., ibid*, 1984, 129, 149).

Partial *O*-tosylation of 1,5-anhydro-L-arabinitol yields mixtures of products. The reactivities of the hydroxyl group at each carbon is found to be C-4>C-2>C-3 with two molar equivalents of tosyl chloride. When one molar equivalent of tosyl chloride is used, the reactivity is C-4>C-3>C-2 (Y. Kondo, *ibid*, 1984, 128, 175). Heating equimolar quantities of pentitols with tosyl chloride in pyridine at 60°C for four hours affords 1,4- or 2,5-monoanhydroalditols, with retention of configuration at the chiral C-2 or C-4 centres. Variation of the reaction conditions by using four molar equivalents of tosyl chloride at 115°C gives mainly 1,4-anhydro-5-chloro-5-deoxypentitols (A. Wisniewski *et al.*, J. Carbohydr. Chem., 1989, 8, 59).

The monooxetanes (88) have been prepared from the appropriate 2,5-anhydroalditols by selective monotosylation of the primary hydroxyl group and subsequent treatment with base. Analogous reactions give the 1,3:2,5:4,6-trianhydrohexitols (89) with L-*ido* and *meso-galacto* configurations (P. Koll *et al.*, Liebigs Ann. Chem., 1987, 205).

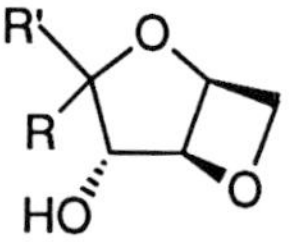

(88) 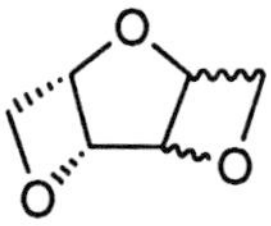 (89)

The reaction pathway for the base transformation of 2,4-*O*-benzylidene-1,6-di-*O*-*p*-toluenesulphonyl-D-glucitol (90) into 1,3-anhydro-2,4-*O*-benzylidene-D-glucitol (93) has been shown to have intermediates in the order (90), (91), (94), (93) and not (90), (91), (92), (93) as previously suggested (H. B. Sinclair, Carbohydr. Res., 1983, 113, 321).

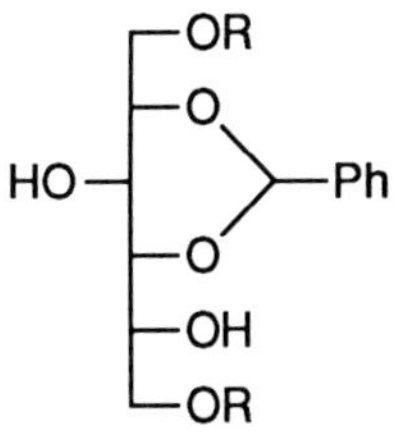

(90) R = Ts
(91) R = H

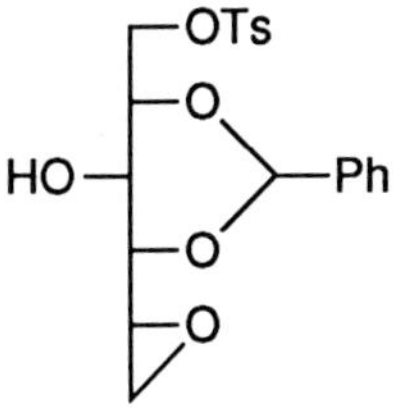

(92)

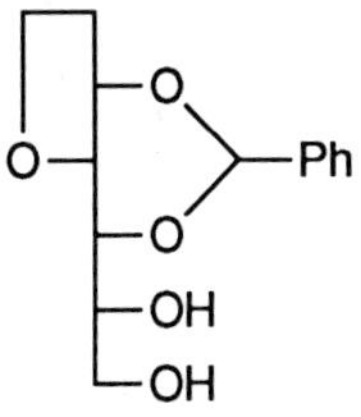

(93)

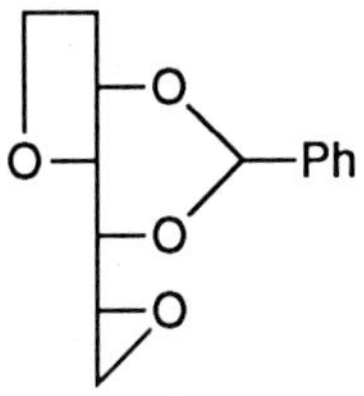

(94)

Other methods of forming anhydroalditols have recently been developed. These include the silylation of glycosides, followed by reductive cleavage in the presence of triethylsilane and trimethylsilyl triflate (J. A. Bennek and G. R. Gray, J. Org. Chem., 1987, 52, 892). 1,5-Anhydro-D-alditols (95) are formed by desulphurisation of glycosyl isothiocyanates (96) with tributyltin hydride, in the presence of the initiator, α,α-azoisobutyronitrile (Z. J. Witczak, Tetrahedron Lett., 1986, 27, 155).

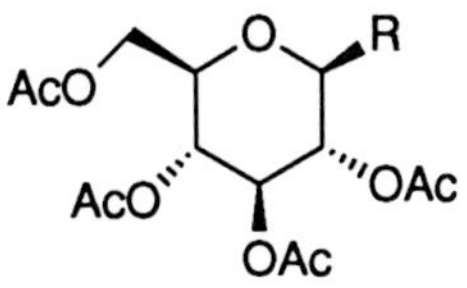

(95) R = H
(96) R = NCS

The previously unknown 2,5-anhydro-D-iditol (97) has been prepared from 2,5-anhydro-D-mannitol *via* the epoxide (98) (D. A. Otero and R. Simpson, Carbohydr. Res., 1984, 128, 79).

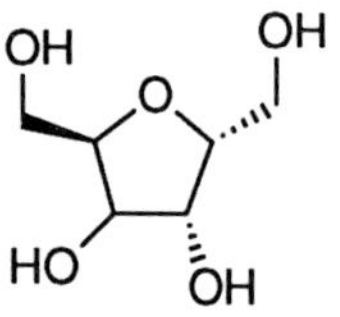

(97) 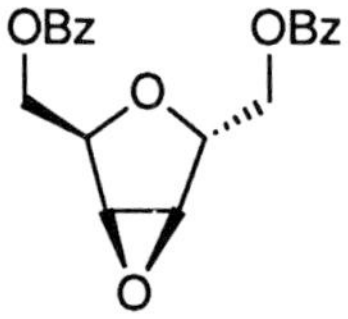 (98)

Payne epoxidation of D-*threo*-(E)-1,2:5,6-dianhydro-3,4-dideoxy-3-hexenitol (99) with hydrogen peroxide in methanol/ acetonitrile leads to a 1:3 mixture of two enantiomerically pure diastereomers of the trianhydrides of hexitols, having D-*manno* (100) and D-*ido* (101) configurations. Similar epoxidation of the *erythro*-(E)-1,2:5,6-dianhydro-3,4-dideoxy-3-hexenitol yields the racemic trianhydroglucitol (102) (P. Koll, M. Oelting, and J. Kopf, Angew. Chem., Int. Edn., 1984, 23, 242).

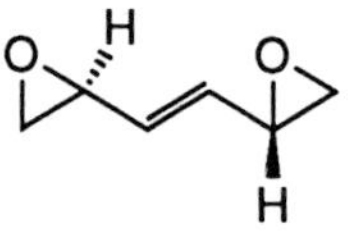

(99)

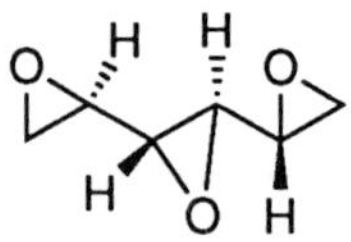

(100) 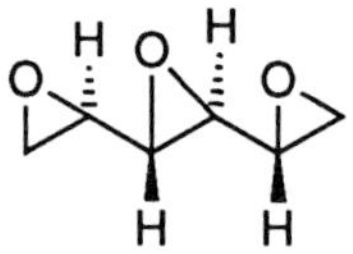 (101)

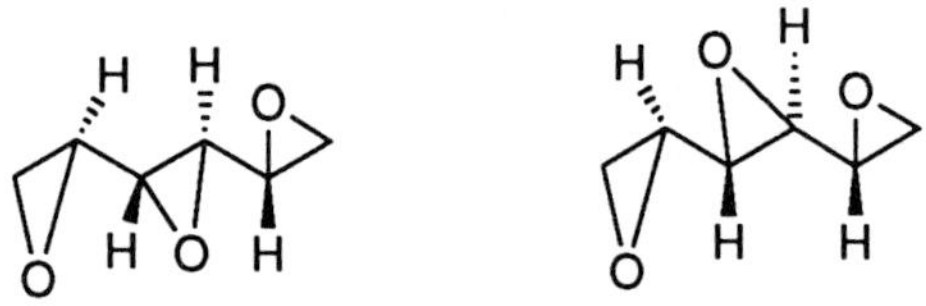

(102)

6. Deoxyalditols

1-Deoxyalditols and 1,2-dideoxyalditols can be prepared by heating the corresponding pentoses, or hexoses, in refluxing hydrazine (J. M. Williams, Carbohydr. Res., 1984, 128, 73).

A stereocontrolled sequence, based on the Sharpless asymmetric epoxidation reaction, has been developed to provide a general entry into the 1,3,5...(2n+1) polyol series (K. C. Nicolaou *et al.*, J. Am. Chem. Soc., 1988, 110, 4672). The methodology is outlined below. It should be noted that the asymmetric epoxidation can be selected to afford either enantiomer at any stage in the reiterative sequence.

Sharpless Asymmetric Epoxidation

(i) Oxidation (ii) Olefination

Selective Reduction

Sharpless A. E.

Reiterate

(2n + 1) Polyol

Vicinal dideoxyalditols can be prepared by reaction of a 3,4-thiocarbonate with triethylphosphite and subsequent hydrogenation of the resulting olefins. Selective benzylation of the primary alcohols of (103), *via* the organotin derivative, followed by tosylation, affords 1,6-dibenzyl-2,5-ditosyl-3,4-dideoxy-D-*threo*-hexitol (104). Reaction of this ditosylate with benzylamine leads to the pyrrolidine derivative (105) in 90% yield (M. Marzi and D. Misiti, Tetrahedron Lett. 1989, 30, 6075).

7. Aminoalditols

Specific activations or protections of 3,4-isopropylidene-D-mannitol, followed by intramolecular S_N2 reactions leads to the chiral epoxides (106) or (107) (X=O), or the chiral diaziridines (X=NH) (Y. Le Merrer *et al.*, Heterocycles, 1987, 25, 541).

1,1 -Thionocarbonyldiimidazole

$(EtO)_3P$, Reflux

(i) H_2, Rh/ Al_2O_3
(ii) pTsOH , aq. MeOH

(i) Bu_2SnO, BnBr, Bu_4NBr
(ii) TsCl , py

(103)

(104)

$BnNH_2$, 80 °C

(105)

N-Tosylation and subsequent nucleophilic ring-opening of the diaziridines (106) by dialkyl copper lithium reagents affords diamino compounds (108). Deacetalisation and oxidative cleavage of these intermediates yield precursors of chiral amino acids (109) (A. Dureault *et al.*, Tetrahedron Lett., 1986, 27, 4157).

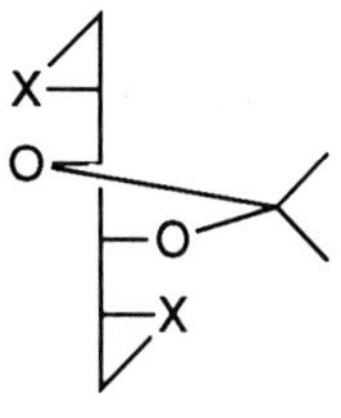

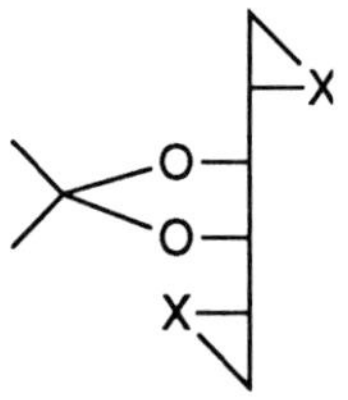

(106) (107)

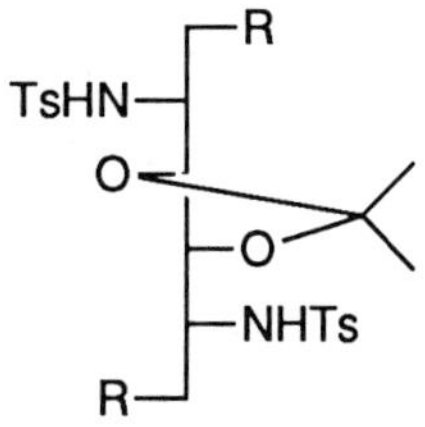

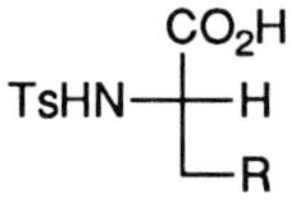

(108) (109)

An intramolecular aminomercuration reaction has been developed, allowing the direct conversion of a sugar into a cyclic aminoalditol, having the same relative and absolute configuration. The method is illustrated by the syntheses of the natural products 1-deoxynojirimycin (110) and 1-deoxymannojirimycin (111) from tri-*O*-benzyl-6-bromo-pyranoside and methyl-α-D-mannopyranoside respectively (R. C. Bernotas and B. Ganem, Tetrahedron Lett., 1985, 26, 1123).

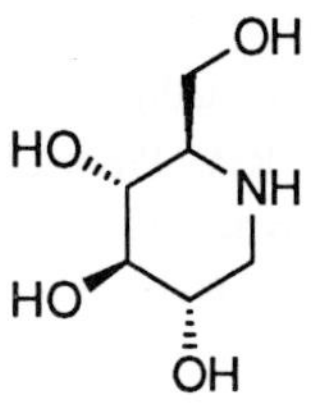

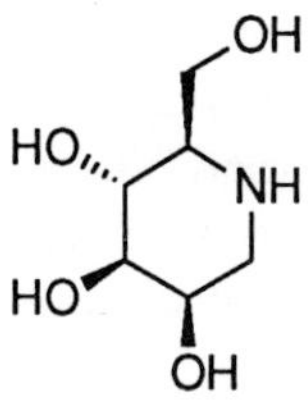

(110) (111)

Reaction of the lactol (112), obtained from D-mannose, with hydrazine, affords the pyrazole (113), an intermediate in the synthesis of 7-amino-3-(β-D-arabinofuranosyl)pyrazolo[4,3-d]pyrimidine (114). This compound is the D-arabino analogue of the antibiotic formycin (J. G. Buchanan, D. Smith, and R. W. Wightman , Tetrahedron, 1984, 40, 119).

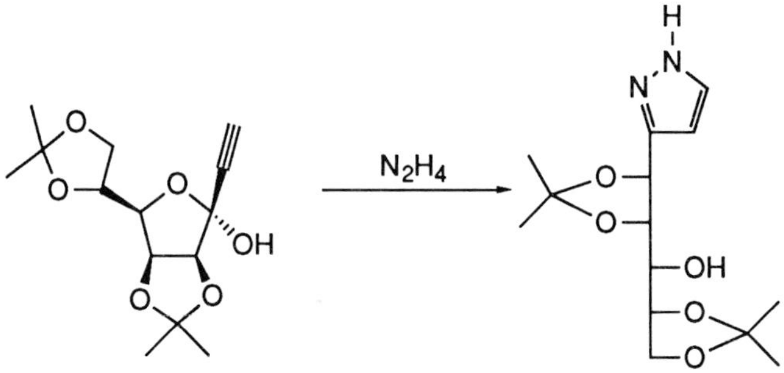

N_2H_4

(112) (113)

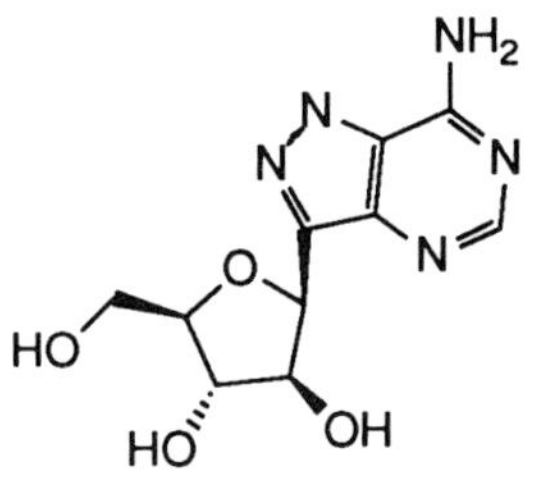

(114)

Other heterocyclic aminoalditols that have been prepared as potential drugs include new acyclic sugar *C*-nucleoside analogues (e.g. 115). These are prepared by reactions of the hydrochlorides of 2-deoxy-2-ethylaminohexoses with 5,5-dimethyl-1,3-cyclohexanedione and sodium carbonate (J. A. Galbis Perez *et al.*, Carbohydr. Res., 1985, 138, 153). 1-Alkyl-(pentitol-1-yl)pyrroles (e.g. 116) have been synthesised *via* the cyclocondensations of 2-(alkylimino)-2-deoxyheptoses and acetylacetone (J. A. Galbis Perez *et al., ibid,* 1984, 132, 153).

Enantiomeric and racemic 2,3,4,5-tetrahydroxypentyl derivatives of adenine, cytosine, and uracil have been prepared, as potential bacteriosides, by condensing a reactive alditol derivative (e.g. 1-*O*-tosyl-2,3:4,5-di-*O*-isopropylidene-DL-ribitol) with the sodium salt of the heterocyclic base, followed by deisopropylidenation (A. Holy, Collect. Czech. Chem.

Commun., 1982, 42, 2786). Alternatively, a reactive aldofuranose derivative can be condensed with the heterocyclic base and subsequent reduction is then employed to afford the alditol derivative.

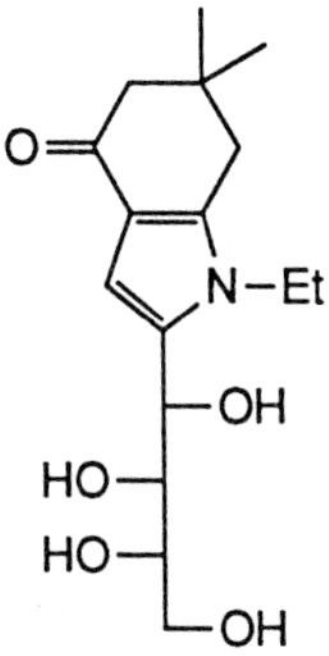

(115) 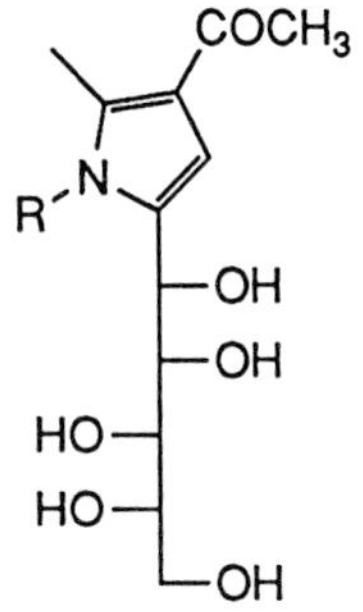 (116)

Polyhydroxylated pyrrolidines (e.g. 117) are receiving a growing amount of interest owing to their properties as potent glycosidase inhibitors, antiviral agents, and as chiral auxilaries in asymmetric synthesis. A quick, stereospecific entry into this class of compounds can be achieved using D-mannitol as the starting material. The synthesis involves five sequential steps, and in the third of these, the pyrrolidine framework is formed in one operation (T. K. M. Shing, Tetrahedron, 1988, 44, 7261).

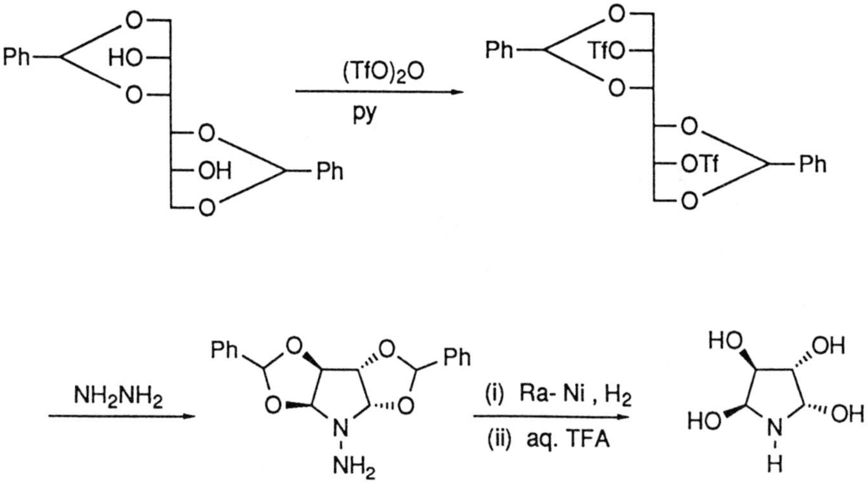

$(TfO)_2O$, py

NH_2NH_2

(i) Ra- Ni , H_2
(ii) aq. TFA

(117)

1,4-Dideoxy-1,4-iminopentitols also show potential as glycosidase inhibitors. The trisubstituted pyrrolidines 1,4-dideoxy-1,4-imino-D-lyxitol (118), 1,4-dideoxy-1,4-imino-D-arabinitol (119), and 1,4-dideoxy-1,4-imino-L-arabinitol (120) have been prepared from either D-mannose or D-lyxose. The compounds exhibit varying degrees of inhibition towards different glycosidases (G. W. J. Fleet *et al.*, Tetrahedron Lett., 1985, 26, 3127).

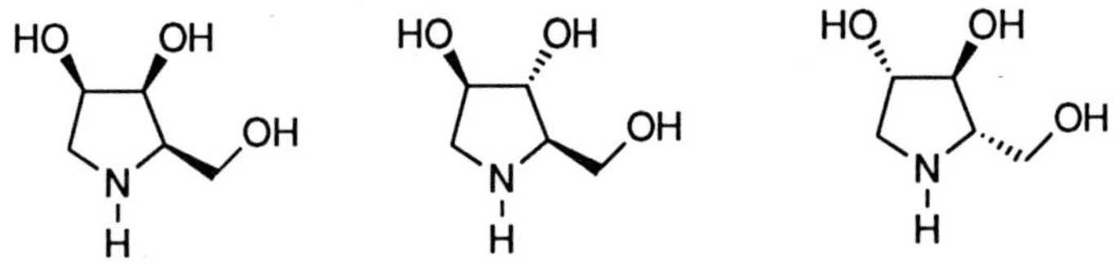

HO OH OH N H (118) HO OH OH N H (119) HO OH OH N H (120)

8. Deoxynitroalditols

A chemoselective reduction of isosorbide-2,5-dinitrate (121) can be achieved by variation of the reducing agent and the conditions used. Thus, reduction with ferrous sulphate affords isosorbide-2-nitrate, whilst zinc/acetic acid yields isosorbide-5-nitrate (O. DeLucchi *et al.*, Gazz. Chim. Ital., 1987, 117, 173).

Reaction of nitroheptenitol(122) with methyl acetoacetate yields the unstable Michael adduct (123) (R. Fernandez-Fernandez, J. Chem. Res. Synop., 1987, 222).

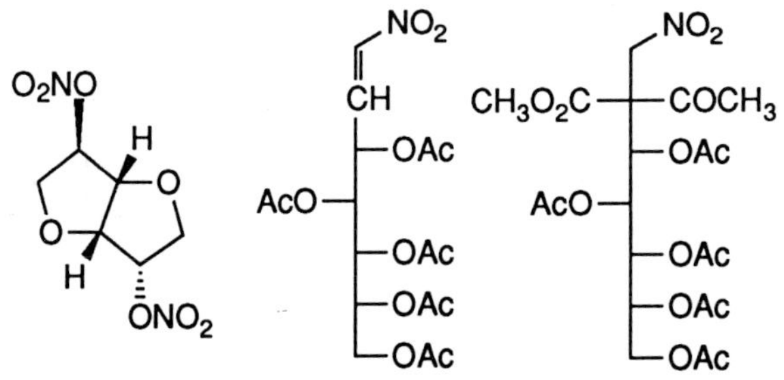

O_2NO H O O H ONO_2 (121) NO_2 CH OAc AcO OAc OAc OAc (122) NO_2 CH_3O_2C $COCH_3$ OAc AcO OAc OAc OAc (123)

9. Sulphur containing derivatives

Thioalditols have been used in an elegant synthesis of all eight L-hexoses (S. Y. Ko *et al.*, Tetrahedron, 1990, 46, 245). A key step in this reiterative two-carbon extension cycle originates from a 6-*S*-phenyl-6-thio-6-deoxyalditol (124). Oxidation of these thioethers to the corresponding sulphoxides and subsequent Pummerer rearrangement yields the *gem*-acetoxysulphides (125). For hexoses with *erythro* stereochemistry at C-4 and C-5, the Pummerer products (125) are treated with diisobutylaluminium hydride (DIBAL-H), giving the *erythro*-aldoses (126). For hexoses with *threo* stereochemistry at C-4 and C-5, potassium carbonate/methanol hydrolyses the geminal acetoxy-sulphide group and simultaneously epimerises the C-2 centre to afford the appropriate *threo*-aldoses (127).

SPh
O
O
R
(124)
MCPBA
O
SPh
+
O
O
R
AcO
SPh
Ac_2O
NaOAc
O
O
R
(125)
DIBAL
CHO
O
O
R
(126)
K_2CO_3,
MeOH
R = Protected
1,2,3 - Triol unit
CHO
O
O
R
(127)

A related development has been reported wherein reduction of an allylic β-ketosulphoxide to an allylic β-hydroxysulphoxide can be made stereoselective to give either diastereomer, according to the nature of the reducing agent (G. Solladie, J. Hutt, and C. Frechou, Tetrahedron Lett., 1987, 28, 61). Asymmetric hydroxylation of the allylic alcohols and subsequent Pummerer rearrangement leads to the formation of three adjacent, chiral hydroxylic centres. The value of this approach was proven in the synthesis of L-penta-*O*-acetylarabinitol (128).

BnO CO$_2$Et; H_3C–S(O)–Ar, LDA / THF → BnO…C(O)CH$_2$S(O)Ar

Ar = p- Tolyl

DIBAL / ZnCl$_2$ → BnO…CH(OH)CH$_2$S(O)Ar; OsO$_4$ → BnO…(OH)(OH)(OH)CH$_2$S(O)Ar

Ac$_2$O / NaOAc → BnO…(OAc)(OAc)(OAc)CH(OAc)SAr; (i) DIBAL (ii) Ac$_2$O → BnO…(OAc)(OAc)(OAc)CH$_2$OAc

(i) Debenzylation (ii) Acetylation → AcO…(OAc)(OAc)(OAc)CH$_2$OAc

(128)

Methyl 6-*S*-phenyl-6-thio-α-D-glucopyranoside is converted into 6-*S*-phenyl-6-thio-D-glucitol pentaacetate by sequential hydrolysis, borohydride reduction, and acetylation. Oxidation with *m*-chloroperbenzoic acid gives the corresponding *S*-epimeric sulphoxides, which undergo a Pummerer rearrangement to yield 1-epimeric L-gulose-*S*-phenyl-monothiohemiacetal hexaacetates (F. Santoyo-Gonzalez and H.H. Beer,

Carbohydr. Res., 1990, 202, 33).

Treatment of 2,3,4,6-tetra-*O*-benzyl-1-mercapto-1-deoxy-D-glucitol (129) with 20% hydrogen peroxide gives a disulphide, which on acetolysis, followed by deacetylation affords the disulphide (130, n=2). Reaction of 2,3,4,6-tetra-*O*-benzyl-1-*O*-tosyl-D-glucitol with (129) gives a monosulphide, which on acetolysis, followed by deacetylation yields the monosulphide (130, n=1) (A. U. Rahman *et al.*, J. Pure Appl. Sci., 1986, 5, 25).

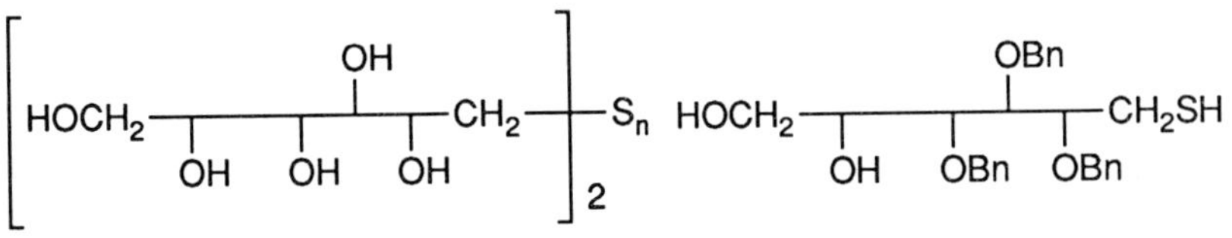

$$\left[HOCH_2-CH(OH)-CH(OH)-CH(OH)-CH(OH)-CH_2-\right]_2 S_n \qquad HOCH_2-CH(OH)-CH(OBn)-CH(OBn)-CH(OBn)-CH_2SH$$

(130) (129)

An interesting discovery was made when reacting the 2,5-bis-dithiocarbonate derivative 1,3:4,6-di-*O*-benzylidene-D-mannitol (131) with tributyltin hydride in toluene, at reflux, containing a catalytic amount of α,α-azoisobutyronitrile. This did not give the expected 2,5-deoxy- product (132). Instead, the reaction furnished (2S, 3S, 4S, 5S)-di-*O*-benzylidene-3,4-dihydroxy-2,5-bishydroxymethylthiolane (133) as the sole product in 80% yield. This radical deoxygenation procedure presents a novel approach to the synthesis of chiral polyhydroxylated thiolanes, pyrrolidines, or tetrahydrofurans, important intermediates in natural product syntheses (A. V. Rama Rao *et al.*,J. Chem. Soc. Chem. Commun., 1988, 1273).

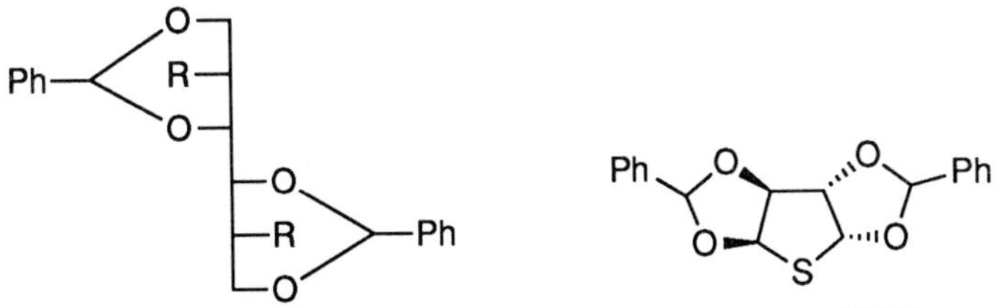

(131) R = OC(S)SMe
(132) R = H

(133)

Second Supplements to the 2nd Edition of Rodd's Chemistry of Carbon Compounds, Vol. IE/F/G, edited by M. Sainsbury

***CHAPTER* 23a**

MONOSACCHARIDES : Synthesis, Chemistry, Structure and Physical properties

Deepthi K. Weerasinghe

1. Introduction

Interest in the chemistry of monosaccharides has grown considerably in the last few years due to an increase in their use as chiral starting materials in synthesis, see chapters 23b and 23c of this supplement, and to the continuing discovery of biologically active compounds that contain monosaccharide units as integral parts of their structures. This has resulted in the selection of certain key compounds as targets for synthesis by internationally renowned research groups. Among a number of review articles, the Specialist Periodical Reports on carbohydrates, published by the Royal Society of Chemistry are recommended for annual overviews of the subject.

The thermal decomposition of monosaccharides, sometimes known as caramelisation, is discussed by Tomasik and his colleagues (*Adv. Carbohydr. Chem. Biochem.,* 1989, *47,* 203) and the use of high pressure liquid chromatography (HPLC) for the separation of monosaccharides is reviewed by Hicks (*ibid.,* 1988, *46,* 17). In this last article different stationary phases, detection systems, and separation parameters are critically compared with the help of numerous examples.

2. Synthesis of Monosaccharides

A recent review by Mukaiyama (*Am. Chem. Soc. Symp. Ser.,* 1989, *386,* 278) provides a discussion of the strategies employed by his group *en route* to important monosaccharides. Over fifty papers are cited in this work, which also deals with mechanistic aspects of the more important reactions.
In this chapter it is not possible to cover all of the achievements of the last decade in the area, so a selection of monosaccharide syntheses will be summarized which embody some important techniques of general applicability.

a). Diels-Alder types

As in the synthesis of many other natural product groups, the Diels-Alder reaction finds useful expression in monosaccharide constructions. It has been used, for example, in the synthesis of racemic *L*-rhodinose (**5**). The key reaction in this work involved the cycloaddition of the nitrone (**1**) and vinyltrimethylsilane (**2**). The resultant isoxazolidine (**3**) on treatment with fluoride ion fragmented to afford the *α,β,*-unsaturated aldehyde (**4**). It was suggested that fluoride ion attack on the silicon atom of the isoxazolidine (**3**) resulted in the formation of a transient aminoaldehyde (**6**). Reduction of (**4**), followed by acid catalyzed deprotection and ring-closure gave the methylglycoside, *L*-rhodinose (P. DeShong and J. M. Leginus, *Tet. Lett.,* 1984, *25,* 5355).

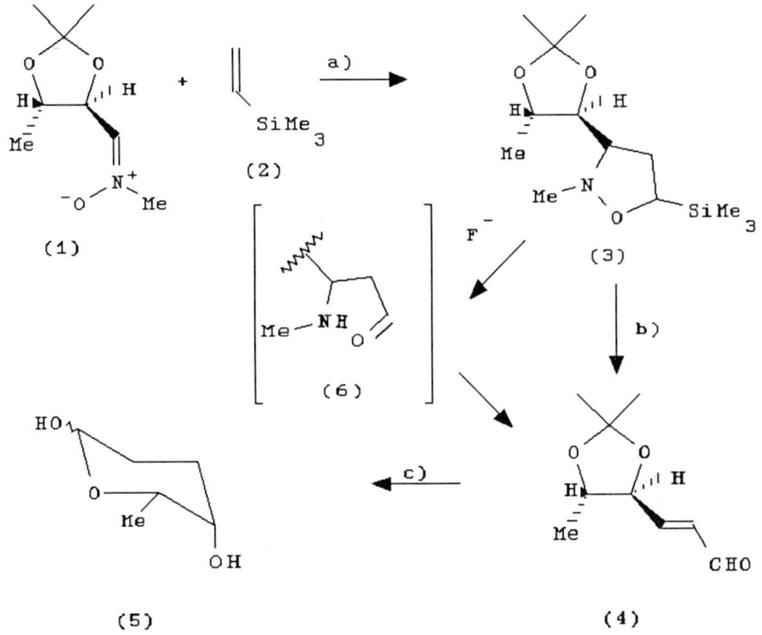

a)80°C/24h. b)HF(50% aq.)/CH_3CN/30min. c) i)[H_2]/(5%Pd/C)/EtOH/2h.
ii)H^+/1h/MeOH. iii)H^+/Acetone

A Diels-Alder reaction between the substituted diene (**7**) and crotonaldehyde was the first step in a synthesis of ($\pm$) β-methyllincosamide (**16**) (E. Larson and S. Danishefsky, *J. Am. Chem. Soc.,* 1983, *105,* 6715). In this case, the cycloaddition reaction was promoted by the Lewis acid boron trifluoride, and the adduct (**8**) was then selectively reduced

(7) R=Bzl (8) (9)

(12) R' = $(MeO)_2P=O$ (11) (10)

(13) (14) R"=Ac (15) R" =Ac

(16) R"=H

a)i)$NaBH_4/CeCl_3$/EtOH -78°. ii) mCPBA/MeOH. iii)PhCOCl/NaH. b)NBS/AcOHaq. /CH_2Cl_2. c)i)$Bu_4NN_3/TMSN_3$. ii)TFA. iii)MsCl. iv) $(MeO)_3P$. d)i)K_2CO_3-MeOH. ii)Im_2CO. iii)MeOH. e)i)AcOH/85°. ii)K_2CO_3-MeOH. iii)Ac_2O/Py

under Luche-type conditions. Further steps involved benzoylation and then oxidation with *m*-chloroperbenzoic acid in the presence of methanol. A second benzoylation afforded the methyl-*β*-galactoside (**9**). From the galactoside a series of reactions were then undertaken which eventually led to (±)-*β*-methyllincosamide (**16**). The racemic natural product was characterized as its peracetate (**15**).

Many variations of this type of chemistry have been described by Danishefsky (*ibid.,* 1985, *107,* 1269, 1274, 7762; *Angew. Chem. Int. Ed. Eng.,* 1987, *26,* 15).

The carbonyl group of crotonaldehyde is the "ene" component in the above work, in addition a carbonyl group may also form part of the "diene" when present as part of a vinyl ether derivative. This is utilized in a synthesis of the triose (**21**) (R. R. Schimdt and M. Maier, *Tet. Lett.,* 1985, *26,* 2065). Here the

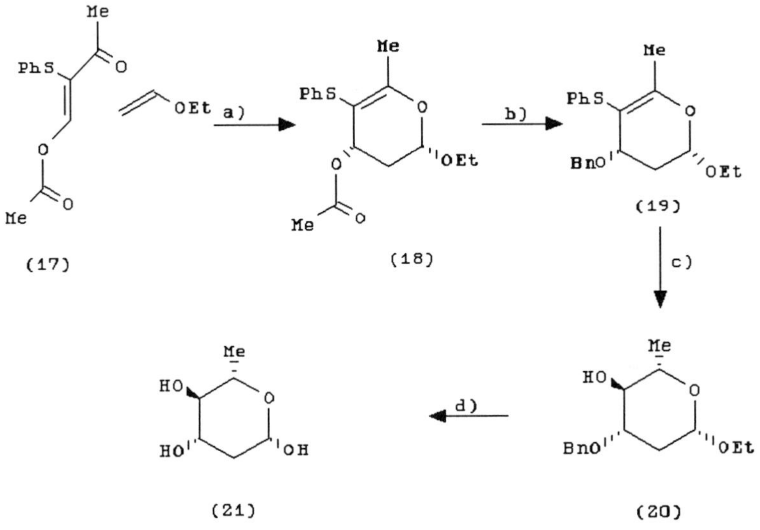

a)70°/72h. b)i)MeOH/K_2CO_3. ii)$PhCH_2Cl$/NaH. c)i)RaNi/THF. ii) $BH_3.Me_2S/H_2O_2$. d)i)Pd/[H]. ii)HCOOH

heterodiene (**17**) and ethylvinyl ether were reacted together to form the pyran (**18**). This cycloaddition reaction proceeded in a stereoselective manner to afford the endo (*cis*) isomer shown, and this was O-deacetylated and then O-benzylated to give the pyran (**19**). The pyran was desulphurised by treatment with Raney nickel and the product was then reacted with borane-dimethylsulphide and oxidized to give the alcohol (**20**). Here the stereochemistry of the hydroxyl group was presumably dictated by the α-orientated benzyloxy group at C-4. Deprotection then completed the preparation of the triose. The use of enones in ($4+2\pi$) cycloaddition reactions has proved to be of value in other monosaccharide syntheses (R. R. Schimdt, *Acc. Chem. Res.,* 1986, *19,* 250; S. Danishefsky and R. R. Webb, *J. Org. Chem.,* 1984, *49,* 1955; K. Sun *et al., ibid.,* 1985, *50,* 4775; P. Herczegh *et al., Heterocyles,* 1989, *28,* 887).

b). Oxirane ring openings

A Japanese group (T. Itoh *et al., Chem. Lett.,* 1985, *(11),* 1679) has prepared optically active *L*-rhodinose (**26**) from (*S*)-glycidyl sulfide (**22**). The oxirane ring of the starting material

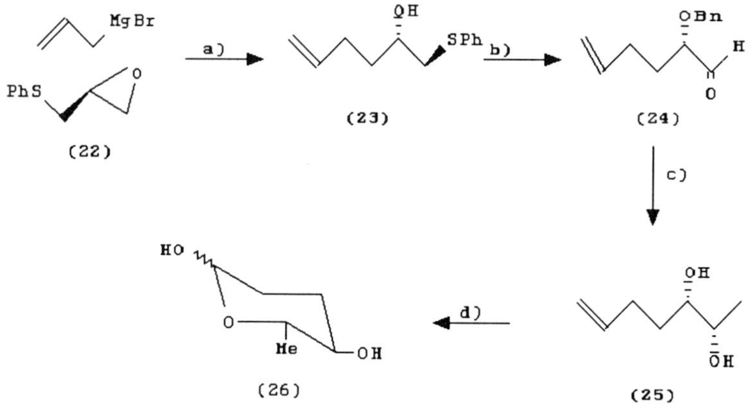

a)CuI/THF/-30°/2h. b)i)BzBr/NaH. ii)$NaIO_4$. iii)DIBAL-H. c)i)MeMgBr/$ZnBr_2$ /THF/0°. ii)Li/NH_3. d)i)O_3. ii)Me_2S

was opened with allylmagnesium bromide in the presence of cuprous iodide. This gave the hydroxysulphide (**23**), which was protected by O-benzylation prior to oxidation, and a Pummerer reaction. Reduction of the product with diisobutylaluminum hydride gave the aldehyde (**24**), treatment of which with methyl magnesium bromide in the presence of zinc bromide, followed by debenzylation yielded the *syn*-diol(**25**). Ozonolysis of the diol and reductive work up with dimethyl sulphide gave *L*-rhodinose (**26**).

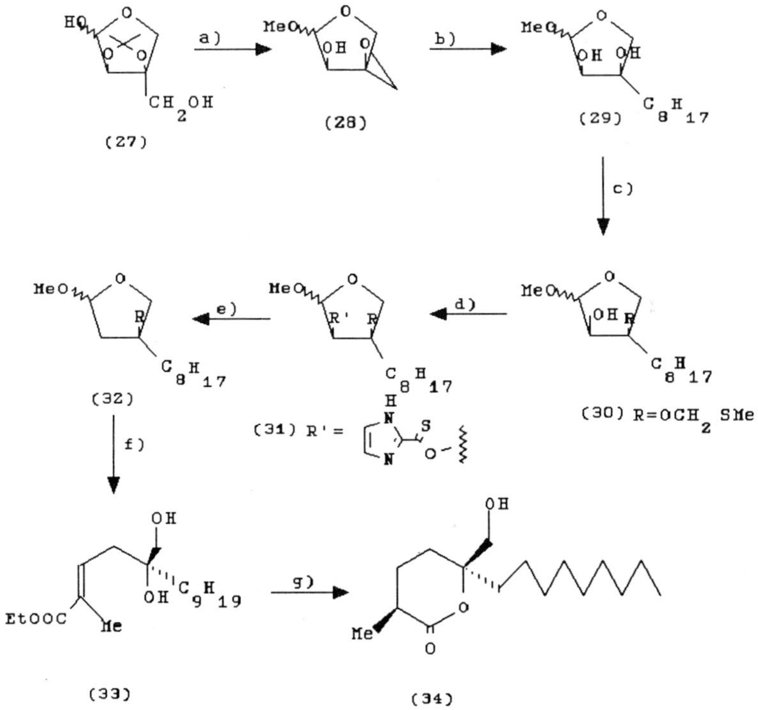

a)i)TsCl. ii)TsOH/MeOH/Δ/2.5d. iii)MeOH/KOH/0°/2h. b) $C_7H_{15}MgBr$/CuI/THF. c)i)Ac_2O/Py. ii)DMSO/Ac_2O/rt/3d. d)thiocarbonyldiimidazole/THF/Δ/1h. e)tBuSnH/AIBN/PhH/Δ/0°/5h. f)i)HCl/acetone/70°/2.5h. ii)$Ph_3P=C(Me)COOEt$ /CH_3CN/Ar/50°/2.5h. g)Pd/C/[H]. ii)MeOH/KOH/H_2O. iii)Ph_3P/Py_2S_2/PhH/rt/3h

The marine antibiotic (-)-malyngolide (**34**) has been synthesized by Ho and Wong (*Can. J. Chem.,* 1985, *63,* 2221) using a variation of earlier work (see P. Ho, *Tet. Lett.,* 1978, 1623). Here 2,3-O-isopropylidene-*D*-apiose (**27**) was initially tosylated then refluxed in the presence of p-toluenesulphonic acid in methanol for 2.5 days, cooled and treated with methanolic potassium hydroxide to afford the epoxide (**28**). A Grignard reaction with octylmagnesium bromide mediated by the presence of cuprous iodide opened the epoxide ring to yield the C-alkylated sugar (**29**). The tertiary hydroxyl group of the alkylated sugar was selectively acetylated before treatment with dimethylsulphoxide to form the methylthiomethyl ether (**31**). The secondary hydroxyl group was next thiocarbonylated, before deoxygenation using tetrabutyltin hydride in the presence of a radical initiator. The resulting deoxy sugar (**32**) was hydrolysed to its furanose derivative, which when reacted under Wittig conditions afforded the unsaturated ester (**33**). Hydrogenation followed by cyclisation, yielded the lactone (**34**), after separation of the epimers.

c). Wittig reactions

Wittig reactions are common place in monosaccharide syntheses, as are Emmons Horner procedures, an example has already been shown in the work of Ho.

In a synthesis of (-)-canadenosolide (**43**) Anderson and Fraser-Reid (*J. Org. Chem.,* 1985, *50,* 4786) reacted the protected sugar aldehyde (**35**) and the ylide from propyltriphenylphosphonium bromide to afford an alkylidene derivative which, upon reduction with Raney nickel, afforded the alkylated sugar (**36**). Acid catalyzed methanolysis, followed by oxidation with pyridinium chlorochromate afforded the ketone (**37**). This was reacted with (ethoxycarbonylmethylene)triphenylphosphorane, and the resulting unsaturated ester was hydrogenated over Raney nickel to cleave the benzyl ether and to reduce the double bond, thereby affording the C-4 hydroxy sugar (**38**). Acid hydrolysis of (**38**) readily formed the hemiacetal (**39**). Reaction of the hemiacetal (**39**) with vinyl ether in the presence of pyridinium-*p*-toluenesulphonate (PypTs), followed by ethoxycarbonylation with lithiumdiisopropylamide and ethyl - chloroformate afforded the γ-lactone (**41**). Acid catalyzed removal of the α-ethoxyethyl group, followed by oxidation with Jones' reagent afforded the bis γ-lactone (**42**), which was converted to the target compound (**43**) by a previously published procedure (M. Kato *et al., ibid.,* 1975, *40,* 1932). Fraser-Reid had used similar strategies to synthesise other bis γ-lactones (*ibid.,* 1985, *50,* 4781).

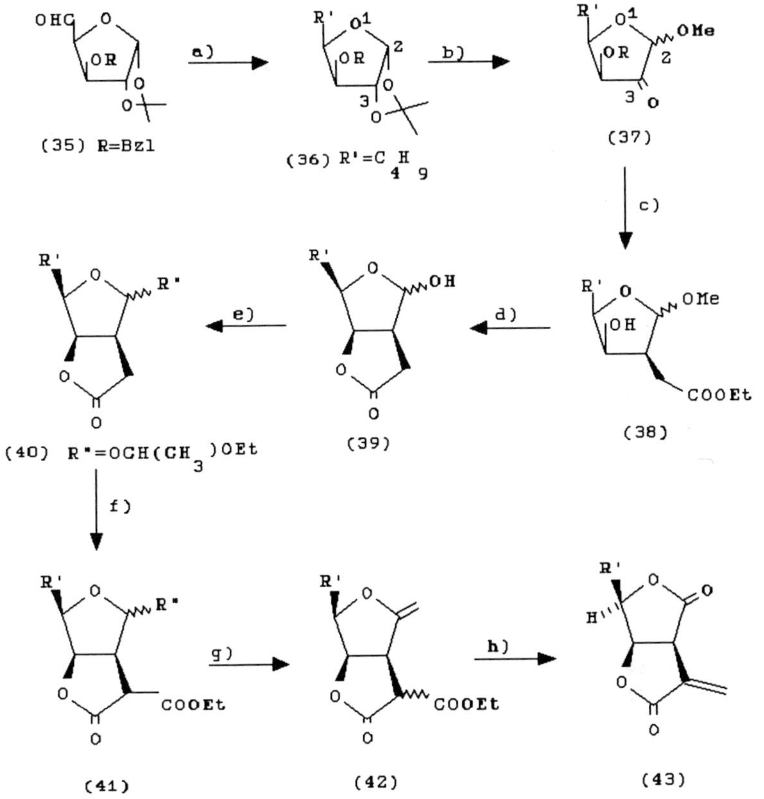

a)i)$Ph_3P=CHC_2H_5$. ii)RaNi/EtOH. iii)[H]/Pd. b)i)MeOH/HCl. ii)PCC. c)i)$Ph_3P=CHCOOEt$/THF. ii)RaNi/EtOH. iii)[H]. d)H_2SO_4/dioxane/H_2O. e)$CH_2=CHOEt$/PypTs/rt/2h. f)LDA/ClCOOEt. g) i)PypTs/EtOH. ii)Jones reagent. h)i)H^+/60°/8h. ii)HAc/Et_3N. iii)formalin/NaAc

d). Organometallic reagents

Following some earlier work on epoxybenzooxocin sugar

analogues, Hauser and his group (*J. Org. Chem.,* 1984, *49,* 2296) have recorded the enantioselective syntheses of the

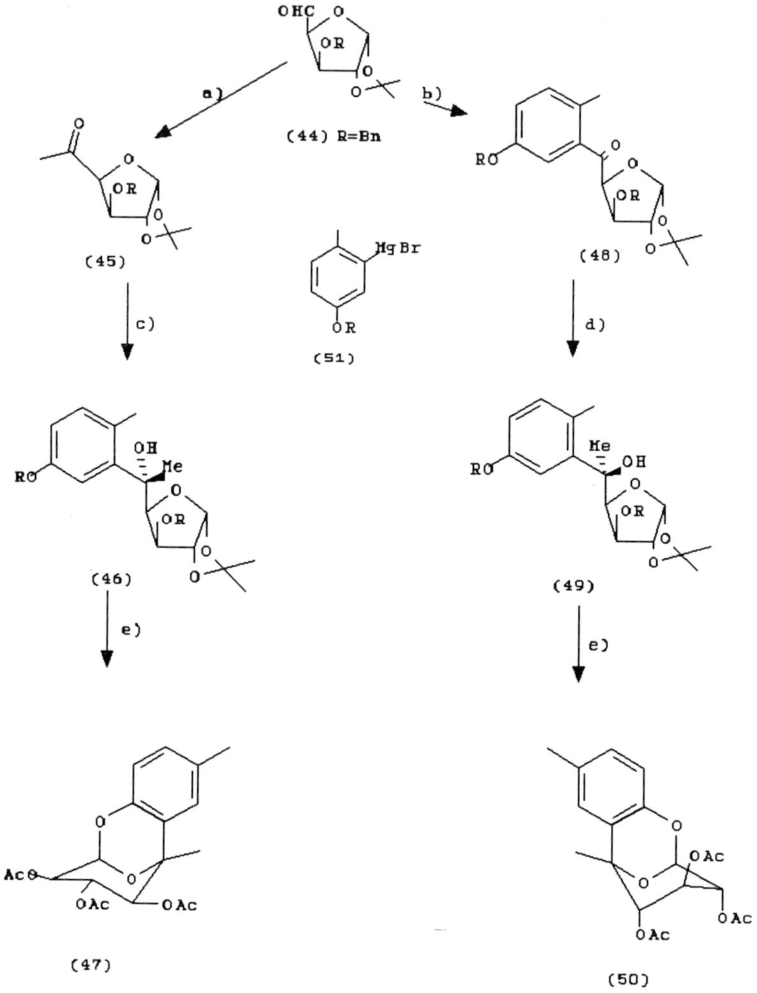

(44) R=Bn

(45)

(51)

(48)

(46)

(49)

(47)

(50)

a)i)MeMgBr. ii)CrO_3.2Py/CH_2Cl_2. b)i)**51**. ii)CrO_3.2Py/CH_2Cl_2. c)b(i). d)MeLi/THF /78°. e)i)Pd/C/EtOH. ii)EtOH/aq/Amberlite IR-120. iii)Ac_2O/Py/rt.

derivatives (**47**) and (**50**), from the common chiral precursor (**44**). This pentose was reacted with 4-benzyloxy-2-methyl phenylmagnesium bromide (**51**) and the secondary alcohol thus formed was then oxidized to the ketone (**48**). A second alkylation, now with methyl lithium gave the tertiary alcohol (**49**), which had the *L-ido* configuration. The benzyl ether protecting groups were cleaved by hydrogenolysis. Hydrolysis in aqueous ethanol catalyzed by amberlite and subsequent O-acetylation then afforded the *L-ido*-2,6-epoxy-2*H*-1-benzoxocin (**50**).

In a parallel series of reactions the pentose (**44**) was reacted

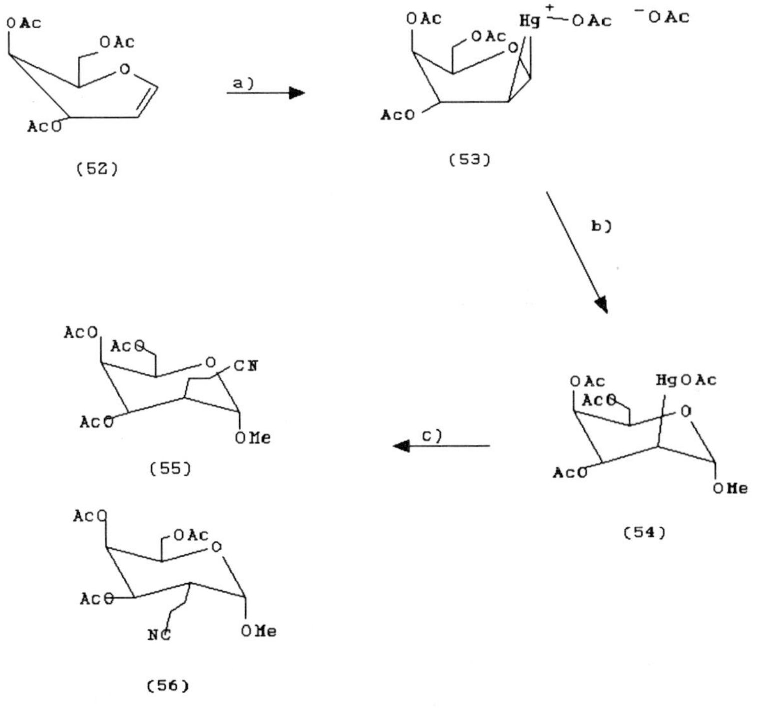

(52) (53) (54) (55) (56)

a)$Hg(OAc)_2$/MeOH. b)MeOH. c)i)Bu_3SnH. ii)$CH_2=CHCN$

with methylmagnesium bromide first and the product oxidised with chromium trioxide. This gave the methyl ketone (**45**) which, when reacted further with the aryl Grignard reagent (**51**), yielded the tertiary alcohol (**46**). This product afforded the *D-gluco*benzoxocin (**47**) when it was reacted in a similar manner to its enantiomer (**49**).

Branched sugars of the type (**55**) and (**56**) have been synthesised through the alkylation of unsaturated monosaccharides. Thus reaction of the acetylated galactal (**52**) with mercuric-acetate afford the adduct (**53**) (B. Giese and K. Groninger, *Tet. Lett.,* 1984, *25,* 2743). As expected the stereochemistry of the addition occurred *trans* to the C-4 acetoxy group. If the reaction was carried out in the presence of methanol, the acetoxy function at C-2 was displaced selectively to afford the organomercuric salt (**54**). (For earlier work on methoxy-mercuration see K. Takiura and S. Honda *Carbohydr. Res.,* 1972, *21,* 379). The mercury bearing substituent at C-3 was removed by reduction with tributyltin hydride to give the 3-deoxysugar, but if the reduction was carried out in the presence of excess of alkene branched deoxysugars of the type (**55**) and (**56**) were formed. The reaction is thought to involve radicals, and the stereoselectivity depends on the alkene substituents. With acrylonitrile a 2:1 mixture of (**56**) to (**55**) was formed.

The reactivity of allyltin derivatives has been exploited in the synthesis of the disaccharide (**60**)(S. Jarosz and B. Fraser-Reid, *J. Org. Chem.,* 1989, *54,* 4011). The allyltin derivative (**58**) was prepared by initially reacting the iodide (**57**) with sodium tosylsulphone and then with tributyltin hydride to yield the allyltin derivative (**58**) (J. E. Baldwin *et al., J. Chem. Soc., Chem. Commun.,* 1986, 1339., Y. Ueno *et al., J. Organomet. Chem.,* 1980, *197,* c1). This compound when reacted with the aldehyde (**59**) yielded two diastereomeric disaccharides. If titanium tetrachloride was added as catalyst a mixture resulted, in which the hydroxyl group at C-6 was either *in* (66%) see (**60**) or *out* (33%) of the plane of the ring. However, upon using a milder Lewis acid zinc chloride as the catalyst, yields were higher, and the isomer ratio was 10:1 in favor of the *in* plane derivative (also see S. Jarosz, *Carbohydr. Res.,* 1987, *166,* 211).

(57) R=Bn

(58)

(59)

(60)

a)i)$NaSO_2Tol/MeOH/\Delta$. ii)$Bu_3SnH/AIBN/PhH$. b)$TiCl_4/CH_2Cl_2/-78°/5min$ or $ZnCl_2/CH_2Cl_2/0°/20min$.

Monosaccharide derivatives such as (**62**) are important as precursors for the synthesis of glycosides and oligosaccharides. Tsui and Gorin (*Carbohydr. Res.,* 1985, *144,* 137) prepared compound (**62**) in high yield from readily available 1,2-*trans* halide (**61**) by reacting it with silver *p*-toluenesulphonate in the presence of methanol and 2,4,6-trimethylpyridine in acetonitrile as a "proton sponge".

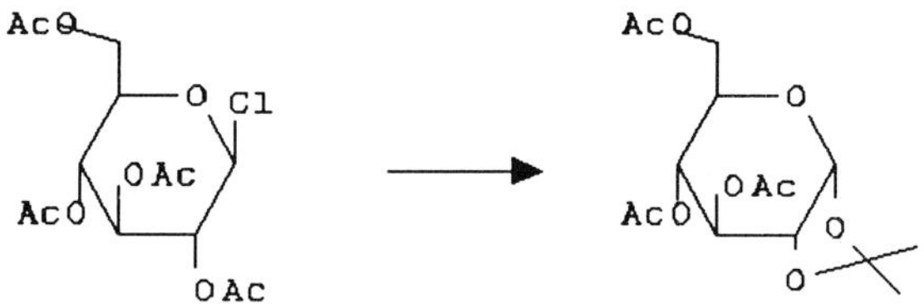

AcO AcO OAc Cl OAc O (61) → AcO AcO OAc O O O (62)

p-TsOAg/CH_3CN/MeOH/Me_3Py/rt/4h

Tetrakis(triphenylphosphine)palladium (0) [$Pd(PPh_3)_4$] effects the stereoselective alkylation of 2-acetoxy-5,6,dihydro-2-pyrans. When stabilised carbanions, such as diethyl sodium acetamidomalonate, were used, the reaction resulted in the formation of alkylated dihydropyrans, with a net retention of stereochemistry at the oxygen bearing carbon atom. On the otherhand, when non-stabilised carbanions were employed *C*-glycosyl compounds were formed with net inversion of the stereochemistry at the oxygen bearing carbon (L. V. Dunkerton *et al., Carbohydr. Res.,* 1987, *171,* 89). [also see M.Brakta *et al., J. Carbohydr. Chem.,* 1987, *6,* 307). Such products are useful starting materials for monosaccharides.

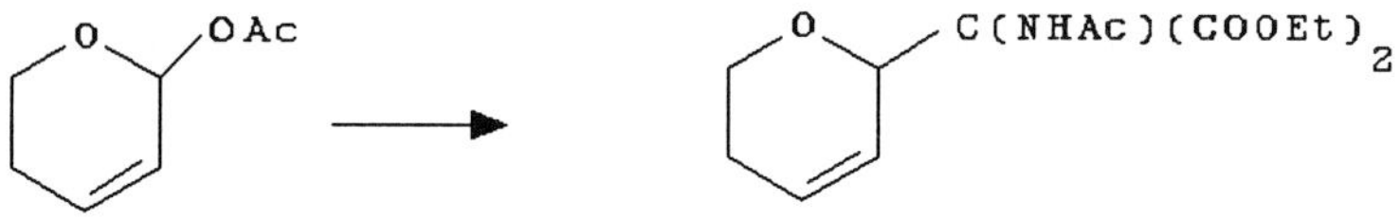

O OAc → O C(NHAc)(COOEt)$_2$

NaC(NHAc)(CO_2Et)$_2$/Ph_3P/$Pd(PPh_3)_4$/DMF/60-70°/12h

Fronza and his colleagues (*Tet. Lett.,* 1982, *23,* 4143) prepared the dideoxy sugar *L*-mycarose (**66**) by the reaction of allylmagnesium bromide upon the methyl ketone (**63**). This gave the adduct (**64**) which was hydrolysed to the triol (**65**). The triol upon ozonolysis in methanol solution containing dimethyl sulphide afforded the cyclic dideoxytriol sugar *L-mycarose*

(**66**). When this reaction was carried out using diallyl zinc reagent, prepared from allylmagnesium bromide, the same clean *erythro* mode of addition was observed. If a formyl group replaced the acetyl function in the starting material the same exclusive erythro preference was observed in a reaction with dialkylzinc in diethylether. But when the aldehyde was reacted with allylmagnesiumbromide a mixture of *erythro* and *threo* adducts was obtained.

A diastereoselective condensation mediated by rhodium has been used in a synthesis of tetroses and pyranoses (M. L Deem and E. A Roman, *Tet. Lett.,* 1988, *29,* 4649). Thus 1,2-di-ethoxyethane was partly converted into ethoxyacetaldehyde in the presence of rhodium. Excess starting material and the aldehyde then underwent a condensation to form a mixture of

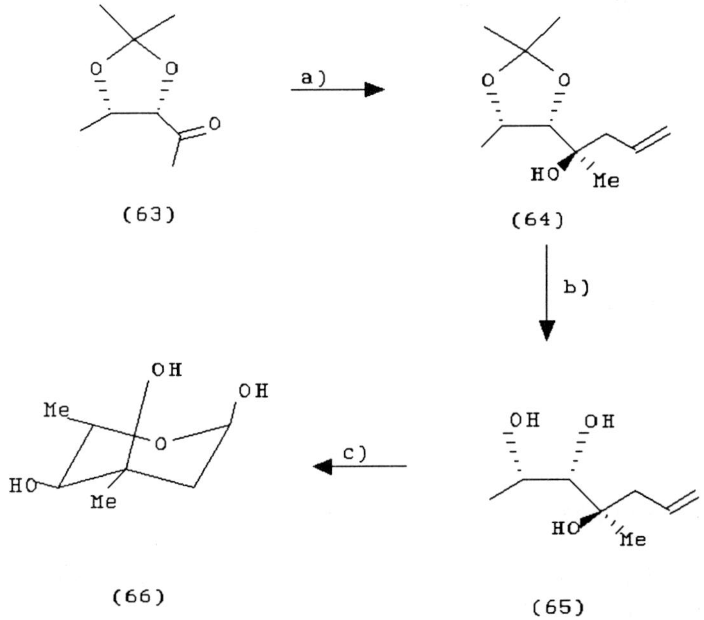

(63) (64) (65) (66)

a)$CH_2=CHCH_2MgBr/Et_2O/-78°$. b)HAc. c)i)$O_3$/MeOH. ii)$Me_2S$

the corresponding tetrose and pyranose derivatives (for other rhodium mediated aldol condensations see S. Sato *et al., ibid.,* 1986, *27,* 5517; M. T. Reetz *et al., ibid.,* 1987, *28,* 793).

e). Miscellaneous reactions

Butadiene was reacted with nitromethane in the presence of trimethylsilylchloride and trimethylamine to give the vinylisoxaline (**67**) (K. B. G Torrel *et al., Tetrahedron,* 1985, *41,* 5569). The isoxazoline on osmylation, followed by oxidative work up yielded the *erythro* diol (**68**) as the major product. The isomer was purified by preparative thin layer chromatography (TLC), before being catalytically hydrogenated in aqueous methanol over palladium to afford dideoxyribose (**69**) in good yield. The authors illustrated the versatility of the synthesis by applying the same sequence of reactions to obtain other deoxysugars.

a) b) c)

(67) (68) (69)

a)$CH_3NO_2/Me_3SiCl/Et_3N$. b)OsO_4/H_2O_2. c)aq.MeOH/Pd/$BaSO_4$/[H]

Chelated hydrazone derivatives are resistant to acetylation, but by the use of a mixture of acetic anhydride and zinc chloride, or trifluoroacetic acid, *D*-arabino-*L-xylo*-(tetra-O-acetyl)hexosulose-1,2-bis(acetylphenyl hydrazone) and other related hydrazone sugars have been prepared (L. Somagyi, *Carbohydr. Res.,* 1985, *145,* 156).

Tri-O-acetyl-1,2-O-(1-cyanoethylidene)-*α*-*D*-galacto-pyranose (**70**) has been stereoselectively converted into the corresponding per-O-acetylated 1,2-*trans*-aldohexopyranosyl cyanide (**71**) using Lewis acid boron trifluoride ethereate (R. W. Meyers and Y. C. Lee, *ibid.,* 1986, *154,* 145). This method is a significant improvement on an earlier preparation that used mercuric

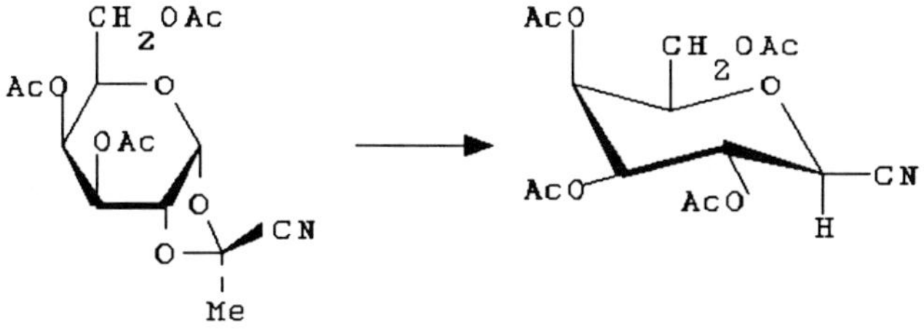

(70) (71)

$BF_3.Et_2O/MeNO_2$/rt/1.5h

cyanide in nitromethane to displace a bromide ion from the corresponding peracetylpyranosyl bromides (R. W Meyers and Y. C Lee, *ibid.,* 1984, *132,* 61).

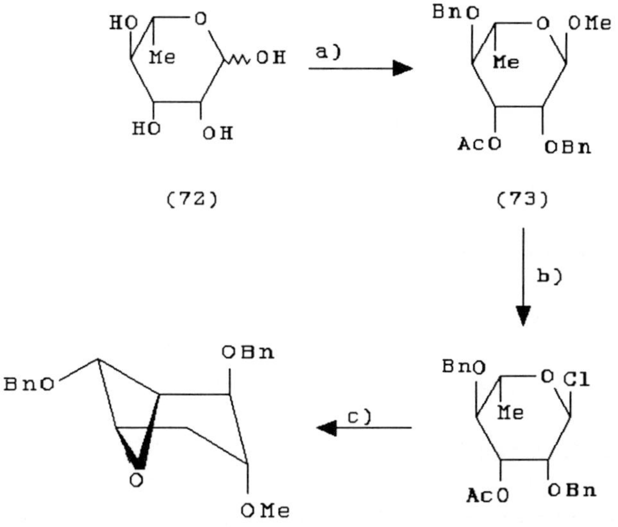

(72) (73)

(75) (74)

a)i)MeOH/H^+. ii)$EtO_3CH/(Me_3)_2CO$. iii)$PhCH_2Br$/NaH. iv)$H^+/PhCH_2Br/Bu_4NBr$. v)Ac_2O/Py. b)$Et_2O/HCl^g/0^o$-->rt/6h. c)^tBuOK/THF/rt.

In the synthesis of the bridged dibenzylated *β-L*-rhamnopyranose (**75**), *L*-rhamnose (**72**) was treated with anhydrous methanol and Amberlite IR-120 resin to form the methyl *α*-and *β-L*-rhamnopyranosides (E. Wu *et al., ibid.,* 1987, *161,* 235). These were next manipulated using published techniques to arrive at the product (**73**). Compound (**73**) was dissolved in ether, saturated with dry hydrochloric acid at $0^{o}C$ and left to warm up to room temperature. This procedure effected *O*-demethylation and the formation of the chloride (**74**). This compound in dry dioxane, when treated with potassium tertiary-butoxide gave the target compound (**75**). (also see R. W Hoffmann *et al., ibid.,* 1983, *123,* 320).

Dimethylfuranosides are available through lithium dicuprate reactions with sugar epoxides (H. Yamamato *et al., ibid.,* 1984, *132,* 287; for general information on the addition of organocopper reagents to epoxides see Y. Nakahara *et al., J. Carbohydr. Chem.,* 1984, *3,* 487).

A technique for deoxygenation of the primary hydroxy group of certain monosaccharides consists of initial *O*-sulphonylation and then nucleophilic displacement of the appropriate sulphinic acid from the product by treatment with metal hydrides (e.g. sodium borohydride, lithium aluminium hydride or lithium triethylborohydride) (R. W. Binkley, *J. Org. Chem.,* 1985, *50,* 5646).

A typical Mitsunobu reaction to reverse the stereochemistry of hydroxyl group is illustrated by the reaction of methyl 2,3-di-*O*-acetyl-*α-D*-galactopyranose with triphenylphosphine and diethyl azodicarboxylate (DEAD) to give the corresponding 6-*O*-benzoate with inversion of stereochemistry (G. Alfredsson *et al., Acta. Chem. Scand.,* 1973, *27,* 724).

3. Amino Monosaccharides

Substantial progress has been made in our knowledge of the chemistry of nitrogen containing sugars. This interest is stimulated because of their occurrence as essential constituents of highly effective biologically active compounds. Recently synthetic aspects of nitrogen containing deoxygenated sugars have been reviewed (H.H. Baer, Pure & Appl. Chem., 1989, **61**,

1217). This review covers the use of organometallic reagents for bond formation (palladium), chain elongation at the non-reducing terminus (iron carbonyl reagents), reduction of sugar sulfonic esters (boron hydrides), amination of sugars (many methods) and also some coverage of the modifications of disaccharides and nitro sugars.

Hauser and Ellenberger (*Chem. Rev.,* 1986, *86,* 35) have reviewed the synthesis of trideoxy amino and nitro hexoses. Their review covers nomenclature and syntheses of many key amino and nitro hexoses from various starting materials.

Of the amino sugars, *L*- daunosamine (**82**), the glycosidic component of several important anthracycline antibiotics, has been the target of many industrial and academic groups. There is a considerable amount of literature on the synthesis of amino sugars. Consequently only a brief cross section of the subject is discussed in this survey.

a). Oxirane ring opening

A racemic and also an asymmetric synthesis of daunosamine (**82**) was accomplished via the *trans*-diaxial opening of the epoxide ring of the key intermediate, 3,4-anhydrohexopyranoside (**80**) with sodium azide (G. Grethe *et al., J. Org. Chem.,* 1983, *48,* 5309). To construct this intermediate *L*-arabinose (**76**) was subjected to a known one-carbon homolagation process. Thus condensation of the natural sugar with nitromethane, followed by a series of established reactions gave (**80**). The azide (**81**) requires an inversion of the hydroxyl group at C-4 if it is to be stereochemically compatible with the target compound. This was effected by mesylation, followed by displacement of the mesylate by benzoate and subsequent hydrolysis of the benzoate to yield the α-hydroxyazide. Finally, the azide group was reduced over palladium and the product hydrolysed to give the hydrochloride of daunosamine (**82**).

In the asymmetric version of the synthesis, an elegant, albeit low yielding, strategy was employed (*idem, ibid.,* 1983, *48,* 5315). Methycyclopentadiene (**83**) was initially subjected to asymmetric hydroboration using (-)-di-3-pinanylborane. This a)i)CH_3NO_2/NaOMe. ii)$BF_3.Et_2O/Ac_2O$. iii)$NaHCO_3$/Tol/Δ. iv)Pd/C/[H]. b)$Ba(OH)_2$ /$H_2SO_4/BaCO_3$. c)i)MeOH/H^+. ii)TsCl/Py. iii)LAH/THF. d)i)TsCl/Py. ii)LAH/THF. e)NaN_3/ MeOEtOH. f)i)MsCl/Py. ii)PhCOONa/DMF. iii)NaOH. iv)Pd/C/[H]. v)HCl

afforded the alcohol (**84**). The β-oriented homoallylic hydroxy group guided the subsequent epoxidation reaction with *m*-chloroperbenzoic acid (mCPBA) synfacially, thus establishing the stereochemistry of the target compound. Jones oxidation of the hydroxy epoxide gave the cyclopentanone (**85**), which on treatment with mCPBA in the presence of sodium bicarbonate underwent Bayer-Villiger ring enlargement to an epoxylactone which was reduced and methylated to afford (**80**). This was converted into daunosamine (**82**), as described before.

(83) (84) (85) (80)

a)i)-Di-3-pinanylborane. ii)H_2O_2/NaOH. b)mCPBA/$NaHCO_3$. ii)CrO_3/Py. c)i)mCPBA/$NaHCO_3$. ii)DIBAL-H/Tol/-78°. iii)MeOH/BF_3.Et_2O.

(86) R' = NMe(COOEt) (87) (88)

a)$Ph_3P=CH_2$/DME. b)i)EtOH/$MeNH_2$/0°. ii)80°/30min. iii)$ClCO_2Et$/$NaHCO_3$/rt. iv)$CH_2=CHOEt$/Py/TsOH. c)$CHCl_3$/I(collidinium)ClO_4/4h/dark. d)i)MeOH/H_2O /KOH/(Pd/C)/[H]/1d. ii)EtOH/Py/TsOH,iii)15%KOH/55°/6h.

In a synthesis of methyl-*α-L*-garosamimide (**88**) Pauls and Fraser-Reid (*Can. J. Chem.,* 1984, *62,* 1532) utilized the opening of the oxirane (**86**) to introduce an amino function. The product was then converted through a series of steps into the fused iodomethyloxazolidinone (**87**). The hydroxyl group at C-5 and the amino substituent at C-4 were incorporated in the oxazolidinone ring, scission of which and reductive deiodination allowed access to the target compound.

b). Condensation type reaction

A bicyclic carba-analogue of the aminosaccharide daunosamine (**82**) was prepared by an ester-imine condensation (J.C. Gallucci *et al., Tetrahedron,* 1989, *45,* 1283). The dianion of ethyl *β*-hydroxybutyrate (**89**) reacted with the imine (**90**) to afford the *β*-lactam (**91**). Sequential ozonolysis and dimethylsulphide treatment afforded the hemiacetal (**92**) in good yield. The purified hemiacetal (**92**) when reacted with an equivalent of

(89) (91) R=C_6H_4-4-OMe (92) (94) (93)

a)i)LDA/THF. ii)PhCH=CHCH=NC_6H_4-4-OMe(**90**). b)i)O_3. ii)Me_2S. c)i)LDA/THF. ii)Ph_3P=CHOMe. d)i)MeOH/H^+. ii)$Ce(NH_4)_2(NO_3)_6/CH_3CN.H_2O$.

lithiumdiisopropylamide, followed by the addition of excess (methoxymethylidine)triphenylphosphorane gave the vinyl ether (**93**). The vinyl ether was cyclised by treatment with an acidic Dowex resin in methanol and the removal of the *p*-methoxyphenyl group was achieved by treatment with ceric ammonium nitrate to yield the pyrano-*β*-lactam (**94**).

c). Diels-Alder type constructions

The N-acetylenaminocarbaldehyde (**95**) served as the oxadiene component in a Diels-Alder type cycloaddition reaction with ethylvinyl ether. The adduct (**96**) embodies the skeleton of certain branched amino sugars (L. F. Tietze *et al., Tet. Lett.,* 1985, *26,* 5273).

MeOOC NHCOMe O + OEt → MeOOC NHCOMe O OEt

(95) (96)

A stereoselective intramolecular cycloaddition was the key step in the preparation of 5-*epi*-desosamine (**102**) (S. W Remiszewski *et al., J. Org. Chem.,* 1984, *49,* 3243). The starting point for this synthesis was the carbamate (**97**) which was converted into the N-sulphinyl derivative (**98**) by reaction with thionyl chloride and pyridine.This compound underwent a spontaneous intramolecular reaction to give the adduct (**99**). This compound was cleaved by reacting it with phenylmagnesium bromide to produce an allyl sulfoxide. This underwent a [2,3]-sigmatropic rearrangement and desulphurisation to yield an allylic alcohol (**100**), as a single stereoisomer, when heated with piperidine in ethanol. The allylic alcohol (**100**) on reaction with paraformaldehyde in the presence of *p*-toluenesulfonic acid, followed by reduction with lithium aluminum hydride gave the amino diol (**101**), which was oxidatively cleaved by ozonization over dry silica gel to give the target compound, **5**-*epi*-desosamine (**102**). The senior author of this work, Weinreb, has used this type of chemistry to synthesize other unusual amino sugars (*ibid.,* 1987, *52,* 1177; *Bull. Soc. Chim. Belg.,* 1986, *95,* 1021).

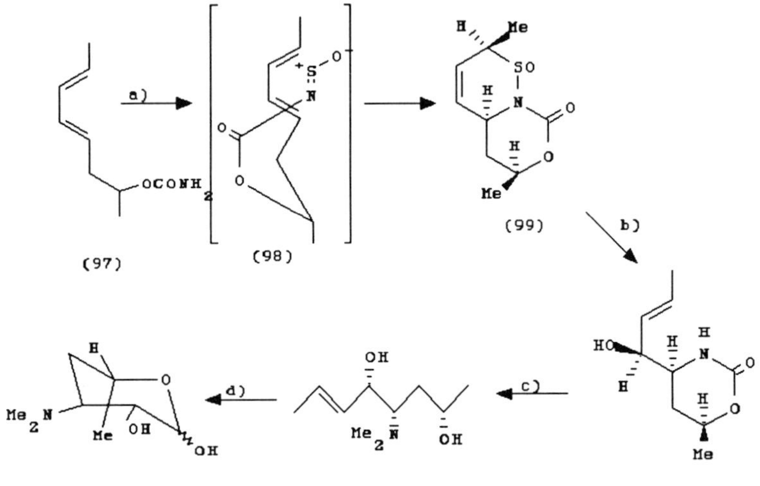

a)$SOCl_2$/Py/Tol/0°-rt. b)i)PhMgBr/THF/-50°. ii)piperidine/EtOH c)i)$(HCHO)_n$/TsOH /PhH/Δ. ii)LAH/THF/12h/Δ. d)i)TFA/CH_2Cl_2. ii)O_3/Silica gel/-78°/Zn/HOAc

Cycloaddition of chlorosulphonyl isocyanate and 1,3-pentadiene, followed by the reductive cleavage of the N-chlorosulphonyl moiety afforded the phenylazetidinone (**103**). This azetidinone on methanolysis gave the *β*-amino ester (**104**) which through a series of oxidation and reduction steps eventually afforded the acetoxyfuranose (**105**). The furanose (**105**) on treatment with ammonia in cold methanol gave N-benzoyldaunosamine (**106**) (F. M. Hauser *et al.,* J. Org. Chem., 1986, *49,* 2236). For other [4+2] cycloaddition reactions see B. J. Fitzsimmons *et al., J. Amer. Chem. Soc.,* 1987,

109, 285; A. K. Forrest *et al., J. Chem. Soc. Perk. Trans. 1,* 1984, 1981).

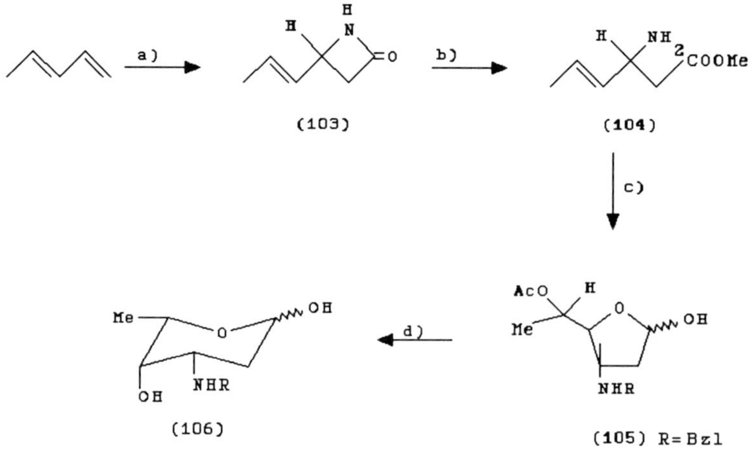

(103) (104)

(105) R=Bzl (106)

a)$ClSO_2NCO/Na_2SO_3$/0-5°/pH7-9. b)MeOH/HCl/2h. c)i)PhCOCl/Py/CH_2Cl_2. ii)OsO_4/TMNO. iii)Ac_2O/Py. iv)DIBAL-H/THF/-100°. d)NH_3/MeOH/0°-rt/2h.

d). Miscellaneous reactions

Sugars bearing terminal 1-cyanovinyl groups were synthesized by Tronchet and colleagues (*J. Carbohydr. Chem.,* 1989, *8,* 217) in their studies to evaluate structure-activity relationships of antiviral agents. The conjugate base of nitromethane added to the formyl sugar (**107**) to afford the β-hydroxy-α-nitro derivative (**108**). Upon acetylation, this product underwent elimination to give the nitroenose (**109**). This compound on treatment with potassium cyanide, in the presence of a phase transfer catalyst, gave the 1-cyanovinyl sugar (**110**). In some related examples the intermediate α-nitromethylnitriles (**111**, R= monosaccharide unit)) were formed, these on gentle heating in the presence of triethylamine afforded the corresponding vinylnitriles.

CHO (107) → a) → CH(OH)CH_2NO_2 (108) → b) → CH=CHNO_2 (109) → c) → CH(CN)=CH_2 (110)

a)CH_3NO_2/MeOH/NaOMe. b)Ac_2O/NaAc/0°-rt/1d. c)i)KCN/Bu_4NBr/THF/Tol /pH7.5/rt/3d. ii)Et_3N/Δ

$$R-CH=CH-NO_2 \xrightarrow{CN^-} R-CH(CN)-CH_2-NO_2 \longrightarrow R-C(CN)=CH_2$$

(111)

Substituted glyceraldehydes have been reacted with α-cyanomethyl acetate under Knoevenagel type reaction conditions. The nucleophilic addition reaction was observed to occur with exclusive *erythro*-selectivity. Ensuing intermolecular cyclisations afforded the corresponding oxazolines which were converted into 2-amino-2-deoxy-*D*-pentoses.(Y. Yamamoto *et al.*, *Agric. Biol. Chem.*, 1985, *49,* 1435).

Sodium azide in the presence of ceric ammonium nitrate added stereoselectively to the *L*-arabinal diacetate (**112**) to give the corresponding azidonitrate (**113**), which was readily reduced to

AcO O OAc → AcO O OAc ONO_2 N_3

(112) (113)

$Ce(NH_4)_2(NO_3)_6/NaN_3/CH_3CN/-10°$

the appropriate 3-aminosugar (H. Hashimoto *et al.*, *Bull. Chem. Soc. Jpn.*, 1986, *59,* 3131). The unusual stereochemistry associated with this type of addition is explained by the steric hindrance of the axially oriented 4-acetoxyl group in the half chair conformation of the substrates. This favours attack of the

MeOOC O OAc AcO a) → MeOOC O OAc AcO OAc N_3 b) → HO O H O OH O NHAc

(114) (115)

a)i)$Ce(NH_4)_2(NO_3)_6/CH_3CN/NaN_3$. ii)NaOAc/HAc. b)i)HCl/MeOH. ii)Pd[H] /MeOH/Ac_2O

azide ion from the opposite side. The generality of this type of approach has been recognized and other related molecules have also been synthesized by similar routes. These include glycocinnamoylspermidines (K. Araki, H.Hashimoto *et al., Carbohydr. Res.* 1982, *109,* 143). Similarly, the 3,6-lactone (**115**) of 2-deoxy-*D*-glucuronic acid was also obtained from a 3-(*β*-)acetoxypyrene derivative (**114**) (E. Darklas *et al., Carbohydr. Res.,* 1982, *103,* 176).

In their synthesis of N-acetylholacosamine (**121**) Fraser-Reid and colleagues (*Can. J. Chem.,* 1984, *62,* 1539) reacted the diol (**116**) under Mitsunobo reaction conditions with epimerisation of the C-4 hydroxyl group to yield the dibenzoate (**117**).

(116) (117) (118) (119) $R=N(Me)CO_2Et$

(120) (121)

a)i)DEAD/Ph_3P/PhCOOH/THF. ii)MeOH/NaOMe. iii)pTsCl/Py. b)i)$MeNH_2$/DMSO /rt. ii)Et_3N/ClCOOEt. c)i)I(collidine)$_2$ ClO_4/dioxane/rt. ii)Me_2CO/NaI. d)i)tBuSnH /PhH. ii)5%KOH. iii)MeOH/Ac_2O. iv)NaH/THF/MeI/nBu_4NI.

Hydrolysis, was followed by sulphonation giving the sulphonate (**118**), which was reacted with methylamine in dimethysulphoxide and ethoxycarbonylated with triethylamine and ethyl chloroformate to afford the trideoxy sugar (**119**). Portionwise addition of iodonium dicollidine perchlorate to a refluxing solution of the urethane (**119**) in dry dioxane afforded an oxazolidinone which was immediately iodinated to give the diiodourethane (**120**). Reduction, followed by a series of reactions gave the target compound (**121**) (also see M. Georges and B. Fraser-Reid, *Tet. Lett.,* 1981, *22,* 4635).

An oxyamino sugar (**125**) was the product when the dibenzylated pyranose (**123**) was reacted with N-hydroxyphthalimide (**122**) in the presence of triphenylphosphine and diethylazodicarboxylate. The product was the O-glycosyl-N-hydroxyphthalimide (**124**), which on refluxing in hydrazine afforded the O-glycosylhydroxylamine (**125**) (E. M. Nashed, E. Grochowski *et al., Carbohydr. Res.,* 1990, *196,* 184; [also see E. Grochowski *et al., ibid.,* 1988, *177,* 244; 1976, *50,* c15; W. A. Szarek *et al., ibid.,* 1977, *57,* c13]).

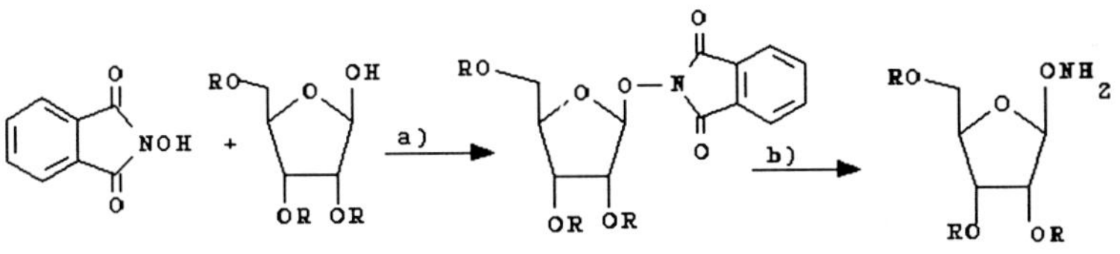

(122) (123) R=Bn (124) (125)

a)Ph_3P/DEAD. b)NH_2NH_2/10min/Δ

3,4,6-Tri-O-acetyl-2-azido-2-deoxy-*α*-*D*-galactopyranosyl bromide has been converted by a series of reactions into the corresponding 4-methoxybenzylidene acetal. This compound was ring opened reductively without destroying the azido group. Furthermore the resulting methoxybenzyl ether was oxidatively cleaved by treatment with DDQ to yield the appropriate azido alcohol (M. Kloosterman *et al., Recl. Trav. Chim. Pays-Bas.,* 1985, *104,* 291).

The free amino function of tertiarybutyl *p*-aminobenzoate was utilized as a nucleophile to displace the trityl group in the 2,3-anhydro-*α*-*D*-ribopyranoside (**126**) to afford the aminodeoxysugar (**127**). Treatment of (**127**) with boron trifluoride in trimethylsilylazide afforded the aziridino sugar (**128**) (F. Latif *et al., Liebigs. Ann. Chem.,* 1987, 717).

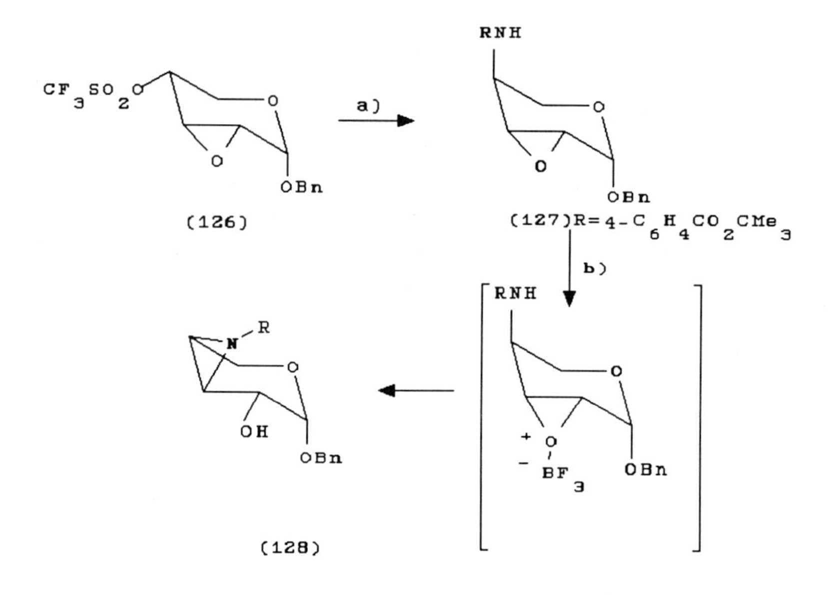

a)4-$NH_2C_6H_4CO_2CMe_3$/CH_3CN/collidine. b)BF_3.Et_2O/Me_3SiN_3.

The ketosugar (**129**) reacted with hydroxylamine hydrochloride in pyridine and ethanol to give the corresponding oxime, which was catalytically reduced to yield the amino sugar (**130**) (J. Yoshimura *et al., Chem. Lett.,* 1985, 69).

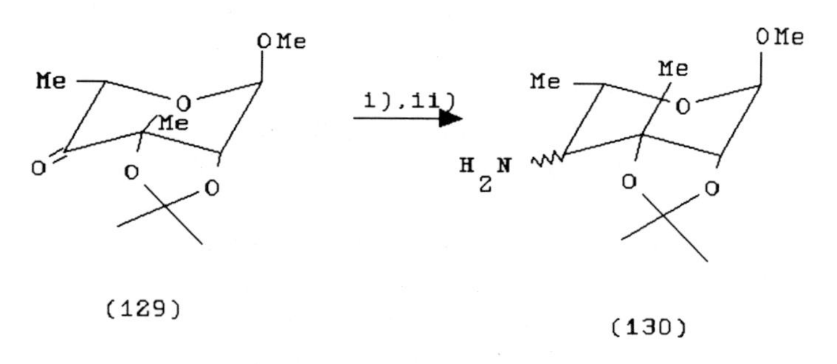

i)NH_2OH.HCl/Py/EtOH. ii)HAc/(Pd/C)/[H].

4. Monosaccharides having a phosphorus atom in the ring

The chemistry of " P sugars " or monosaccharides that have the oxygen atom of the hemiacetal ring replaced by a phosphorus atom, has been the subject of a recent review (H. Yamamoto and S. Inokawa, *Adv. Carbohydr. Chem. Biochem.,* 1984, *42,* 135). In this report a survey of the literature up to 1983, with emphasis on structural, physical properties, synthesis and the biological activities of phosphino and phosphinyl pyranoses and phosphinyl furanoses is discussed. The structural analysis covered nuclear magnetic resonance spectroscopy, mass spectrometry, and X-ray crystallography.

Following on from earlier routes to hydroxyphosphinyl sugars,

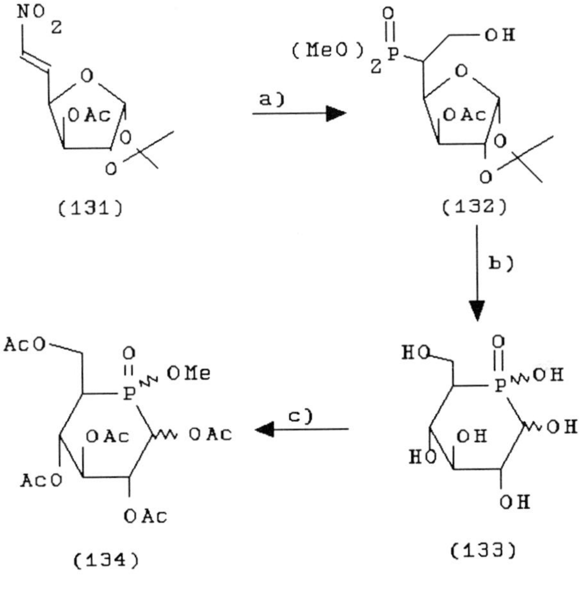

(131) (132) (133) (134)

a)i)dimethylphosphonate/100°/N_2/2h. ii)PtO_2/[H]/HCl. iii)HAc/$NaNO_2$. b)i)Ph_3CCl /Py. ii)$NaH_2Al(OCH_2CH_2OCH_3)_2$(SDMA)/iPrOH/$H^+$. iii)$H_2O_2$. c)i)$CH_2N_2$/$Me_2SO$ /MeOH. ii)Ac_2O/Py

Yamamoto and colleagues (*J. Org. Chem.,* 1985, *50,* 3517) have extended their methodology to the preparation of 5-deoxy-

5-C-(hydroxyphosphinyl)-aldohexopyranose (**133**). The 3-O-acetylnitrofuranose (**131**) on heating with dimethylphosphonate afforded a dimethoxylphosphinyl derivative that was catalytically hydrogenated. The ensuing 6-aminoglucofuranose was diazotized and hydrolysed to afford the primary alcohol (**132**). A series of reactions effected a ring expansion of this molecule to generate the phosphacyclohexane (**133**) which was characterized as the 1-(O-methyl)pentaacetate (**134**) (for similar work see S. Inokawa *et al., Carbohydr. Res.,* 1982, *106,* 31; K. Seo *et al., ibid.,* 1986, *148,* 168; H.Yamamoto *et al., ibid.,* 1984, *127,* 35).

The first synthesis of a hexoheptanose (**138**) having a phosphorus atom in the heptanoid ring has been accomplished (K. Seo, *Carbohydr. Res.,* 1983, *123,* 201). A solution of the

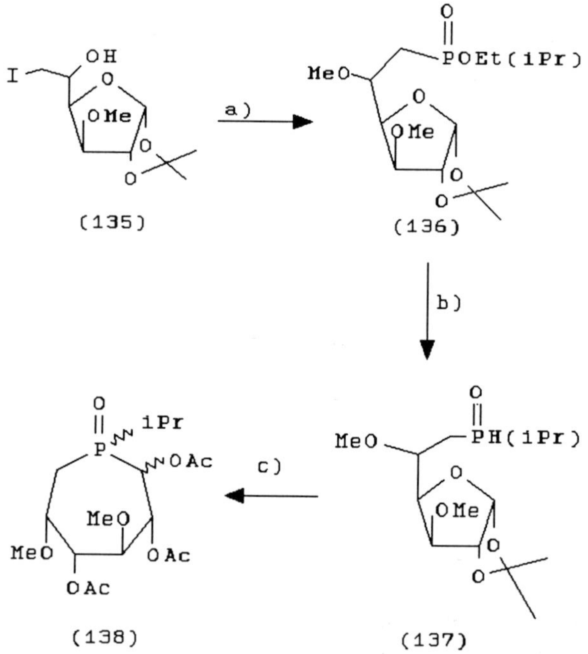

(135) (136) (137) (138)

a)i)iPrP(OEt)$_2$/110°. ii)MeI/AgO. b)70%SDMA/PhH/THF/0°/Ar. c)i)HCl/40°/3h/Ar. ii)Ac$_2$O/Py.

iodide (**135**) in diethylisopropyl phosphonite was heated to effect a Michaelis-Arbuzov reaction. The C-5 hydroxyl group of the resultant phosphinyl sugar, was methylated by treatment with methyl iodide in the presence of silver oxide affording the O-methyl analogue (**136**). The phosphinyl group was reduced with sodium dihydrobis(2-methoxyethoxy)aluminate (SDMA) and the phosphinoxide (**137**) was then cyclised to the hepta-noid sugar by treatment first with hydrogen chloride and then with acetic anhydride. This gave the peracetate (**138**).
Wroblewski (*Tetrahedron,* 1983, *39,* 1809) has investigated the ring closure of aminophosphites in aqueous media. This reaction type has value in the synthesis of phospha sugars and finds expression in the work of Mukahametov and colleagues (*Zhur. Obsch. Khim.,* 1979, *49,* 1756). These workers condensed 4-hydroxy-4-methylpentan-2-one (**139**) with dimethyl

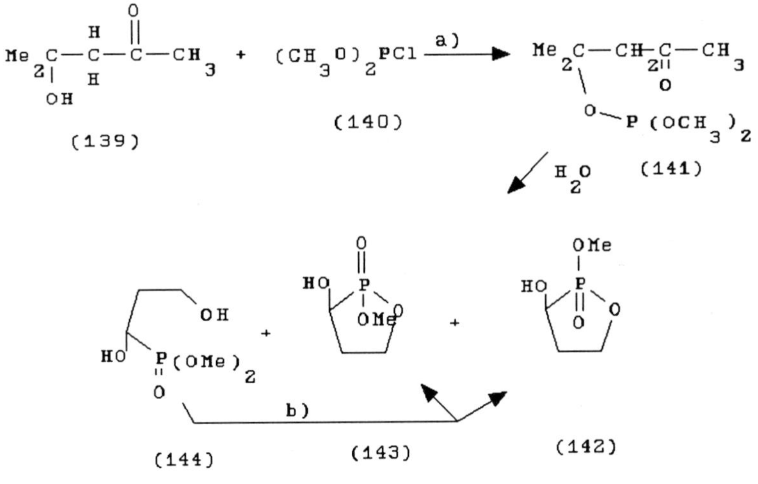

a)Et_3N/PhH. b)MeOH/HCl

phosphorochloridite (**140**) in the presence of triethylamine to obtain the phosphite (**141**). This on hydrolysis afforded a mixture of compounds (**142, 143, 144**) in the proportion 1:3:6. The acyclic phosphonate (**144**) was cyclised to a mixture of (**142**) and (**143**) by the treatment of triethylamine (for other examples see, Wroblewski *et al., Tetrahedron,* 1986, *42,* 13., *Carbohydr. Res.,* 1984, *125,* c1)

5. Cyano Monosaccharides

Glycosyl cyanides are among the most useful carbon-linked derivatives of carbohydrates and have been utilized in the synthesis of various derivatives. The reaction of alkaline cyanide solution, or of hydrogen cyanide, with keto sugars has been the method of choice to introduce a nitrile function, but ring opening of sugar oxiranes, and also the displacement of labile sugar triflates with cyanide ions have been used (T. T. Thang *et al., Tet. Lett.,* 1980, *21,* 4495; N. R. Williams, *J. Chem. Soc., Chem. Commun.,* 1967, 1012; A. Knierzinger and A. Vasella, *ibid.,* 1984, 9; P. L. Durette, *Carbohydr. Res.,* 1982, *100,* c27).

As an example of a oxirane ring opening reaction which results in a cyano sugar, the epoxide (**145**) was treated with

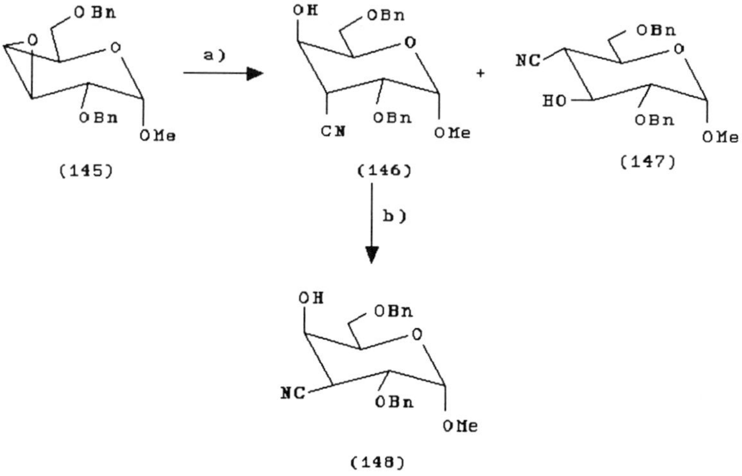

a)Et_2AlCN/PhH. b)NaOMe/MeOH.

diethylaluminium cyanide in benzene solution. Both the 2- and the 3- cyano isomers were formed in the ratio 1:5, and as expected in both cases the cyanide group was shown to be *α*-oriented. If the 3-isomer is treated with sodium methoxide, epimerisation occurs to give the corresponding *β*-nitrile (**148**) (A. Mubarak and B. Fraser-Reid, *J. Org. Chem.* 1982, *47,* 4265).

Brimacombe *et al.,*(*Carbohydr.Res.,*1982, *110,* 207) in their synthesis of methyl-N-acetyl-*α*-*L*-vanucosaminide (**151**) reacted the keto sugar (**149**) with an excess of potassium cyanide to form the kinetically preferred cyanohydrin (**150**), albeit in moderate yield, but unreacted (**149**) was then recycled to show a good net productivity. The cyanohydrin was converted to the target compound (**151**) by reduction of its mesylate, further reduction, and finally N-acetylation.

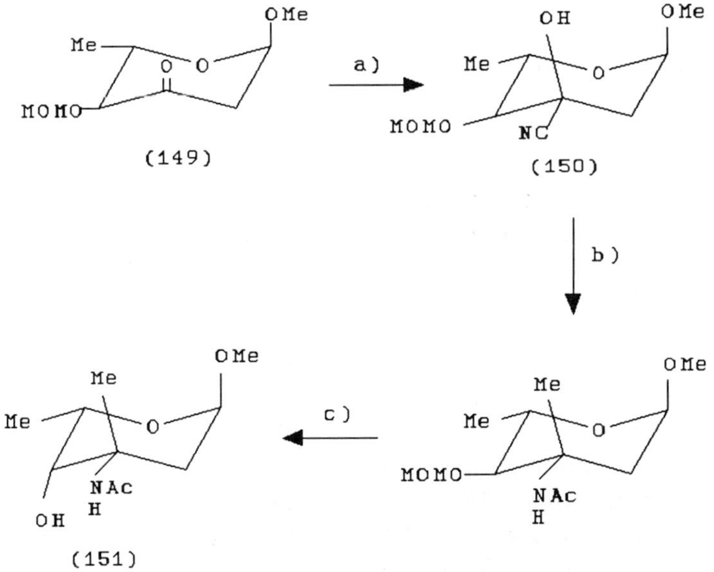

a)KCN/$NaHCO_3$/CH_2Cl_2. b)i)MsCl. ii)LAH/Ether/Δ. iii)RaNi[H]/rt/70h. iv)Ac_2O/Py. c)H^+/MeOH.

Kini and colleagues (*Carbohydr. Res.,* 1987, *159,* 81) reacted the pentaacetate (**152**) with trimethylsilylcyanide in the presence of borontrifluoride in nitromethane to obtain the cyano sugar (**153**) the stereochemistry of the reaction being determined by the orientation of the acetoxy group at C-2 (also see F. G. Delas Heras *et al., Tetrahedron,* 1985, *41,* 3867).

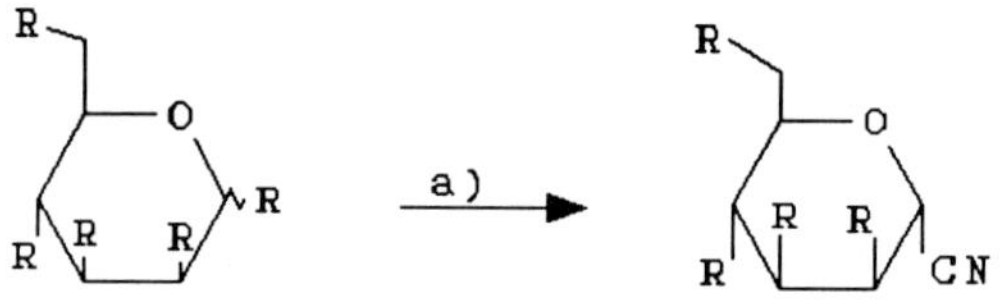

(152) R=OAc

(153)

a) $Me_3SiCN/BF_3.Et_2O/CH_3CN/\Delta$

6. Halides of Monosaccharides

Halogen substituents are known to enhance the sweetness of sugars (L. Hough and S. Phadnis, *Nature,* 1976, *263,* 800; C. K. Lee, *Adv. Carbohydr. Chem. Biochem,* 1987, *45,* 199), whilst fluorinated carbohydrates have found utility as *in vivo* imaging agents for carbohydrate metabolism. Interest in these compounds are demonstrated by the number of reviews on the subject (A. A. E. Penglis, *Adv. Carbohydr. Chem. Biochem.,* 1981, *38,* 195; T. Tsuchiya, *ibid.,* 1990, *48,* 91; R. Csuk and B. I. Glanzer, *ibud.,* 1988, *46,* 73; P. W. Kent, Fluorinated Carbohydrates, N. F. Taylor (ed.), *A.C.S. Symp. Ser. 374,* 1988, 1; J. Mann, *Chem. Soc. Rev.,* 1987, *16,* 381 and J. T. Welch, *Tetrahedron,* 1987, *43,* 3123).

Su and fellow workers (*J. Org. Chem.* 1981, *46,* 1791) have synthesized 2-deoxy-2-halogeno-arabinofuranoses (**156**) from the ribofuranoses (**154**),via their triflates (**155**), through displacement with a lithium halide in the presence of pyridine (also see *idem, ibid.,* 1982, *47,* 1507).

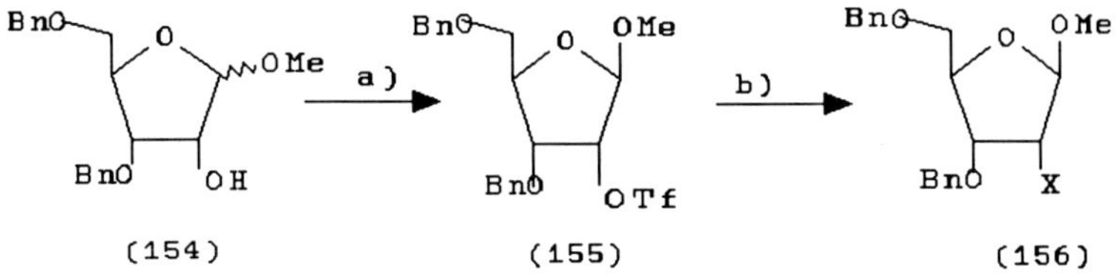

(154) (155) (156)

a)Py/CH_2Cl_2/Tf_2O/-15°/N_2. b)DMSO/HMPA/LiX (X=Cl,Br or I).

Lee (*Carbohydr. Res.,* 1988, *177,* 247) has also approached the synthesis of halogeno-sugars via the displacement of the

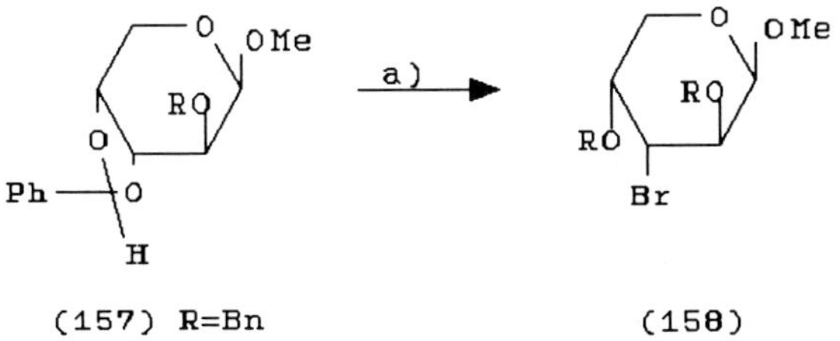

(157) R=Bn (158)

a)$BrCCl_3$/UV/1.5h.

corresponding mesylates, or sulphonates. Benzylidene acetals of pyranosides (**157**) have been regio and stereospecifically opened to the corresponding bromo sugars (**158**) by treatment with trichloromethyl bromide under photochemical conditions (T. G. Richmond *et al., J. Chem. Soc. Chem. Commun.,* 1988, 94). Typically such reactions are effected by exposure of the substrates to light from a mercury lamp and are claimed to be cleaner and more efficient than those which use N-bromo-succinimide (also *idem., ibid.,* 1988, 272).

Phosphoryl chloride, or bromide, in the presence of dimethyl-formamide can be reacted with the uridine derivative (**159**) to afford the corresponding halide (**160**) (K. Hirota *et al., Synthesis,* 1983, 121).

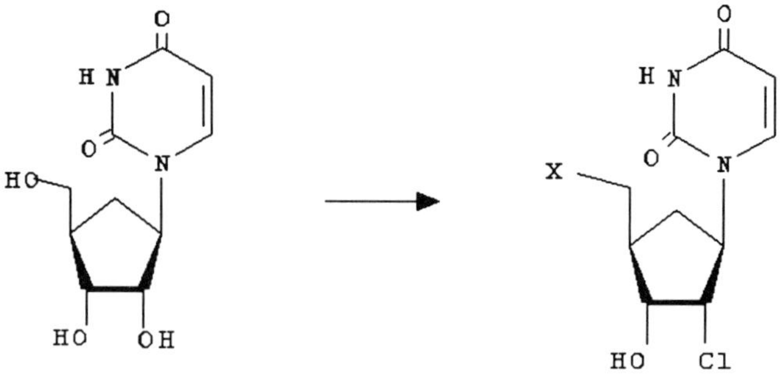

(159) (160)

i)DMP/POX_3. ii)MeOH/reflux X=Br or Cl

Bis(*sym*-collidine)iodine tetrafluoroborate, formed by the reaction of silver(I)tetrafluoroborate, 2,4,6-trimethylpyridine and iodine [$I^+(collidine)_2BF_4^-$], has been used to iodofluorinate sugars. For example, a reaction of diacetyl deoxyrhamnal (**161**) with this reagent afforded the iodofluorinated sugar (**162**) (R. D. Evans, *ibid.,* 1987, 551).

Lichtenthaler and colleagues (*Tet. Lett.,* 1990, *31,* 71) formed the ulosylbromide (**164**) by the action of N-bromosuccinimide upon the O-benzylated alkene (**163**). Bromides of this type show a variety of useful reactions, which include

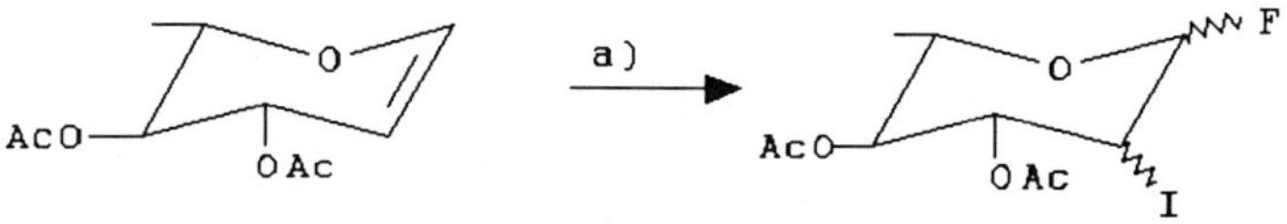

(161) (162)

a)$I(collidine)_2BF_4/CH_2Cl_2$/rt.

selective glycosidations, and selectivity during reduction. They also enter into cycloaddition reactions and can be C-homologated to give disaccharides.

(163) R=Bn

(164)

a)NBS/MeOH/CH_2Cl_2

The benzoylated sugar (**165**) on reaction with hydrogen bromide in acetic acid afforded the 6-deoxy-*D*-allofuranosyl bromide (**166**)(H. S. El Khadem and V. Nelson, *Carbohydr. Res.,* 1981, *98,* 195; also see K. Bock *et al., ibid.,* 1981, *90,* 7, 17).

(165) $R=CH_2C_6H_4$-4-NO_2

(166)

a)HBr/HAc/0-5°/5d.

Diethylaminosulphur trifluoride (DAST, F_3SNEt_2), has been used to fluorinate the primary hydroxyl group of methyl *α-D*-glucopyranoside. Prolonged treatment caused a second substitution to occur at C-4 (P. J. Card, *J. Org. Chem.,* 1983, *48,* 393; *J. Carbohydr. Chem.,* 1985, *4,* 453). Glycosyl fluorides have been also synthesised by the reaction of DAST upon suitable substrates (G. H. Posner and S. R. Haines, *Tet. Lett.,* 1985, *26,* 5).

Twai and colleagues (*Kaigi. Daigakko. Kenkyu. Hokuko.,* 1985, *28,* 69; *Chem. Abstr.,* 1986, *105,* 79241) reacted titanium bromide ($TiBr_4$) with alkylated glucose to afford glycosyl bromides. The authors claim that with N-acetylglucosamine a N ——> O acyl rearrangement occurred.

α-Nitroepoxide sugars (**167**) were denitrified by halogeno nucleophiles to afford keto halogeno sugars of the type (**168**) (T. Nakagawa *et al., ibid.,* 1987, *163,* 227). Typically the nitroepoxide is treated with the lithium halide in tetrahydrofuran, or dimethylformamide, at room temperature.

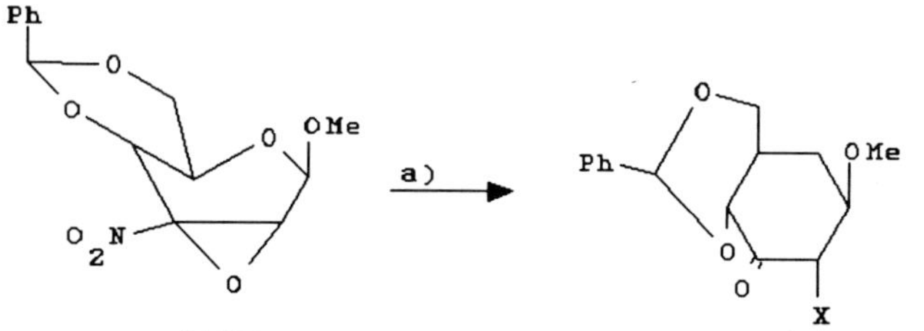

(167)

(168) X=Br, Cl or I

a)LiX/THF/4h/rt (X=Cl, Br or I).

3-Iodosugars can be formed by the displacement of tosyl groups by sodium iodide in dimethylformamide. Thus tolylsulphonyl pyranose (**169**) afforded 3-deoxy-3-iodo-*D*-glucose (**170**) in high yield. Acetylated and benzylated iodo derivatives have been similarly prepared. Retention of stereochemistry in these reactions suggests the intermediacy of epoxides (M. Yamamada, *ibid.,* 1981, *96,* 121).

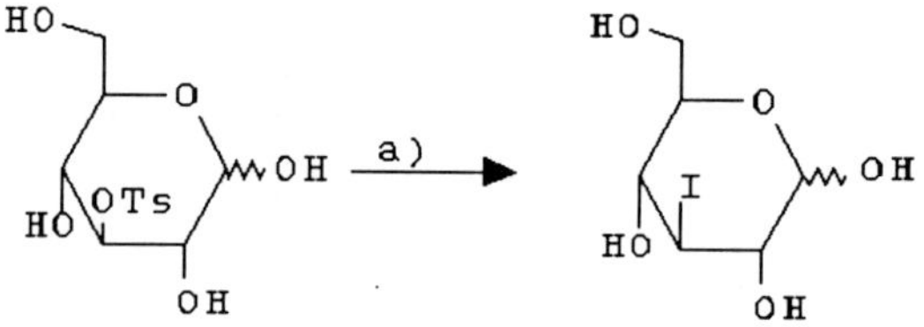

(169)

(170)

a)NaI/DMF/120°/4h

In another approach this time to iodochloro compounds, 3,4,6-tri-O-acetyl-*D*-galactal (**171**) was stirred in the dark with silver imidazolate and mercury(II)chloride, in the presence of powdered molecular sieves in acetonitrile at room temperature. When iodine was added this gave the mixed isomers (**172**) and (**173**) in the ratio 5:1 (P. J. Garegg, *ibid.,* 1981, *92,* 157).

(171) R=OAc (172) (173)

a)i)Ag Imidazolate/$HgCl_2$ ii)I_2/CH_3CN

For a discussion of the chlorination of D-glucals of the type (**171**) see Sakakibara (*Chem. Lett.,* 1987, 7). Samuelson (*Can. J. Chem.,* 1981, *59,* 339) recommended a mixture of triphenylphosphine, tribromoimidazole and imidazole in toluene for the bromination of sugar hydroxyl groups (*idem, J. Chem. Soc. Perkin Tran 1.* 1980, 2866).

Ferrier and Haines (*J. Chem. Soc. Perkin Trans 1.,* 1984, 1675) showed that furanose derivatives (**174**) on irradiation in the presence of bromine gave the bromides (**175**). This seems to be a reaction of general applicability.

(174) R=Bzl (175)

a)CCl_4/Br_2/hv

The steric bulk of the hindered base 2,4,6-tributylpyridine (TTBP) was used in the preparation of the triflate (**177**). Previous attempts to sulphonate, the primary hydroxyl group of the glucopyranose (**176**) using other bases were unsuccessful. Subsequent displacement of the triflate group with halide ions afforded the deoxyhalogeno sugars (**178**) in good yield (M. G. Ambrose and R. W. Brinkley, *J. Org. Chem.,* 1983, *48,* 674).

(176) R=OAc (177) (178)

a)Tf_2O/CH_2Cl_2/TTBP. b)nBu_4NX/CH_2Cl_2 (X=Br,Cl,I or F)

The diisopropylidine sugar (**179**) on treatment with diethyl bromomethyleneammonium bromide in the presence of triethylamine afforded the bromo sugar (**180**) directly without recourse to any activation of the C-1 hydroxyl group (D. P. McAdam, R. V. Stick *et al., Aust. J. Chem.,* 1988, *41,* 563). In an earlier paper the same authors used diethyl chloromethyleneammonium chloride to obtain the corresponding chloro sugars (*ibid.,* 1983, *36,* 1239).

(179) (180)

a)$Et_2N^+=CHBr_2Br^-/CH_2Cl_2/Et_3N$/2h/rt

7. Thio and Thia Monosaccharides

Capon and MacLeod (*J. Chem. Soc., Chem. Commun.,* 1987, 1200) recorded the isolation of the first naturally occurring thia sugar, 5-thia-*D*-mannose (**181**), from a marine sponge.

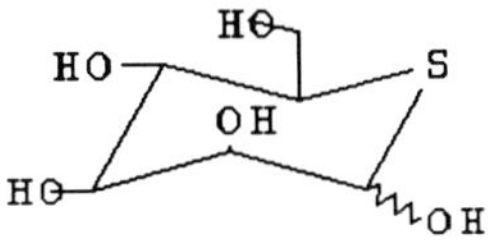

(181)

For monosaccharides in which the heterocyclic oxygen atom has been replaced by sulphur a variety of synthetic techniques have evolved. For example, 5-thia-*D*-glucose (**185**) has been

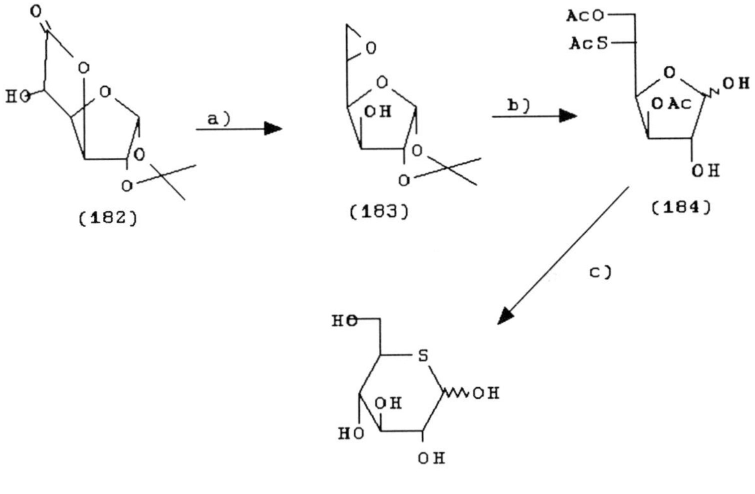

(182) (183) (184) (185)

a)i)TsCl. ii)$LiBH_4$/THF. iii)NaOMe/MeOH/-15 °. b)i)MeOH/H_2NCSNH_2. ii)HAc/Ac_2O /NaAc/1.5d. iii)TFA/rt/20min. c)NaOMe/MeOH/N_2

prepared from the hydroxylactone (**182**). This compound was O-tosylated, before the lactone function was reduced and ring opened with sodium methoxide generating the epoxide (**183**). When this was treated with thiourea the epoxide unit cleaved regioselectively affording an intermediate hydroxythiol, which was acetylated to afford the triacetate (**184**). Ring expansion of this product was effected by successive treatment first with trifluoroacetic acid and then with sodium methoxide (H. Drignes and B.Henrissat, *Tet. Lett.,* 1981, *22,* 5061).

A similar strategy was adopted in the synthesis of the corresponding allose (**189**). Here the diacetal (**186**), already containing the sulphur atom, was oxidized by acetic anhydride-methylsulphoxide and the resulting ketone was isolated as the crystalline hydrate (**187**). Sodium borohydride, reduction, followed by acid catalyzed ring enlargement afforded the 5-thia-*D*-allose (**189**) (N. A. L. Al-Masoudi and N. A. Hughes, *Carbohydr. Res.,* 1986, *148,* 25).

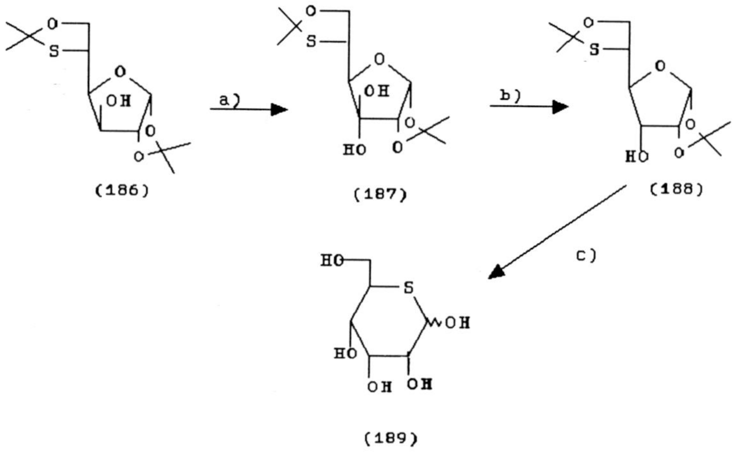

(186) (187) (188) (189)

a)Ac_2O/DMSO/rt/2d. b)$NaBH_4$/EtOH. c)HAc/90°/2h.

The same authors using minor modifications of the above synthesis, prepared the analogous 5-thia-*D*-altrose (**190**) (*ibid.,* 1986, *148,* 39., also see E. Tanahashi *et al., ibid.,* 1983, *117,* 304., A. Hasegawa *et al., ibid.,* 1983, *122,* 168).

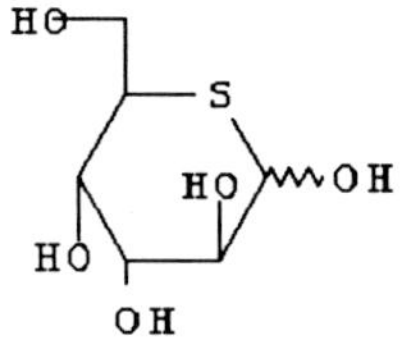

(190)

D-Glucose pentaacetate when fused with thiophenol in the presence of zinc chloride yielded the corresponding 1-thio-*β*-glycoside. Other Lewis acids were also studied for use in such conversions (T. Shimadate *et al., Bull. Chem. Soc. Jpn.,* 1982, *55,* 3552).

For monosaccharides bearing thiol or sulphur bearing substituents the usual method is nucleophilic displacement of an acetate or a similar group. For example, glycosyl acetates when heated under reflux in dry dichloromethane containing thiolactic acid and aluminum chloride afforded the corresponding 1-thioglycosyl acetates in good yields (B. Raja-nikanth and R. Seshadri, *Tet. Lett.,* 1987, *28,* 2295).

Similarly, 1-thio-*β*-*D*-glucopyranose (**192**) was prepared by reacting the *α*-chloride (**191**) with potassium thioacetate (A. Hasegawa, *Carbohydr. Res.,* 1983, *123,* 183).

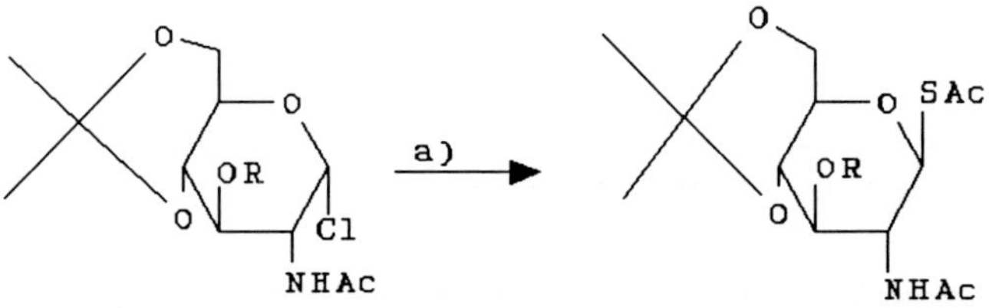

(191) R=CH(Me)COOMe (192)

a)$CH_2Cl_2/Me_2CO/KSAc$

Another technique for the preparation of thioglycosides requires the reaction of methoxy pyranosides with trimethylmethylthiosilane (or phenylthiosilane) in dichloromethane containing zinc chloride or tetrabutylammonium iodide [see (**193**) ——> (**194**)]. Many examples of the procedure are known, illustrating the generality of the reaction (S. Hanessian and Y.Guindon, *ibid.,* 1980, **86**, c3).

HO O OH HO OMe OH (193) —i,ii→ HO O OH SPh HO OH (194)

i)Me_3Si-SMe/Py. ii)PhS-$SiMe_3$/ZnI_2 or Bu_4NI/CH_2Cl_2/60°/8h

Phosphorodithioic acid adds across the double bond of acetylated glucals to afford the corresponding phosphorodithioates in quantitative yield (J. Borowiecka *et al., Tetrahedron.,* 1988, *44,* 2067). For the synthesis of C-2 deuterium labelled monosaccharides deuterated dialkylphosphorodithioic acid is recomended. The *α*-dithiophosphates of 2-deoxy sugars are known to undergo substitution very easily, thus the phosphorodithioates may be reacted with appropriate nucleophiles to yield branched sugars (*idem, J. Carbohydr.,* 1983, *2,* 99).

Witczak (*Heterocycles,* 1983, *20,* 1435) has reviewed the synthesis of thiocyanates of monosaccharides and also the chemistry of their isothiocyanate analogues (*Adv. Carbohydr. Chem. Biochem.,* 1986, *44,* 91).

8. Carbocyclic analogues of monosaccharides

" Pseudo" sugars i.e. compounds where the ring oxygen atom of the cyclic forms of sugars have been replaced by a methylene group are found in Nature as constituent parts of antibiotics and related compounds. The term carba-sugars is also in use. The synthesis of pseudo sugars has been reviewed by Suami and Ogawa (*Adv. Carbohydr. Chem. Biochem.,* 1990, *48,* 21; *Synth. Org. Chem. Jpn.,* 1985, *43,* 26).

Some illustrations of this type of endeavor include the synthesis of the pentaacetyl derivative of pseudo-*β*-*L*-mannopyranose (**200**) (Tadano *et al., J. Org. Chem.,* 1987, *52,* 1946). In this construction the thioacetal (**195**), from *D*-ribose and thioethanol, was protected and converted into the aldehyde (**196**). A condensation with diethyl malonate, followed by reduction of the double bond of the product then gave the primary alcohol (**197**). This was cyclised to the cyclohexane (**198**), by conversion first to the formyl derivative and an intramolecular aldol type reaction. Hydrolysis and thermal decarboxylation, prior to reduction and dehydration gave the cyclohexene (**199**). Finally

a)i)TrCl/DMAP/Py. ii)$PhCH_2Br$/NaH/DMF. iii)TsOH/MeOH. iv)$tBuPh_2SiCl$/Imi/DMF. v)$HgCl_2/CH_3CN$. b)i)$CH_2(COOMe)_2/Py/Ac_2O$. ii)RaNi[H]. iii)Bu_4NF. c)i)PCC /CH_2Cl_2. ii)Ac_2O/Py. d)i)DMSO/NaCl/Δ. ii)LAH/THF. e)i)BH_3/THF. ii)$NaOH/H_2O_2$. iii)Ac_2O/Py. iv)Pd/C[H]/Ac_2O/Py.

oxidative hydroboration, acetylation and reduction gave the acetate (**200**). Using the same approach the authors used other naturally occurring sugars, e.g. D-xylose and D-arabinose, to effect the synthesis of the corresponding carbocyclic analogues (*idem, et al., Chem. Lett.,* 1984, 1919; R. Blattner and R. J. Ferrier *J. Chem. Soc., Chem. Commun.,* 1987, 1008).

Of the naturally occurring carbohydrates, D-fructose is considered to have the sweetest taste due to the adoption of the β-D-fructopyranose form which is assumed to fit well on the natural receptor sites of the tongue. Therefore the pseudofructopyranose (**204**) was an attractive target for synthesis. Interestingly, when it was prepared this pseudo sugar was found to be organoleptically equal to natural D-fructose (T. Suami *et al., Chem. Lett.,* 1985, 719). The route chosen for its

a)DBU. b)i)mCPBA. ii)NaAc/Ac_2O/DMAP/Py. iii)NaOMe/mCPBA. iv)Ac_2O/Py /DMAP. c)i)LAH. ii)Ac_2O/Py/DMAP. iii)NaOMe/MeOH/H^+

construction was as follows: the diacetylbromo cyclohexene (**201**) was debrominated with 1,8-diazabicycloundec-7-ene to give the diene (**202**), which underwent preferential epoxidation of the exocyclic double bond with mCPBA. This epoxide was ring opened with sodium acetate and the product immediately acetylated to fix the stereochemistry of the C-1 center. Subsequent epoxidation of the endocyclic double bond and acetylation afforded the oxirane (**203**). Reductive opening of the epoxide and hydrolysis then produced the target molecule (**204**).

The cellular second messenger *D-myo*-inositol triphosphate (**205**) has been the target of many synthetic strategies (H. Streb *et al., Nature.,* 1983, *306,* 67; M. J. Berridge and R. F. Irvine *ibid,* 1984, *312,* 315; M. J. Berridge *Annu. Rev. Biochem.,* 1987, *56,* 159).

OH OH
OP
OP
PO
OH

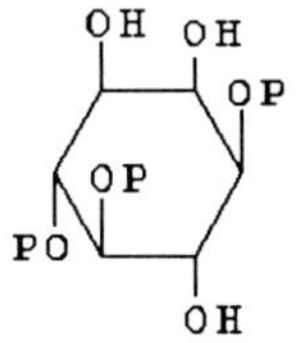

(205) P=PO(OH)$_2$

In one approach the partially protected *myo*-inositol (**210**) has been synthesized from the lactone (**206**) (Y.Watanabe *et al., Chem. Lett.,* 1987, 123). In this work the acetonide was O-tosylated, reduced and treated with potassium carbonate in methanol to afford the *L-ido* compound (**207**). Methanolysis to remove the acetonite protection, benzylation, followed by acid hydrolysis of the methylether gave the diol (**208**).

Reduction of the product yielded the ring opened iditol with two primary hydroxyl groups. These were selectively tritylated before O-benzylation of the secondary hydroxyl groups. Hydrolysis of the trityl groups gave the diol (**209**), which when treated under Swern conditions gave the corresponding dialdehyde. Finally the dialdehyde was reduced and cyclised using a low valent titanium reagent. This gave the *myo*-inositol derivative (**210**), together with other isomers, differing in stereochemistry at the hydroxylated centers. The *myo*-inositol derivative (**210**) has previously been reported as the key intermediate in the synthesis of *myo*-inositol (**205**) (S. Ozaki *et al., Tet. Lett.,* 1986, *27,* 3157).

(206) (207) (208) R=Bn

(210) (209)

a)i)TsCl/Py. ii)DIBAL-H. iii)K_2CO_3/MeOH/0°-rt. b)i)HCl/MeOH. ii)NaH/$PhCH_2Cl$ /DMF.iii)H_2SO_4/AcOH. c)i)$NaBH_4$/MeOH. ii)TrCl/Py. iii)NaH/$PhCH_2Cl$/DMF. iv)HCl/Dioxane. d)i)$(COCl)_2$/DMSO/Et_3N. ii)$TiCl_4$/Zn/Cu/THF

D-Glucose when subjected to Ferrier rearrangement conditions (see below) formed a tribenzyloxyhydroxy-cyclohexanone which was then converted to the pseudo-hexopyranosides (**211, 212** and **213**) (D.H.R.Barton *et al., J. Chem. Soc., Chem. Commun.,* 1988, 1184)

The Ferrier rearrangement (*J. Chem. Soc. Perkin Trans. 1,* 1979, 1455) has been utilized in many other syntheses, one such is the synthesis of aminocyclitols (**217**) (D. Semeria *et al., Synthesis,* 1983, 710). In this route methyl-α-D-glucopyranoside on treatment with triphenylphosphine, iodine and imidazole, followed by acetylation afforded the iodoacetate (**214**), which on reaction with silver fluoride in the presence of pyridine

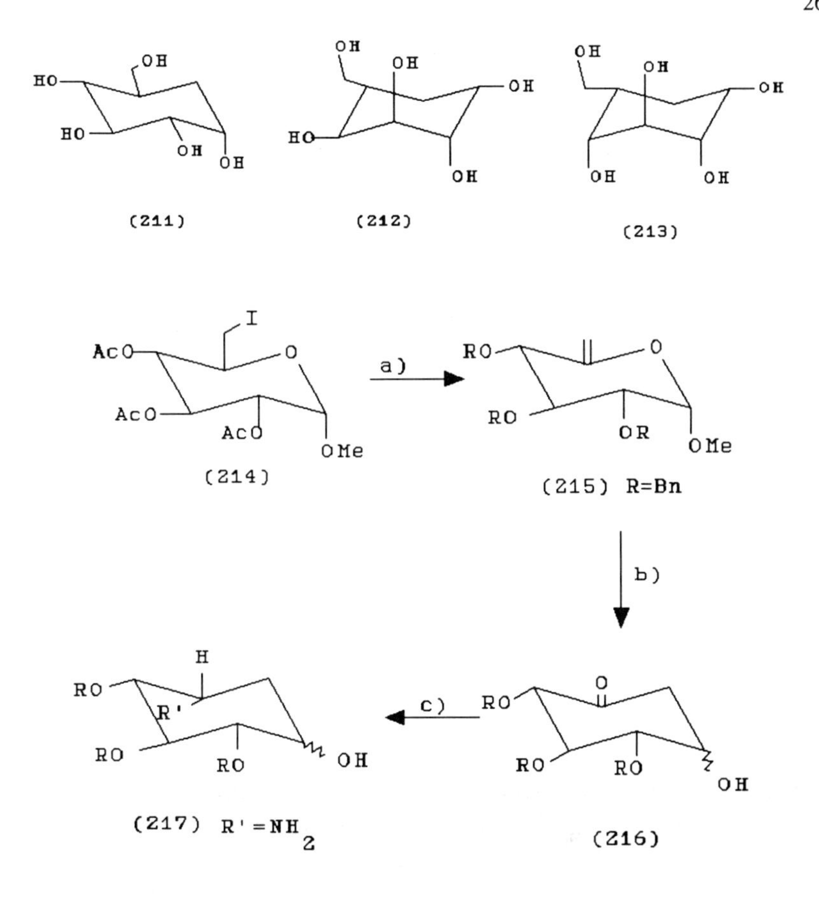

a)i)AgF/Py. ii)NH_3/MeOH. iii)NaH/DMF/$PhCH_2Br$. b)$HgCl_2$/Me_2CO/H_2O.
c)i)NH_2OH.HCl/MeOH/Py. ii)LAH/THF

eliminated hydroiodic acid to afford tri-O-benzyl-5-eno-pyranoside (**215**). After O-benzylation of the hydroxyl groups at C-2, C-3 and C-4, the pyranoside (**215**) was then subjected to the Ferrier rearrangement to give the cyclohexanone (**216**). This on reduction with sodium borohydride gave the protected cyclitols (**217, R=OH**), or when it reacted with hydroxylamine and the product oxime reduced with lithium aluminum hydride afforded the aminocyclitols (**217, R=NH_2**) [for similar preparations of amino pseudosugars see I. Pelyvas , F. Sztaricskai and R. Bognair (*J. Chem. Soc., Chem. Commun.,* 1984, 104; F. Chretien and Y. Chapleur (*ibid.,* 1984, 1268); A. S. Machado, A Olesker, S Castillion and G. Lukacs (*ibid,* 1985, 330)].

9. Chemistry of Monosaccharides

Araki and fellow workers (*Tet. Lett.,* 1987, *28,* 5853) showed that bromosugars react with olefines in the presence of a radical initiator and tributyltin hydride to afford branched sugars (**219**).

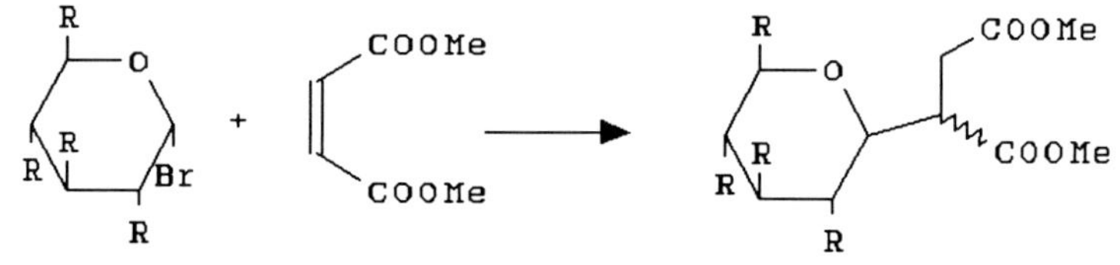

(218) R=OAc

(219)

Bu_3SnH/AIBN/PhH/Δ.

Glucose acetals have been cleaved in a one pot oxidation-esterification procedure which operates *via* the hemiacetal. Typically,

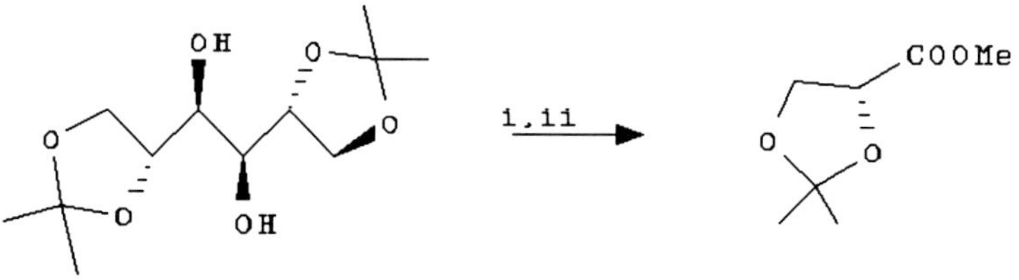

(220)

(221)

i)$NaIO_4/H_2O$/MeOH. ii)Br_2/H_2O/MeOH

sodium metaperiodate is added to a solution of the "hydroxy compound" in aqueous methanol, followed by oxidation with bromine in aqueous methanol (F.W. Lichtenthaler *et al., Synthesis,* 1988, 790) (**220** ——> **221**).

Pyranoid glycals were converted to their 2-deoxy-glycosidic orthoesters in very high yield, by treatment with methanol and palladium(II)acetate. These reactions are illustrated by a number of examples (C. W. Holzapfel *et al., Heterocycles,* 1989, *28,* 433) (**222** ——> **223**).

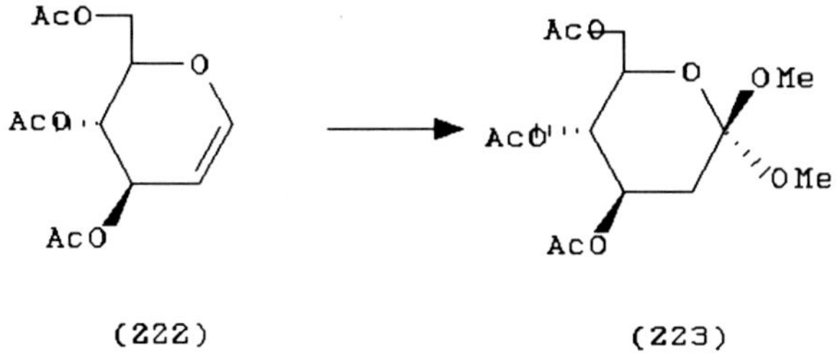

(222) (223)

MeOH/Pd(II)Ac

Following on from their earlier work on the regioselective acylation reactions of monosaccharides, Horton and colleagues (*Carbohydr. Res.,* 1985, *144,* 317, 332) have shown that 6-deoxy-glucals can be selectively silylated at the allylic hydroxyl group. Silyl base-stable protecting groups are invaluable in the manipulation of monosaccharides (**224** ——> **225**).

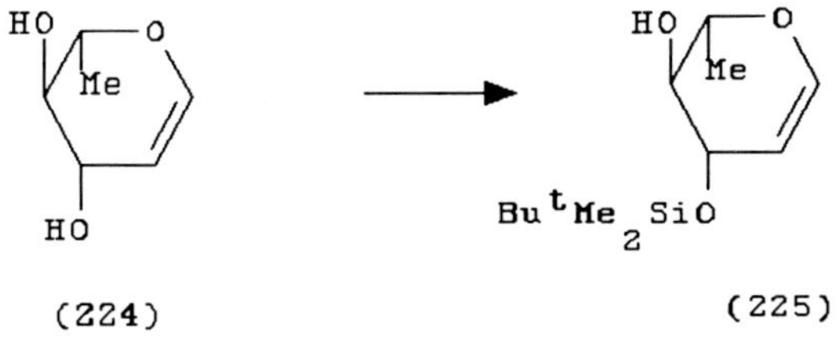

(224) (225)

DMF/Imidazole/tBuMe$_2$SiCl/rt/2h

1,1-Bis(benzamido)-1-deoxy-D-glucitol reacted with acetic anhydride in nitromethane to yield 1-acetamido-1-benzamido-1-

deoxy-D-glucitol. Essentially this amounts to the replacement of one benzamido group with an acetamido group (R.A. Cadenas *et al., An. Asoc. Quim. Argent.,* 1986, *74,* 251).

O-Benzylated monosaccharides when treated with Lewis acids, such as stannous chloride or titanium chloride, in methylene

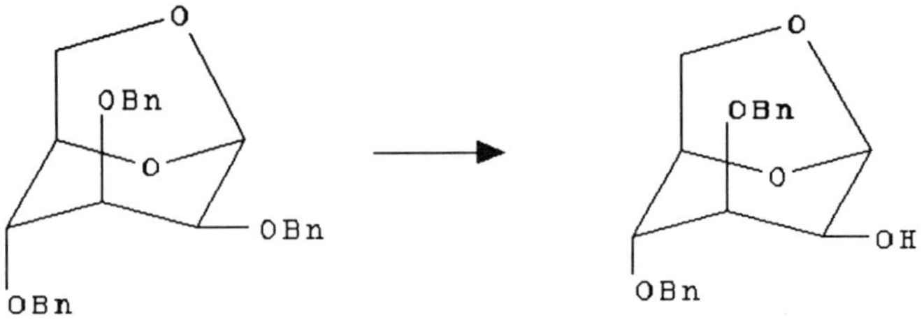

(226) (227)

$SnCl_4/CH_2Cl_2$/rt/0.5h

chloride at room temperature undergo regioselective de-O-benzylation at C-3 (H. Hori *et al., J. Org. Chem.,* 1989, *54,* 1346) (**226** ——> **227**).

Similarly, Watanabe and fellow workers (*Carbohydr. Res.,* 1986, *154,* 165) have demonstrated that when fully acetylated monosaccharides are heated under reflux in toluene containing bis(tributyl tin)oxide these compounds afford the 1-O-deacetylated analogues (**228** ——> **229**).

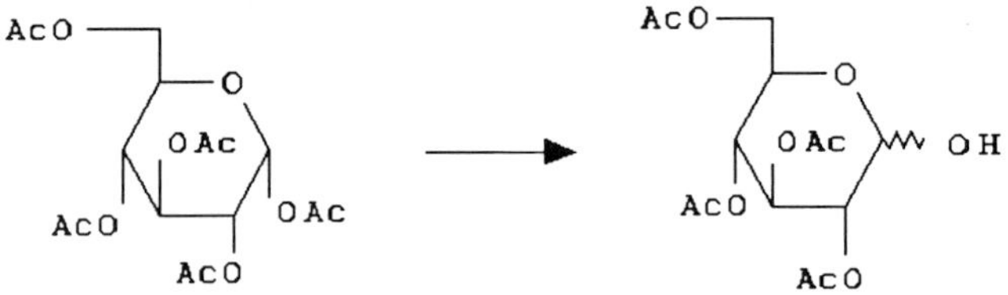

(228) (229)

$(Bu_3Sn)_2O$/Tol/reflux/3.5

1,2-Dichloro-1,1,3,3-tetraisopropyldisiloxane has been used to protect the 3,4- hydroxyl groups of *β*-glucosyldiglyceride. Then the primary 6′-hydroxyl function of the products can be phosphorylated (C.A.A. van Boeckel *et al., Tetrahedron,* 1985, *41,* 4557) (**230** ——> **231**).

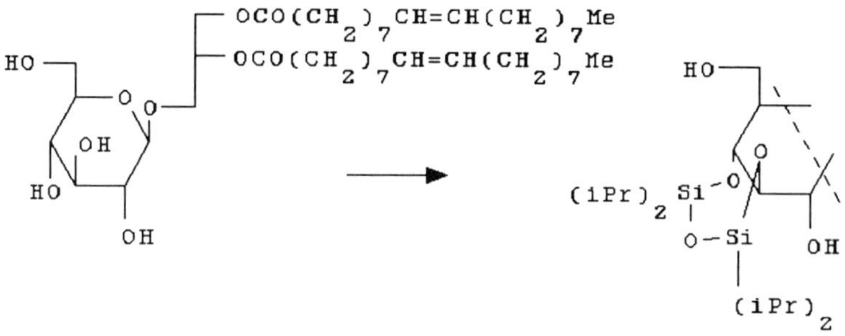

(230) (231)

$[Cl(iPr)_2Si]_2O$/DMF/Py.HCl

Brimacombe and Kabir (*Carbohydr. Res.,* 1986, *150,* 35) have shown that the hydroxypropenylfuranose (**233**) when osmylated affords the triol (**234**). The starting material for this procedure was the formyl derivative (**232**) which was subjected

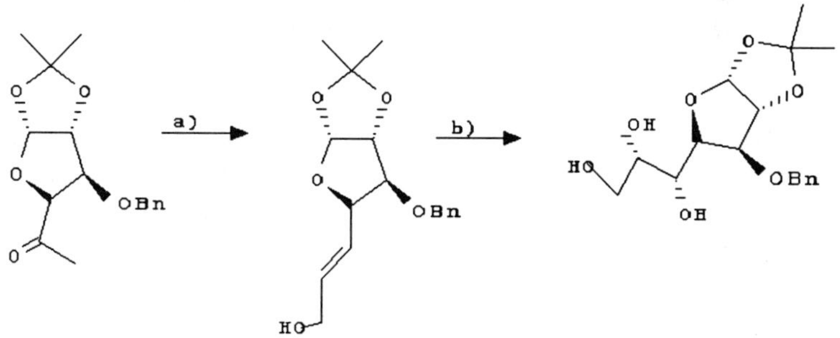

(232) (233) (234)

a)i)$Ph_3P=CHCHO$/PhH/ ▵. ii)DIBAL-H. b)OsO_4/Me-morphine-N-oxide/Me_2CO/H_2O

to a Wittig reaction with (formylmethylene)triphenylphosphorane prior, to selective reduction of the aldehyde group. The osmylation step was observed to follow Kishi's empirical rules, (*Tetrahedron,* 1984, *40,* 2247), resulting in the specific stereochemistry denoted in the structure (**234**).

The availability of enantiometrically pure D-glucose, D-fructose, D-mannitol *etc.* makes monosaccharides viable starting materials in many synthetic strategies towards chiral natural products. The subject has attracted the attention of many leading chemists and is exemplified by numerous reviews (F. W. Lichtenthaler *Nat. Prod. Chem., R. Atta-Ur, ed., Springer, Heidelberg, NY.,* 1986, 227; *New Aspects in Organic Chemistry, Z. Yoshida, ed., Elsevier, Amsterdam,* 1989, 351; J. Jurczak *et al., Tetrahedron,* 1986, *42,* 447; C. Hubschwerlen. *Synthesis,* 1986, 962; T. Rosen and C. Heathcock, *Tetrahedron,* 1986, *42,* 4909; B. Fraser-Reid *Acc. Chem. Res.,* 1985, *18,* 347; T. D. Inch, *Tetrahedron,* 1984, *40,* 3161). See the following chapters.

10. Structure and Physical properties of Monosaccharides

Increased information on the conformation of carbohydrates has prompted the integration of recommendations on the conformational nomenclature for these compounds (*I.U.P.A.C., Pure and Applied Chem.,* 1981, 1902., *Eur. J. Biochem.,* 1980, *3,* 295). Anomeric and exo-anomeric effects in carbohydrates in general are discussed by Tvarsoka and Blehal (*Adv. Carbohydr. Chem. Biochem.,* 1989, *47,* 45). Mathlouthi and Koenig (*ibid.,* 1986, *44,* 7) have reported on the vibrational spectra of carbohydrates, discussing the computational calculation of vibrational frequencies, Fourier-transform infrared and laser-Raman data. Hydrogen bonding in sugars has been hypothesized to be correlated to the degree of sweetness (R. S. Shallenberger *et al., Nature,* 1967, *216,* 480; G. G. Birch *C.R.C. Critical Rev. Food Sci. Nutr.* 1976, *8,* 54). Szarek and colleagues (*Can. J. Chem.,* 1984, *62,* 1512) studied the relationship between the vibrational spectra and the structures of some monosaccharides to evaluate the validity of the existence of a correlation between hydrogen bonding and sweetness.

Egli and Djerassi (*Helv. Chem. Acta,* 1982, *65,* 1898) have described a computer programme to assist structure determination of unknown compounds. In this they also discuss the possibilities of using this programme to evaluate the structures of carbohydrates based on ^{1}H N.M.R spectral data. Recent advances in microcomputers have made spectroscopical data accumulation by Fourier-transform techniques possible. This has opened the way to many ^{1}H, ^{13}C N.M.R applications.

^{13}C N.M.R assignments, sampling techniques, and the applications of N.M.R spectroscopy in interpreting the conformational and structural integrity of monosaccharides are surveyed by Bock and Pederson (*Adv. Carbohydr. Chem. Biochem.,* 1983, *41,* 27).

Barber, Sedgwick and colleagues (*J. Chem. Soc., Chem. Commun.,* 1981, 325) revolutionized the field in mass spectrometry by the fast atom bombardment (F.A.B) technique for the analysis of non volatile substances. Here an accelerated beam of atoms are fired towards a viscous liquid containing the sample to be analyzed. Upon impact of the atom beam on the sample matrix, kinetic energy is transformed to the surface molecules sputtering the liquid into the high vacuum of the ion source. The "gas phase" ions thus generated without prior volatilization are analyzed. Recently , Dell reported the use of this technique in carbohydrates analysis (*Adv. Carbohydr. Chem. Biochem.,* 1987, *45,* 19).

The biochemical pathways in *Escherichia coli* have been monitored *in vivo* using D-[U-^{13}C]glucose and ^{13}C N.M.R. spectroscopy (N. E. Mackenzie, *J. Chem. Soc., Chem. Commun,* 1983, 145). The diastereomers of hexa and tetra alditols (**236** and **237**) were converted to their peracetates and it was then demonstrated that individual isomers, within each group, could be identified by comparing vicinal spin-spin coupling constants determined by ^{1}H N.M.R. spectroscopy with those predicted from a generalized Karplus equation.(S. Masumune *et al., ibid.,* 1986, 261). In a similar ^{1}H NMR technique the *threo* and *erythro* isomers of 1-C-substituted glycerols have been identified

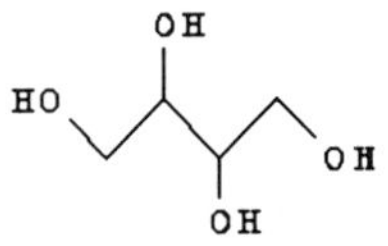

(236) 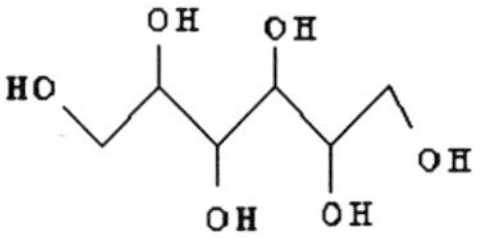 (237) 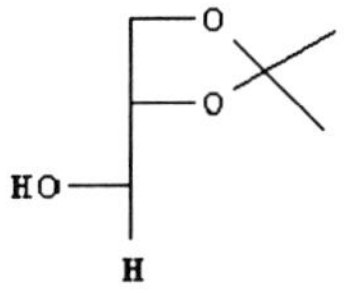 (238)

using the difference between the chemical shifts of the two methyl resonances of the isopropylidine groups.(**238**). Many other examples are given (E. H. El Ashry *ibid.,* 1986, 1025).

Solutions of D-idose in deuterated water were analyzed by one and two dimensional ^{13}C N.M.R spectroscopy. α- and β-Pyranose forms (**239** and **240**) along with the two furanoses, the hydrate form of the open chain, and the carbonyl forms were been detected (J. R. Snyder and A. S. Serianni, *J. Org. Chem.,* 1986, *51,* 2694., also see A. S. Serianni, *J. Amer. Chem. Soc.,* 1982, *104,* 4037). Reuben (*J. Amer. Chem. Soc.,* 1985, *107,* 5867) studied the effects of solvation on the ^{13}C N.M.R spectrum of D-idose. The relative abundance of α-, β-pyranoses to the α-, β-furanoses was found to be solvent dependent. The same author had previously shown that partially deuterated hydroxyl groups in monosaccharides have unique chemical shifts (*ibid.,* 1984, *106,* 6180). Praly and Lemieux (*Can. J. Chem.,* 1987, *65,* 213) applied ^{13}C N.M.R resonance to study

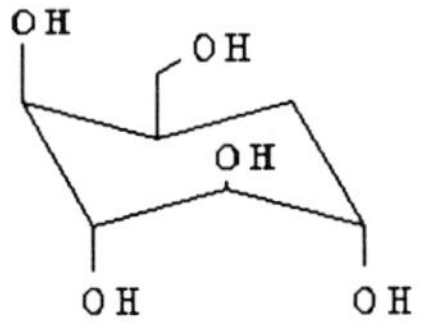

(239) 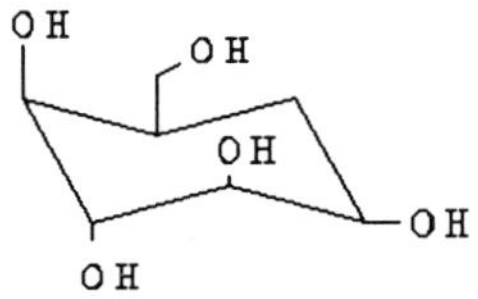 (240)

the effects of solvent polarity and hydrogen-bond formation on the conformational equilibria for some 2-substituted tetrahydropyrans. Water was found to have a strong exo-anomeric effect. This indicates that this solvent hydrogen bonds to the ring oxygen atom. Further it was shown that anomeric hydroxyl groups prefer the equatorial orientation in this solvent.

Duin and fellow workers (*J. Org. Chem.,* 1986, *51,* 1298) reported empirical force field calculations upon a series of diols. Diols are converted to borate esters and are of importance in the configurational analysis of monosaccharides, they are also utilized in separation techniques. The inherent flexibility of the furanose ring introduces structural modifications in the naturally occurring nucleic acids, thus playing an important role in biological functions. Olson and Sussmann (*J. Amer. Chem. Soc.,* 1982, *104,* 270) compared theoretical and experimental evidence to determine the degree of flexibility of the furanose ring. The potential energy of the pseudorotational movements of the furanose ring were also estimated. The puckering dependence of ribose in the conformation of nucleic acids has stimulated interest in theoretical calculations to obtain the dynamics and energetics of the pseudorotation cycle of D-ribofuranose. The relative energies of the respective conformations were found to be dependent on torsional energy (S. C. Harvey and M. Prabhakaran, *J. Amer. Chem. Soc.,* 1986, *108,* 6128).

Double Fourier-transform ("**2D**") NMR methods were employed to obtain ^{13}C and ^{1}H chemical shifts for mono and disaccharides (G. A. Morris and L. D. Hall, *J. Amer. Chem. Soc.,* 1981, *103,* 4703). In pyranoses relative orientations of substituents and their electronegativity have been used to predict coupling constants, thus a additivity rule for *anti* and *gauche* vicinal ^{1}H-^{1}H coupling constants has been developed (C. Altone and C. A. G. Haasnoot, *Org. Magn. Reson.,* 1980, *13,* 417; *Chem. Abstr.,* 1981, *94,* 15999x).

The tautomerisation or mutarotation reaction that occurs upon the addition of "crystalline reducing" sugars to an appropriate solvent has been of interest from a chemical and biochemical viewpoint. The kinetics of this phenomena in water has been studied (P. W. Wertz *et al., J. Amer. Chem. Soc.,* 1981, *103,* 3916). Back and Polavaragm (*Carbohydr. Res.,* 1987, *165,* 173) reported on the time dependent Fourier-transform infrared spectra of galactose, fucose and glucose substrates. The changes observed in the case of α-fucose were found to adhere to first order reaction kinetics.

Second Supplements to the 2nd Edition of Rodd's Chemistry of Carbon Compounds, Vol. IE/F/G, edited by M. Sainsbury

Chapter 23b

MONOSACCHARIDES: USE IN SYNTHESIS AS CHIRAL TEMPLATES

KARL J. HALE

Introduction

Early work on the use of carbohydrate derivatives for asymmetric induction has been reviewed (T.D. Inch, *Adv. Carbohydrate Chem. and Biochem.*, 1972, **27**, 191). The present chapter deals with developments that have occurred subsequently.

1. Asymmetric reduction

(a) Asymmetric homogenous hydrogenation with transition metal complexes of monosaccharide phosphines

Several comprehensive review articles have appeared that discuss aspects of this topic (V. Čaplar, G. Comisso, and V. Šunjić, *Synthesis*, 1981, 85; M. Nógrádi, *"Stereoselective Synthesis"*, VCH, Verlagsgesellschaft, Weinheim, 1987; H. Brunner, *Topics in Stereochem.*, 1988, 18; H. Brunner, *Synthesis*, 1988, 645).

(i) Homogenous hydrogenation of didehydroamino acids

Homogenous hydrogenation of didehydroamino acids, catalysed by rhodium (I)-monosaccharide phosphine complexes, has emerged as a highly effective method for the synthesis of optically pure α-amino acids. The best enantioselectivities are attained with rhodium (I) catalysts derived from phenyl 4,6-*O*-(*R*)-benzylidene-2,3-*O*-bis(diphenylphosphino)-β-D-glucopyranoside (**1**) (R. Selke and H. Pracejus, *J. Mol. Catal.*, 1986, **37**, 213; W. Vocke, R. Hanel, F.U. Flother, *Chem Tech.(Leipzig)*, 1987, **39**, 123; E.I. Klabunovskii, B.D. Polnikov, *Khim. Prom.-st (Moscow)*, 1986, **9**, 518, *CA* 1987, **106**, 17578), and (*3R,4R*)-3,4-bis-diphenylphosphinoxytetrahydropyran (**2**) (I. Habuš, Z. Raza and V. Šunjić, *J. Mol. Catal.*, 1987, **42**, 173). These provide *N*-acyl α-amino acids with the 2*S* configurat-

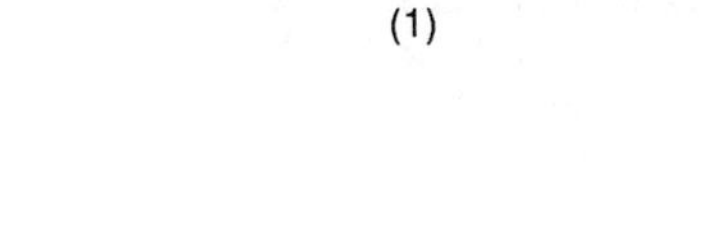

Ph O O O OPh Ph_2PO $OPPh_2$

(1)

Ph_2PO Ph_2PO O

(2)

TABLE 1
Enantioselective hydrogenation of didehydroamino acid derivatives to *S*-amino acids

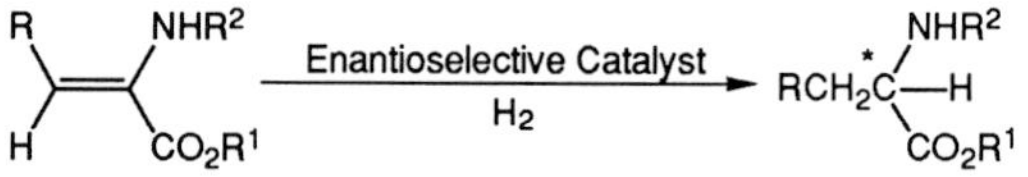

(1) R = Ph, R^1 = H, R^2 = Ac
(3) R = Ph, R^1 = Me, R^2 = Ac
(5) R = H, R^1 = H, R^2 = Ac
(7) R = H, R^1= Me, R^2 = Ac
(9) R = Ph, R^1 = H, R^2 = Bz

(2) R = Ph, R^1 = H, R^2 = Ac
(4) R = Ph, R^1 = Me, R^2 = Ac
(6) R = H, R^1 = H, R^2 = Ac
(8) R = H, R^1 = Me, R^2 = Ac
(10) R = Ph, R^1 = H, R^2 = Bz

Enantioselective Catalyst		Substrate	Conditions	Yield (%)	e.e. (%)	Ref.
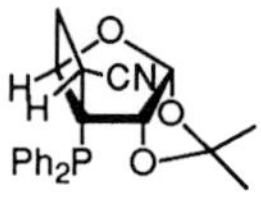	$Rh(C_8H_{14})_2Cl$, C_6H_6, MeOH	(1)	1 atm, 72 h, r. temp	84	91.6 (**S**)	1
		(1)	15 atm, 10 h, r. temp	100	85.6 (**S**)	
		(3)	1 atm, 72 h, r. temp	51	70.6 (**S**)	
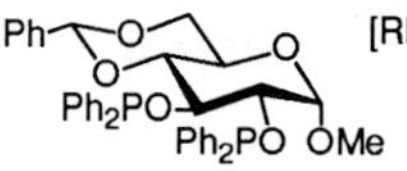	$[Rh(C_7H_8)Cl]_2$, $AgPF_6$, EtOH	(1)	1 atm, 90 min, 0°C	100	75.0 (**S**)	2
		(5)	1 atm, 60 min, -20°C	100	80.0 (**S**)	
		(7)	1 atm, 30 min, 0°C	100	78.0 (**S**)	
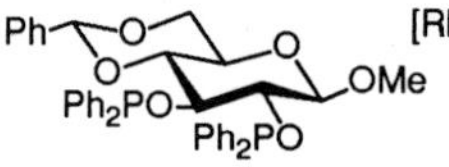	$[Rh(C_7H_8)Cl]_2$, $AgPF_6$, EtOH	(3)	1 atm, 30 min, r. temp	100	80.0 (**S**)	3
		(5)	1 atm, 5 min, r. temp	100	91.0 (**S**)	
		(7)	1 atm, 10 min, r. temp	100	88.0 (**S**)	
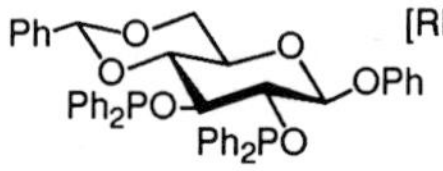	$[Rh(C_7H_8)Cl]_2$, $AgBF_4$, MeOH	(1)	0.1 MPa,-27°C		99.3 (**S**)	4
		(3)	0.1 MPa,-20°C		95.4 (**S**)	
		(9)	5.0 MPa, 25°C		95.0 (S)	
	$([Rh(C_7H_8)_2]ClO_4)$, C_6H_6, EtOH	(1)	1.48 atm, 24h, -15°C, Et_3N	79	90.4 (**S**)	5

1. S. Saito, Y. Nakamura, Y. Morita, *Chem. Pharm. Bull.*, 1985, **33**, 5284.
2. W.R. Cullen, Y. Sugi, *Tetrahedron Lett.*, 1978, **19**, 1635.
3. Y. Sugi, W.R. Cullen, *Chem. Lett.*, 1979, 39.
4. R. Selke, H. Pracejus, *J. Mol. Catal.*, 1986, **37**, 215.
5. I. Habuš, Z. Raza, V. Šunjić, *J. Mol. Catal.*, 1987, **42**, 173.

TABLE 2
Enantioselective hydrogenation of didehydroamino acid derivatives to *R*-amino acids

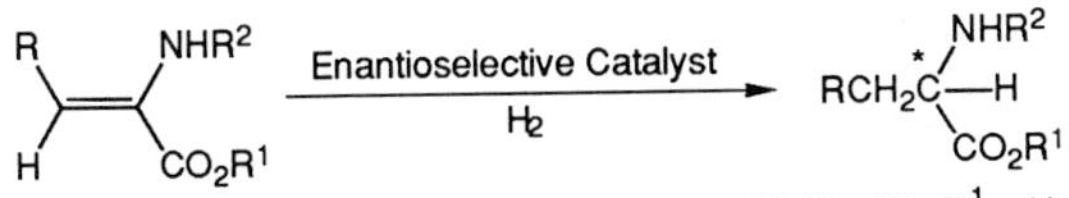

$$\text{RCH=C(NHR}^2\text{)CO}_2\text{R}^1 \xrightarrow[\text{H}_2]{\text{Enantioselective Catalyst}} \text{RCH}_2\text{C}^*\text{H(NHR}^2\text{)CO}_2\text{R}^1$$

(1) R = Ph, R^1 = H, R^2 = Ac
(3) R = Ph, R^1 = Me, R^2 = Ac
(5) R = H, R^1 = H, R^2 = Ac
(7) R = Ph, R^1 = H, R^2 = Bz

(2) R = Ph, R^1 = H, R^2 = Ac
(4) R = Ph, R^1 = Me, R^2 = Ac
(6) R = H, R^1 = H, R^2 = Ac
(8) R = Ph, R^1 = H, R^2 = Bz

Enantioselective Catalyst	Substrate	Conditions	Yield (%)	e.e. (%)	Ref.
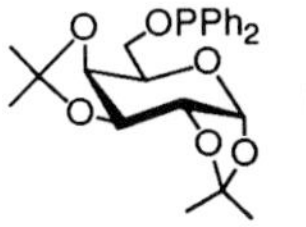 $Rh_2(C_8H_{12})_2Cl_2$, EtOH, C_6H_6	(1)	1 atm, 24h, r. temp, Et_3N	99	67.0 (**R**)	1
$Rh_2(C_8H_{12})_2Cl_2$, EtOH, C_6H_6	(1)	1 atm, 24h, r. temp, Et_3N	100	48.0 (**R**)	2
	(3)	1 atm, 24h, r. temp, Et_3N	96	41.0 (**R**)	
	(7)	1 atm, 24h, r. temp, Et_3N	100	78.0 (**R**)	
$[Rh(C_8H_{12})]ClO_4$, EtOH	(1)	10^5 Pa, 20 min, r. temp	100	55.0 (**R**)	3
	(1)	10^5 Pa, 1 h, 0°C	100	70.0 (**R**)	
	(3)	10^5 Pa, 1 h, r. temp	100	52.0 (**R**)	
 $[Rh(C_7H_8)_2]ClO_4$, EtOH, C_6H_6	(1)	1.48 atm, 7 h, -15 to -20°C	53	62.7 (**R**)	4
$Rh_2(C_8H_{12})_2Cl_2$, EtOH, C_6H_6	(1)	1 atm, 24h, r. temp, Et_3N	100	10.0 (**R**)	2
	(3)	1 atm, 24h, r. temp, Et_3N	100	7.0 (**R**)	

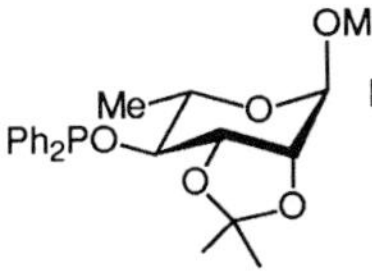

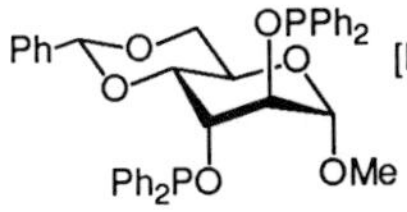

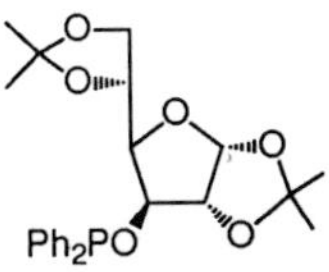

1. M. Yamashita et al., *Bull. Chem. Soc. Japan*, 1982, **55**, 2917.
2. M. Yamashita et al., *Bull. Chem. Soc. Japan.*, 1986, **59**, 175.
3. D. Sinou, G. Descotes, *React. Kinet. Catal. Lett*, 1980, **14**, 463.
4. I. Habuš, Z. Raza, V. Šunjić, *J. Mol. Catal.*, 1987, **42**, 173.

ion in ee's ranging from 90.4 to 99.3% (Table 1). The corresponding *2R*-enantiomers can be synthesised in somewhat lower optical yield using the Rh(I) complexes shown in Table 2.

The factors controlling enantioselectivity in didehydroamino acid hydrogenations with monosaccharide phosphine-Rhodium (I) catalysts tend to be rather complex and variable. Aside from catalyst and substrate structure, a number of other parameters can influence the stereochemical outcome of reduction. For instance, varying the temperature of a particular hydrogenation can often afford products of different optical purity. An example of this behavior can be found in the hydrogenation of α-acetamidocinnamic acid with [(**3**)Rh(COD)]BF_4 in THF. Lowering the reaction temperature from 25°C to 0°C caused an increase in product ee

Scheme 1

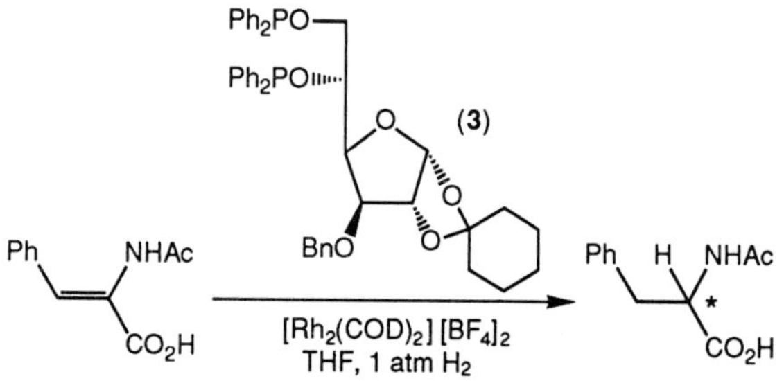

$[Rh_2(COD)_2][BF_4]_2$, THF, 1 atm H_2

temp, °C	ee
25	40% (*R*)
0	52% (*R*)
-78	10% (*R*)

from 40 to 52%, whereas cooling it to -78°C provided a product with only 10% ee (Scheme 1) (T.H.Johnson and G.Rangarajan, *J. Org. Chem.*, 1980, **45**, 62).

Changing the hydrogenation solvent can likewise adjust product enantioselectivity. For instance, when methyl α-acetamidocinnamate was hydrogenated with the cationic complex derived from (**1**), $[Rh(C_7H_8)Cl]_2$, and $AgBF_4$ (Scheme 2), use of benzene as the reaction solvent gave a product with 20%

Scheme 2

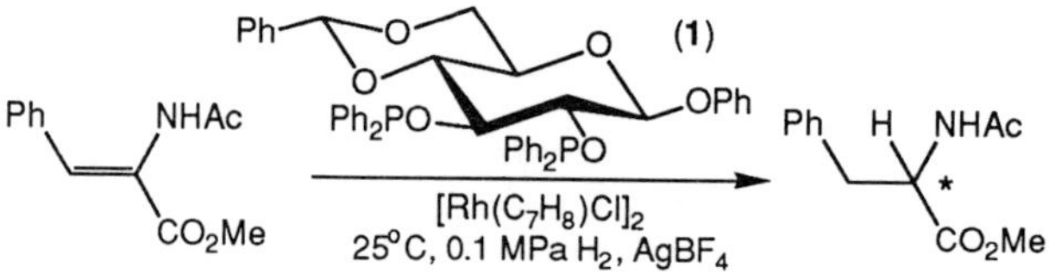

$[Rh(C_7H_8)Cl]_2$, 25°C, 0.1 MPa H_2, $AgBF_4$

Solvent	ee
C_6H_6	69% (*S*)
EtOH	89% (*S*)

ee lower than that obtained from ethanol (R. Selke et al., *loc cit.*).

The choice of catalyst counterion can further modulate enantioselectivity, as illustrated by the hydrogenation of α-acetamidocinnamic acid with the cationic complexes [(Glucophinite)Rh(COD)]Y in tetrahydrofuran. Here, alteration of the metal counterion from tetraphenylborate to tetrafluoroborate led to a 10% improvement in optical induction (Scheme 3) (T.H. Johnson et al., *loc cit.*).

Scheme 3

Ph_2PO Ph_2PO (3) BnO

Ph NHAc CO_2H → $[Rh_2(COD)_2][Y]_2$ 25°C, 1 atm H_2 → Ph H NHAc * CO_2H

Counterion (Y)	% ee (R) THF	EtOH
BF_4^-	40	30
PF_6^-	32	30
BPh_4^-	30	35
ClO_4^-	36	28

Catalyst stoichiometry can have a substantial impact on the stereochemical course of a hydrogenation. Such was the case for the reduction of α-acetamidocinnamic acid with the chiral catalyst prepared from 2 mol of (**4**) and 1 mol of Rh(cyclooctene)$_2$Cl (**5**). This delivered the *S*-amino acid in 92% ee, while an

Scheme 4

(4) CN H Ph_2PO

Ph NHAc CO_2H → $Rh_2(COD)_2$ Cl (**5**) H_2, 1 atm 25°C, C_6H_6-MeOH → Ph H NHAc * CO_2H

Mole Ratio [**4**]:[**5**]	Time	% yield	% ee
2:1	72 h	84	92 (*S*)
1:1	1 week	100	31 (*S*)

alternative catalyst constructed from 1 mol of (**4**) and 1 mol of (**5**), provided a product with only 31% optical yield (Scheme 4) (S. Saito, Y. Nakamura, Y. Morita, *Chem.*

Pharm. Bull., 1985, **33**, 5284).

While an increase in hydrogenation pressure often enhances the rate of reduction with rhodium-monosaccharide phosphinites, its effect on enantioselectivity is usually detrimental, as shown in Scheme 5 (S. Saito et al., *loc cit.*).

Scheme 5

Ph, NHAc, CO_2Me → (4) Ph_2PO; $Rh_2(COD)_2$ Cl (**5**), 25°C, C_6H_6-MeOH → Ph H NHAc * CO_2Me

H_2Pressure (atm)	Time	% yield	% ee
1	72 h	51	71(*S*)
30	10 h	100	43(*S*)

(ii) Homogenous hydrogenation of acrylic acid derivatives

A variety of acrylic acid systems have been asymmetrically hydrogenated using monosaccharide phosphine-Rh(I) complexes (Table 3). As with didehydroamino acid derivatives, enantioselectivities are often maximal when the substrates are capable of strong bidentate coordination to a conformationally rigid chiral catalyst. Thus, dimethyl itaconate was reduced in 100% optical yield using the bicyclic ligand (**6**) and $[(C_8H_{12})RhCl_2Rh(C_8H_{12})]$ as the homogenous catalyst (Scheme 6) (M. Yamashita et al., *Bull. Chem. Soc. Japan*, 1986, **59**, 175).

Scheme 6

CO_2Me, CH_2CO_2Me → (6) OMe, Ph_2PO; $[Rh_2(COD)_2]$ $[Y]_2$, 25°C, 1 atm H_2, C_6H_6-EtOH → Me, H, CO_2Me, *, CH_2CO_2Me

Counterion (Y)	% ee
Cl^-	100 (*S*)
ClO_4^-	73 (*S*)

Again, many of the factors that determine enantioselectivity in homogenous hydrogenations of didehydroaminoacid derivatives also apply to acrylic acid systems.

TABLE 3
Enantioselective hydrogenation of acrylic acid derivatives with monosaccharide phosphine/phosphinite-rhodium (I) catalysts

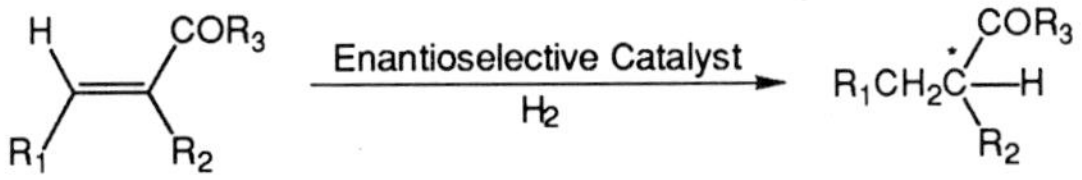

(1) $R_1 = H, R_2 = CH_2CO_2H, R_3 = OH$
(3) $R_1 = H, R_2 = CH_2CO_2Me, R_3 = OH$
(5) $R_1 = H, R_2 = CH_2CO_2Me, R_3 = OMe$
(7) $R_1 = Ph, R_2 = Me, R_3 = OH$

Enantioselective Catalyst, H_2

(2) $R_1 = H, R_2 = CH_2CO_2H, R_3 = OH$
(4) $R_1 = H, R_2 = CH_2CO_2Me, R_3 = OH$
(6) $R_1 = H, R_2 = CH_2CO_2Me, R_3 = OMe$
(8) $R_1 = Ph, R_2 = Me, R_3 = OH$

Enantioselective Catalyst	Substrate	Conditions	Yield (%)	e.e. (%)	Ref.
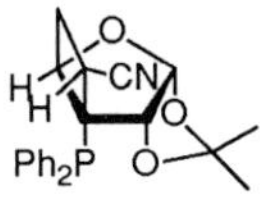 $Rh(C_8H_{14})_2Cl$, C_6H_6, MeOH	(1) (3)	1 atm, 18 h, r. temp 1 atm, 72h, r. temp	100 84	70 (**R**) 90 (**R**)	1
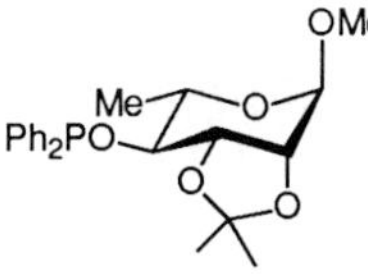 $Rh_2(C_8H_{12})_2Cl_2$, C_6H_6, EtOH	(5)	1 atm, 24h, r. temp, Et_3N	100	100 (**S**)	2
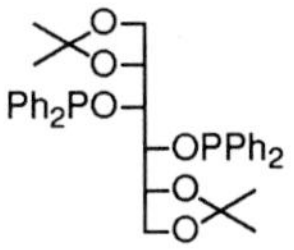 $Rh_2(C_8H_{12})_2Cl_2$, C_6H_6, EtOH	(1)	1 atm, 24h, r. temp, Et_3N	50	78 (**S**)	3
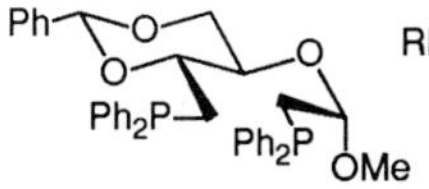 $Rh_2(C_8H_{12})_2Cl_2$, $NaClO_4$, EtOH	(1) (7)	1 atm, r. temp, Et_3N 1 atm, 24h, r. temp, Et_3N	100 100	34 (**S**) 58 (**S**)	4

1. S. Saito, Y. Nakamura, Y. Morita, *Chem. Pharm. Bull.*, 1985, **33**, 5284.
2. M. Yamashita et al., *Bull. Chem. Soc. Japan*, 1986, **59**, 175.
3. M. Yamashita et al., *Bull. Chem. Soc. Japan*, 1989, **62**, 942.
4. D. Lafont, D. Sinou, G. Descotes, *J. Organometal. Chem.*, 1979, **169**, 87.

For instance, in the reduction of dimethyl itaconate (Scheme 6), changing the metal counterion from chloride to perchlorate lowered the ee of the (*S*)-enantiomer to 73% (M. Yamashita, *Bull. Chem. Soc. Japan*, 1989, **62**, 942). Solvent variations and addition of base can similarly influence asymmetric induction (Scheme 7) (D. Sinou, G. Descotes, *loc cit.*).

Scheme 7

(**7**)

[Rh(COD) (**7**)] ClO_4
25°C, 10^5 Pa H_2

Solvent	% ee
EtOH + 0.1 eq Et_3N	51 (*S*)
EtOH + 2.0 eq Et_3N	28 (*S*)

In addition to the chiral catalysts derived from rhodium complexes of monosaccharide phosphines and phosphinites, a small number of ruthenium chelates have been used for asymmetric hydrogenation. Thus, the combination of dodecacarbonyltriruthenium with chiral phosphinites (**3**) and (**8**) catalysed the reduction of α,β-unsaturated acids, but the optical yields were disappointing (8-38%) (T.H. Johnson, L.A. Siegle, V.K. Chaffin, *J. Mol. Catal.*, 1980, **9**, 307).

Asymmetric reduction of α,β-unsaturated ketones can be accomplished in the

(**3**) (**8**)

presence of 1,2-O-isopropylidene α-D-glucofuranose and $RuCl_2PPh_3$, but again the product ee's are low (Scheme 8) (G. Descotes, D. Sinou, *Tetrahedron Lett.*, 1976, 4083).

(b) Asymmetric heterogenous hydrogenation

Asymmetric heterogenous hydrogenation with monosaccharide-based transition metal catalysts has received comparatively little attention, despite the potential advantages for product isolation and development of a reusable catalyst. A cationic Rh-complex of PhO-β-glup (**1**) bound to a commercial sulphonated styrene-divinylbenzene copolymer has been prepared. This catalyst improved enantioselectivi-

Scheme 8

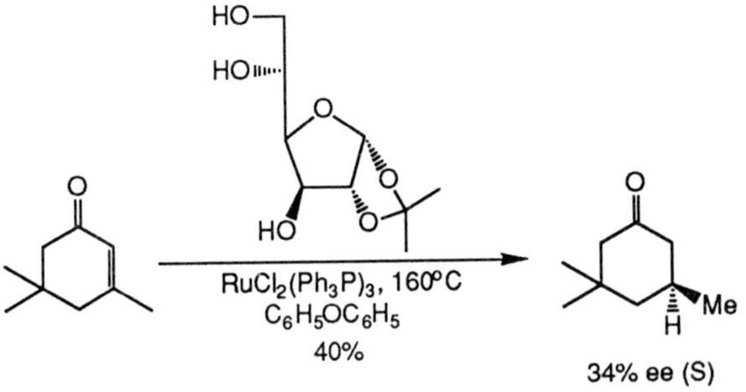

$RuCl_2(Ph_3P)_3$, 160°C
$C_6H_5OC_6H_5$
40%

34% ee (S)

ty in the hydrogenation of didehydroamino acid derivatives by 3% relative to its homogenous counterpart (R. Selke, *J. Mol. Catal.*, 1986, **37**, 227). More recent studies, however, suggest that the benzylidene acetal is actually cleaved by the acid functionality of such polymeric supports (R. Selke, K. Haeupke, H.W. Krause, *J. Mol. Catal.*, 1989, **56**, 315).

(c) Asymmetric hydrosilylation

Asymmetric reduction of aromatic alkyl ketones with silanes and chiral rhodium complexes frequently delivers alcohols of low optical purity. However, with complexes derived from monosaccharide phosphinites, ee's up to 65% have been achi-

Scheme 9

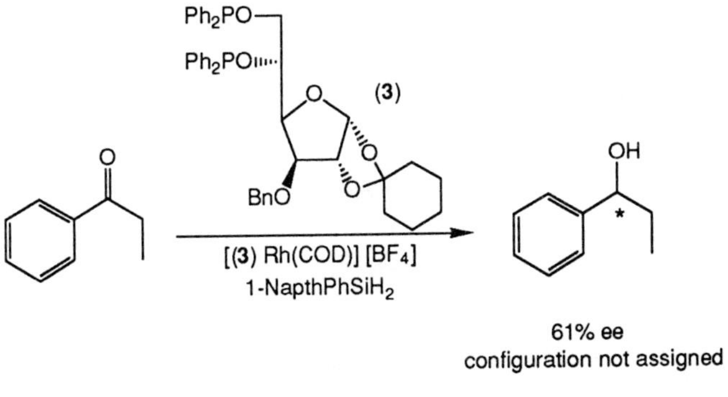

(**3**)

[(**3**) Rh(COD)] [BF_4]
1-NapthPhSiH_2

61% ee
configuration not assigned

eved (T.H. Johnson, K.C. Klein, and S. Thomen, *J. Mol. Catal.*, 1981, **12**, 37).

(d) Chiral reducing agents derived from monosaccharides

The chiral complex formed from lithium aluminium hydride and 3-*O*-benzyl-1,2-*O*-cyclohexylidene-α-D-glucofuranose (Landor's reagent) has been used to asymmetrically reduce ketones, ketone oximes, their *O*-ether derivatives, and *N*-phenylazomethines to optically active alcohols, primary amines, and secondary amines of the *S*-configuration (Scheme 10) (S.R Landor, B.J. Miller, and A.R. Tatchell, *J. Chem. Soc.*, 1966, 1822; *ibid.*, 1966, 2280; *ibid.*, 1967, 197; S.R. Landor et al., *J. Chem Soc., Perkin Trans. 1*, 1974, 1902; *ibid.*, 1978, 605; S.R Landor et al., *Bull. Chem. Soc. Japan*, 1984, **57**, 1658). In the case of alcohols and secondary amines, the product stereochemistry is thought to arise through preferential transfer of H_2 from complex (**9**) to the ketone or azomethine, when the substrate has its highest priority substituent oriented away from the benzyloxy group. For oximes

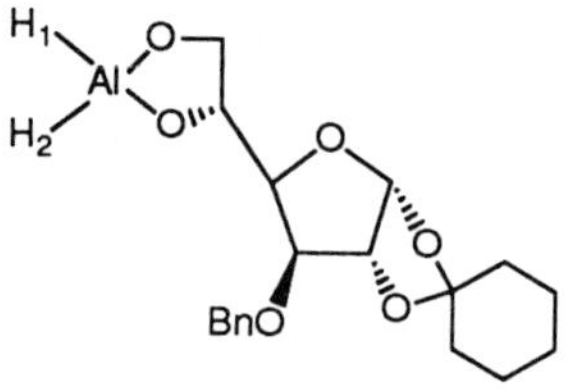
H1
H2
Al
O
BnO

(**9**)

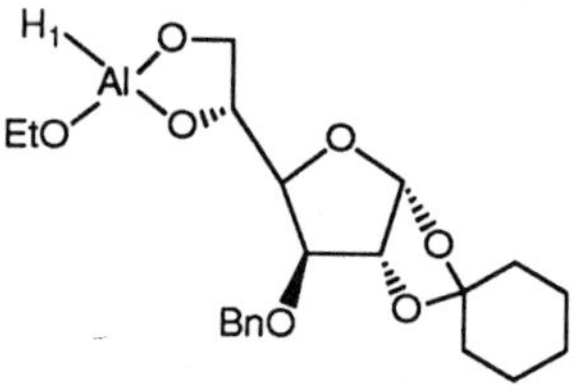
H1
EtO
Al
O
BnO

(**10**)

a chelate is believed to be initially formed with (**9**), which then undergoes reduction by H_2 from a second molecule of (**9**). Alcohols and amines of the *R*-configuration, are available using the ethanol modified complex (**10**), which effects reduction by transfer of H_1. Changing the 3-*O*-benzyl ether to a cyclohexyl methyl ether did not

Scheme 10

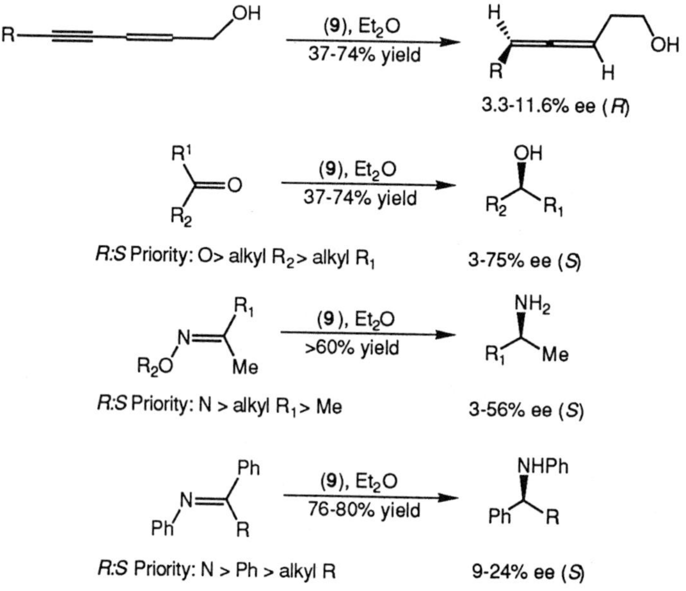
R
OH
(9), Et2O
37-74% yield
H
R
H
OH
3.3-11.6% ee (R)
R1
O
R2
(9), Et2O
37-74% yield
OH
R2
R1
R:S Priority: O> alkyl R2> alkyl R1
3-75% ee (S)
R1
N
R2O
Me
(9), Et2O
>60% yield
NH2
R1
Me
R:S Priority: N > alkyl R1> Me
3-56% ee (S)
Ph
N
Ph
R
(9), Et2O
76-80% yield
NHPh
Ph
R
R:S Priority: N > Ph > alkyl R
9-24% ee (S)

significantly improve the enantioselectivity of these reductions (S.R. Landor et al., *J. Chem. Soc. Perkin Trans. 1*, 1984, 493). Reagent (**9**) could also be used to convert alkene-yn-ols to *R*-allenic alcohols but the optical yields were low (Scheme 10).

Complexes of lithium tetrahydridoaluminate with C_2 symmetrical diols, such as 1,4:3,6-dianhydro-D-mannitol and 1,3:4,6-di-*O*-benzylidene-D-mannitol, have been examined for the asymmetric reduction of ketones, but did not prove very effective (N. Baggett, P. Stribblehill, *J. Chem. Soc. Perkin Trans. 1*, 1977, 1123).

Comprehensive accounts are available on the use of monosaccharide modified lithium aluminium hydride reagents as chiral reductants, which discuss these applications in more detail (H. Haubenstock, *Topics in Stereochem.*, 1983, **14**, 231; E.R. Grandbois, S.I. Howard, J.D. Morrison, *"Asymmetric Synthesis"*, Academic Press, New York, 1983, **2**, Chapter 3).

Alcohols with an *S*-configuration can be obtained in modest ee (28-68%) when aromatic alkyl ketones are treated with a combination of sodium borohydride, zinc chloride and 1,2:5,6-di-*O*-isopropylidene-α-D-glucofuranose in tetrahydrofuran (Scheme 11) (A. Hirao et al., *J. Chem. Soc. Chem. Commun.*, 1979, 807; *Bull. Chem. Soc. Japan.*, 1981, **54**, 1424).

Scheme 11

(1) $NaBH_4$ (1 equiv), $ZnCl_2$ (0.3 equiv), THF
(2 equiv), Ketone (1 equiv)
30°C, 48h, 98-100%.

R = Me (50% ee, ***S***)
R = *n*-Et (68% ee, ***S***)
R = *n*-Pr (58% ee, ***S***)
R = *i*-Pr (28% ee, ***S***)

Replacing the Lewis acid component with a suitable carboxylic acid gives access to the corresponding *R*-enantiomers in 57-85% ee (Scheme 12). The optimum reduction conditions employ a 0.5: 1.0: 1.2: 2.0 mixture of ketone: sodium borohydride: isobutyric acid: glucofuranose, and "age" the system for 1 hour prior to addition of the ketone (J.D. Morrison, E.R. Grandbois, S.I. Howard, *J. Org. Chem.*,

Scheme 12

(1) $NaBH_4$ (1 equiv),
$(CH_3)_2CH_2CO_2H$ (1.2 equiv),
THF, 0°C for 2 min;
warm to 30°C for 1h.
(2) Add (2 equiv) 30°C, 1h
(3) Ketone, 0°C 2h, 50-62%.

R = Me (78% ee, ***R***)
R = *n*-Et (85% ee, ***R***)
R = *n*-Pr (74% ee, ***R***)
R = *i*-Pr (57% ee, ***R***)

TABLE 4
Representative ketones reduced by K-glucoride

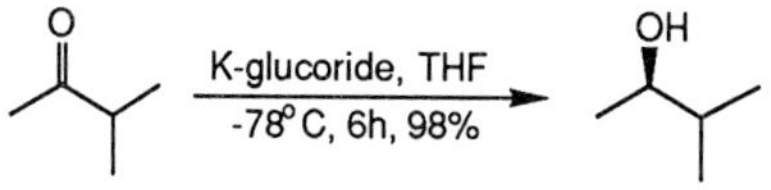

K-glucoride, THF, -78°C, 6h, 98%	39% ee (***R***)	(Eq. 1)
K-glucoride, THF, -78°C, 10h, 95%	20% ee (***R***)	(Eq. 2)
K-glucoride, THF, -78°C, 48h, 88%	84% ee (***R***)	(Eq. 3)
K-glucoride, THF, -78°C, 16h, 95%	70% ee (***R***)	(Eq. 4)
K-glucoride, THF, -78°C, 12h, 97%	70% ee (***R***)	(Eq. 5)
K-glucoride, THF, -78°C, 12h, 97%	42% ee (***R***)	(Eq. 6)
K-glucoride, THF, -78°C, 10h, 95%	61% ee configuration not assigned	(Eq. 7)
K-glucoride, THF, -78°C, 10h, 92%	60% ee (***R***)	(Eq. 8)
K-glucoride, THF, -78°C, 12h, 82%	77% ee (***S***)	(Eq. 9)
K-glucoride, THF, -78°C, 12h, 95%	48% ee (***S***)	(Eq. 10)

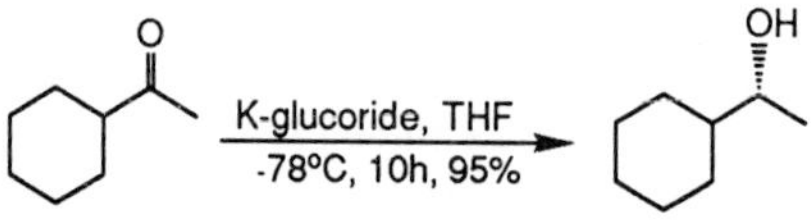

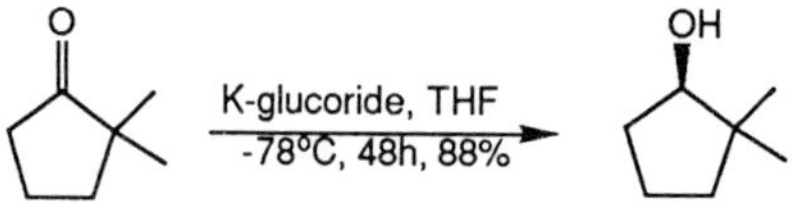

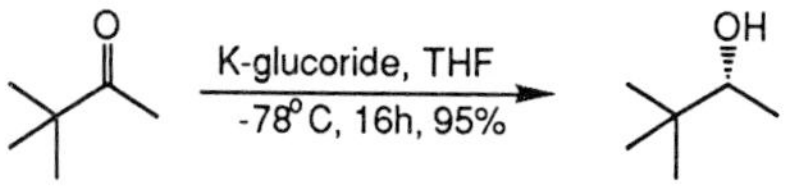

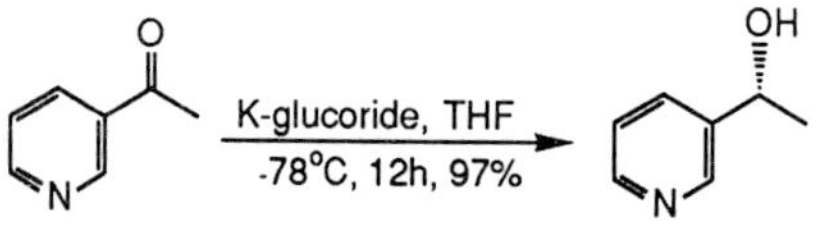

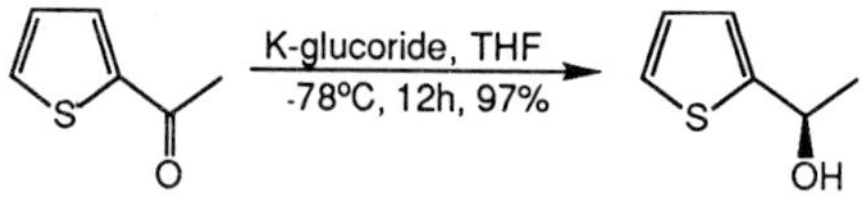

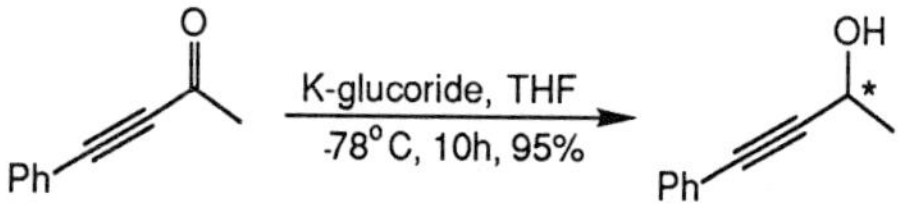

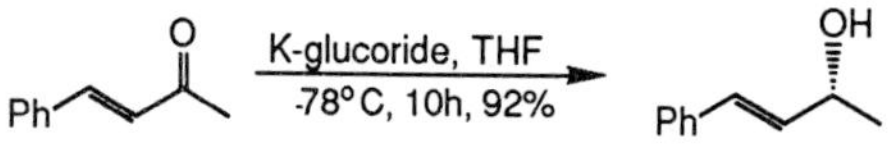

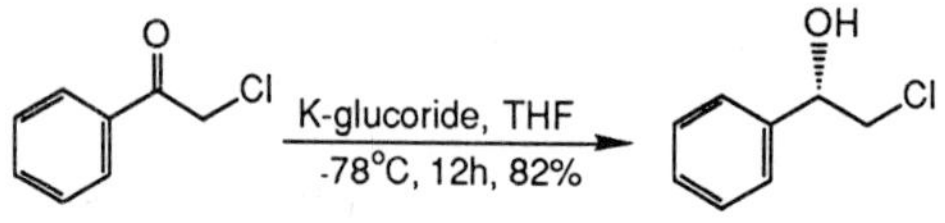

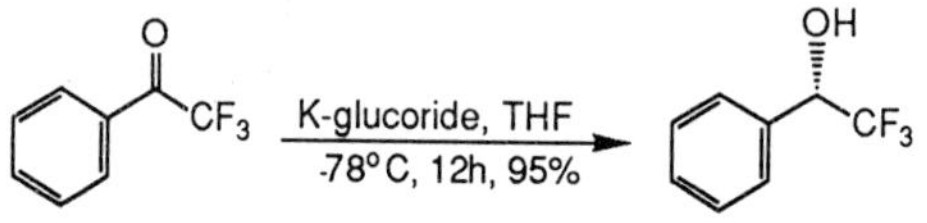

1980, **45**, 4229; A. Hirao et al., *ibid.*, 1980, **45**, 4231; A. Hirao et al., *J. Chem. Soc. Perkin Trans. 1*, 1981, 900). Asymmetric reduction of aromatic ketones has also been accomplished with chiral acyloxyborohydrides derived from sodium borohydride and the carbohydrate acids (**11**) and (**12**) (A. Hirao et al., *Agric. Biol. Chem.*, 1981, **45**, 693).

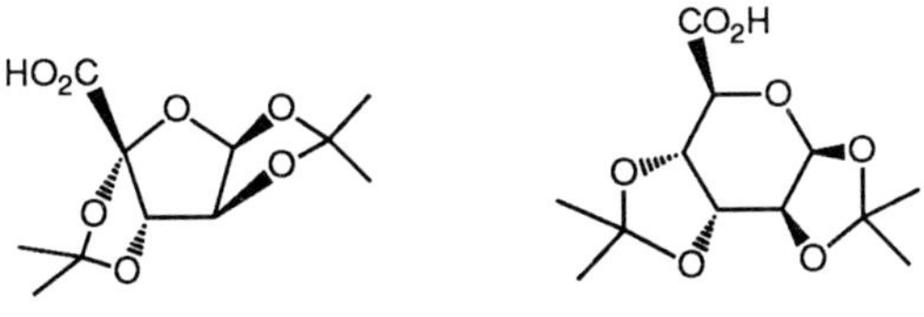

HO_2C CO_2H

(**11**) (**12**)

A noteworthy chiral borohydride reagent is K-Glucoride (Potassium 9-*O*-(1,2:5,6-di-*O*-isopropylidene-α-D-glucofuranosyl)-9-boratobicyclo[3.3.1]nonane). It is synthesised from the borinic ester 9-*O*-(1,2:5,6-di-*O*-isopropylidene-α-D-glucofuranosyl)-9-boratabicyclo[3.3.1]nonane by addition of potassium hydride in tetrahydrofuran (H.C. Brown, W.S. Park, B.T. Cho, *J. Org. Chem*, 1986, **51**, 1934; 3278; 3398). K-Glucoride reduces alkyl phenyl ketones, and hindered alkyl ketones to

Scheme 13

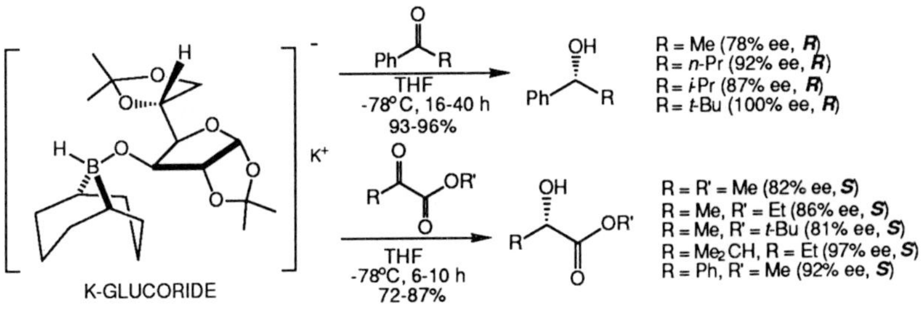

K-GLUCORIDE K^+

THF, -78°C, 16-40 h, 93-96%

R = Me (78% ee, *R*)
R = *n*-Pr (92% ee, *R*)
R = *i*-Pr (87% ee, *R*)
R = *t*-Bu (100% ee, *R*)

THF, -78°C, 6-10 h, 72-87%

R = R' = Me (82% ee, *S*)
R = Me, R' = Et (86% ee, *S*)
R = Me, R' = *t*-Bu (81% ee, *S*)
R = Me_2CH, R = Et (97% ee, *S*)
R = Ph, R' = Me (92% ee, *S*)

the corresponding *R*-alcohols in excellent ee, although with aliphatic, α,β-unsaturated, and sterically undemanding ketones the optical purities are less satisfactory. K-Glucoride is one of the best reagents currently available for the conversion of an α-keto ester into an α-hydroxy ester of the *S*-configuration. Indeed, in substrates possessing bulky alkyl groups attached to the ketone, it delivers products with almost 100% ee (Scheme 13).

Asymmetric reductions with the closely related chiral borohydride reagent (**13**) derived from 1,2:5,6-di-*O*-cyclohexylidene-α-D-glucofuranose have been described (B.T. Cho, *Taehan Hwahakhoe Chi*, 1990, **34**, 313; *CA* 1990, **113**, 131,663t). In the case of acetophenone, reduction with (**13**) gave the *R*-alcohol in 78% ee.

A reagent of more restricted utility is potassium 1,2:5,6-di-*O*-isopropylidene-D-mannitol thexylhydridoborate (**14**). It sometimes gives good results in systems where K-glucoride proves ineffective. For instance, in the reduction of 2-cyclohexen-

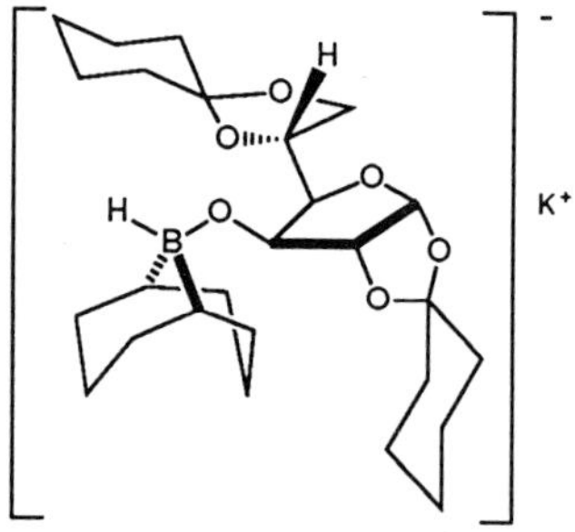

(13)

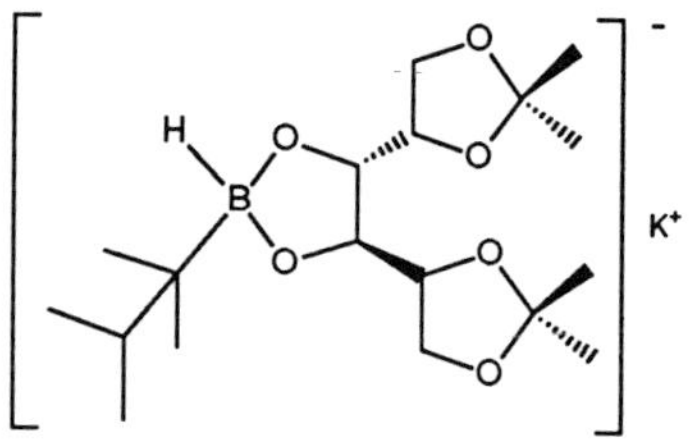

(14)

1-one with K-glucoride, only the products of 1,4-reduction were detected, whereas with (**14**) the *S*-allylic alcohol was produced in 71% ee (Scheme 14) (H.C. Brown, B.T. Cho, W.S. Park, *J. Org. Chem.*, 1987, **52**, 4020).

Scheme 14

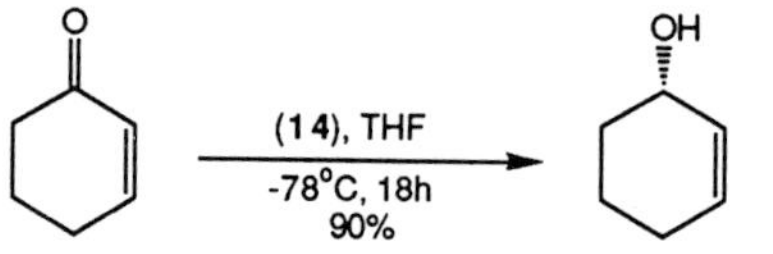

(14), THF
-78°C, 18h
90%

71% ee (*S*)

1,4-Dihydronicotinamide sugar pyranosides have been evaluated for the asymmetric 1,4-reduction of α,β-unsaturated iminium salts. However, the product ketones that

Scheme 15

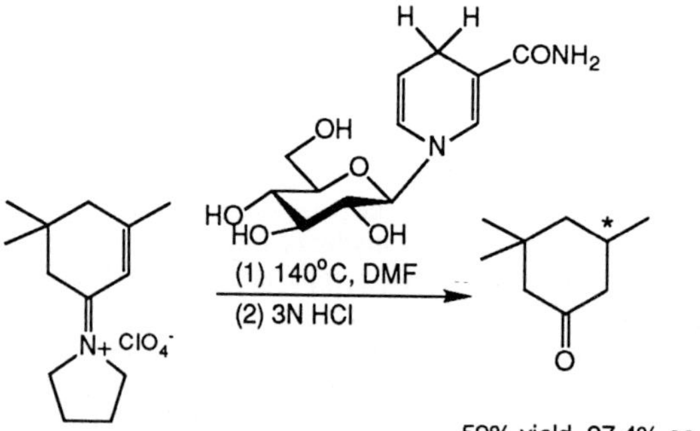

(1) 140°C, DMF
(2) 3N HCl

59% yield, 27.4% ee

are obtained after enamine hydrolysis, generally have low ee (3-31% ee) (Scheme 15) (T. Makino, et al., *Chem. Ind.*, 1977, 277; *Can. J. Chem.*, 1980, **58**, 387).

1,4-Dihydronicotinamide derivatives such as (**15**) and (**16**) have proven useful for the asymmetric reduction of α-keto esters. In the case of methyl phenyl glyoxalate, reagent (**15**) delivered the *R*-α-hydroxy ester in 79% ee, while reagent (**16**) furnished the *S*-enantiomer in 77% ee (Scheme 16) (S. Zehani, J. Lin, G. Gelbard, *Tetrahedron*, 1989, **45**, 733).

Scheme 16

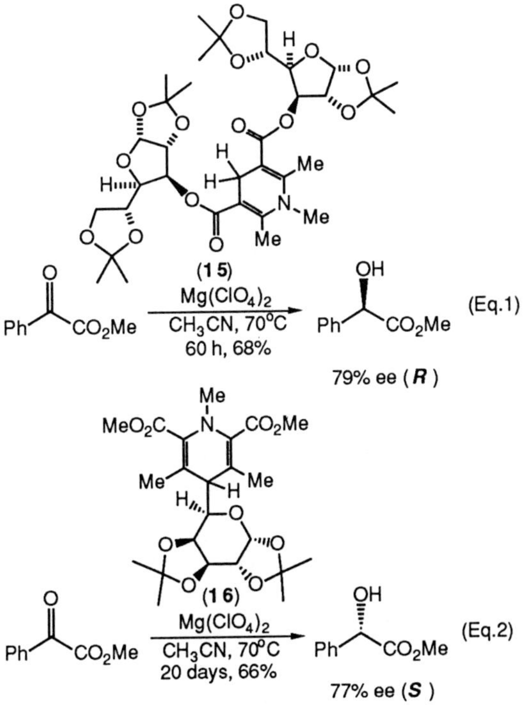

(**15**)
$Mg(ClO_4)_2$
CH_3CN, 70°C
60 h, 68%
(Eq.1)
79% ee (***R***)

(**16**)
$Mg(ClO_4)_2$
CH_3CN, 70°C
20 days, 66%
(Eq.2)
77% ee (***S***)

2. Enantioselective alkylation reactions

(a) Monosaccharide derived enolates in enantioselective aldol reactions

The low temperature aldol reactions between benzaldehyde and the lithium enolates of propionates (**17-23**) proceed with low diastereofacial selectivity (1 to 4), and with *threo* diastereoselectivity ranging from 1 to 6 (C.H. Heathcock, C.T. White, J.J. Morrison, D. VanDerveer, *J. Org. Chem.*, 1981, **46**, 1296).

In contrast, the chiral titanium enolate (**25**) undergoes highly stereoselective aldol reactions with aldehydes to give β-hydroxy esters in good yield and in 91-96% ee (Scheme 17) (R.O. Duthaler et al., *Angew. Chem. Int. Ed. Engl.*, 1989, **28**, 495; R.O. Duthaler, A. Hafner, M. Riediker, *Pure Appl. Chem.*, 1990, **62**, 631). Interestingly, enantioselectivity improved slightly when these reactions were conducted between -15°C and room temperature, although the yields suffered. A key step in a

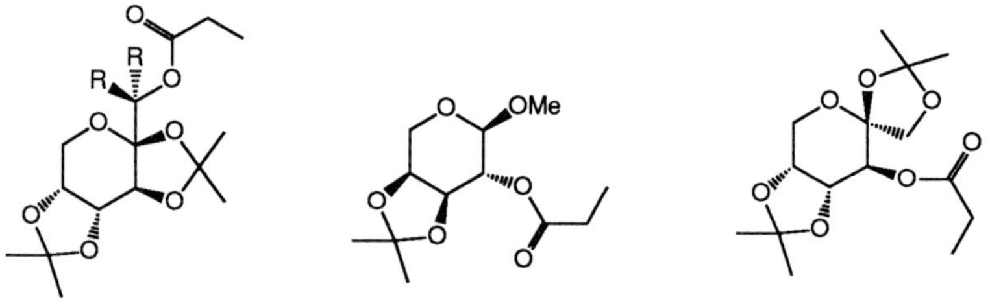

(17) R = H
(18) R = Me

(19)

(20)

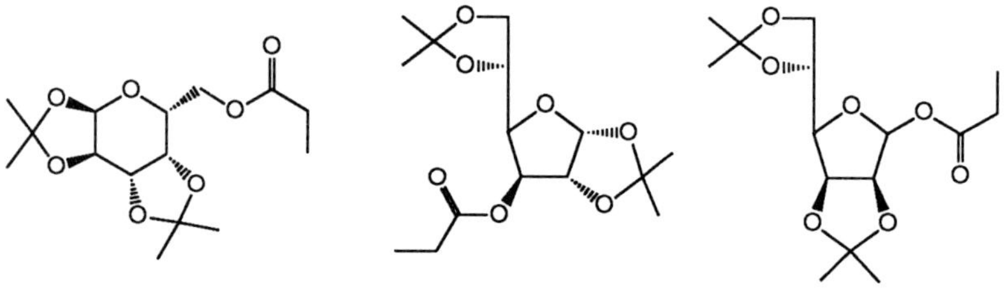

(21)

(22)

(23)

relatively facile synthesis of optically pure (-)-(*S*)-ipsenol utilised this methodology on a 2 mole scale (K. Oertle et al., *Helv. Chim. Acta*, 1990, **73**, 353).

Scheme 17

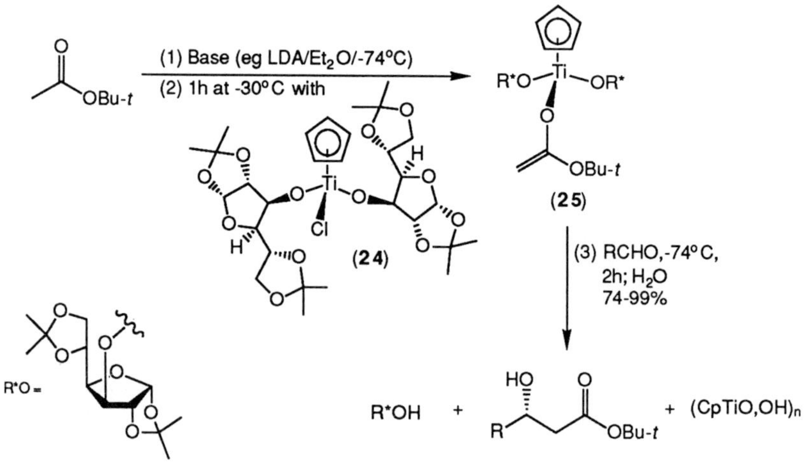
OBu-t
(1) Base (eg LDA/Et2O/-74°C)
(2) 1h at -30°C with
(24)
R*O-Ti-OR*
OBu-t
(25)
(3) RCHO,-74°C, 2h; H2O 74-99%
R*O =
R*OH + HO O R OBu-t + (CpTiO,OH)n
90-96% ee

Extension of these ideas to an *N*-bis(silylated)glycine enolate opened up an expeditious route to D-*threo*-β-hydroxy-α-amino acids (Scheme 18) (G. Bold, R.O. Duthaler, M. Riediker, *Angew. Chem. Int. Edn. Engl.*, 1989, **28**, 497). In all the cases examined, except for alkyl glyoxalates, the aldol addition products had ee's and de's greater than 96%. As in the acetate-aldol reaction, the chiral glycine titanium enolate adds preferentially to the *re*-face of the aldehyde. A notable feature of this technology is the ease with which the monosaccharide component can be recovered at the end of the reaction.

Scheme 18

(1) Base (eg LDA/THF/-78°C)
(2) 17h at -78°C with
(3) RCHO, inverse addition -78°C to room temp
(4) AcOH, THF, H_2O
(5) Boc_2O, $NaHCO_3$, dioxane H_2O, 2.5h, room temp

R*O =

R*OH + [product: R, HO, NHBoc, OEt] + $(CpTiO,OH)_n$

Yields between 43-70% 78-98% ee 87-99% de

(b) Enantioselective alkylation of monosaccharide derived carbanions

The chemical behavior of monosaccharide lithium ester enolates toward alkyl halides is often capricious, depending largely on the degree of intramolecular complexation that exists between the carbohydrate framework and the lithium counte-

Scheme 19

(1) LDA, THF -78°C
(2) C_2H_5I, -70°C

(26) **(27)** (***2R:2S***) = 67:33 **(28)**

30% 50%

rion (H. Kunz, J. Mohr, *J. Chem. Soc. Chem Commun.*, 1988, 1315). For example, in the alkylation of (**26**) with iodoethane, the desired 2-methylbutyrate (**27**) could only be obtained in 30% yield and with disappointing diastereoselectivity. The preferred reaction pathway actually involved decomposition of the enolate into the ketene, which then reacted with the starting enolate, to deliver (**28**) in 50% yield. In this particular enolate, strong chelation between the lithium and the monosaccharide was thought to increase the leaving group properties of the sugar alkoxide, and facilitate the decomposition process (Scheme 19). Chelation phenomena were again invoked to explain the inertness of the enolate derived from (**29**) towards various electrophiles (Scheme 20).

Scheme 20

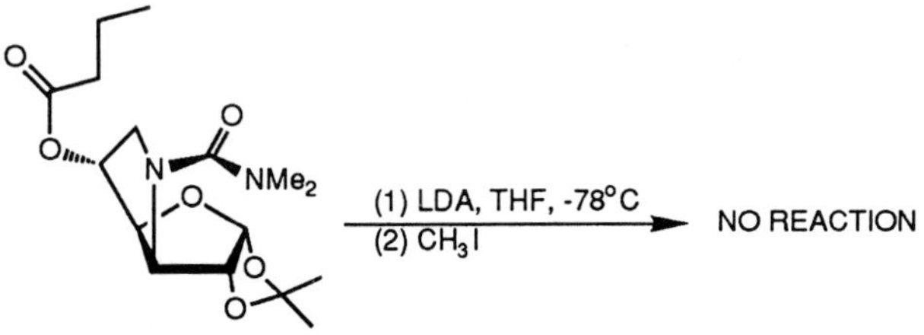

(1) LDA, THF, -78°C
(2) CH_3I
NO REACTION

(**29**)

In stark contrast to these results, ester enolates from 1,2:5,6-di-*O*-isopropylidene-α-D-allofuranose (**30**) undergo alkylation reactions with alkyl iodides with reasonable facility. In these systems, the enolates exist exclusively as their *Z*-isomers, and are

Scheme 21

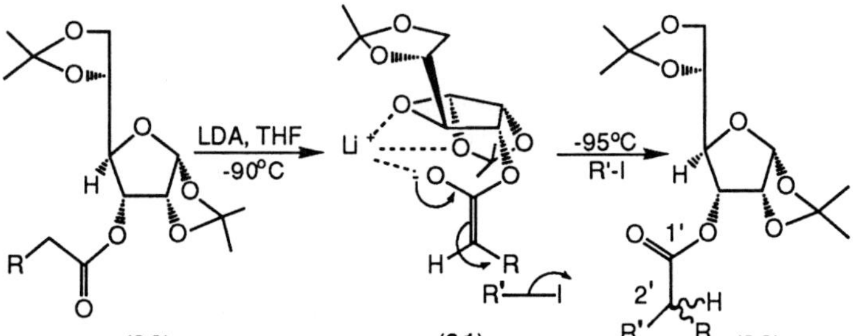

LDA, THF, -90°C → (**31**) → -95°C, R'-I

(**30**) (**31**) (**32**)

Z-enolates formed exclusively except for entry 7

ENTRY	R	R'	YIELD (%)	Diastereoselectivity (2'*R*):(2'*S*)
1	Me	Et	27	1:5
2	Et	Me	65	4.5:1
3	*n*-Pr	Me	58	4:1
4	*i*-Pr	Me	75	6:1
5	*t*-Bu	Me	100	10:1
6	$Ph(CH_2)_2$	Me	55	4:1
7	*o*-MeOPh	Me	84	1:5

thought to react via a chelated transition state such as (**31**) (Scheme 21). For entries 2-6, the *2'R*-diastereomer was the major product formed. However, when the R group had a lower priority than the R'-group being transferred from the iodoalkane, then the *2'S*-diastereoisomer was the favoured product. In entry 7, the *E*-enolate was formed, this is presumably due to complexation with the *o*-methoxy group; it reacted to give mainly the *2'S*-stereoisomer.

A range of chiral methylated α-amino acids are accessible from *N*-(6-deoxy-1,2;3,4-di-*O*-cyclohexylidene-α-D-galactopyranosylidene)-alanine methyl ester, after deprotonation at -78°C, addition of HMPT, and alkylation with an activated alkyl

Scheme 22

(1) LDA, THF, -78°C; HMPT
(2) R-Br, -70°C
(3) aq. HCl; aq. NH_4OH

S-configuration
formed in 44-85% ee

bromide at -70°C (Scheme 22). The products obtained, after removing the chiral auxiliary, typically had optical yields in the 44-85% range (I. Hoppe, U. Schöllkopf, R. Tölle, *Synthesis*, 1983, 789).

Scheme 23

(1) Li, NH_3 -78°C
(2) 3 eq. LDA Et_2O, -50°C, 4h
(3) 2 eq. E^+ -78°C, 4h

90-94% de
78-90%

E^+ = MeI, Ethylene oxide, (***R***) and (***S***)-Propylene oxide

A powerful method has appeared for the regio- and diastereoselective alkylation of 1-alkoxy-1,4-cyclohexadienes, that employs the per-*O*-methyl-β-D-glucopyranosyl group as the chiral inductor (Scheme 23) (D. Stanssens, D. De Keukeleire, M. Vandewalle, *Tetrahedron Lett.*, 1987, **28**, 4195; *Tetrahedron:Asymmetry*, 1990, **1**, 547). The starting 1,4-cyclohexadienes are conveniently prepared from the appropriate phenyl-β-D-glucopyranoside, by permethylation and Birch reduction. They undergo selective deprotonation at the C-6 position of the cyclohexadienyl moiety after treatment with 3 equiv. of lithium diisopropylamide at -50°C. The chelated carbanions that result, undergo alkylation with a range of carbon electrophiles at either -78°C or -90°C, with good to excellent diastereocontrol. Removal of the chiral auxiliary is accomplished by either Lewis acid catalysed transketalisation, or acid hydrolysis with hydrochloric acid in THF (Scheme 24). Alternatively, the alkylated 1,4-cyclohexadiene moiety can be further modified prior to removal of the sugar.

Scheme 24

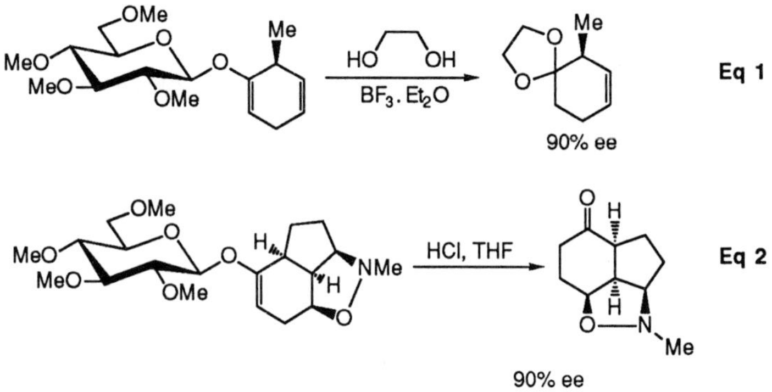

An outstanding method for enantioselectively transferring an allyl synthon to aldehydes and ketones utilises the chiral titanium complex shown in Scheme 25. It smoothly allylates a range of hindered and unhindered prochiral aldehydes at -74°C, to

Scheme 25

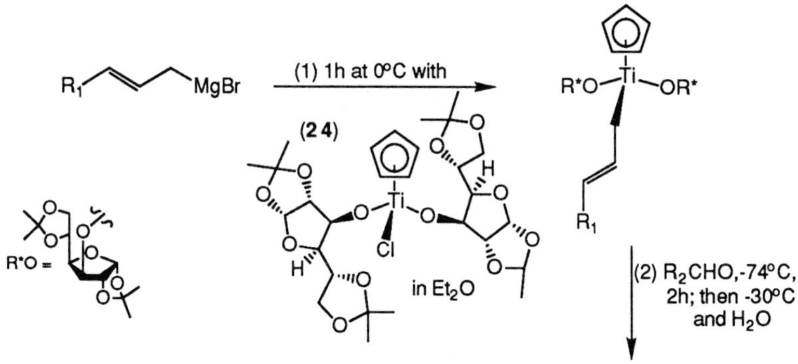

ENTRY	R_1	R_2	% YIELD
1	H	Ph	85 (90% ee)
2	H	Et	67 (93% ee)
3	H	C_6H_{11}	78 (92% ee)
4	H	*t*-Bu	58 (88% ee)
5	Me	*i*-Pr	46 (83% ee)
6	Ph	*i*-Bu	59 (88% ee)
7	Vinyl	*i*-Bu	68 (90% ee)

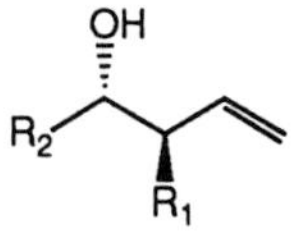

provide chiral homoallylic alcohols in 88-94% ee (M. Riediker, R.O. Duthaler, *Angew. Chem. Int. Ed. Engl.*, 1989, **28**, 494). In the case of ketones, addition only proceeds at higher reaction temperatures (*ca.* 0°C), and as one would expect this leads to a reduction in product ee (typically *ca.* 50%: although, 80% ee for acetophenone). Diminished enantioselectivity was also encountered with chiral allyltitanium complexes having only one monosaccharide ligand present (R.O. Duthaler, A. Hafner, W. Riediker, *Pure Appl. Chem.*, 1990, **62**, 631).

Modest levels of asymmetric induction can occasionally be found in the addition of Grignard reagents to alkyl phenyl ketones, provided they are conducted with 1,2:5,6-di-*O*-isopropylidene-α-D-glucofuranose or methyl 4,6-*O*-benzylidene-2,3-di-*O*-methyl-α-D-glucopyranoside as additives. The extent of optical induction

Scheme 26

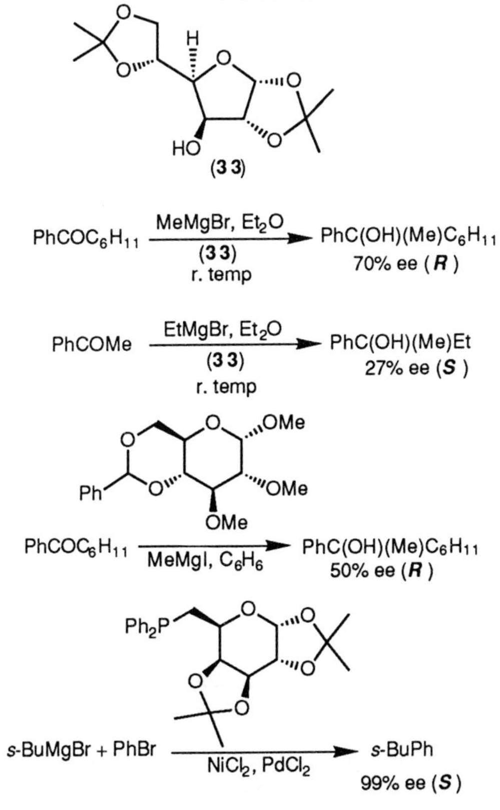

(**33**)

$PhCOC_6H_{11}$ → [MeMgBr, Et_2O, (**33**), r. temp] → $PhC(OH)(Me)C_6H_{11}$ 70% ee (***R***)

PhCOMe → [EtMgBr, Et_2O, (**33**), r. temp] → PhC(OH)(Me)Et 27% ee (***S***)

$PhCOC_6H_{11}$ → [MeMgI, C_6H_6] → $PhC(OH)(Me)C_6H_{11}$ 50% ee (***R***)

s-BuMgBr + PhBr → [$NiCl_2$, $PdCl_2$] → *s*-BuPh 99% ee (***S***)

depends not only on the nature of the Grignard reagent being used, but also on the structure of the substrates (T.D. Inch, et al., *Tetrahedron Lett.*, 1969, 3657). A later study investigated the potential of 1,2:5,6-Di-*O*-isopropylidene-α-D-allofuranose as a chiral conditioning agent for Grignard reactions, but this gave lower optical yields of products (N. Baggett, R.J. Simmonds, *J. Chem. Soc. Perkin Trans. 1*, 1982, 197). 6-*C*-(Diphenylphosphino)-1,2:3,4-di-*O*-isopropylidene-α-D-galactopyranose has been employed as a chiral additive for palladium catalysed Grignard cross-coupling reactions between aryl bromides and *sec*-butyl magnesium bromide, but the chemical yields are extremely low (Scheme 26) (A. Iida, M. Yamashita, *Bull. Chem. Soc. Japan*, 1980, **61**, 2365). Only one report has appeared concerning the application of chiral cuprate reagents derived from monosaccharides for asymmetric induction, and again the results are disappointing (J.S. Zweig et al., *Tetrahedron Lett.*, 1975, 2355).

Asymmetric Michael addition of 4-*tert*-butyl thiophenol to 2-cyclohexenone delivers the *R*-adduct in up to 77% ee, if chiral amino-alcohol derivatives of D-glucose are incorporated into the reaction mixture (Scheme 27) (F.Wang, M. Tada, *Agric. Biol. Chem.*, 1990, **54**, 2989). The stereochemistry of addition can be rationalised by invoking intermolecular hydrogen bonding between the pyran hydroxy

Scheme 27

Catalytic

room temp.
Toluene

77% ee (***R***)

Via:

group and the enone, and simultaneous ion pairing of the chiral ammonium species with the thiolate anion. Both these events conspire to favour nucleophilic attack via the transition state shown in Scheme 27.

(c) Enantioselective alkylation with monosaccharide electrophiles

(i) Monosaccharides as chiral leaving groups

The trapping of prochiral carbanions with monosaccharide electrophiles was first enunciated for Schiff bases derived from glycine (Scheme 28) (P. Duhamel, J.Y. Valnot, J. Jamal Eddine, *Tetrahedron Lett.*, 1982, **23**, 2863; *ibid.*, 1984, **25**, 2355). These undergo deprotonation at -70°C with lithium diisopropylamide in THF, and successfully react with mixed sulphate esters of D-glucose to give amino acid esters in up to 76% ee (P. Duhamel et al., *ibid.*, 1987, **28,** 3801; *Tetrahedron*, 1988, **44**, 5495).

Scheme 28

(1) LDA, THF, -70°C
(2) (34)
(3) 0.2N, HCl/Ether; 6N HCl reflux

$R_1 = R_2 = Ph$, 56% ee (***S***)
R_1 = t-Bu, R_2 = 4-Me_2NPh 76% ee (***S***)

In analogous fashion, monosaccharides can be utilised for the enantioselective protonation of ester enolates, although with much less success. For example, when the cyclic ester (**35**) was treated with base, and quenched with either 1,2:5,6-di-*O*-isopropylidene-α-D-glucofuranose (**33**), or methyl 4,6-*O*-benzylidene-α-D-glucopyranoside, the *S*-enantiomer was formed in 31% and 15% ee respectively (Scheme 29) (U. Gerlach, S. Hunig, *Angew. Chem. Int. Ed. Engl.*, 1987, **26**, 1283).

Scheme 29

(1) LDA, THF, -78°C
(2) (33)

(**35**) (***R,S***)

(**36**) 31% ee (***S***)

Birch reduction of 2- or 3-furoic acids, followed by enantioselective protonation with (**33**), delivered products of even lower optical purity (typically *ca.* 3%) (T. Kinoshita, T. Miwa, *J. Chem. Soc. Chem. Comm.*, 1974, 181)

Scheme 30

$SnCl_4$
C_6H_6

(**37**)

(**38**) 21% ee

Partial asymmetric cyclisation of the monosaccharide-ester derivative of homogeranic acid (**37**) has been achieved by treatment with stannic chloride in anhydrous benzene (Scheme 30). The desired *cis*-tetrahydroactinidiolide (**38**) was isolated in 21% ee (Its absolute stereochemistry was not assigned) (T. Kato, S. Kumuzawa, Y, Kitahara, *Synthesis*, 1972, 573).

Scheme 31

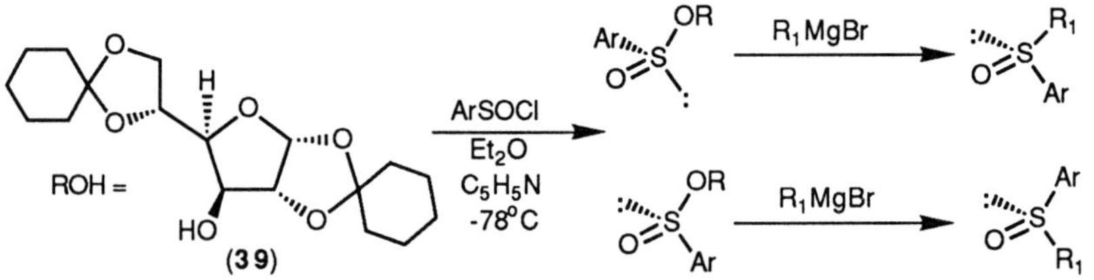

The diastereomeric sulphinate esters formed from arene sulphinyl chlorides and 1,2:5,6-di-*O*-cyclohexylidene-α-D-glucofuranose afford almost optically pure sulphoxides after reaction with Grignard reagents, and recrystallisation (Scheme 31). The method allowed (*S*)-1-napthyl phenyl sulphoxide to be obtained in 95% ee (29% overall yield), after one recrystallisation of a crude reaction mixture in which it was present in only 40% ee (D.D. Ridley, M.A. Smal, *J. Chem. Soc. Chem. Comm.*, 1981, 505).

Chiral phosphate esters can be conveniently prepared in enantiomerically pure form by ring opening of monosaccharide 1,3,2-oxazaphosphorinanes with metal alkoxides and treatment with acid (Scheme 32) (T.D. Inch et al., *J. Chem. Soc. Chem. Comm.*, 1975, 720; *J. Chem. Soc. Perkin Trans. 1*, 1975, 1892). For a detailed discussion of this and related work, see C. Richard Hall and T.D. Inch, *Tetrahedron*, 1980, **36**, 2059.

Scheme 32

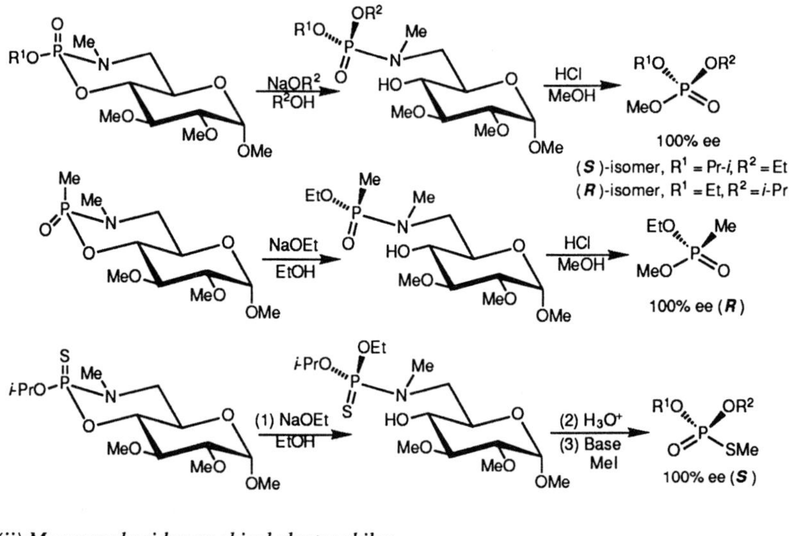

(ii) Monosaccharides as chiral electrophiles

Excellent levels of diastereoselection have been observed in the nucleophilic additions of lithium dialkylphosphite anions (R. Huber et al., *Helv. Chim. Acta.*,

1985, **68**, 1730; B. Bernet, E. Krawczyk, A. Vasella, *ibid.*, 1985, **68**, 2299) and tris(trimethylsilyl)phosphite ($P(OSiMe_3)_3$) (R. Huber, A. Vasella, *ibid.*, 1987, **70**,

Scheme 33

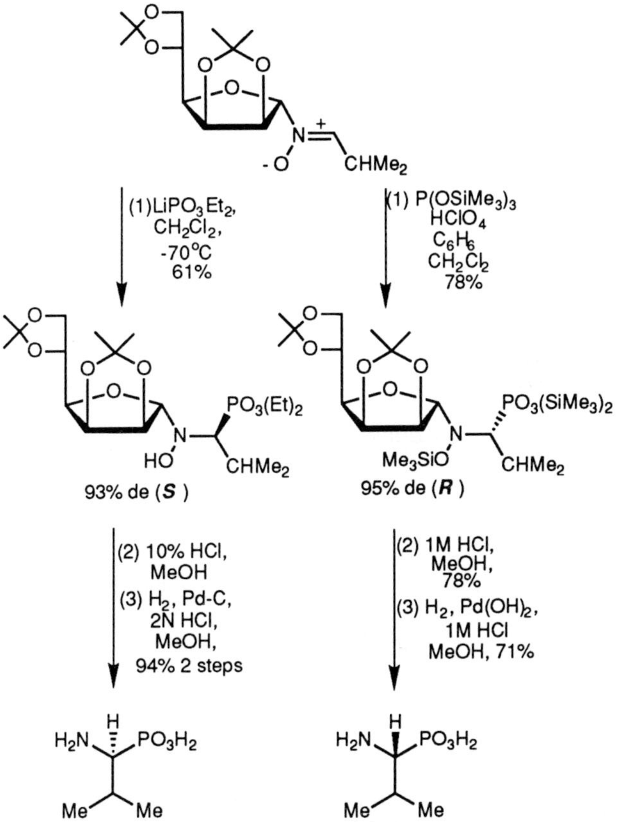

1461) to Z-mannofuranosyl nitrones, due to the combined operation of a kinetic anomeric effect and steric phenomena (R. Huber, A. Vasella, *Tetrahedron*, 1990, **46**, 33). The resulting adducts can be easily transformed, without loss of optical activity, into chiral α-aminophosphonic acids, which are precursors of phosphonopeptides (Scheme 33) (P. Kafarski, B. Leijczak, P. Mastalerz, *"Phosphonopeptides:Synthesis and Biological Activity"*, Beitrage zur Wirkstofforschung, Heft 25, Berlin, 1985).

An asymmetric version of the Strecker synthesis of α-aminonitriles has been devised which utilises 2,3,4,6-tetra-*O*-pivaloyl-β-D-galactopyranosylamine as a chiral auxiliary (H. Kunz, W. Sager, *Angew. Chem. Int. Ed. Engl.*, 1987, **26,** 557). After conversion into the appropriate Schiff base, treatment with trimethylsilyl cyanide, under Lewis acid catalysis, leads to the preferential formation of the α*R*-diastereomer in high yield (Scheme 34). This stereochemical outcome can be attributed to chelate formation between the Lewis acid, O-2, and the imine nitrogen, which favours attack by the cyanide ion on the same side of the imine as the ring oxygen. Acid hydrolysis of the resulting α-amino nitriles liberates optically pure D-amino acids (Scheme 34).

Scheme 34

Me_3SiCN, $ZnCl_2$, $CHCl_3$

2'S:2'R = 3-9:1

Me_3SiCN, $SnCl_4$, THF

2'R:2'S = 7-13:1

AUXILIARY REMOVAL

HCl, HCO_2H, 94%

100% ee

2,3,4,6-Tetra-*O*-pivaloyl-β-D-galactopyranosylamine has proven a very effective auxiliary for an asymmetric version of the Ugi-four-component-condensation (H. Kunz, W. Pfrengle, *J. Am. Chem. Soc.*, 1988, **110**, 651; *Tetrahedron*, 1988, **44**, 5487). Typically, the diastereoselectivity of this process ranges from 15-35:1, with the (2*R*, β-D) diastereoisomers predominating (Scheme 35). Acid hydrolysis of these intermediates affords enantiomerically pure *R*-amino-acids in high yield. Use of 2,3,4-tri-*O*-pivaloyl-α-D-arabinopyranosylamine as the chiral template allows the sense of stereochemical induction to be reversed (H. Kunz, W. Pfrengle, W. Sager, *Tetrahedron Lett.*, 1989, **30**, 4109).

Under the influence of Lewis acids, galactosyl Schiff bases such as (**40**) undergo a stereoselective Mannich reaction with alkoxy dienes, which if followed by intramolecular Michael ring closure, can be used as an effective entry into chiral piperidine alkaloid systems (H. Kunz, W. Pfrengle, *Angew. Chem. Int. Edn. Engl.*, 1989, **28**, 1067). Using this methodology a synthesis of *S*-anabasine was achieved (Scheme 36).

One very effective route to aryl β-amino acid esters of the 3*S*-configuration is through Lewis acid mediated addition of silyl ketene acetals to galactosyl Schiff bases (Scheme 37). (H. Kunz, D. Schanzenbach, *Angew. Chem. Int. Edn. Engl.*, 1989,

examined.

Under Lewis acid catalysis, 3-substituted acrylate derivatives of dihydro-D-glucal undergo highly selective Diels-Alder reactions with an assortment of dienes to deliver *R*-cycloadducts in up to 96% de. The opposite enantiomers are available using 4-*O*-pivaloyl-dihydro-L-rhamnal as the chiral auxiliary (Scheme 42) (W. Stahle, H. Kunz, *Syn. Lett.*, 1991, 260).

The chiral α-chloronitroso ether (**49**) undergoes facile [4+2] cycloaddition at -70°C with a variety of dienes to deliver *N*-unsubstituted 3,6-dihydro-*2H*-1,2-oxazines in essentially optically pure condition (H. Felber et al., *Tetrahedron Lett.*, 1984, **25**, 5381; *Helv. Chim. Acta.*, 1986, **69**, 1137). Since the absolute configuration of the

Scheme 43

NH. HCl
OMe
OMe
OMe
OMe
NH. HCl
CH_2Cl_2, EtOH
-70°C, 15 min
72%
(**51**)[b] >96% e.e.
N=O
Cl
CH_2Cl_2, EtOH
-70°C, 15 min
70%
(**50**)[a]
>96% e.e.
(**49**)
R
CH_2Cl_2, EtOH
-20°C
63-69%
R
O
NH
Me
Me
(**52**) R = H[b]
(**53**) R = CO_2Et[b]
>96% e.e.
a) *1 S, 4 R*
b) Absolute configuration not determined

ENDO ADDITION

CONFORMATION A

Cl
(**49**)
$C_{(2)}$
$O(C_4)$
N
O
Cl

dihydrooxazine (**50**) was (*1S,4R*), it was suggested that (**49**) added *endo* to the diene when the nitroso group was in conformation A (Scheme 43). The enantiomeric dihydrooxazines can be synthesised from an α-chloronitroso derivative of D-ribose (H. Braun et al., *Tetrahedron:Asymmetry*, 1990, **1**, 403).

Monosaccharide *O*-thioformates and 1,1-dithiooxalates have proven willing participants in Diels-Alder reactions, but only display poor diastereoselectivity. However, with some combinations of reagents, diastereomeric ratios as high as 9:1

Scheme 44

(1) (55), C_6H_6, rt
(2) MeOH
90% overall

(54)

(56) 9:1
Configuration not assigned

can be achieved (Scheme 44) (P. Herczegh et al., *Tetrahedron Lett.*, 1986, **27**, 1509; *Heterocycles*, 1989, **28**, 887).

Reasonable selectivity in favour of the *S*-adducts has been observed in the aza-Diels-Alder synthesis of *N*-galactosyldehydropiperidine derivatives with unactivat-

Scheme 45

(58)
$ZnCl_2.Et_2O$
THF
4°C, 12 h
96%

(57)

(59)

S:*R* = 87:13

ed dienes, a typical example being shown in Scheme 45. It is important to conduct these cycloadditions at 4°C otherwise diminished diastereoselectivity results (W. Pfrengle, H. Kunz, *J. Org. Chem.*, 1989, **54**, 4261).

(ii) Reactions of chiral dienes derived from monosaccharides

The initial studies on Diels-Alder reactions of monosaccharide dienes have been reviewed (L.A. Paquette, *"Asymmetric Synthesis"*, Ed. J.D. Morrison, Academic Press, New York, 1984, **3**, Chapter 7).

Monosaccharides have served as sources of numerous chiral dienes for use in asymmetric [4+2] cycloadditions. The first reactions were conducted between substituted glycosyl dienes and alkyl glyoxalates, and were attempted with a view to preparing oligosaccharides (S. David et al., *J. Chem. Soc. Perkin Trans. 1*, 1974, 2274; *ibid.*, 1976, 1831; *ibid.*, 1979, 1795, 2230; *J. Chem. Soc. Chem. Comm.*, 1975, 701; *ibid.*, 1978, 535; *Nouv. J. Chem.*, 1977, **1**, 375; *Tetrahedron*, 1978, **34**, 299). A representative case is provided the addition of benzyl 2,3,6-tri-*O*-benzyl-4-*O*-(but-1-en-3-yl)-α-D-glucopyranoside to butyl glyoxalate (Scheme 46). Although this diene reacted with quite high diastereofacial selectivity, its *endo-exo* selectivity was poor, as was generally the case with most of the early dienes studied.

Scheme 46

In contrast, (*E*)-3-trimethylsiloxybuta-1,3-dienyl 2,3,4,6-tetra-*O*-acetyl-β-D-glucopyranoside (**59a**) and its congeners (**59b-e**) displayed noteworthy diastereofacial reactivity in their reactions with dienophiles (Schemes 47 and 48) (R.J. Stoodley et al., *J. Chem. Soc. Chem. Comm.*, 1983, 754; *ibid.*, 1986, 668, 1116; *J.Chem. Soc. Perkin Trans 1.*, 1985, 525; *ibid.*, 1988, 1773, *ibid.*, 1989, 739, 1841; *Tetrahedron*, 1984, **40**, 4657). The selectivities were optimal when the the diene component was heavily functionalised. Thus, diene (**59a**) reacted with *N*-phenylmaleimide in benzene at room temperature to afford an 86:14 mixture of cycloadducts (**60a**) and (**61a**) (R.J. Stoodley et al., *J. Chem. Soc. Perkin Trans. 1*, 1988, 1773) whereas its more highly substituted counterpart (**59b**) reacted to give a >19:1 mixture of (**60b**) and (**61b**) (Scheme 48) (R.J. Stoodley et al., *ibid.*, 1989, 1841). All the dienophiles studied showed a marked preference for *endo* addition to the less hindered top face of dienes (**59a-e**), when they were in the conformation depicted in Scheme 49. The effect of removing the 2- or 6-acetoxy substituents from 2,3,4,6-tetra-*O*-acetyl β-linked glucosyl dienes (**59a-e**) was usually to lower the diastereoselectivity of their cycloaddition reactions. Changing the anomeric configuration of the dienyl glucoside from β to α had similar consequences (R.J. Stoodley et al., *J. Chem. Soc. Perkin Trans. 1*, 1990, 1339, 3113). The latter effect was again manifest with 2-deoxy-glucosyl dienes. Thus, (**62α**) reacted with *N*-phenylmaleimide to give a 1:1 mixture of cycloadducts (**63α**) and (**64α**), whereas the corresponding *β*-anomer (**62β**) afforded a 4:1 mixture of (**63β**) and (**64β**), in which the former isomer predominated (Scheme 50). For (*E*)-3-(*tert*-butyldimethylsilyloxy)buta-1,3-dienyl 2,3,4-tri-*O*-acetyl α- and β-

Scheme 47

C_6H_6, 5°C
46-73%
ISOLATED YIELDS FOR MAJOR CYCLOADDUCTS

R	Diene	Major		Minor
R = H	(**59a**)	9	:	1
R = CO_2Me		7.33	:	1
R = Ac		3	:	1

Scheme 48

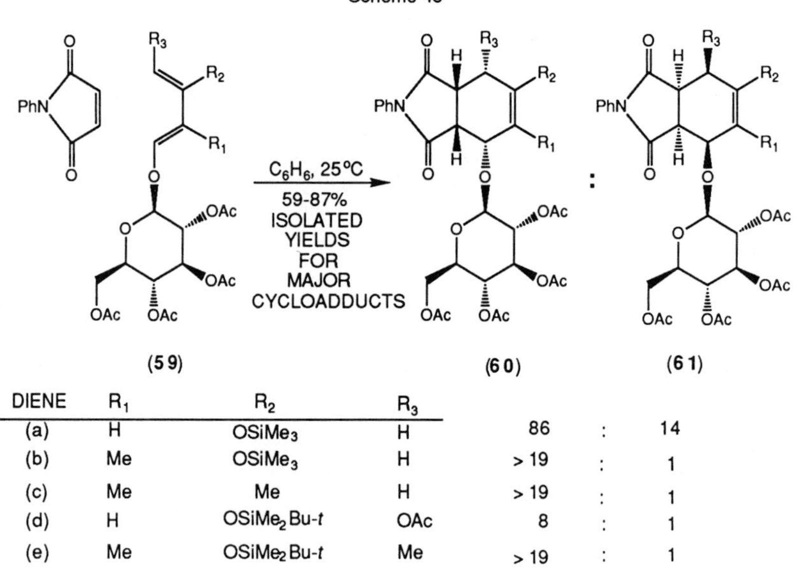

DIENE	R_1	R_2	R_3	(60)		(61)
(a)	H	$OSiMe_3$	H	86	:	14
(b)	Me	$OSiMe_3$	H	> 19	:	1
(c)	Me	Me	H	> 19	:	1
(d)	H	$OSiMe_2Bu$-*t*	OAc	8	:	1
(e)	Me	$OSiMe_2Bu$-*t*	Me	> 19	:	1

Scheme 49

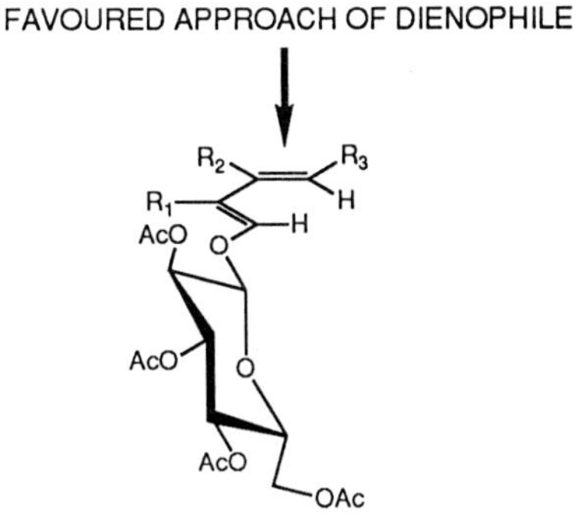

FAVOURED APPROACH OF DIENOPHILE

xylopyranosides the diastereofacial reactivities were found to be equal and opposite (Scheme 51).

The Manchester workers have elegantly exploited their methodology to synthesise the (+)-enantiomer of the antibiotic Bostrycin (Scheme 52) (R.J. Stoodley et al., *J. Chem. Soc. Chem. Comm.*, 1989, 17). Acid hydrolysis cleanly detached the sensitive aglycone from the chiral auxiliary, enabling the natural product to be prepared in good yield.

Scheme 50

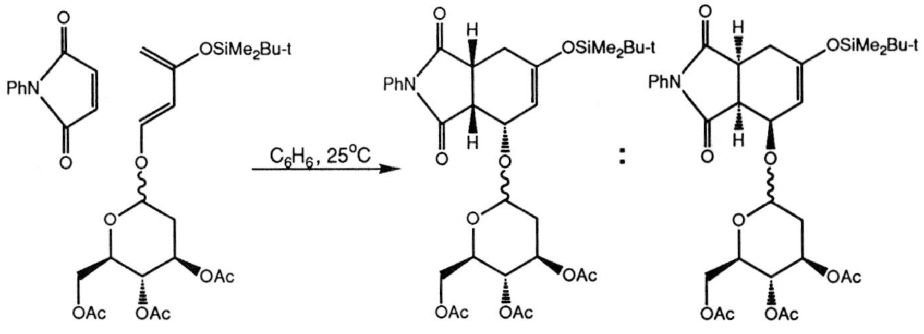

C_6H_6, 25°C

(62) **(63)** **(64)**

	(63)		(64)
(62) α-ANOMER	1	:	1
(62) β-ANOMER	4	:	1

Scheme 51

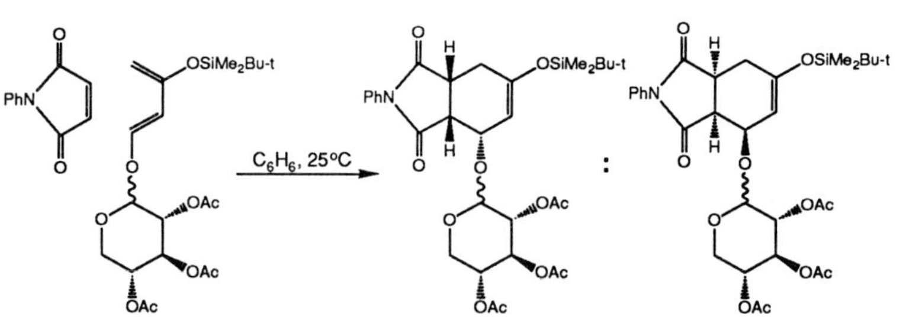
OSiMe2Bu-t
PhN
C6H6, 25°C
OAc
H
:

(6 5) (6 6) (6 7)

	(6 6)		(6 7)
(6 5) α-ANOMER	4	:	1
(6 5) β-ANOMER	1	:	4

Scheme 52

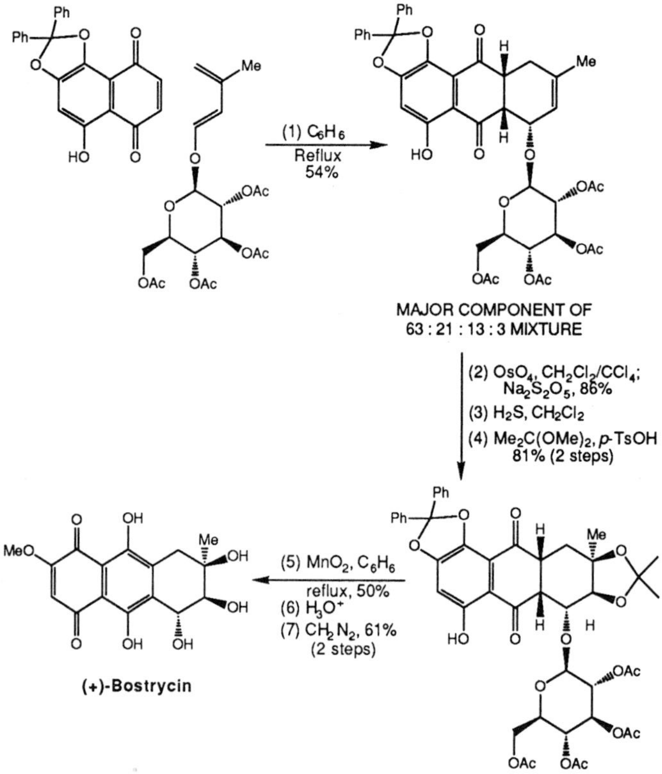
Ph
Me
HO
OAc
(1) C6H6
Reflux
54%
MAJOR COMPONENT OF
63 : 21 : 13 : 3 MIXTURE
(2) OsO4, CH2Cl2/CCl4;
Na2S2O5, 86%
(3) H2S, CH2Cl2
(4) Me2C(OMe)2, p-TsOH
81% (2 steps)
(5) MnO2, C6H6
reflux, 50%
(6) H3O+
(7) CH2N2, 61%
(2 steps)
MeO
OH
(+)-Bostrycin

Extension of these ideas to dienes derived from 3-*O*-trifluoroacetyl-α-D-daunosamine has given access to a plethora of anthracycline analogues (M.G. Brasca, S. Penco, *Eur. Pat. Appl.* EP 287,353, 19 Oct, 1988; *CA*, 1989, **111**, 7749v).

Scheme 53

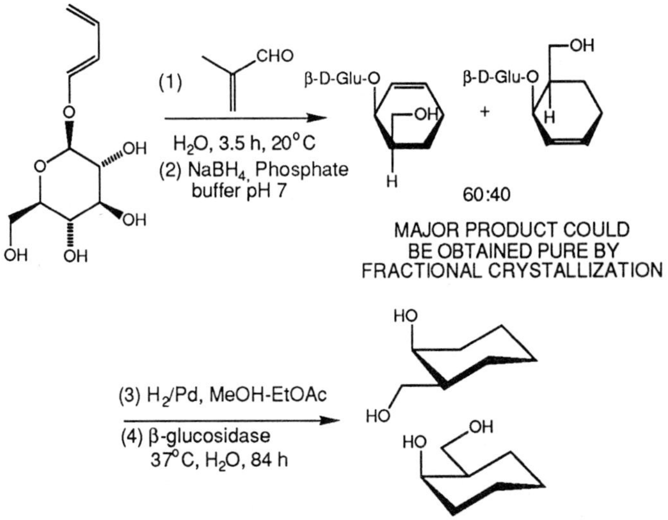

(1) CHO; H_2O, 3.5 h, 20°C (2) $NaBH_4$, Phosphate buffer pH 7

β-D-Glu-O ... OH ... H + β-D-Glu-O ... OH ... H

60:40

MAJOR PRODUCT COULD BE OBTAINED PURE BY FRACTIONAL CRYSTALLIZATION

(3) H_2/Pd, MeOH-EtOAc
(4) β-glucosidase 37°C, H_2O, 84 h

Details have appeared on the uncatalysed cycloadditions of unprotected 1,3-butadienyl α- and β-glucosides with methacrolein in aqueous media (Scheme 53) (A. Lubineau, Y. Queneau, *Tetrahedron Lett.*, 1985, **26**, 2653; *J. Org. Chem.*, 1987, **52**, 1001; *Tetrahedron*, 1989, **45**, 6697). These are reactions which display a strong preference for *endo* addition, yet which exhibit poor facial selectivity. Nevertheless, the marked crystallinity of the cycloadducts frequently allows purification through fractional crystallisation. The sugar component can usually be removed from the chiral products by the action of α- or β-glucosidases.

Scheme 54

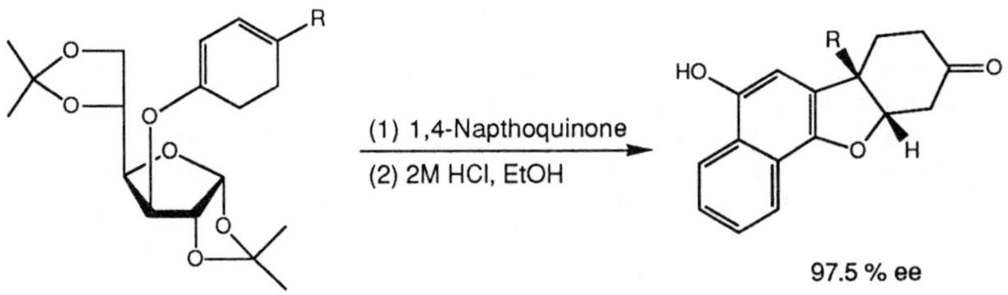

(1) 1,4-Napthoquinone
(2) 2M HCl, EtOH

97.5 % ee

R = H or Me

Excellent diastereofacial discrimination has been reported in the addition of 1,4-napthoquinone to alkoxycyclohexa-1,3-dienes derived from 1,2:5,6-di-*O*-isopropylidene-α-D-glucofuranose (Scheme 54). After acid hydrolysis and rearrangement, optically pure β-alkoxycyclohexanones can be obtained in 97.5% ee (C.W. Bird, A. Lewis, *Tetrahedron Lett.*, 1989, **30**, 6227).

(b) Asymmetric [3+2] cycloaddition reactions with monosaccharide auxiliaries

All the published work on this topic has focussed on the 1,3-dipolar cycloadditions of *N*-glycosyl nitrones to olefins. Typically, these reactions proceed

Scheme 55

(68)

(69a) R = Me
(69b) R = H

"O-endo"

Anti

(70) R = Me (5'S) 90% de
(71) R = H (5'S) 75% de

(72) R = Me (5S) >90% ee
(73) R = H (5S) 80% ee

with high stereocontrol, as exemplified by the additions of methyl methacrylate to 2,3:5,6-di-*O*-isopropylidene-D-mannofuranosylnitrones (**69a**) and (**69b**) (Scheme 55). These yielded isoxazolidines (70) and (71) with the *5'S*-configuration in 95% and 75% de respectively (A. Vasella, *Helv. Chim. Acta.*, 1977, **60**, 426; *ibid.*, 1273; R. Huber and A. Vasella, *Tetrahedron*, 1990, **46**, 33). The product stereochemistry was rationalised by either *syn* or *anti* addition of the dipolarophile to the glycosyl nitrone when it was in the "*O-endo*" conformation. However, later studies on a conformationally rigid *N*-mannosylspironitrone indicated that *anti*-attack was the preferred mode of addition, possibly because steric considerations outweighed stereoelectronic effects (B. Bernet, E. Krawczyk, A. Vasella, *Helv. Chim. Acta*, 1985, **68**, 2299). The corresponding *5'R*-isoxazolidines are accessible from nitrones derived from D-ribosyl hydroxy oxime (**74**). *N*-Glycosyl nitrones prepared *in situ*

Scheme 56

TrO— OH CHNOH (74) → (1) CH_2=CH_2 $CHOCO_2Bu$-*t* / $CHCl_3$, 75°C, 17h 65 bar → TrO— N— H CO_2Bu-*t* (75)(5 *R*) d.e. 72%

(2) $FeCl_3$ / 1% MeOH-CH_2Cl_2 (3) HCl, aq. MeOH → H N O *t*-BuO_2C H

43% overall yield
(76) (5*R*) 100% ee

from alkyl glyoxalates can participate in [3+2] cycloadditions but tend to react with lower selectivity (Scheme 56) (A. Vasella, R. Voeffray, *J. Chem. Soc. Chem. Comm.*, 1981, 97)

The chiral isoxazolidines are usually liberated from the parent monosaccharide by either mild acid hydrolysis (eg, $HClO_4$/MeOH or 1M HCl, 6% aq. MeOH) or through hydrogenolytic cleavage, although the latter protocol also cleaves the N-O linkage.

N-Glycosyl nitrones are becoming increasingly popular starting materials for the preparation of enantiomerically pure target molecules, as illustrated by the following syntheses of (+)-acivicin (Scheme 57) (S. Mzengeza, R.A. Whitney, *J. Am. Chem. Soc.*, 1987, **109**, 276; *J. Org. Chem.*, 1988, **53**, 4074), and (+)-nojirimycin (Scheme 58) (A. Vasella, R. Voeffray, *Helv. Chim. Acta*, 1982, **65**, 1134). In the former, a 19:1 mixture of the *5'S:5'R* isomers arose from the reaction of nitrone (**77**) with chiral vinylglycine (**78**). Presumably, this was because both reactants constituted a "matched" pair and double asymmetric induction could occur. In the sequence to nojirimycin (Scheme 58), three of the four contiguous asymmetric centres present in the target structure were created in a single step, through [3+2] cycloaddition between a glyoxalate-derived mannofuranosylnitrone and furan. The bicyclic nature of the adduct subsequently allowed all the remaining chirality to be installed in a predictable way using routine chemistry.

Scheme 57

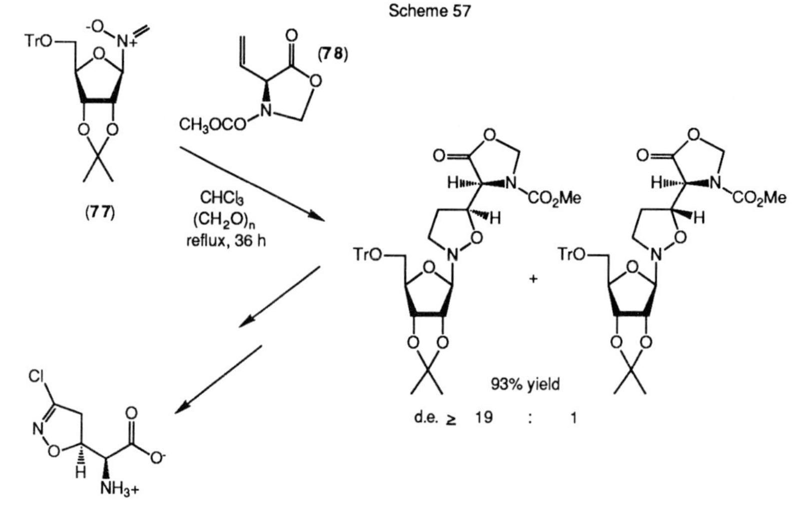

(77) (78)

$CHCl_3$
$(CH_2O)_n$
reflux, 36 h

93% yield

d.e. ≥ 19 : 1

(+)-Acivicin

Scheme 58

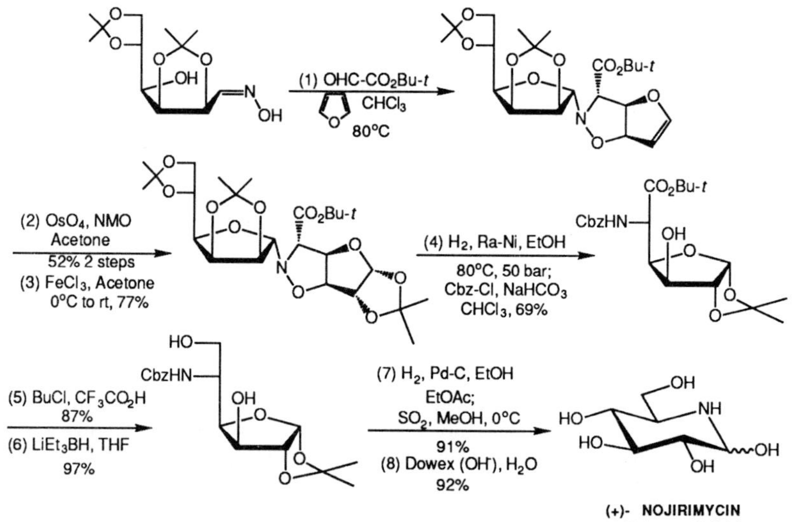

(1) OHC-CO_2Bu-*t*
$CHCl_3$
80°C

(2) OsO_4, NMO
Acetone
52% 2 steps
(3) $FeCl_3$, Acetone
0°C to rt, 77%

(4) H_2, Ra-Ni, EtOH
80°C, 50 bar;
Cbz-Cl, $NaHCO_3$
$CHCl_3$, 69%

(5) BuCl, CF_3CO_2H
87%
(6) $LiEt_3BH$, THF
97%

(7) H_2, Pd-C, EtOH
EtOAc;
SO_2, MeOH, 0°C
91%
(8) Dowex (OH^-), H_2O
92%

(+)- **NOJIRIMYCIN**
19.5% overall yield

(c) Asymmetric [2+2] cycloadditions with monosaccharide auxiliaries

The ketene-imine cycloaddition process has been successfully applied to a carbohydrate-based ketene to give, after removal of the sugar with mild acid, a *cis*-3-

Scheme 59

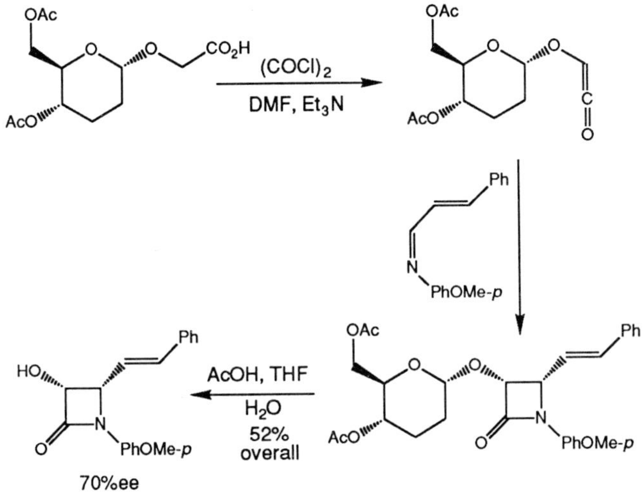

hydroxy-β-lactam in 70% ee and in moderate yield (B.C. Borer, D.W. Balogh, *Tetrahedron Lett.*, 1991, **32**, 1039) (Scheme 59).

4. Asymmetric carbenoid additions with monosaccharide bearing copper catalysts

Ethyl diazoacetate adds to a variety of halogenated olefins to afford chiral cyclopropane carboxylates in modest ee, provided copper (II)-Schiff base catalysts derived from amino-sugars (**79-82**) are incorporated into the reaction mixture (D. Holland, D.A. Laidler, D.J. Milner, *J. Mol. Catal.*, 1981, **11**, 119).

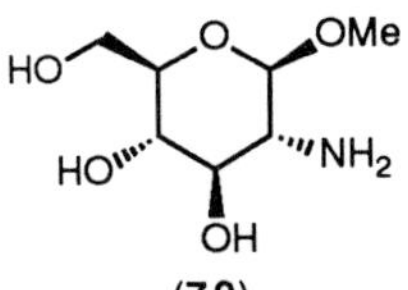

(**79**) 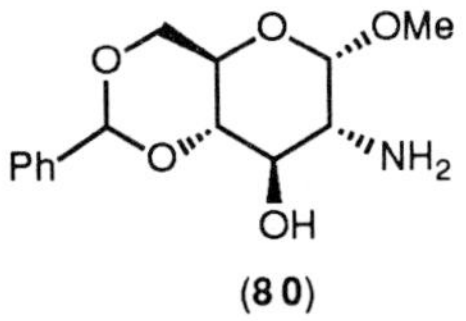 (**80**)

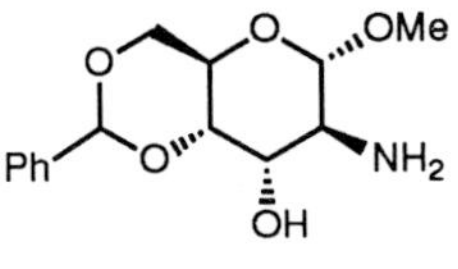

(**81**) 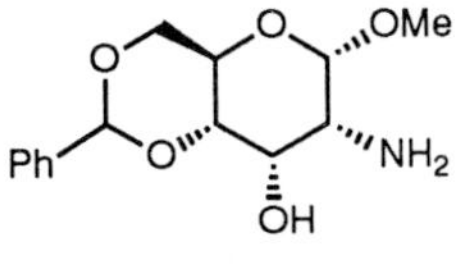 (**82**)

Second Supplements to the 2nd Edition of Rodd's Chemistry of Carbon Compounds, Vol. IE/F/G, edited by M. Sainsbury

Chapter 23c

MONOSACCHARIDES: USE IN THE ASYMMETRIC SYNTHESIS OF NATURAL PRODUCTS

KARL J. HALE

Introduction

A number of excellent monographs are available on the utilisation of monosaccharides as chiral starting materials for the synthesis of enantiomerically pure natural products. These have dealt with all the major developments that were reported prior to 1984 (B. Fraser-Reid and R.C. Anderson, *Fortschr. Chem. Org. Naturst.*, Springer-Verlag, Wien 1980, 39, 1; A. Vasella, *"Modern Synthetic Methods 1980"* R. Scheffold (Ed.), Salle und Sauerlander. Aurau 1980, 174; S. Hanessian, *"Total Synthesis of Natural Products: The Chiron Approach"*, Pergamon, Oxford, 1983; T.D. Inch, *Tetrahedron,* 1984, 40, 3161; N.K. Kochetkov, A.F. Sviridov, M.S. Ermolenko, *et al.*, *"Carbohydrates in the Synthesis of Natural Compounds"*, Nauka: Moscow, USSR, 1984). The topic is also reviewed comprehensively on an annual basis in the Specialist Periodical Reports of The Royal Society of Chemistry, *"Carbohydrate Chemistry"* Vols 9-23 (covering the literature in years 1977-1989). The present chapter discusses noteworthy synthetic achievements that have occurred since 1984, and in view of the vast amount of material that is available for coverage, it is necessarily selective. A deliberate attempt has been made to highlight syntheses which reflect the important methodological advances that have taken place in the last decade, and wherever appropriate these developments have been referenced in the text.

1. Use of monosaccharides for the assembly of natural products with carbocyclic ring systems

(a) Radical cyclisation strategies

Monosaccharide derivatives are being used increasingly for the

Scheme 1

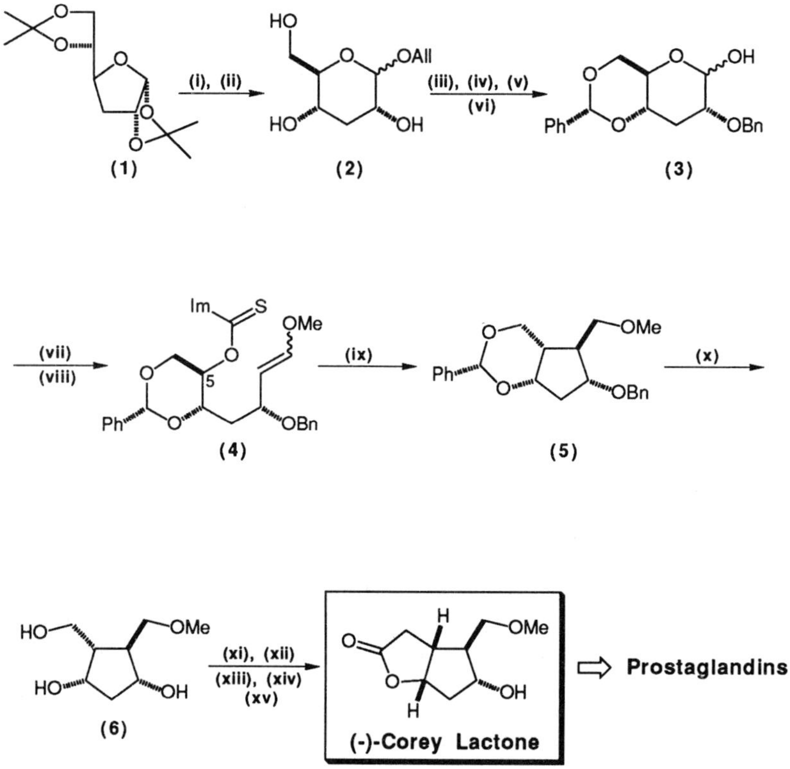

(1) —(i), (ii)→ (2) —(iii), (iv), (v), (vi)→ (3)

—(vii), (viii)→ (4) —(ix)→ (5) —(x)→

(6) —(xi), (xii), (xiii), (xiv), (xv)→ **(-)-Corey Lactone** ⇨ **Prostaglandins**

Reagents: (**i**) AG 50W-X8 H^+ resin, EtOH, H_2O, 4h, Δ; (**ii**) AG 50W-X8 H^+ resin, Allyl Alcohol, C_6H_6, Δ, 2 h, ; (**iii**) $PhCHBr_2$, Py, Δ, 90 min.,(32%, 3 steps); (**iv**) NaH, BnBr, DMF, -10°C to 0°C, 4h, (87%); (**v**) $[Ir(COD)(PPh_2Me)]^+PF_6^-$,THF, 2 h, (100%); (**vi**) HgO, $HgCl_2$, 9:1 Acetone-H_2O, (98%); (**vii**) $Ph_3P^+CH_2OCH_3Cl^-$, THF, *n*-BuLi, -20°C to r. temp.,16 h, (84%); (**viii**) Thiocarbonylbis(imidazole), $(ClCH_2)_2$, Δ, 4 h, (74%); (**ix**) *n*-Bu_3SnH, AIBN, PhMe, (58%); (**x**) H_2, Pd-C, 1:1 EtOAc-AcOH, 50 p.s.i.; (**xi**) TsCl, CH_2Cl_2, Py,; (**xii**) Ac_2O, Py, DMAP, (95%, 3 steps); (**xiii**) NaCN, DMF, 85°C, (83%); (**xiv**) K_2CO_3, MeOH; (**xv**) conc. HCl, 100°C, 2h, (38%, 2 steps).

construction of five and six-membered chiral carbocycles by free radical cyclisation reactions (C.S. Wilcox, L.M. Thomasco, *J. Org. Chem.*, 1985, 50, 546; C.S. Wilcox, J.J. Gaudino, *J. Am. Chem. Soc.*, 1986, 108, 3102; T.V. RajanBabu, *J. Am Chem. Soc.*, 1987, 109, 609; T.V. Rajanbabu, T. Fukunaga, and G.S. Reddy, *J. Am. Chem. Soc.*, 1989, 111, 1759). A number of recent natural product syntheses have incorporated such strategies.

(-)-Corey Lactone

The Corey lactone is an important intermediate in the synthesis of many naturally-occuring prostaglandins. Its synthesis (Scheme 1) in asymmetric form has recently been achieved from 1,2:5,6-di-*O*-isopropylidene-3-deoxy-α-D-glucofuranose by T.V. RajanBabu (*J. Org. Chem.*, 1988, 53, 4522). Compound (**1**) was initially subjected to acid hydrolysis to remove the isopropylidene protecting groups, and Fischer glycosidation then afforded a mixture of allyl glycosides (**2**) in almost quantitative yield. A benzylidene acetal was used to block the C(4) and C(6) hydroxyls, and the remaining C(2) alcohol was benzylated, prior to unmasking the anomeric centre. Wittig reaction on lactol (**3**) allowed the necessary one carbon homologation to be made at C(1), and led to a mixture of *E* and *Z* geometrical isomers which were not separated but converted directly into thiocarbonylimidazolides (**4**). The key radical ring-closure was executed in 58% yield by treatment with tri-*n*-butyltin hydride and a free radical initiator (D.H.R. Barton and S.W. McCombie, *J. Chem. Soc., Perkin Trans.* 1, 1975, 1574; W. Hartwig, *Tetrahedron*, 1983, 39, 2609). The *trans*-1,5-cyclopentane (**5**) was formed almost exclusively, as is generally the case for cyclic hex-5-enyl radicals derived from 4,6-*O*-benzylidenated hexopyranoses. Product (**5**) was thought to arise from the radical adding to the olefin in a boat-like transition state (**7**) that minimised *A1,3* strain. This would place the 4,6-*O*-benzylidene acetal in a chair conformation with the phenyl and but-3-enyl groups both in an equatorial orientation.

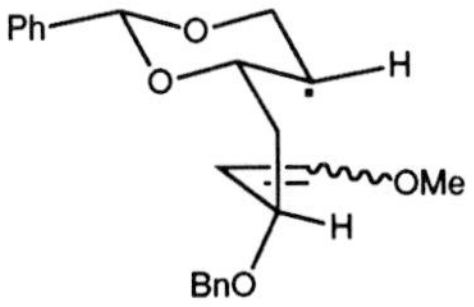

Ph O O H OMe H BnO

(**7**)

The final stages of the route involved hydrogenation of (**5**) under mildly acidic conditions to produce (**6**). The primary alcohol of (**6**) was tosylated, the remaining hydroxy groups acetylated, and nucleophilic displacement of the tosyloxy substituent effected in good yield with cyanide ion. After *O*-deacetylation with base, lactonisation to the Corey lactone was accomplished with strong acid.

(-)-Allosamizoline

A free radical cyclisation also featured in the synthesis of (-)-allosamizoline from D-glucosamine hydrochloride by N.S. Simpkins, S. Stokes, and A.J. Whittle (*Tetrahedron Lett.*, 1992, 33, 793). The critical reaction in this strategy (Scheme 2) was the free radical-mediated ring closure of thiocarbonylimidazolide oxime ether (**11**) to compound (**12**) (P.A. Bartlett, K.L. McLaren, P.C. Ting, *J. Am. Chem. Soc.*, 1988, 110, 1633). This proceeded in 54% yield and gave a mixture of diastereoisomers at C(1). Conversion of the *O*-benzylhydroxylamino substituent in (**12**) to a secondary alcohol proved troublesome, but was nevertheless achieved by ozonolysis of oxime (**13**) and reductive work-up with sodium borohydride. This furnished alcohol (**14**), which was transformed into (-)-allosamizoline hydrochloride in four steps.

(-)-Silphiperfolene

(-)-Silphiperfolene is an architecturally novel sesquiterpenoid triquinane isolated from roots of *Silphium perfoliatum* (F. Bohlmann, and J. Jakupoviv, *Phytochemistry,* 1980, 19, 259). It was first synthesised in optically active form by L.A. Paquette, R.A. Roberts, and G.J. Drtina (*J. Am. Chem. Soc.*, 1984, 106, 6690) starting from (R)-(+)-pulegone. Despite the fact that (-)-silphiperfolene bears no apparent resemblance to a monosaccharide, its synthesis was recorded from keto-ester (**16**) (J.K. Dickson, Jr., and B. Fraser-Reid, *J. Chem. Soc., Chem. Comm.*, 1990, 1440) (Scheme 3). The route commenced with a noteworthy regio- and stereo-selective axial methylation at C(2) instigated by reaction with potassium hydride and methyl iodide (Y. Chapleur, *J. Chem. Soc., Chem. Comm.*, 1983, 141). The product ketone was chemoselectively reacted with vinylmagnesium bromide, and the ethyl ester reduced to a primary alcohol that was protected as a *tert*-butyldimethylsilyl ether. Treatment of (**17**) with thionyl chloride was accompanied by chlorination and double bond rearrangement, and after displacement with potassium acetate, transesterification led to an exocyclic allylic alcohol. Oxidation to the α,β-unsaturated aldehyde followed by addition of methyl magnesium chloride gave (**18**) along with its epimer. The undesired alcohol was converted to (**18**) by Mitsunobu reaction with benzoic acid (O. Mitsunobu, *Synthesis*, 1981, 1) followed by treatment with methyllithium to remove the benzoate ester. Claisen rearrangement of (**18**) then furnished (**19**) in 78% yield. Addition of ethyl magnesium bromide to (**19**), followed by Swern oxidation (A.J. Mancuso, S.L. Huang, D. Swern, *J. Org. Chem.*, 1978, 43, 2480) and ozonolysis provided a keto-aldehyde which was

Scheme 2

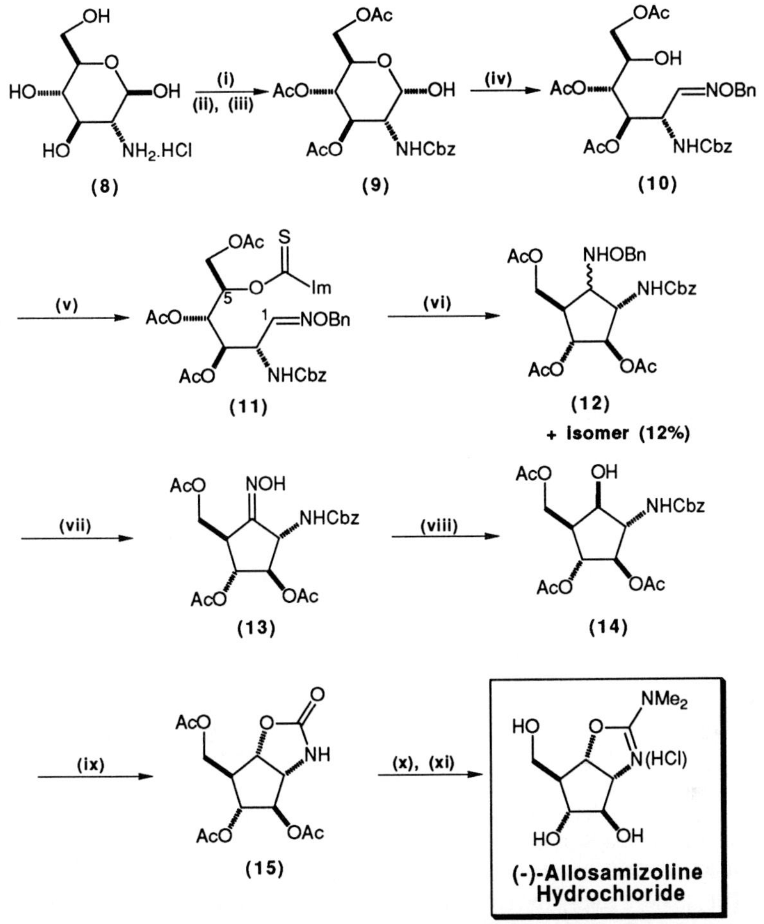

(8) → (i), (ii), (iii) → (9) → (iv) → (10) → (v) → (11) → (vi) → (12) + isomer (12%) → (vii) → (13) → (viii) → (14) → (ix) → (15) → (x), (xi) → (-)-Allosamizoline Hydrochloride

Reagents: (**i**) CbzCl, $NaHCO_3$, H_2O, (96%); (**ii**) Ac_2O, Et_3N, THF, DMAP, (82%); (**iii**) $(NH_4)_2CO_3$, MeOH, THF, (88%); (**iv**) $NH_2OBn.HCl$, Py, CH_2Cl_2, (88%); (**v**) $Im_2C{=}S$, C_6H_6, (82%); (**vi**) Bu_3SnH, AIBN, C_6H_6, Δ, (54%); (**vii**) *m*-CPBA, Na_2CO_3, EtOAc, (79%); (**viii**) O_3, CH_2Cl_2, -40°C; $NaBH_4$, MeOH, -40°C to rt, (20-40%); (**ix**) $SOCl_2$, (82%); (**x**) Et_3OBF_4, CH_2Cl_2; Me_2NH, CH_2Cl_2, (80%); (**xi**) NaOMe, MeOH, HCl, (98%).

Scheme 3

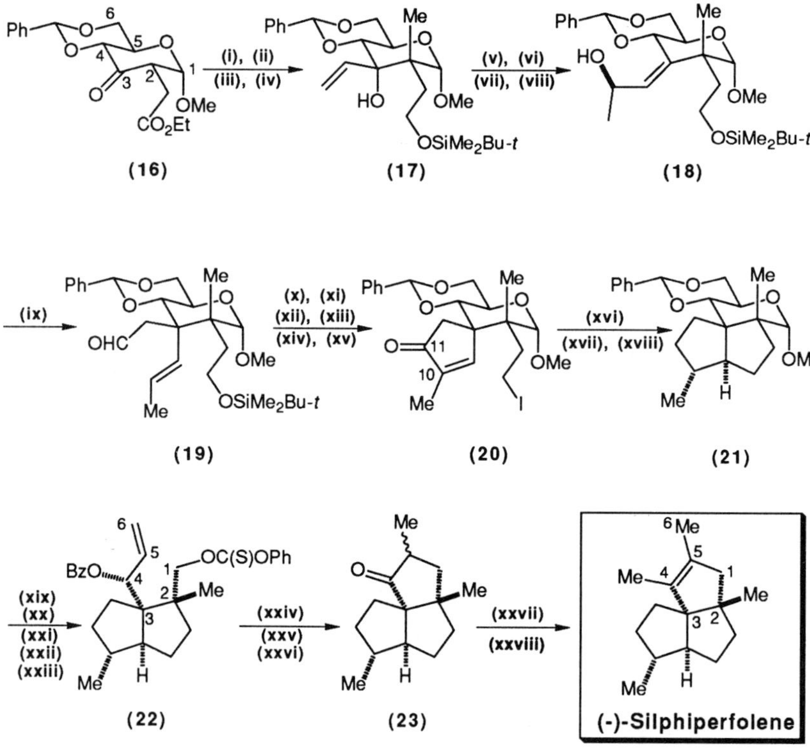

Reagents: (**i**) KH, THF, MeI (78%); (**ii**) H_2C=CHMgBr, THF, 20 min.; (**iii**) $LiAlH_4$, THF, 0°C, 45 min.; (**iv**) *t*-$BuMe_2SiCl$, imidazole, THF, (73%, 3 steps); (**v**) $SOCl_2$, Py, THF, 0°C, 1h; (**vi**) KOAc, DMF, then NaOMe, MeOH (62%, 2 steps); (**vii**) PCC, CH_2Cl_2; (**viii**) MeMgCl, THF, (83%, 2 steps); (**ix**) EtOCH=CH_2, $Hg(OCOCF_3)_2$, Et_2O, 12 h; Xylene, Δ, 24 h, (78%); (**x**) EtMgBr, THF; (**xi**) $(COCl)_2$, Me_2SO, CH_2Cl_2, -78°C; (**xii**) O_3, MeOH, CH_2Cl_2; (**xiii**) KOBu-*t*, THF, (45%, 4 steps); (**xiv**) *n*-Bu_4NF, THF; (**xv**) Ph_3P, I_2, C_6H_6, (77%, 2 steps); (**xvi**) *n*-Bu_3SnH, AIBN, C_6H_6, Δ, (81%); (**xvii**) $(Me_3Si)_2NLi$, $PhN(SO_2CF_3)_2$, THF; (**xviii**) H_2, Pd, MeOH, (52%, 2 steps); (**xix**) *N*-Bromosuccinimide, $BaCO_3$, CCl_4, Δ, 0.5 h; (**xx**) NaI, MeCOEt, Δ, 18h; (**xxi**) Zn(Hg), EtOH, Δ, 2 h, (82%, 2 steps); (**xxii**) $NaBH_4$, MeOH, r. temp., 5 min.; (**xxiii**) PhOC(S)Cl, DMAP, CH_2Cl_2,(96%, 2 steps); (**xxiv**) *n*-Bu_3SnH, C_6H_6, AIBN, Δ; (**xxv**) $LiAlH_4$, Et_2O, 0°C, 30 min.; (**xxvi**) PDC, 4A Mol. Sieves, Et_2O, 1h, (54%, 3 steps); (**xxvii**) MeLi, Et_2O, -78°C, 30 min.; (**xxviii**) $POCl_3$, Py, (34%, 2 steps).

converted to the desired iodo spiro-cyclopentenone (**20**), by treatment with potassium *tert*-butoxide, *O*-desilylation, and iodination. Conjugate addition of the primary radical generated from (**20**) by reaction with tri-*n*-butyltin hydride and AIBN led to the desired methyl epimer at C(10) being formed exclusively in excellent yield. Deoxygenative removal of the C(11) keto group was then achieved by palladium-catalysed hydrogenation (V.B. Jigajinni and R.H. Wightman, *Tetrahedron Lett.*, 1982, 23, 117) of the enol-triflate (J.E. McMurry and W.J. Scott, *Tetrahedron Lett.*, 1983, 24, 979). The benzylidene acetal in the resulting pyranoside diquinane (**21**) was cleaved with *N*-bromosuccinimide and after halide exchange, reductive acetal ring opening furnished an alkenyl-aldehyde (B. Bernet and A. Vasella, *Helv. Chim. Acta.*, 1979, 62, 1990, 2400, 2411; *ibid.*, 1984, 67, 1328), that was reduced and thioacylated to give (**22**). Radical mediated ring-closure of (**22**) proceeded smoothly to give a mixture of methyl epimers at C(5) (M.J. Robins and J.S. Wilson, *J. Am. Chem. Soc.*,1981, 103, 932). This stereochemical outcome was inconsequential since the C(5) position was later to be incorporated into the olefinic system. The sequence required for this conversion involved reduction of the benzoate ester, oxidation of the alcohol to ketone (**23**), addition of methyllithium, and dehydrative elimination with phosphorus oxychloride. These last five steps proceeded in 18% overall yield.

(-)-Pipitzol

The strategy of tandem radical cyclisation on multiply branched monosaccharide derivatives has now emerged as very useful methodology for obtaining, highly functionalised, chiral polyquinane systems (R. Tsang and B. Fraser-Reid, *J. Am. Chem. Soc.*, 1986, 108, 2116) An illustration of how this approach can be tailored to a specific target molecule is provided by the D-mannose-based route to (-)-α-pipitzol developed by H. Pak, I.I. Canalda, and B. Fraser Reid (*J. Org. Chem.*, 1990, 55, 3009). As with (-)-silphiperfolene, the idea of generating a pyranoside diquinane was central to this strategy, but in this approach the requisite intermediate (**27**) was assembled by tandem radical cyclisation of the branched-sugar (**26**) (Scheme 4). A key step in the preparation of (**26**) was the Johnson ortho-ester spiro-Claisen rearrangement (D.B. Tulshian, R. Tsang, B. Fraser-Reid, *J. Org. Chem.*, 1984, 49, 2347; W.S. Johnson *et al.*, *J. Am. Chem. Soc.*, 1970, 92, 741) on allylic alcohol (**25**). The pyranoside system of diquinane (**27**) was then unravelled by conversion to iodide (**28**), and reductive elimination with zinc amalgam. This provided aldehyde (**29**), which was α-hydroxylated (G.M. Rubottom, M.A. Vasquez, D.R. Pelegrina, *Tetrahedron Lett.*, 1974, 4319) and reduced to the 1,2-diol. Hydroboration of the olefin, followed by two-step oxidation furnished

Scheme 4

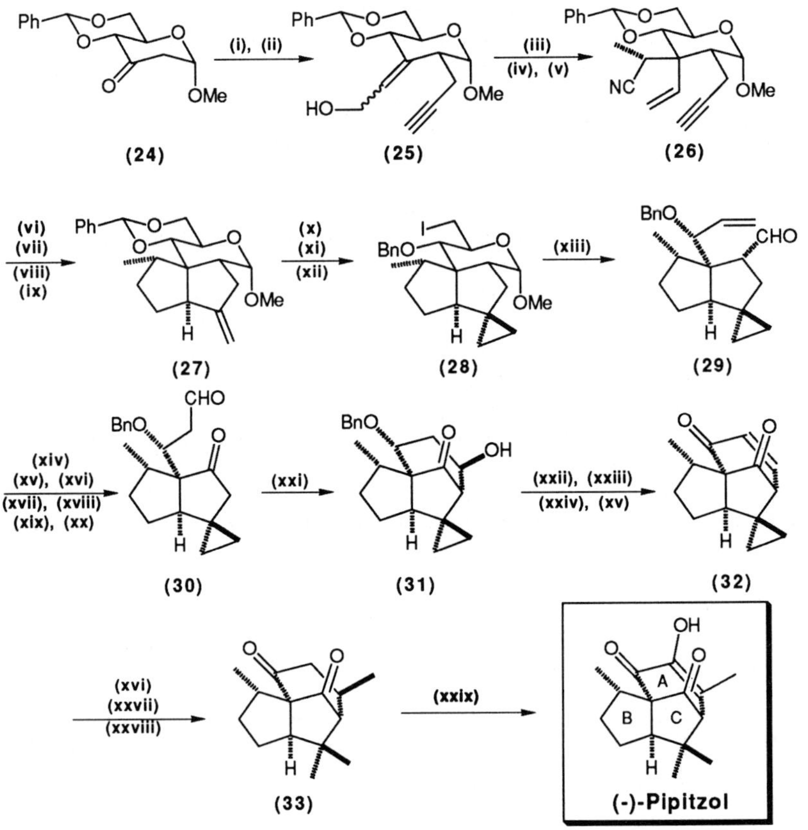

Reagents: (**i**) KH, THF, Propargyl Bromide, 0°C, (50%); (**ii**) $(EtO)_2P(O)CH_2CO_2Et$, NaH, THF, 0°C then DIBAL-H, (53%); (**iii**) $EtC(OEt)_3$, $EtCO_2H$, Xylene, Δ, (85%); (**iv**) DIBAL, then PCC, then NaOMe, (85%); (**v**) $NH_2OH.HCl$, NaOAc, AcOH, then $(CF_3CO)_2O$, (65%); (**vi**) *n*-Bu_3SnH, AIBN, C_6H_6, Δ; (**vii**) Silica gel, CH_2Cl_2, (65%); (**viii**) DIBAL, (88%); (**ix**) PhOCSCl then *n*-Bu_3SnH (70%); (**x**) Et_2Zn, CH_2I_2, PhMe, 60°C, (94%); (**xi**) DIBAL, (60%); (**xii**) Ph_3P, I_2, Imidazole, PhMe, (90%); (**xiii**) Zn(Hg) (70%); (**xiv**) $KN(SiMe_3)_2$, *t*-$BuMe_2SiCl$, DMF, -60°C; (**xv**) *m*-CPBA; (**xvi**) *n*-Bu_4NF; (**xvii**) $NaBH_4$ (64%, 4 steps); (**xviii**) $BH_3.Me_2S$; (**xix**) $NaIO_4$; (**xx**) PDC, (25%, 3 steps); (**xxi**) 0.25% Na_2CO_3, 4:3 MeOH, H_2O, (90%); (**xxii**) TsCl; (**xxiii**) H_2, Pd-C; (**xxiv**) LiBr, DMF, Δ; (**xxv**) Dess-Martin Periodinane, CH_2Cl_2; (**xxvi**) Me_2CuLi, THF, -78°C; (**xvii**) H_2, PtO_2 AcOH, 3-4 atm; (**xxviii**) Dess-Martin Periodinane, (24% from **xxii**); (**xxix**) SeO_2, H_2O, Dioxane, Δ, 18 h, (47%).

Scheme 7

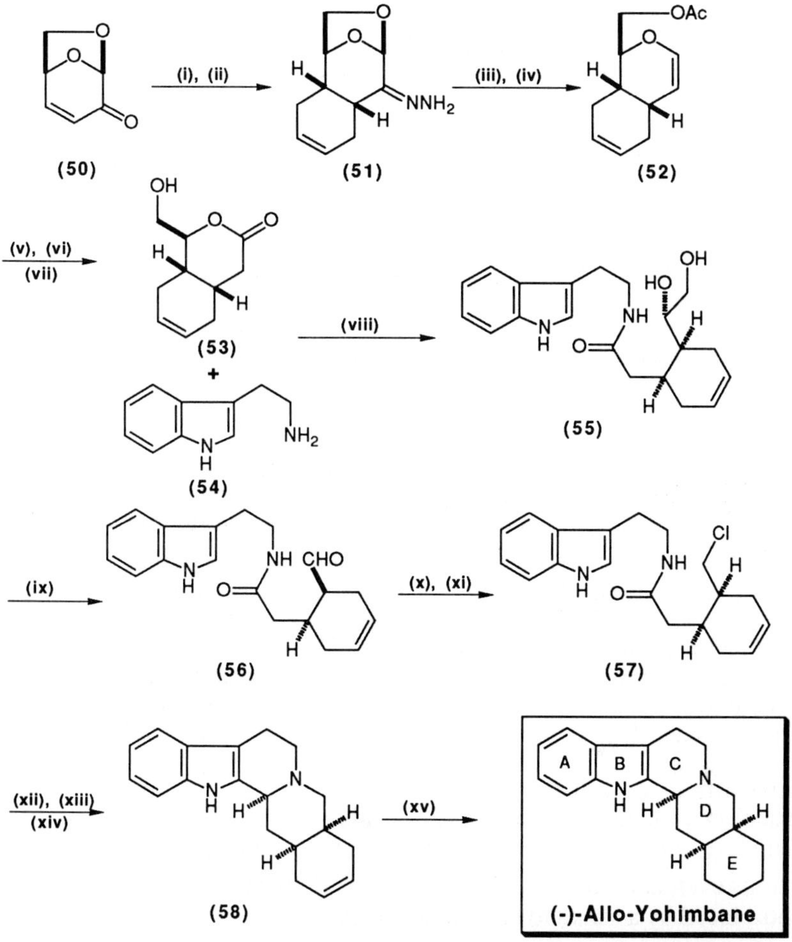

(50) (i), (ii) → (51) (iii), (iv) → (52)

(v), (vi) (vii) → (53) + (54) (viii) → (55)

(ix) → (56) (x), (xi) → (57)

(xii), (xiii) (xiv) → (58) (xv) → (-)-Allo-Yohimbane

Reagents: (**i**) 1,3-Butadiene, 140°C, 10 h, (98%); (**ii**) NH_2NH_2-H_2O (5.0 equiv), Et_3N, (1.0 equiv), EtOH, 60°C, 20 min.; (**iii**) NaH, Me_2SO, r. temp., 1.5 h, (93%, 2 steps); (**iv**) Ac_2O (3.0 equiv), Py (7.0 equiv), CH_2Cl_2, 4 h, (100%); (**v**) 1 N HCl, THF, 50°C, 2h ; (**vi**) Jones reagent, 0°C, 5 min., (90%, 2 steps); (**vii**) NaOMe, MeOH, 5 min., (86%); (**viii**) Tryptamine (2.0 equiv), THF, *i*-Pr_2NEt,(96%); (**ix**) HIO_4 (1.2 equiv), THF-H_2O, (3:1); (**x**) $NaBH_4$, MeOH, r. temp., 5 min., (93%, 2 steps); (**xi**) Ph_3P, CCl_4 (3.0 equiv), $CHCl_3$-Py (2:1), (51%, 75% conversion yield); (**xii**) $LiN(SiMe_3)_2$, THF, -78°C to r. temp., (91%); (**xiii**) $POCl_3$, 100°C, 1.5 h; (**xiv**) $NaBH_4$, MeOH, H_2O (9:1), pH 7, 0°C, (73%, 2 steps); (**xv**) H_2, Pd-C, MeOH, CH_2Cl_2, (1:1), (92%).

and butadiene also played a pivotal role in an interesting asymmetric total synthesis of (-)-allo-yohimbane (M. Isobe, N. Fukami, T. Goto, *Chemistry Lett.*, 1985, 71). This reaction yielded only one cycloadduct (**51**), and thus allowed the DE ring junction stereochemistry to be set in a single step (Scheme 7). Application of an eliminative Wolff-Kishner reduction to (**51**) followed by acetylation delivered glycal (**52**). Reaction with Jones reagent and selective deesterification then provided lactone (**53**) which condensed with tryptamine (**54**) to give (**55**). The latter underwent diol cleavage with periodic acid to furnish aldehyde (**56**). After borohydride reduction, chlorination with triphenylphosphine and carbon tetrachloride led to (**57**). Closure of the D-ring was best achieved by exposing (**57**) to lithium hexamethyldisilazide. The C-ring of (**58**) was established by treatment with phosphorus oxychloride, and reduction with sodium borohydride. Hydrogenation of the double bond completed this total synthesis of (-)-allo-yohimbane.

β-Santalene

β-Santalene is a component of highly prized East Indian sandalwood oil. It is characterised by a norbornyl ring system in which there is a quaternary chiral centre and an exocyclic olefin. A stereocontrolled synthesis of β-santalene has been reported by S. Takano *et al.* (*J. Chem. Soc., Chem. Comm.*, 1987, 1720) from *(S)*-5-hydroxymethylbuten-2-olide (**61**) (Scheme 8). The preparation of this material had previously been described from D-mannitol (S. Takano *et al.*, *Synthesis*, 1986, 403). The bicyclic framework of the target compound was created in a Diels-Alder reaction between (**61**) and cyclopentadiene. This produced a crystalline *endo*-adduct (**62**) along with some minor stereoisomers. Hydrogenation of (**62**) followed by saponification, periodate cleavage, borohydride reduction and acidification yielded intermediate (**63**). This was then alkylated on its least hindered face with 4-methyl-pent-3-enyl bromide, and a series of reductions eventually resulted in (**66**). The olefin present in β-santalene was introduced by a standard elimination sequence.

(-)-Cryptosporin

The "inverse"-electron-demand (IED) Diels-Alder reaction has received only limited attention from chemists involved in natural product synthesis. An IED [4+2] cycloaddition did, however, feature in a beautifully crafted asymmetric synthesis of (-)-cryptosporin that began from L-fucal (R.B. Gupta and R.W. Franck, *J. Am. Chem. Soc.*, 1989, 111, 7668). The pathway commenced with a Bradsher cycloaddition reaction (C.K. Bradsher, and F.H. Day, *Tetrahedron Lett.*, 1971, 409) between isoquinolinium salt (**68**) and L-fucal (**69**)

Scheme 8

(59) (60) (61)

(62) (63) (64)

(65) (66)

(67) **(+)-β-Santalene**

Reagents: (**i**) $NaIO_4$, 5% aq. $NaHCO_3$, 0°C to r. temp., 1 h; $(EtO)_2P(O)CH_2CO_2Et$, K_2CO_3, H_2O, r. temp.,1 pot procedure, (86%, 97:3 *E*:*Z*); (**ii**) $(i\text{-}Pr)_2NEt$, PhSH, C_6H_6 42 h, (100%); (**iii**) conc. HCl, EtOH, 22 h, (100%); (**iv**) H_2O_2, AcOH, 2 h, (81%); (**v**) $CaCO_3$, PhMe, Δ, (69%); (**vi**) Cyclopentadiene (10 equiv), Sealed Tube, 140°C, 9h, (60%); (**vii**) H_2, Pd-C, MeOH, (98%); (**viii**) 20% aq. NaOH, $NaIO_4$, $NaBH_4$, 0°C to r. temp., (1 pot); (**ix**) 10% aq. HCl, (66%, 2 steps); (**x**) LDA (2.1 equiv), THF, -78°C then 4-Methyl-pent-3-enyl bromide (61%); (**xi**) DIBAL, (2 equiv), THF, -78°C; (**xii**) NH_2NH_2 $-H_2O$ (3.5 equiv.), KOH (3.5 equiv), $(CH_2OH)_2$, 130 to 200°C, (73%, 2 steps); (**xiii**) TsCl, Et_3N, DMAP, 0°C to r. temp.; (**xiv**) NaI, MeCOEt, 70°C; (**xv**) DBU, C_6H_6, 70°C, (87%).

Scheme 9

(68) (69) (70) (71) (72) (73) (74) (75) (-)-Cryptosporin

Reagents: (**i**) $CaCO_3$, MeOH, 55-60°C, 72 h; (**ii**) 1 N HCl, MeCN, 24 h, (95%, 2 steps); (**iii**) 2,2-$(MeO)_2CMe_2$, Me_2CO, *p*-TsOH, 4 A Sieves, 0-25°C, 1 h; (**iv**) *t*-$BuMe_2SiOTf$, Et_3N, CH_2Cl_2, 0-25°C, 1 h, (91%, 2 steps); (**v**) $RuCl_3$, $NaIO_4$, MeCN, H_2O, CCl_4, 5-25°C, 0.5 h, (78%); (**vi**) *t*-$BuMe_2SiOTf$, Et_3N, CH_2Cl_2, 0-25°C, 4 h, (95%); (**vii**) $LiBH_4$, EtOH, THF, H_2O, 25°C, 20 h, (90%); (**viii**) MeI, K_2CO_3, MeOH, 25°C, 30 h, (56%); (**ix**) 3 N HCl, MeCN, 45°C, 4 h, (92%); (**x**) Salcomine, MeCN, O_2, 45 min., (72%); (**xi**) BCl_3, CH_2Cl_2, -78°C to -40°C, 6 h, (77%).

(Scheme 9). This gave a tetracyclic adduct that underwent ring-opening in aqueous acid to furnish amino-aldehyde (**70**). Ketalisation of the diol, followed by conversion to silyl enol ether (**71**), and oxidative cleavage, supplied ketone (**72**). This was transformed into napthol (**74**) by formation of silyl enol ether (**73**), reductive cleavage of the dinitrophenyl group with lithium borohydride, and execution of a modified Hofmann elimination followed by acid hydrolysis. The next step was salcomine oxidation of (**74**) in acetonitrile. This regimen allowed a 72% yield of (**75**) to be isolated. Low temperature deprotection of quinone (**75**) with boron trichloride was the finale of this very elegant approach to (-)-cryptosporin, which confirmed that the natural material was the (+)-enantiomer.

(+)-Diplodiatoxin

The intramolecular Diels-Alder reaction has now emerged as a powerful technique for the construction of polycylic natural products, since it can combine high regioselectivity with excellent stereocontrol. A perfect demonstration of these properties can be found in the route to (+)-diplodiatoxin from D-glucose by A. Ichihara *et al.* (*Tetrahedron Lett.*, 1986, 27, 1347) (Scheme 10). The synthesis began by deoxygenation of known alcohol (**76**) (S. Hanessian and G. Rancourt, *Can. J. Chem.*, 1977, 55, 1111) and debenzylidenation. Selective tritylation of the resulting diol, oxidation with pyridinium dichromate (E.J. Corey and G. Schmidt, *Tetrahedron Lett.*, 1979, 399), and Wittig reaction produced (**78**). Reduction of the exocyclic olefin with diimide gave a 2:1 mixture of methyl epimers at C(4), in which the desired isomer (**79**) predominated. After opening of the pyranoside ring with 1,3-propanedithiol, a very unusual eliminative method was used for obtaining aldehyde (**81**). A series of Wittig-Horner reactions were then employed for elaborating (**81**) into the chiral *(E,E,E)*-trienone (**83**). Intramolecular Diels-Alder reaction of (**83**) proceeded in 85% yield to afford a single stereoisomer (**84**). Demethylation with aluminium trichloride in tetrahydrothiophene produced (+)-diplodiatoxin in 64% yield.

Hexahydronapthalene fragment of kijanolide and tetronolide

A similar monosaccharide initiated IMDA strategy has been recorded by E. Yoshii *et al.*, (*J. Org. Chem.*, 1988, 53, 1092) for the construction of the hexahydronapthalene fragment in kijanolide and tetronolide (Scheme 11).

(-)-Forskolin intermediate

Forskolin is a labdane-diterpenoid isolated from roots of *Coleus forskohlii* (S.V. Bhat *et al.*, *Tetrahedron Lett.*, 1977, 19, 1669). It has powerful hypotensive activity and can greatly reduce intraocular

Scheme 10

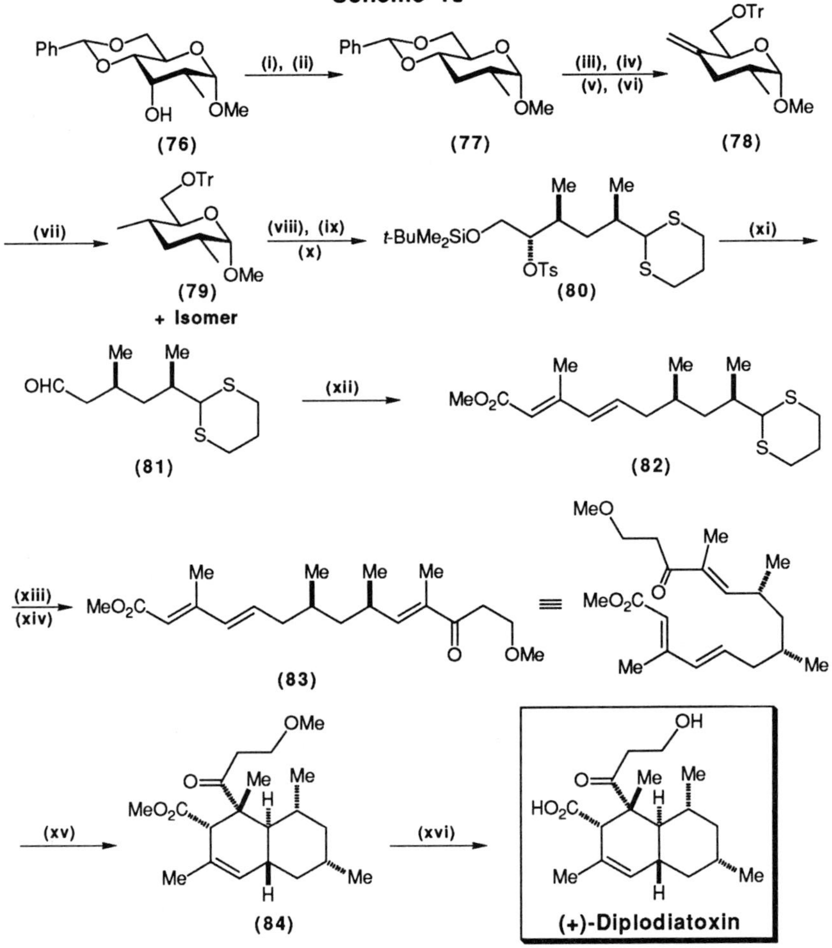

Ph (76) OH OMe —(i), (ii)→ (77) OMe —(iii), (iv), (v), (vi)→ (78) OTr OMe

—(vii)→ (79) OTr OMe + Isomer —(viii), (ix), (x)→ (80) t-BuMe$_2$SiO OTs Me Me S S —(xi)→

(81) OHC Me Me S S —(xii)→ (82) MeO$_2$C Me Me Me S S

—(xiii), (xiv)→ (83) MeO$_2$C Me Me Me Me O OMe ≡ MeO MeO$_2$C O Me Me Me Me

—(xv)→ (84) OMe O MeO$_2$C Me Me H Me H Me —(xvi)→ OH O HO$_2$C Me Me H Me H Me

(+)-Diplodiatoxin

Reagents: (**i**) NaH, THF, CS_2, MeI; (**ii**) Bu_3SnH, AIBN, Toluene, Δ, 48 h, (75%, 2 steps); (**iii**) *p*-TsOH, MeOH, r. temp., 3h; (**iv**) TrCl, Py, r. temp.; (**v**) PDC, DMF, r. temp., (95%, 2 steps); (**vi**) $Ph_3P^+CH_3Br^-$, *n*-BuLi; PhMe, 0°C to r. temp., 3h, (87%); (**vii**) N_2H_4, H_2O, O_2, EtOH, Δ, 6h, (100%, 2:1 Equatorial Me-4: Axial Me-4); (**viii**) 1,3-Propanedithiol, $BF_3.Et_2O$, 0°C, 4h, (100%); (**ix**) *t*-BuMe$_2$SiCl, Imidazole, -10°C, 4h; (**x**) TsCl, DMAP, $CHCl_3$, r. temp., 12 h, (80%, 2 steps); (**xi**) DBU, Py, 100°C, 19 h, (40%); (**xii**) LDA, THF, HMPT, -60°C, *(E)*-$MeO_2CCH{=}C(Me)CH_2P(O)(OEt)_2$, 6 h, (95%); (**xiii**) $Hg(ClO_4)_2$. $3H_2O$, $CHCl_3$-THF, r. temp., 10 min.; (**xiv**) *n*-BuLi, THF, DMF, $(EtO)_2P(O)CH(Me)CO(CH_2)_2OMe$, -1.5°C, (57%, 2 steps, 95:5 *E:Z*); (**xv**) 140°C, PhMe, Sealed Tube, 37 h, (85%); (**xvi**) $AlCl_3$, r. temp., 15 h, Tetrahydrothiophene, (64%).

Scheme 11

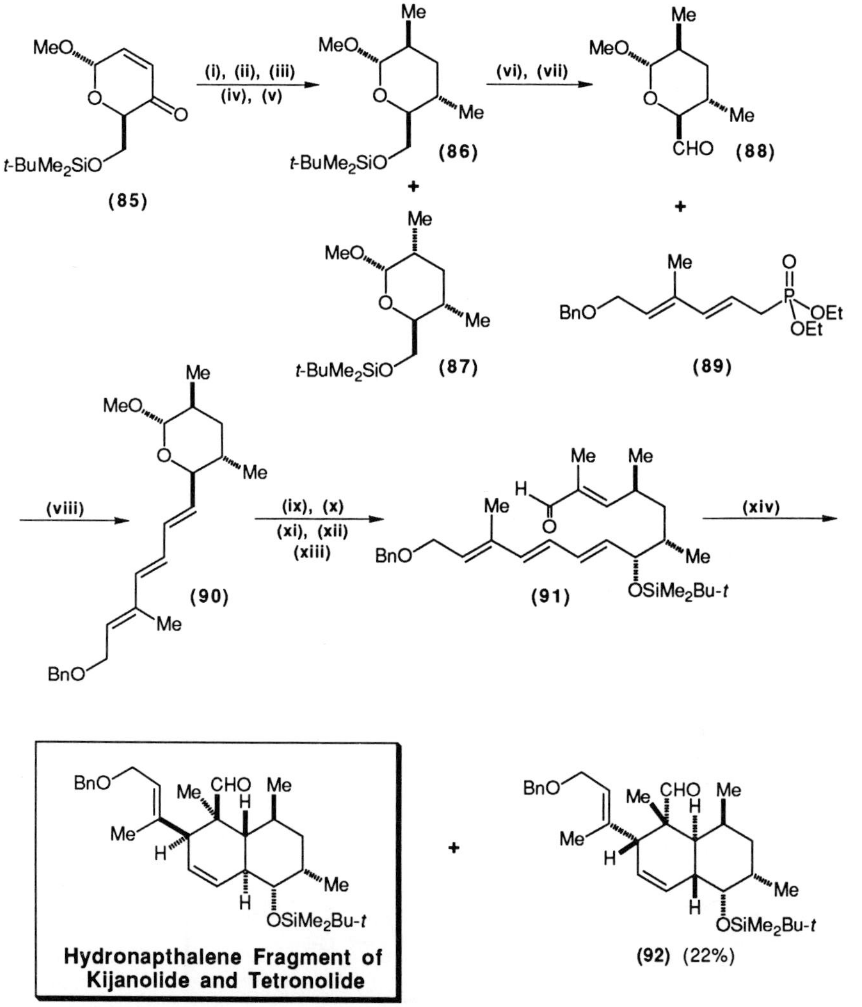
(i), (ii), (iii)
(iv), (v)
(vi), (vii)
(85)
(86)
(87)
(88)
(89)
(viii)
(ix), (x)
(xi), (xii)
(xiii)
(xiv)
(90)
(91)
Hydronapthalene Fragment of Kijanolide and Tetronolide
(92) (22%)

Reagents: (**i**) MeLi; (**ii**) PCC, CH_2Cl_2; (**iii**) H_2, Pd-C; (**iv**) $Ph_3P{=}CH_2$; (**v**) H_2, Ni, (22%, 5 steps, (**86**):(**87**) = 9:1); (**vi**) *n*-Bu_4NF, THF; (**vii**) Me_2SO, $(COCl)_2$, CH_2Cl_2, Et_3N; (**viii**) KOBu-*t*, THF, (55%, 3 steps); (**ix**) AcOH, H_2O, THF; (**x**) $Ph_3P{=}C(Me)CO_2Et$, MeCN, (46%, 2 steps); (**xi**) *t*-$BuMe_2SiCl$, imidazole, DMF,(70%); (**xii**) DIBAL-H, Et_2O -70°C; (**xiii**) PCC, AcONa, CH_2Cl_2, (45%, 2 steps); (**xiv**) Me_2AlCl, -80°C to -40°C, 5.5 h, (65%).

Scheme 12

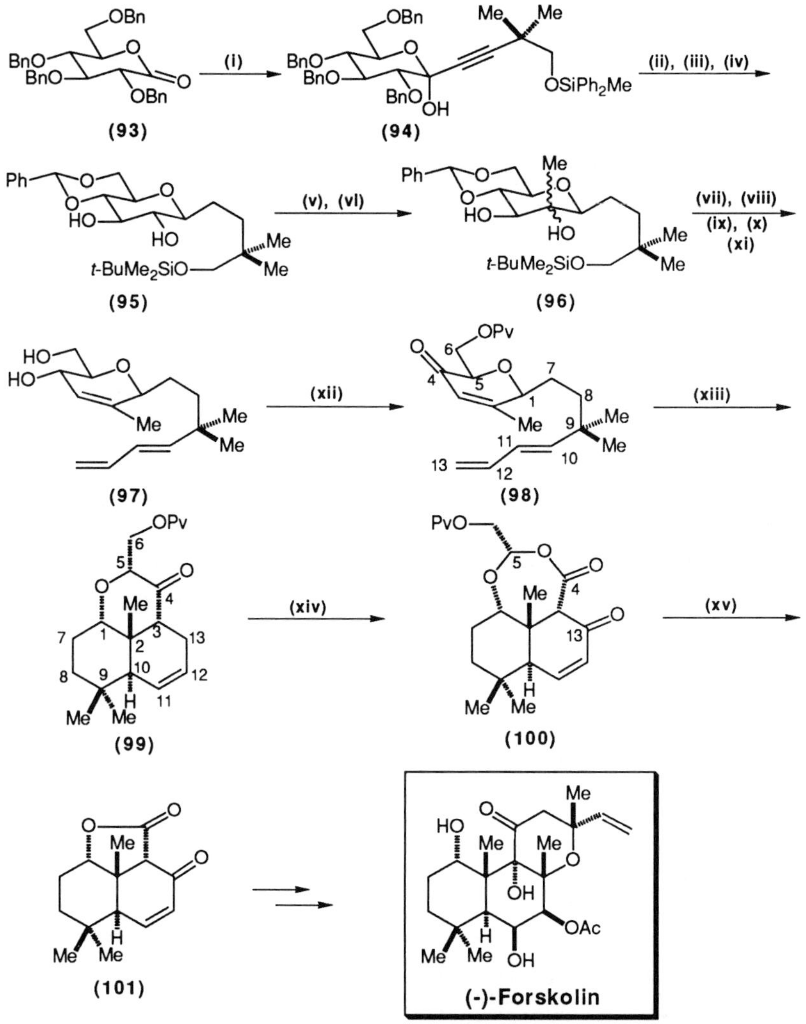

Reagents: (**i**) $Ph_2MeSiOCH_2C(Me)_2CC$-Li; (**ii**) Et_3SiH, BF_3-Et_2O, CH_2Cl_2, (55%, 2 steps); (**iii**) H_2, Pd-C, MeOH, $HClO_4$, then $PhCHBr_2$, Py; (**iv**) *t*-$BuMe_2SiCl$, Pyridine, DMAP, CH_2Cl_2, (76%, 3 steps); (**v**) *n*-Bu_2SnO, PhMe, then Br_2/CCl_4, (72%); (**vi**) MeMgBr, THF, (82%); (**vii**) NaH, CS_2, *n*-Bu_4NI, MeI; (**viii**) $(MeO)_3P$, (63%); (**ix**) *n*-Bu_4NF, THF, then PCC, (65%); (**x**) Allylmagnesium Bromide, then $POCl_3$, Py, (64%); (**xi**) CSA, MeOH, (85%); (**xii**) Me_3CCOCl, Py, DMAP, CH_2Cl_2, then PCC (50%); (**xiii**) Hydroquinone, Xylene, Δ, (75%); (**xiv**) Na_2CrO_4, Ac_2O, AcOH; (**xv**) NaOMe, MeOH, (67%, 2 steps).

pressure, and thus is of potential therapeutic benefit (K.B. Seamon, *Ann. Rep. Med. Chem.*, 1984, 19, 293). R. Tsang and B. Fraser-Reid (*J. Org. Chem.*, 1992, 57, 1067) have described a synthetic approach to forskolin that utilises D-glucose as the chiral starting material. As shown in Scheme 12, *O*-benzylated 1,5-glucuronolactone (**93**) was initially converted into alkynyl alcohol (**94**), and this deoxygenated at the anomeric centre by ionic hydrogenation (D. Rolf and G.R. Gray, *J. Am. Chem. Soc.*, 1982, 104, 3539; M.D. Lewis, J.K. Chao, Y. Kishi, *ibid.*, 1982, 104, 4976). The product was debenzylated, and an *O*-benzylidene acetal used to protect C(4) and C(6). This furnished (**95**), which was regioselectively oxidised at C(2) by bromine treatment of the stannylene (S. David and S. Hanessian, *Tetrahedron*, 1985, 41, 643). Reaction of the resulting ketone with methylmagnesium bromide resulted in an epimeric mixture of diols (**96**) which were converted to the 2,3-trisubstituted olefin by way of the cyclic thionocarbonates. A sequence involving desilylation, oxidation, Grignard addition, elimination and deacetalation eventually yielded triene (**97**). The primary alcohol in (**97**) was pivaloylated and subsequent oxidation of the remaining hydroxyl served to activate the dienophilic portion of (**98**). Intramolecular Diels-Alder reaction of (**98**) afforded starting material and a single adduct (**99**) in 2:1 ratio, despite heating at 140°C for two days. A concurrent Baeyer-Villager/allylic oxidation on (**99**) was achieved with sodium chromate. This inserted an oxygen into the C(4)-C(5) bond while also oxygenating C(13). Upon methanolysis, the acylal function in (**100**) was cleaved to give (**101**); this had previously been converted into forskolin by M.I. Columbo *et al.* (*J. Org. Chem.*, 1990, 55, 5631; F.E. Ziegler, B.H. Jaynes, M.T. Saindane, *J. Am. Chem. Soc.*, 1987, 109, 8115). An alternative intermolecular [4+2] approach to the AB ring system of forskolin has been reported from D-glucose by D.S. Bhakuni, (*Pure & Appl. Chem.*, 1990, 62, 1389).

(+)-Compactin

One of the more unusual variants of the intramolecular Diels-Alder reaction is where a vinylallene is employed as the diene component for the cycloaddition (E.A. Deutsch and B.B. Snider, *J. Org. Chem.*, 1982, 47, 2682; B.B. Snider and B.W. Burbaum, *J. Org. Chem.*, 1983, 48, 4370). The often improved stereoselectivity that results in such cyclisations can be attributed to the greater rigidity of a vinylallene compared to a 1,3-diene, which reduces the number of conformations that can be adopted in the transition state. An especially striking example of the exploitation of this reaction in total synthesis can be found in the route that was developed to (+)-compactin from D-glucose by G.E. Keck and D.F. Kachensky (*J. Org. Chem.*, 1986, 51, 2487; see also E.A. Deutsch and B.B. Snider, *loc. cit.*) (Scheme 13) This exploited earlier work on the preparation of diol (**104**) from tri-*O*-

Scheme 13

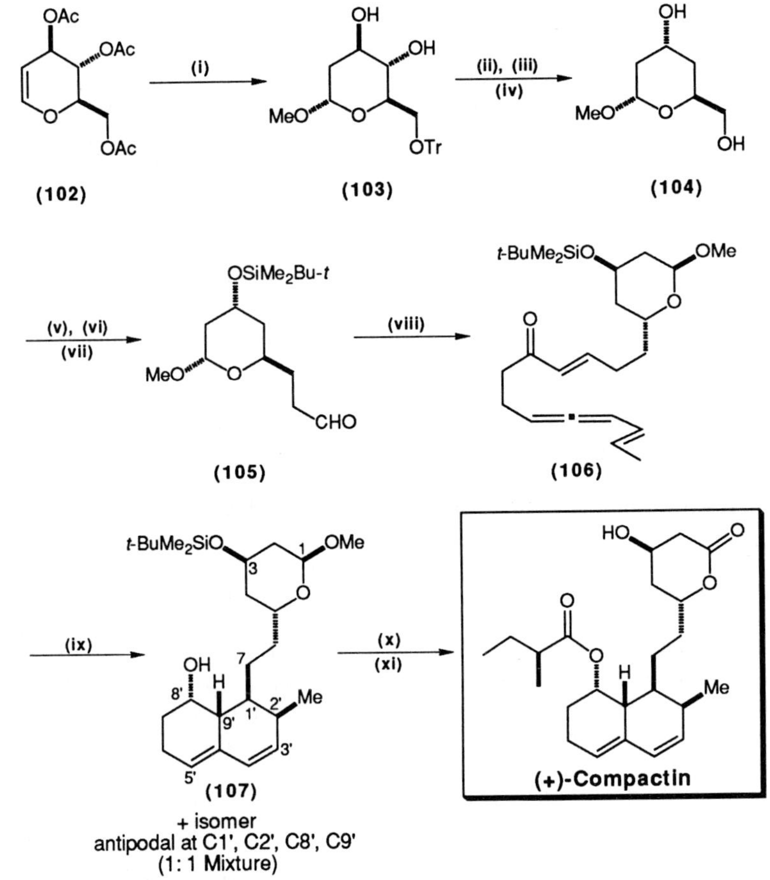
OAc
OAc
O
OAc
(102)
(i)
OH
OH
MeO
O
OTr
(103)
(ii), (iii)
(iv)
OH
MeO
O
OH
(104)
(v), (vi)
(vii)
OSiMe2Bu-t
MeO
O
CHO
(105)
(viii)
t-BuMe2SiO
OMe
O
O
(106)
(ix)
t-BuMe2SiO
1
OMe
3
O
7
OH
H
8'
2'
Me
1'
9'
3'
5'
(107)
+ isomer
antipodal at C1', C2', C8', C9'
(1: 1 Mixture)
(x)
(xi)
HO
O
O
O
O
H
Me
(+)-Compactin

Reagents: (**i**) NaOMe, MeOH, 2h, r. temp. then add $Hg(OAc)_2$; NaCl, $NaBH_4$, 30 min.; TrCl, Py, 18 h, (56%); (**ii**) NaH, HMPA, 0°C to r. temp., 2h, THF, 1-[(2,4,6-triisopropyl-phenyl)sulphonyl]imidazole, -23°C to 0°C, 3h, (86%); (**iii**) $LiAlH_4$, Et_2O, -15°C, 2h, (95%);(**iv**) Li, NH_3, -78°C to -33°C, (90%); (**v**) PhOCSCl, Py, 0°C to r. temp., (87%); (**vi**) *n*-$Bu_3SnCH_2CH{=}CH_2$, Toluene, *hv*, 7h; *t*-$BuMe_2SiCl$, DMF, imidazole, (82%); (**vii**) O_3, MeOH, Me_2S; (**viii**) (±)-1-(Triphenylphosphoranylidene)-2-oxo-5,6,8-decatriene, THF, (73%); (**ix**) PhMe, 140°C, Sealed Tube, 1.25 h; LiB(*s*-Bu)$_3$H, THF, 0°C, 30 min.; (**x**) *(S)*-(+)-2-methylbutyric anhydride, DMAP, Et_3N, CH_2Cl_2, (84%, 3 steps); (**xi**) 10% HCl, THF, 45-55°C, 20 min., Sealed Tube; Ag_2CO_3, Celite, PhMe, 95°C, 2h, (77%, 2 steps).

acetyl-D-glucal (E.J. Corey *et al.*, *J. Am. Chem. Soc.*, 1980, 102, 1439; Y.L. Yang and J.R. Falck, *Tetrahedron Lett.*, 1982, 23, 4305; P.A. Grieco *et al.*, *J. Am. Chem. Soc.*, 1983, 105, 1403; T. Rosen, M.J. Taschner and C.H. Heathcock, *J. Org. Chem.*, 1984, 49, 3994). Diol (**104**) underwent selective acylation with phenyl chlorothiocarbonate in pyridine to give a product that was chain extended in 82% yield by irradiation in the presence of excess allyltri-*n*-butylstannane (G.E. Keck and J.B. Yates, *J. Am. Chem. Soc.*, 1982, 104, 5829; G.E. Keck *et al.*, *Tetrahedron*, 1985; 41, 4079). This produced aldehyde (**105**) after silylation and ozonolysis. Wittig reaction of (**105**) with racemic 1-(triphenylphosphoranylidene)2-oxo-5,6,8-decatriene afforded a 1:1 mixture of diastereomers (**106**), which underwent facile intramolecular [4+2] addition at 140°C under high-dilution conditions (0.05 M). Immediate reduction of the 1:1 mixture of cycloadducts furnished (**107**) along with its diastereomer, which could not be separated until after acylation with *(S)*-(+)-2-methylbutyric anhydride. Concurrent acid hydrolysis of both the methyl glycoside and the *tert*-butyldimethylsilyl ether was performed according to the conditions set out by Grieco *et al.* (*loc. cit*). Selective oxidation of the product hemiacetal with Fetizon's reagent gave the target lactone.

(ii) [3+2] Cycloaddition strategies

For a long time, the five-membered carbocyclic ring systems of many natural products resisted attempts at their construction through [3+2] cycloaddition. However, recent advances in synthetic methodology have dramatically changed this position, and the following section deals with carbocylic natural product syntheses that have begun from monosaccharides which have exploited this approach.

(+)-Brefeldin A

(+)-Brefeldin A is a fungal metabolite obtained from a variety of sources. It displays antibiotic, antimitotic, cytostatic and antiviral activity. Its structure was secured by X-ray crystallographic analysis (H.P. Weber, D. Hauser, and H.P. Sigg, *Helv. Chim. Acta.*, 1971, 54, 2763), which revealed a 13-membered macrolactone ring *trans*-fused to a five-membered carbocycle, in which there is a remote hydroxyl group. B.M. Trost *et al.* (*J. Am. Chem. Soc.*, 1986, 108, 284) have provided full details of a synthesis of (+)-brefeldin A in which the issue of relative and absolute stereochemical control in the cyclopentane ring was resolved through a judicious choice of D-mannitol as a chiral starting material. Their route, outlined in Scheme 14, centres around the use of a palladium-catalysed [3+2] cycloaddition reaction (B.M. Trost, *Chem. Soc. Rev.*, 1982, 11, 141; *Angew. Chem. Int. Edn. Engl.*, 1986, 25, 1) between 2-[(trimethylsilyl)methyl]allyl acetate and known acrylate (**108**) (H. Matsunaga *et al.*, *Tetrahedron Lett.*, 1983, 24, 3009). This

Scheme 14

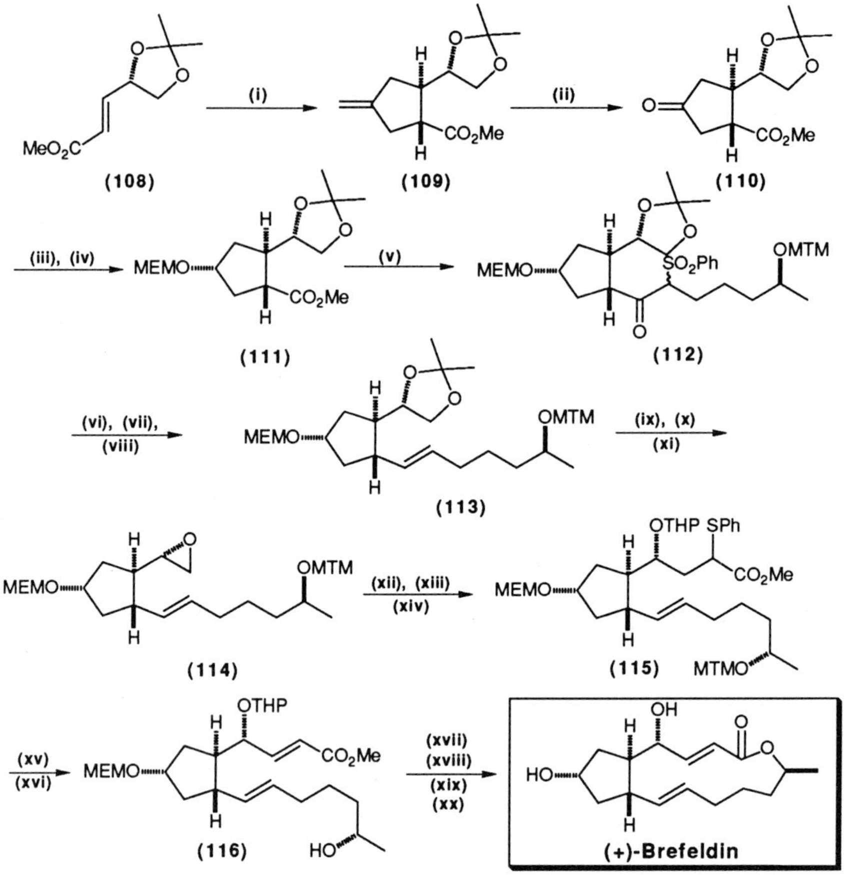

Reagents: (**i**) $AcOCH_2C(=CH_2)CH_2SiMe_3$, 33 mol % $(i\text{-}PrO)_3P$, 7 mol % $Pd(OAc)_2$, PhMe, 100°C, (87%); (**ii**) O_3, MeOH, -78°C, Me_2S, (84%); (**iii**) DIBAL-H, 2,6-Di-*tert*-butyl-4-methyl -phenol, PhMe, -78°C to 0°C; (**iv**) MEM-Cl, $(i\text{-}Pr)_2NEt$, CH_2Cl_2, (65%, 2 steps); (**v**) *n*-BuLi, THF, *(S)*-$PhSO_2(CH_2)_4CH(OMTM)Me$, -78°C to r. temp (89%); (**vi**) $NaBH_4$, MeOH, r. temp, (97%); (**vii**) Ac_2O, DMAP, Py, r. temp., (97%); (**viii**) 6% Na-Hg, MeOH, Na_2HPO_4, -20°C, (64%); (**ix**) aq. HCl, THF, r. temp, (64%); (**x**) TsCl, py, 0°C to r. temp., (89%); (**xi**) NaOMe, MeOH, Et_2O, 0°C, (93%); (**xii**) $PhSCH_2CO_2H$ (1.0 equiv.), LDA, (2.0 equiv), THF, 0°C; (**xiii**) CH_2N_2, Et_2O; (**xiv**) Dihydropyran, PPTS (81%, 3 steps); (**xv**) NCS, $AgNO_3$, MeCN-H_2O, $CaCO_3$; (**xvi**) *N,O*-Bis(trimethylsilylacetamide), PhMe, 95°C; K_2CO_3, MeOH, (92%); (**xvii**) Alkaline Hydrolysis; (**xviii**) $(PyS)_2$, Ph_3P, Xylene; Xylene, Δ; (**xix**) AcOH, THF, H_2O, 3:3:1, 50°C, 4 h; (**xx**) $TiCl_4$, (2.5 equiv.), CH_2Cl_2, 0°C (approx. 85%, last 3 steps).

led to a 3-4:1 mixture of cycloadducts in which isomer (**109**) predominated. After ozonolysis, ketone (**110**) was chemoselectively reduced, probably under chelation-controlled conditions, with the modified DIBAL reagent described by S. Iguchi *et al.*, (*J. Org. Chem.*, 1979, 44, 1363). This gave a 6:1 mixture enriched in the desired alcohol, which was separated and protected as a MEM ether. Conversion of (**111**) into β-keto sulphone (**112**) was followed by reduction, acetylation, and treatment with sodium-amalgam to provide the *(E)*-olefin (**113**) in 65% yield (M. Julia, J.M. Paris, *Tetrahedron Lett.*, 1973, 4833; P.J. Kocienski, B. Lythgoe, and S. Rushton, *J. Chem. Soc., Perkin Trans. 1*, 1978, 829). Removal of the acetonide with acid, tosylation of the primary alcohol, and treatment with base furnished epoxide (**114**). This underwent facile ring-opening with the dianion derived from phenylthioacetic acid (B.M. Trost and K.K. Leung, *Tetrahedron Lett.*, 1975, 4197; K. Iwai *et al.*, *Bull. Chem. Soc. Jpn.*, 1977, 50, 242). Methylation of the resulting acid with diazomethane, protection of the remaining hydroxyl as a tetrahydropyranyl ether, and elimination of the sulphoxide eventually generated (**116**). This was saponified to the hydroxy acid which underwent ring-closure via the thiopyridyl ester method of E.J. Corey and K.C. Nicolaou (*J. Am. Chem. Soc.*, 1974, 96, 5614). Treatment of the cyclised material with aqueous acid removed the tetrahydropyranyl ether, and titanium tetrachloride deprotected the MEM ether to provide (+)-brefeldin.

(-)-Specionin

(-)-Specionin is a recently discovered iridoid, with potent antifeedant activity against the Eastern spruce budworm, a pest of North American forests (C.C. Chang and K. Nakanishi, *J. Chem. Soc. Chem Comm.*, 1983, 605). An elegant synthesis of (-)-specionin (Scheme 15) has been developed from D-xylal by D.P. Curran *et al.* (*J. Am. Chem. Soc.*, 1987, 109, 5280; *Tetrahedron*, 1988, 44, 3079). It incorporated an Ireland-Claisen rearrangement on alkene (**117**) (R.E. Ireland, R.H. Mueller and A.K. Willard, *J. Am. Chem. Soc.*, 1976, 98, 2868), and an intramolecular nitrile oxide [3+2] cycloaddition reaction (A.P. Kozikowski, *Acc. Chem. Res.*, 1984, 17, 410) on (**118**) to create the 6,5-bicyclic ring system present in the target structure. After some routine manipulations this combined strategy stereospecifically delivered the Δ^2-isoxazoline (**119**), which underwent epoxidation and hydrogenolytic N-O bond cleavage to give β-hydroxyketone (**120**). Mesylation of (**120**) was accompanied by β-elimination. It delivered an enone that was reduced by lithium aluminium hydride at low temperature to give a 3:1 mixture of allylic alcohols (**121**), in which the β-epimer

Scheme 15

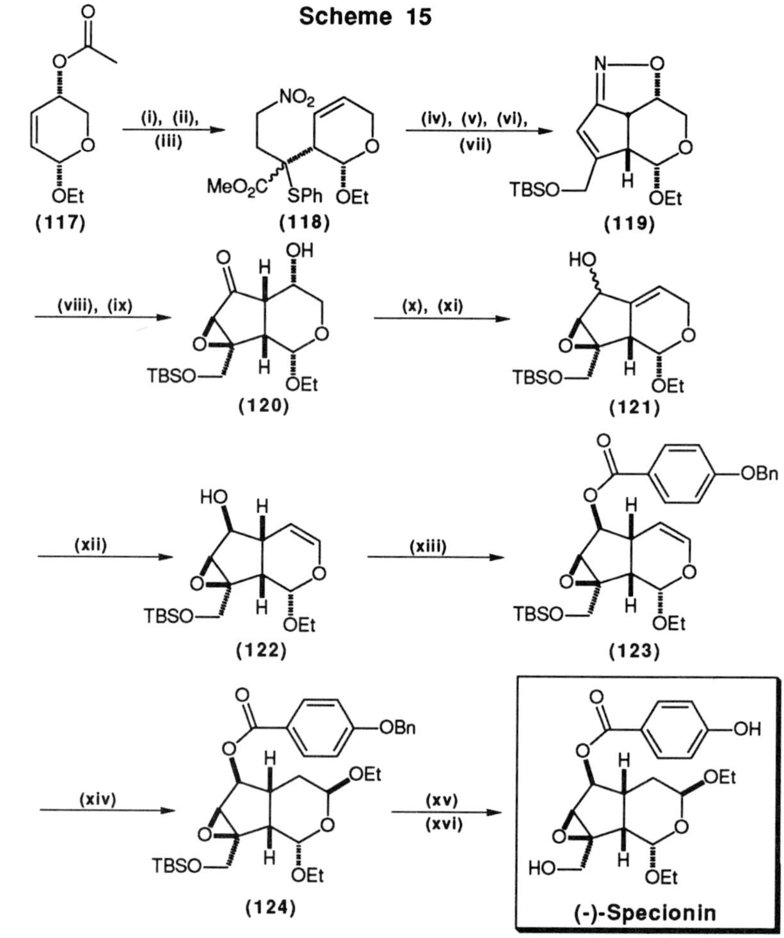

Reagents: (**i**) LDA, THF, -78°C, *t*-BuMe$_2$SiCl, HMPA, warm to 0°C; aqueous work-up; PhMe, 70°C, 20 h; HMPA, KF, MeI, K_2CO_3, 12h, (75%); (**ii**) LDA, THF, -78°C, $(PhS)_2$, (81%); (**iii**) LDA, THF, -78°C, 30 min., $H_2C{=}CHNO_2$, 1h, warm to r. temp., 30 min. (85%); (**iv**) MeNCO, Et_3N, C_6H_6, 25°C to 85°C; H_2O, 2 h, (89%); (**v**) *m*-CPBA, CH_2Cl_2, $NaHCO_3$, 0°C for 4h, then r. temp.; C_6H_6, Δ, (54%); (**vi**) *n*-BuLi, DIBAL, THF, -78°C to 0°C, 1h (92%); (**vii**) *t*-BuMe$_2$SiCl, Imidazole, DMF, 45°C, 16h, (72%); (**viii**) 3,5-Dinitroperoxy-benzoic acid, 4,4'-thiobis(2-*t*-butyl-6-methylphenol), NaH_2PO_4, CH_2Cl_2, (77%); (**ix**) H_2, Rh/Al_2O_3, MeOH, H_2O, (5:1), 1.5 h, (57%); (**x**) MsCl, Et_3N, CH_2Cl_2, 25°C, (70%); (**xi**) $LiAlH_4$, Et_2O, -78°C to r. temp., (93%); (**xii**) $(Ph_3P)_3RuH_2$, EtOH, Δ, 5h, (64%); (**xiii**) *p*-BnOC$_6$H$_4$CO$_2$H, DCC, DMAP, CH_2Cl_2, 72 h (61%); (**xiv**) $Hg(OAc)_2$, EtOH, 3 M NaOH, $NaBH_4$, -78°C, (70%); (**xv**) *n*-Bu$_4$NF, THF; (**xvi**) H_2, Pd-C, EtOH, 3h, (90%, 2 steps).

predominated. The double bond in (**121**) was then migrated with dihydrotetrakis(triphenylphosphine)ruthenium (II) in ethanol, to provide a 3:1 mixture of enol acetals (M.Takahashi *et al.*, *Chem. Lett.*, 1981, 1435). The major product (**122**) was esterified with *p*-benxyloxybenzoic acid and dicyclohexylcarbodiimide (DCC) to produce (**123**), and this converted to (**124**) by ethoxymercuration and reduction. Removal of the *O*-silyl and *O*-benzyl protecting groups concluded this pathway to (-)-specionin.

The oxahydrindene component of the avermectins and milbemycins

The intramolecular nitrile oxide cycloaddition (INOC) reaction also played a central role in the approach of M. Prashad and B. Fraser-Reid, (*J. Org. Chem.*, 1985, 50, 1564) to the oxahydrindene portion of the avermectins and milbemycins (Scheme 16). Starting from a derivative (**125**) of diacetone-D-glucose, an eight step sequence was followed for the production of compound (**126**). Treatment of (**126**) with phenyl isocyanate generated an intermediate nitrile oxide which spontaneously added to the pendant vinyl group to yield (**127**). Cleavage of the isoxazoline unit with Raney nickel furnished β-hydroxy ketone (**128**). This was reduced with lithium aluminium hydride, tritylated at the primary position, and mesylated to give (**129**). This underwent a noteworthy regiospecific elimination to install the C(3)-C(4) olefin present in the oxahydrindene sub-unit. A series of protecting group adjustments eventually liberated the hemiacetal, which was reduced to triol (**131**) by sodium borohydride. Selective tosylation of the primary hydroxyl group in (**131**) led to tetrahydrofuran formation, and oxidation finalised the route.

(iii) [2+2] Cycloaddition strategies

(-)-Echinosporin

(-)-Echinosporin (XK-213) is a fermentation product of *Streptomyces echinosporus* (T. Saro *et al.*, *J. Antibiot.*, 1982, 35, 266). It is of pharmacological interest on account of its promising anticancer profile in several rodent tumour models. It has recently been prepared in enantiomerically pure form from L-ascorbic acid by A.B. Smith, III, G.A. Sulikowski, and K. Fujimoto (*J. Am. Chem. Soc.*, 1989, 111, 8039; A.B.Smith, III *et al.*, *J. Am. Chem. Soc.*, 1992, 114, 2567). Their approach was founded on a novel asymmetric [2+2]-photocycloaddition reaction between dihydrofuran (**133**) and cyclopentenone. This generated a 4:1.2:1 mixture of three photoadducts in 79% yield, from which the major isomer (**134**) was isolated in 50% yield. The preference for head-to-tail regiochemistry in this addition was expected on the basis of previous photoreactions between enol ethers and enones (P. de Mayo, J.-P. Pete, M. Tchir, *J. Am. Chem. Soc.*, 1967, 89, 5712; P.E. Eaton, *Acc. Chem. Res.*, 1968, 1, 50; D.

Scheme 16

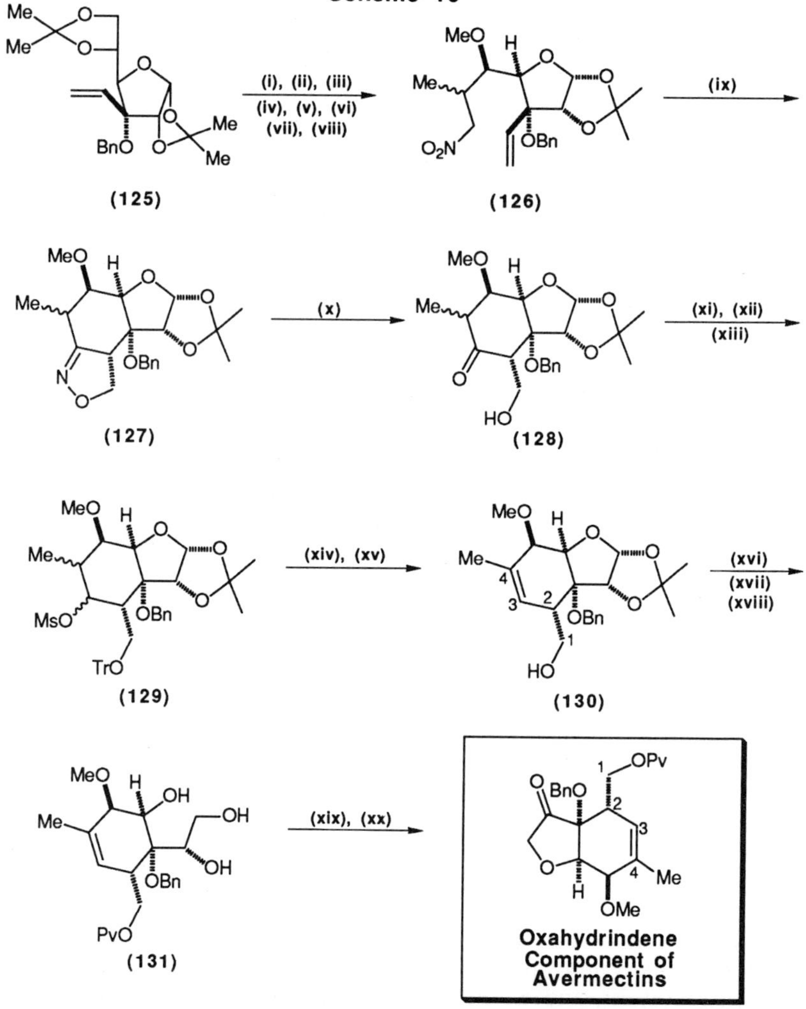

Reagents: (**i**) H_3O^+, (95%); (**ii**) *t*-$BuPh_2SiCl$, Et_3N, (98%); (**iii**) NaH, MeI, *n*-Bu_4NI, (98%); (**iv**) *n*-Bu_4NF, (98%); (**v**) $(COCl)_2$, Me_2SO, CH_2Cl_2, -78°C, Et_3N, (90%); (**vi**) CH_3NO_2, KOBu-*t*, (77%); (**vii**) MsCl, (77%); (**viii**) MeLi, (60%); (**ix**) PhNCO, Et_3N, C_6H_6, Δ, (74%); (**x**) H_2, Ni, (67%); (**xi**) $LiAlH_4$, (65%); (**xii**) Ph_3CCl, (90%); (**xiii**) MsCl, (86%); (**xiv**) NaOAc, HMPA,100°C, 3-4 days, (50-55%); (**xv**) Camphorsulphonic acid, MeOH, r. temp., (94%); (**xvi**) *t*-BuCOCl, (80%); (**xvii**) 0.5% H_2SO_4, (85%); (**xviii**) $NaBH_4$, (60%); (**xix**) TsCl, (70%); (**xx**) $(COCl)_2$, Me_2SO, CH_2Cl_2, Et_3N, -78°C, (70%).

Scheme 17

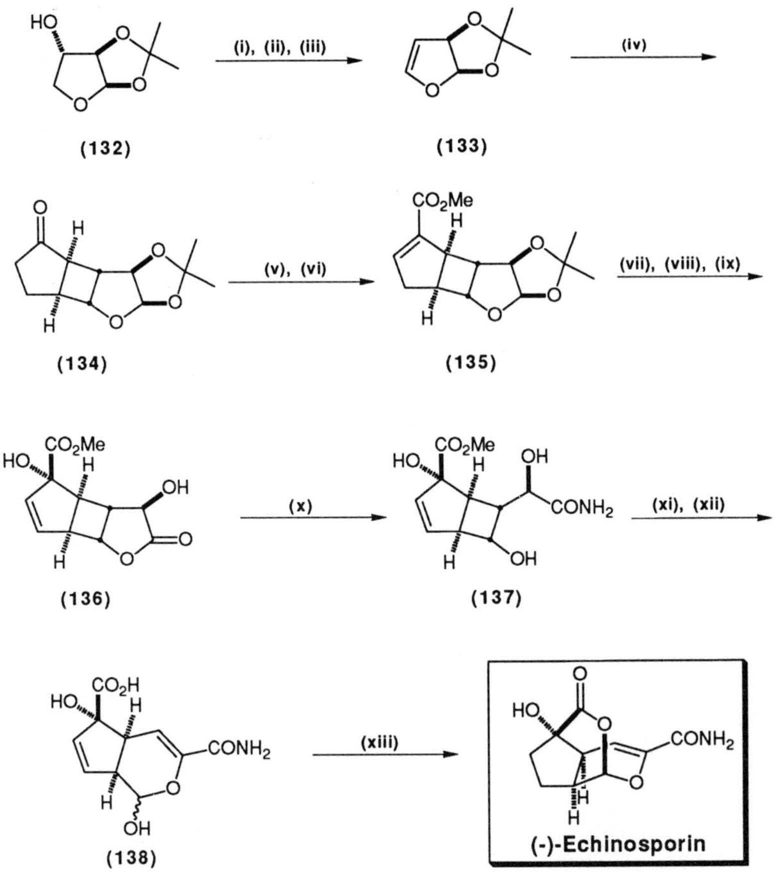

(132) —(i), (ii), (iii)→ (133) —(iv)→

(134) —(v), (vi)→ (135) —(vii), (viii), (ix)→

(136) —(x)→ (137) —(xi), (xii)→

(138) —(xiii)→ **(-)-Echinosporin**

Reagents: (**i**) $(COCl)_2$, Me_2SO, CH_2Cl_2, Et_3N, -78°C; (**ii**) $TsNHNH_2$, EtOH; (**iii**) Na, $(CH_2OH)_2$, 135°C, (83%, 3 steps); (**iv**) *hv*, pentane, (0.1 equiv), Cyclopentenone, 22 h, (50%); (**v**) LDA, THF, -78°C to -20°C, $PhN(SO_2CF_3)_2$, 20 h, (76%); (**vi**) $Pd(OAc)_2$, Ph_3P, Et_3N, CO, MeOH, DMF, (83%); (**vii**) $KN(SiMe_3)_2$,20% HMPA in THF, -78°C, (+)-(Camphorsulphonyl)oxaziridine, (90-94%); (**viii**) Bio-Rad AG 50W-X2 H^+ Resin, 50% aq. MeCN, (60-70%); (**ix**) $Pd_2(DBA)_3.CHCl_3$, MeCN(Allyl-O)$_2$CO, 80°C, (50-55%); (**x**) NH_4OH, MeOH, (86%); (**xi**) SO_3.Py, DMSO, Et_3N, CH_2Cl_2, 0°C (46%); (**xii**) 3.6 N HCl, 48 h, (100%); (**xiii**) Bu_3P (2.5 equiv), DEAD (2.5 equiv), THF, -15°C, 1 h; Bu_3P-DEAD (2.5 equiv), -15°C, then r. temp overnight, (28-31%).

Ermont, D. De Keukeleire, M. Vandewalle, *J. Chem. Soc., Perkin Trans. 1*, 1977, 2349). The stereochemical outcome was also predictable since the cyclopentenone would clearly prefer to add to the much less hindered *exo*-face of the bicyclic dihydrofuran (**133**). Kinetic deprotonation of ketone (**134**) and trapping with *N*-phenyltriflimide afforded an enol triflate which underwent carbonylation, in the presence of methanol and palladium (0), to give methyl ester (**135**) (S. Cacchi, E. Morera, and G. Ortar, *Tetrahedron Lett.*, 1985, 26, 1109). Conversion of (**135**) into the corresponding potassium-dienolate and oxidation with 2-(phenylsulphonyl)-3-phenyloxaziridine installed the C(8)-hydroxyl group with the correct α-stereochemistry (F.A. Davis *et al.*, *J. Org. Chem.*, 1984, 49, 3241), the concave-convex shape of the tricyclic ring system governing the direction of approach by the oxygen electrophile. Deprotection of the 1,2-*O*-acetal with acid, and palladium-mediated dehydrogenation then afforded (**136**) (I. Minami and J. Tsuji, *Tetrahedron*, 1987, 43, 3903). Ammonolysis of (**136**) delivered (**137**), which underwent retro-aldol fragmentation in the oxidation step to give (**138**) after acid-hydrolysis of the methyl ester. Intramolecular glycosyl ester formation via a Mitsunobu reaction delivered (-)-echinosporin in 28-31% yield (O. Mitsunobu, *Synthesis*, 1981, 1; A.B. Smith, III, K.J. Hale, and R.A. Rivero, *Tetrahedron Lett.*, 1986, 27, 5813; A.B. Smith, III, *et al.*, *J. Am. Chem. Soc.*, 1991, 113, 2092).

(iv) [2+1] Cycloaddition strategies

(+)-Bicyclohumulenone

A good example of an asymmetric [2+1] cycloaddition reaction being employed in the synthesis of a natural product from a monosaccharide, can be found in the route to (+)-bicyclohumulenone by Y. Fukuyama, M. Hirono, and M. Kodama (*Chem. Lett.*, 1992, 167). (+)-Bicyclohumulenone is a humulane-type sesquiterpenoid isolated from the liverwort *Plagiochila siophila* (A. Matsuo *et al.*, *J. Chem. Soc., Chem. Comm.*, 1979, 174). It consists of a 10-membered carbocyclic ring fused to a disubstituted cyclopropane, in which there is a quaternary chiral centre. A key step in this particular route (Scheme 18), which began from D-mannitol, was the stereoselective Simmons-Smith cyclopropanation of allylic ether (**139**) to afford a 10:1 mixture of diastereomers. Subsequent acid hydrolysis of the isopropylidene group from the major product, and 1,2-diol cleavage afforded aldehyde (**141**), which was homologated to unsaturated diester (**142**), reduced to (**143**) and alkylated. Further reduction of the ester groups in (**144**), afforded a diol which was converted to ditriflate (**145**), and then deoxygenated with lithium triethylborohydride. After removal of the *p*-methoxybenzyl ether, oxidation gave an aldehyde, which was transformed into oxirane (**146**) by addition of dimethylsulphoxonium methylide. Unfortunately, the intramolecular carbanionic cyclisation of (**146**) proceeded in low

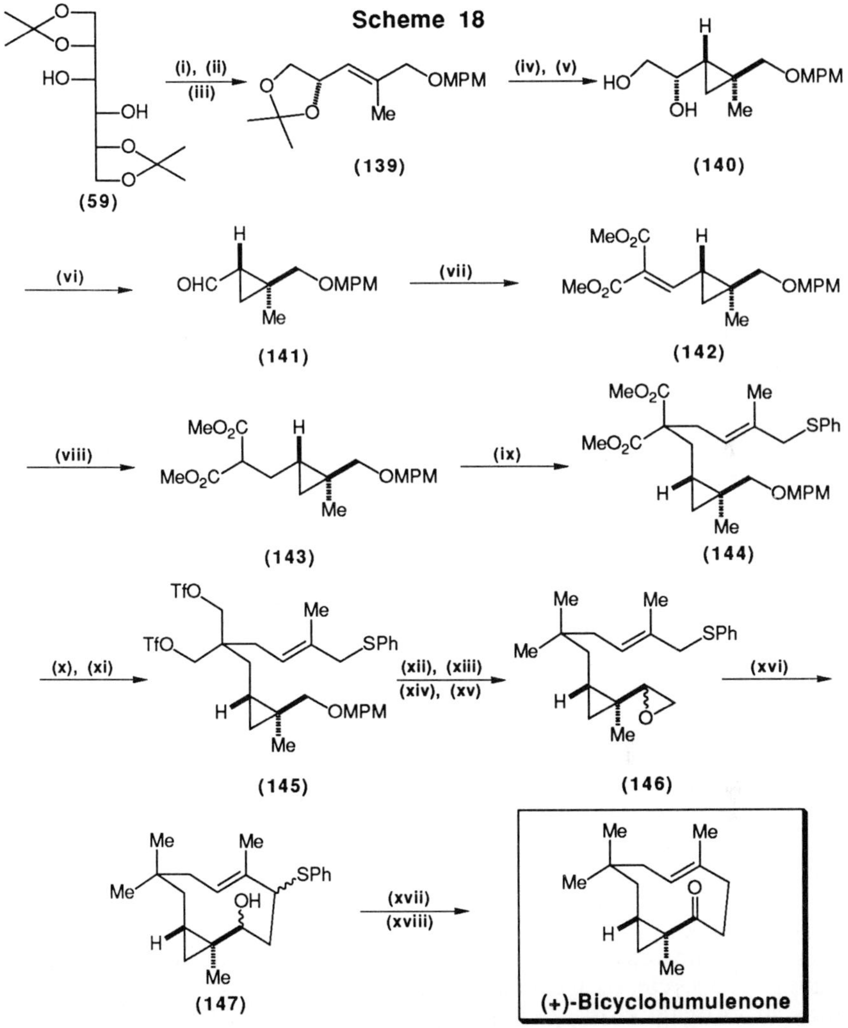

Scheme 18

Reagents: (**i**) $NaIO_4$, 5% Aq. $NaHCO_3$, 1 h then add $(EtO)_2P(O)CH(Me)CO_2Et$, K_2CO_3, H_2O, (58%, 3:1 *E:Z*); (**ii**) DIBAL-H, PhMe, 0^oC, (81%); (**iii**) NaH, DMF, *p*-$MeOC_6H_4CH_2Cl$, (96%); (**iv**) CH_2I_2, Zn-Cu, Et_2O, (89%); (**v**) 2 M HCl, MeOH; (**vi**) $NaIO_4$, THF, H_2O (3:1); (91%, 2 steps); (**vii**) $(MeO_2C)_2CH_2$, Pyridinium Acetate, CH_2Cl_2,(96%); (**viii**) L-Selectride, THF, -78^oC; (**ix**) NaH, DMF, *(E)*-$PhSCH_2C(Me)$=$CHCH_2Cl$, (86%, 2 steps); (**x**) $LiAlH_4$, THF; (**xi**) $(CF_3SO_2)_2O$, Py; (**xii**) $LiBHEt_3$, THF, (81%, 3 steps); (**xiii**) DDQ, CH_2Cl_2-H_2O (18:1); (**xiv**) PCC, CH_2Cl_2, (64%, 2 steps); (**xv**) $Me_2S(O)$=CH_2, NaH, DMSO; (**xvi**) *n*-BuLi, HMPA-THF, -78^oC to r. temp., (17%); (**xvii**) Na, *t*-BuOH; (**xviii**) $(COCl)_2$, Me_2SO, Et_3N, CH_2Cl_2, -78^oC.

yield. Nevertheless, desulphurisation of (**147**) and Swern oxidation eventually delivered the target compound.

(c) Nucleophilic cyclisation strategies for the synthesis of carbocyclic ring systems in natural products

(i) Intramolecular aldol reactions

(-)-Prostaglandin E_2

P. Belanger and P. Prasit (*Tetrahedron Lett.*, 1988, 29, 5521) have recorded a simple route to oxygenated cyclopentenone (**150**) from D-ribonolactone (Scheme 19). This is a key intermediate in the three component coupling approach to (-)-prostaglandin E_2 developed by C.R. Johnson and T.D. Penning (*J. Am. Chem. Soc.*, 1986, 108, 5655). The first stages in the synthesis of (**150**) involved isopropylidenation of the 2,3-diol system in (**148**), introduction of an iodo group at C(5), and base-induced elimination to yield enol lactone (**149**). This was then reduced with lithium tri-*tert*-butoxyaluminium hydride to the enolic hemiacetal. The latter underwent ring-opening and intramolecular aldol reaction to generate a β-hydroxy cyclopentanone, which eliminated after mesylation to afford cyclopentenone (**150**). The phosphine stabilised organocopper reagent (M. Suzuki *et al., Isr. J. Chem.*, 1984, 24, 118) obtained from vinylstannane (**152**) smoothly added to (**150**) at low temperature. The resulting enolate was then captured by iodide (**151**) in the presence of HMPA after warming the reactants. The *trans* vicinally disubstituted product (**153**) was obtained in 53% yield. Desilylation was effected with aqueous hydrofluoric acid in acetonitrile, and reduction of the α-alkoxy ketone accomplished with aluminium amalgam. Enzymatic deprotection of PGE_2 methyl ester has previously been reported by C.J. Sih *et al.* (*J. Am. Chem. Soc.,* 1975, 97, 865).

(+)-Mannostatin A

(+)-Mannostatin A is an unusual sulphur- and nitrogen-containing carbocycle, indigenous to *Streptoverticillium vercillus* var. *quintum* (T. Aoyagi *et al., J. Antibiot.*, 1989, 42, 883). It is a very potent and selective inhibitor of Golgi processing mannosidase II, and is therefore of use as a biochemical probe (A.D. Elbein *et al., Biochemistry*, 1990, 29, 10062). S. Knapp and T.G. Murali Dhar (*J. Org. Chem.*, 1991, 56, 4096) have recently completed a stereocontrolled total synthesis of (+)-mannostatin A from intermediate (**154**) (Scheme 20). The latter was available after slight modifications of the route developed to (**150**) (Scheme 19) by P. Belanger and P. Prasit (*loc. cit.*). Thus, cyclopentenone (**154**) was cleanly reduced on its less hindered face to provide alcohol (**155**) (A.L. Gemal and J.L. Luche, *J.*

Scheme 19

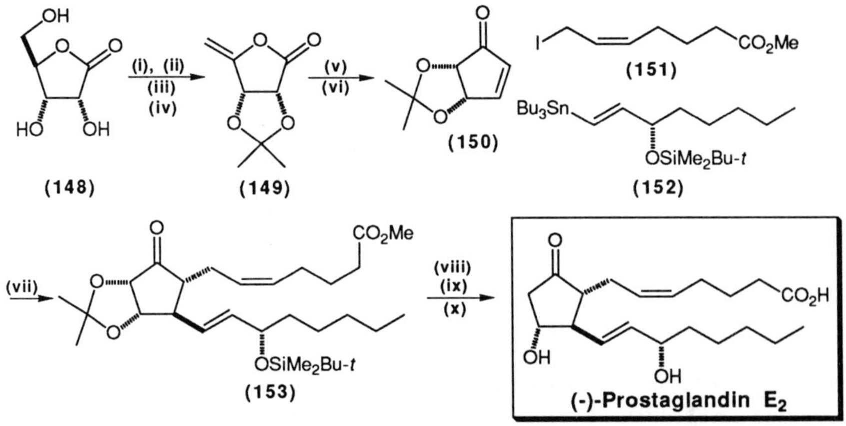

Reagents: (**i**) Acetone, H^+; (**ii**) TsCl, Py; (**iii**) NaI, Acetone, Δ; (**iv**) DBU, C_6H_6, r. temp., (67%, 4 steps); (**v**) (1 equiv) $LiAlH(OBu\text{-}t)_3$, THF, 0°C to r. temp, aq. NH_4Cl; (**vi**) MsCl, Py, CH_2Cl_2, (74%, 2 steps); (**vii**) *n*-BuLi, (**152**), THF, -78°C, Bu_3P-CuI, add (**150**), HMPA, (**151**) (1.5 equiv), warm to -30°C, 3 h, (53%); (**viii**) aq. HF, MeCN, (89%); (**ix**) Al(Hg), THF, H_2O, (98%); (**x**) *Rhyzopus oryzae*, 0.1 M Phosphate Buffer, pH 7, H_2O, EtOH, (76%).

Scheme 20

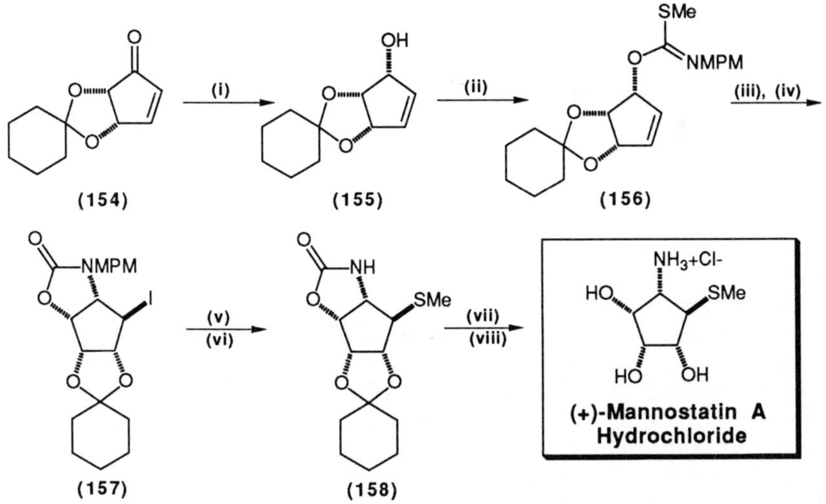

Reagents: (**i**) $NaBH_4$, $CeCl_3$, MeOH, 0°C, 20 min., (83%); (**ii**) NaH, *p*-$MeOC_6H_4CH_2NCS$ MeI; (**iii**) I_2, THF, sieves, 3h; (**iv**) aq. Na_2SO_3, 12 h, (85%, 3 steps)); (**v**) CAN, aq. MeCN, 1h, (92%); (**vi**) NaSMe, DMF, 2 h, (90%); (**vii**) 2 N KOH, Δ, 2 h; (**viii**) 6 N HCl, 10 h, (95%, 2 steps).

Am. Chem. Soc., 1981, 103, 5454; D.R. Borcherding, S.A. Scholtz, and R.T. Borchardt, *J. Org. Chem.*, 1987, 52, 5457). Controlling the chirality at this centre was essential at this stage, since it was subsequently to be used for dictating the stereochemical outcome of iodooxazolidinone formation. Interestingly, the *N-p*-methoxybenzyl group in product (**157**) was removed prior to the thiomethyl substituent being introduced. This tactic enabled nucleophilic displacement to proceed with overall retention of configuration due to participation by the oxazolidinone nitrogen. Compound (**158**) was transformed into (+)-mannostatin A hydrochloride in two more steps.

Paniculide B

K. Tadano, A. Miyake, and S. Ogawa (*Tetrahedron*, 1991, 47, 7259) have succeeded in preparing a key intermediate (**166**) for the synthesis of enantiomerically pure paniculide B from D-glucose (Scheme 21). Two noteworthy reactions in their sequence were the Robinson annulation of keto-aldehyde (**160**) for six-membered carbocycle construction (K. Tadano *et al., Bull. Chem. Soc. Jpn.*, 1987, 60, 1727), and the formation of the furanoside in (**162**) by exposure of (**161**) to palladium (II) acetate in ethyl vinyl ether (K. Fugami, K. Oshima, and K. Utimoto, *Tetrahedron Lett.*, 1987, 28, 809; R.C. Larock and D.E. Stinn, *ibid.*, 1989, 30, 2767). The latter reaction not only resulted in carbon-carbon bond formation, but also caused migration of the endocyclic double bond towards the original sugar ring, so that an enol ether was generated. This was then subjected to hydroboration and oxidative work-up to give (**162**) after silylation. Subsequent treatment with Jones reagent selectively hydrolysed the ethyl glycoside, but left the 1,2-*O*-isopropylidene unit intact. The liberated lactol was oxidised *in situ* to the corresponding lactone. More vigorous treatment of this product with acid also led to the 1,2-acetal being clipped, and the resulting diol was then degraded to the α,β-unsaturated aldehyde (**163**) with sodium periodate and then base. The final stages of the route were conversion of (**163**) into (**164**) and inversion of the hydroxyl group. The latter task was accomplished by an oxidation-reduction sequence. Unfortunately, a significant amount of lactone migration was also observed in this process. The crude mixture was then epoxidised, the desired product (**165**) being isolated as the major component in 42% yield. Protection of the remaining hydroxyl group furnished (**166**), which had previously been converted to paniculide B by A.B. Smith, III, and R.E. Richmond (*J. Org. Chem.*, 1981, 46, 4814; *J. Am. Chem. Soc.,* 1983, 105, 575).

The C(28)-C(49) segment of rapamycin

Rapamycin is a metabolite of *Streptomyces hygroscopicus* that has aroused considerable synthetic and medical interest on account of its

Scheme 21

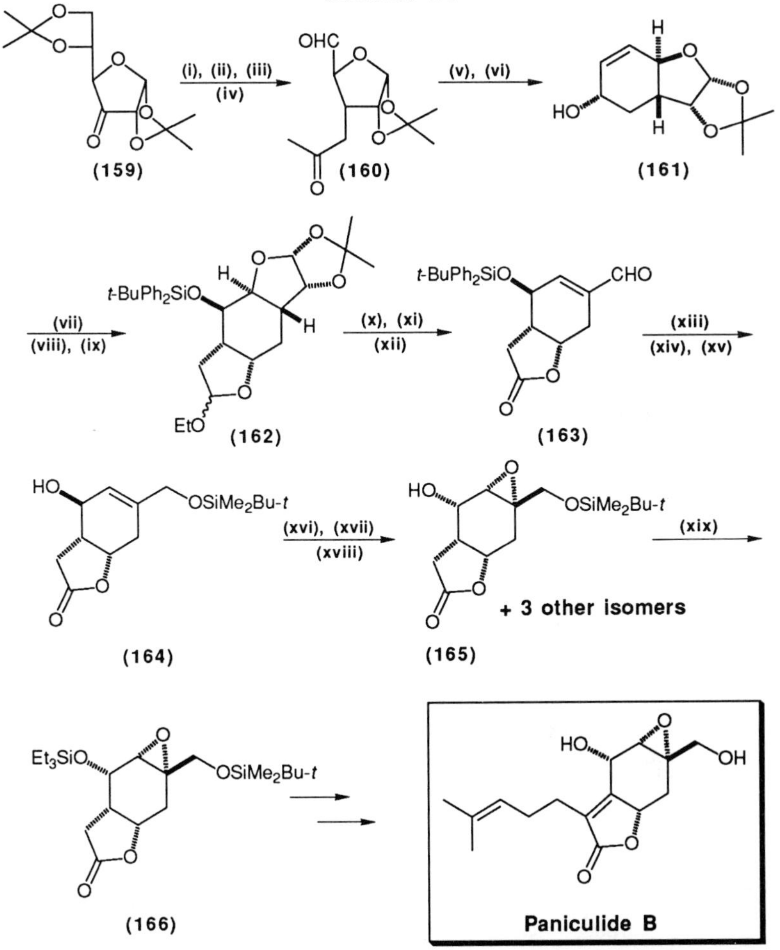

(159) —(i), (ii), (iii) / (iv)→ (160) —(v), (vi)→ (161)

—(vii) / (viii), (ix)→ (162) —(x), (xi) / (xii)→ (163) —(xiii) / (xiv), (xv)→

(164) —(xvi), (xvii) / (xviii)→ (165) + 3 other isomers —(xix)→

(166) → → **Paniculide B**

Reagents: (i) $Ph_3P{=}CHCOMe$, C_6H_6, Δ, 2h, (96%, *Z:E* = 3:1); (**ii**) Raney Ni (T4), MeOH, H_2, 20 h, (88%); (**iii**) 60% aq. AcOH, (98%); (**iv**) $NaIO_4$, aq. MeOH, (100%); (**v**) DBU, C_6H_6, Δ; Ac_2O, Py, (45%); (**vi**) $NaBH_4$, $CeCl_3$. 7 H_2O, -10°C; (**vii**) Ethyl Vinyl Ether, $Pd(OAc)_2$; then Py, Hexanes (62%, 2 steps); (**viii**) B_2H_6, THF, 0°C; H_2O_2, NaOH, 0°C, (74%); (**ix**) *t*-$BuPh_2SiCl$, imidazole, DMF, (68%); (**x**) Jones Reagent, Acetone, 0°C, (73%); (**xi**) 60% Aq. AcOH, 90°C; (**xii**) $NaIO_4$, aq. MeOH; DBU, C_6H_6, 55°C; (**xiii**) $NaBH_4$, $CeCl_3$.7 H_2O, MeOH, (59%, 4 steps); (**xiv**) *n*-Bu_4NF, THF; (**xv**) *t*-$BuMe_2SiCl$, DMAP, Et_3N, CH_2Cl_2, (76%); (**xvi**) PCC, 4A Sieves, CH_2Cl_2; (**xvii**) $NaBH_4$, $CeCl_3$. 7 H_2O, (79%, 2 steps); (**xviii**) *m*-CPBA, $NaHCO_3$, CH_2Cl_2, (93%); (**xix**) Et_3SiCl, Et_3N, CH_2Cl_2, (95%).

Scheme 22

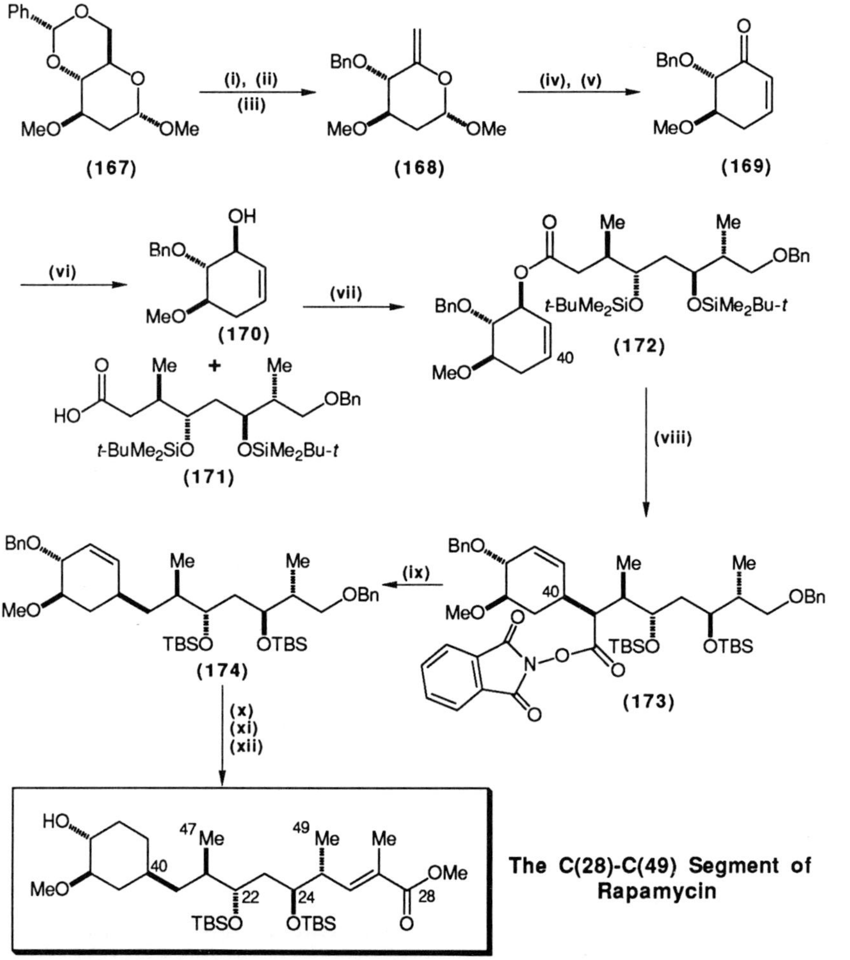

The C(28)-C(49) Segment of Rapamycin

Reagents: (**i**) *N*-Bromosuccinimide, $BaCO_3$, CCl_4, Δ, (93%); (**ii**) NaOMe, MeOH, 12 h, (81%); (**iii**) NaH, BnBr, *n*-Bu_4NI, DMF, (90%); (**iv**) $HgCl_2$, Acetone, H_2O, Δ, (85%); (**v**) MsCl, Py, (91%); (**vi**) $LiBH_4$, $CeCl_3.7\ H_2O$, THF-MeOH (1:1), -78°C, 0.5 h, (67%); (**vii**) [3-(Dimethylamino)propyl]-3-ethylcarbodiimide] (EDCI), CH_2Cl_2, (75%); (**viii**) LDA, THF, HMPA, -78°C, 15 min.; *t*-$BuMe_2SiCl$, THF, warm to r. temp.; PhMe, Δ, 2h; aq. LiOH, THF, 2 h; *N*-Hydroxyphthalimide, CH_2Cl_2, EDCI, DMAP, 8 h; (**ix**) *i*-PrOH, H_2O, *t*-BuSH, *N*-Methylcarbazole, *hv* (Pyrex), 2 h, (54%, 4 steps); (**x**) Raney Ni, EtOH, 30 min.; H_2, PtO_2, EtOAc; H_2, Pd-C, EtOAc, (66%); (**xi**) $Ru(PPh_3)_2Cl_2$, C_6H_6, (73%); (**xii**) $Ph_3P{=}C(Me)CO_2Me$, PhMe, (64%).

marked immunosuppressant activity. S. Danishefsky *et al.* (*J. Org. Chem.*, 1991, 56, 5826) have disclosed an approach to the C(28)-C(49) region of this molecule which exploited a mercuric chloride induced Ferrier reaction (R.J. Ferrier, *J. Chem. Soc., Perkin Trans. 1*, 1979, 1455) on enol ether (**168**) to build the cyclohexane ring (Scheme 22). This led to enone (**169**), after treatment of the initially formed aldol product with methanesulphonyl chloride and pyridine. Enone (**169**) was reduced to allylic alcohol (**170**) in 67% yield using the Luche conditions (*loc cit.*), and the latter esterified with acid (**171**), prepared from *(R)*-3-(benzyloxy)-2-methylpropanal. This furnished (**172**), which was subjected to an Ireland silyl ketene acetal Claisen rearrangement (*loc. cit.*) to attach the acyclic side chain at C(40) in a completely stereoselective manner. After hydrolysis of the silyl ester, the resulting carboxylic acid was converted to its *N*-(acyloxy)pthalimide (**173**), which underwent decarboxylation upon irradiation (Pyrex) in aqueous isopropanol, in the presence of *N*-methylcarbazole and *tert*-butylthiol (K. Okada, K. Okamoto, and M. Oda, *J. Am. Chem. Soc.*, 1988, 110, 8736). This delivered (**174**) which was hydrogenated, selectively oxidised by a palladium-mediated dehydrogenation reaction (H. Tomioka *et al., Tetrahedron Lett.*, 1981, 22, 1605), and chain extended to the target α,β-unsaturated ester by means of a stabilised Wittig reaction.

(+)-Lycoricidine

N. Chida, M. Ohtsuka, and S. Ogawa (*Tetrahedron Lett.*, 1991, 32, 4525) have developed a catalytic version of the Ferrier reaction (N. Chida *et al.*, *Bull. Chem. Soc. Jpn.,* 1991, 64, 2118) and incorporated it into their stereoselective route to the polyhydroxylated phenanthridone alkaloid (+)-lycoricidine from D-glucose (Scheme 23). Thus, treatment of enol ether (**175**) with 1 mol% of mercuric trifluoroacetate in acetone and water provided a cyclohexanone derivative, which underwent spontaneous dehydration to (**176**), upon reaction with methanesulphonyl chloride and triethylamine. The keto group was then stereoselectively reduced to the alcohol with sodium borohydride and cerium trichloride, and this protected with *p*-methoxybenzyl chloride. The azido group was then reduced with lithium aluminium hydride to supply (**177**) as a single isomer. This underwent condensation with 6-bromopiperonylic acid via a mixed anhydride protocol (S. Yamada, Y. Kasai, T. Shiori, *Tetrahedron Lett.*, 1973, 1595). The next stage of the synthesis was assembly of the complete phenanthridone skeleton by palladium-catalysed intramolecular Heck reaction on amide (**178**), using the thallium (I)-modified conditions developed by R. Grigg *et al.* (*Tetrahedron Lett.*, 1991, 32, 687). This produced (**179**) after selective deprotection of the *p*-methoxybenzyl ether. Mitsunobu reaction with benzoic acid was used to invert the stereochemistry of allylic alcohol

Scheme 23

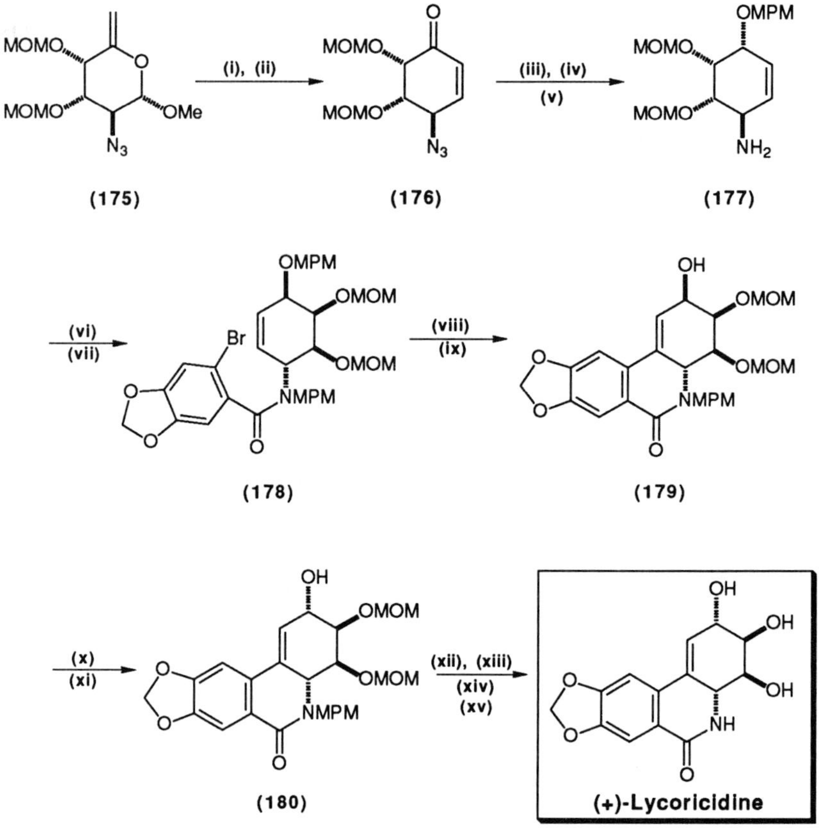

MOMO, MOMO, OMe, N_3 (175) —(i), (ii)→ (176) —(iii), (iv), (v)→ (177) —(vi), (vii)→ (178) —(viii), (ix)→ (179) —(x), (xi)→ (180) —(xii), (xiii), (xiv), (xv)→ (+)-Lycoricidine

(175) (176) (177)

(178) (179)

(180) **(+)-Lycoricidine**

Reagents: (**i**) $Hg(OCOCF_3)_2$ (1 mol %), Me_2CO-H_2O (2:1), r. temp., 20 h; (**ii**) MsCl, Et_3N, CH_2Cl_2, (69%, 2 steps); (**iii**) $NaBH_4$, $CeCl_3.7H_2O$, MeOH, (86%); (**iv**) NaH, DMF, *p*-$MeOC_6H_4CH_2Cl$, (60%); (**v**) $LiAlH_4$, Et_2O; (**vi**) 6-Bromopiperonylic acid, $(EtO)_2P(O)CN$, Et_3N, DMF, (89%, 3 steps); (**vii**) NaH, *p*-$MeOC_6H_4CH_2Cl$, DMF, (100%); (**viii**) $Pd(OAc)_2$ (20 mol%), 1,2-bis(diphenylphosphino)ethane (40 mol %), Tl(OAc) (2 equiv), DMF, 140°C, (68%); (**ix**) DDQ, CH_2Cl_2-H_2O (19:1), (53%); (**x**) Ph_3P, DEAD, $PhCO_2H$, THF, (68%); (**xi**) NaOMe, MeOH, r. temp.; (**xii**) 1M HCl, THF-H_2O, (1:1), 50°C, 2 h; (**xiii**) Ac_2O, Py, (49%, 3 steps); (**xiv**) CF_3CO_2H-$CHCl_3$, (1:1), r. temp.,(53%); (**xv**) NaOMe, MeOH, (100%).

(**179**), and the *O*-benzoate saponified with base. Further protecting group manipulations eventually delivered (+)-lycoricidine

Interestingly, the first chiral synthesis of (+)-lycoricidine was also recorded from D-glucose. In this approach an intramolecular nitroaldol strategy was used for assembling the cyclohexane ring, but unfortunately this step was non-stereoselective (H. Paulsen and M. Stubbe, *Tetrahedron Lett.*, 1982, 23, 3171; *Liebigs Ann. Chem.*, 1983, 535).

The oxahydrindene segment of the avermectins and milbemycins

Intramolecular aldol reactions on monosaccharide derivatives have featured in several carbocylisation pathways to the hexahydrobenzofuran sector of the avermectins and mibemycins. In the approach of M. Hirama *et al.* (*J. Org Chem.*, 1988, 53, 706), the pivotal intermediate for aldol ring closure was prepared through addition of the kinetic enolate derived from (**182**) onto enal (**183**) (Scheme 24). This gave an 8.5:1 mixture of two adducts, enriched in the desired 5R,6R-isomer. The latter was separated in 47-56% yield, and gave homoallylic alcohol (**184**) after further processing. Interestingly, the intramolecular aldolisation of (**184**) occurred under the Swern oxidation conditions, and supplied aldehyde (**185**) in 41-50% yield. Reduction and protecting group manipulation eventually provided tetraol (**187**). This underwent selective *O*-tripsylation at the primary alcohol followed by spontaneous tetrahydrofuran formation. Swern oxidation then furnished the desired precursor of the avermectin southern hemisphere in 45% yield, along with a 30% yield of the tertiary methylthiomethyl ether.

(ii) Intramolecular Claisen condensation

The hexahydrobenzofuran unit of the avermectins/milbemycins

D.R. Williams, F.D. Klingler, and U. Dabral (*Tetrahedron Lett.*, 1988, 29, 3415) have reported a synthesis of a fragment corresponding to the hexahydrobenzofuran region of the avermectin/milbemycin antibiotics. Again, enolate chemistry was exploited for construction of the carbocyclic ring. These workers chose 1,4-anhydrosorbitol as their chiral starting material (Scheme 25). This was subjected to selective mono-*O*-isopropylidenation, silylation on the less-hindered hydroxyl at C(2), and benzylation at OH(3). Removal of the 5,6-*O*-acetal followed by selective benzoylation and methylation furnished (**190**) in 51% overall yield. Routine transformations led to methyl ketone (**191**), which reacted with 3-butenylmagnesium bromide to give a 3:1 mixture of diastereomers, in favour of (**192**). A series of oxidations interrupted by a debenzylation step was used to obtain keto-lactone (**194**). It was selectively deprotonated α- to the lactone carbonyl upon addition of lithium diisopropylamide. The resulting ester enolate

Scheme 24

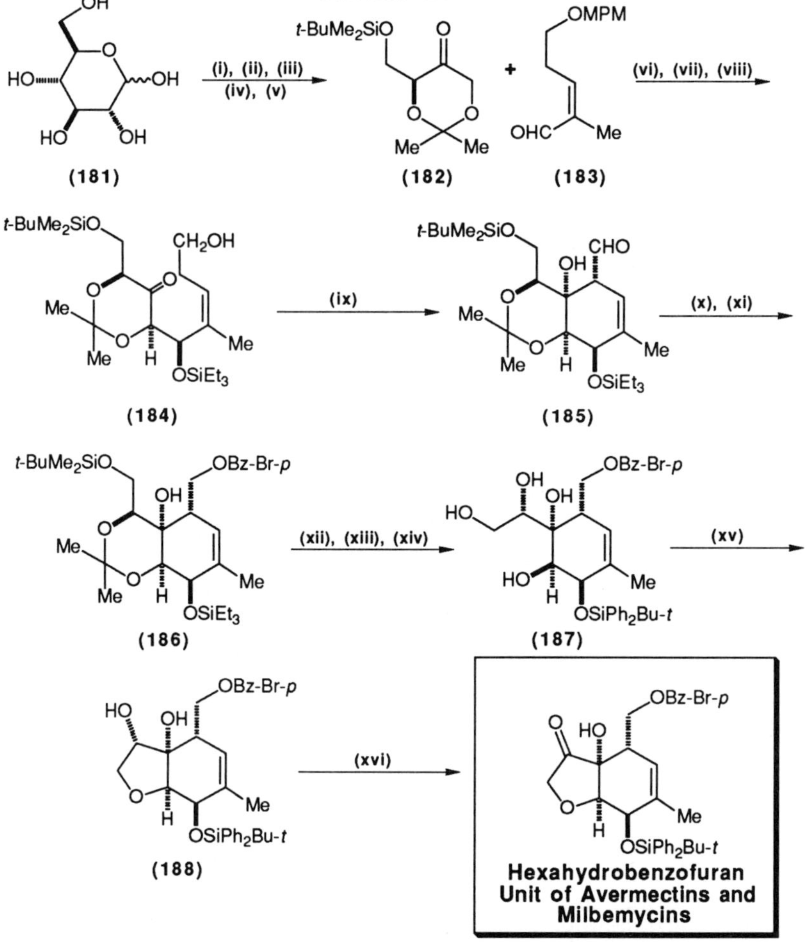

Reagents: (**i**) MeO(Me)C=CH_2, H^+; (**ii**) $NaIO_4$; (**iii**) $NaBH_4$; (**iv**) *t*-$BuMe_2SiCl$, DMAP, Et_3N; (**v**) Me_2SO, $(COCl)_2$, Et_3N, -78°C (46%, 5 steps); (**vi**) (**182**), LDA, -78°C, 30 s, then (**183**), 5 s, (47-56%); (**vii**) Et_3SiCl, imidazole, DMF, r. temp., 5 min, (83%); (**viii**) DDQ (1.5 equiv), CH_2Cl_2, H_2O, (20:1), (76%); (**ix**) Me_2SO (1.5 equiv), $(COCl)_2$ (1.5 equiv), *i*-Pr_2NEt, (10 equiv), -60°C, then 22°C, 10 h, (40-51%); (**x**) $NaBH_4$, MeOH, 0°C, (85%); (**xi**) Py, $ClCOC_6H_4Br$-*p*, 22°C, (73%); (**xii**) Py-HCl (5 equiv), Py, 22°C, 28 h, (75%); (**xiii**) DMF, *t*-$BuPh_2SiCl$, Imidazole, 22°C, (93%); (**xiv**) CF_3CO_2H-MeOH (1:2), 22°C, 30 h, (82%); (**xv**) 2,4,6-(*i*-Pr)$_3C_6H_2SO_2Cl$, Py, 22°C, 23 h, (71%); (**xvi**) Me_2SO (6 equiv), $(COCl)_2$ (3 equiv), Et_3N (10 equiv), -60°C, 20 min.,(45%, along with 30% of 3° Hydroxyl MTM ether).

Scheme 25

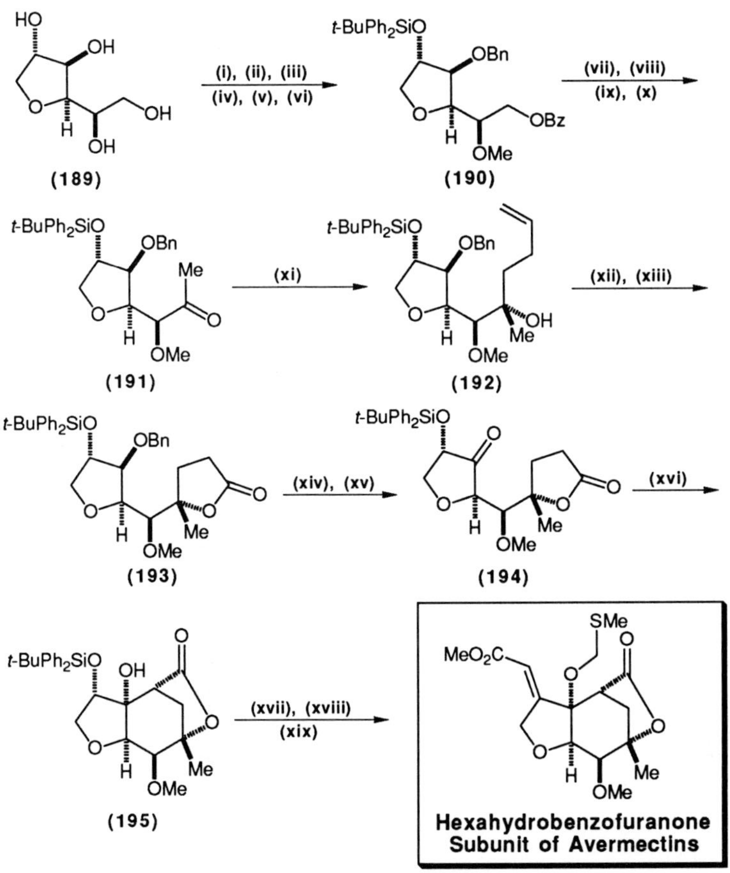

(189) —(i), (ii), (iii), (iv), (v), (vi)→ (190) —(vii), (viii), (ix), (x)→ (191) —(xi)→ (192) —(xii), (xiii)→ (193) —(xiv), (xv)→ (194) —(xvi)→ (195) —(xvii), (xviii), (xix)→ Hexahydrobenzofuranone Subunit of Avermectins

Reagents: (i) $Me_2C{=}O$, H^+; (ii) t-$BuPh_2SiCl$, Et_3N, DMAP, CH_2Cl_2,-10°C to 22°C; (**iii**) NaH, THF, BnBr, 0°C to 22°C; (**iv**) Cat. p-TsOH, MeOH, 14 h; (**v**) PhCOCl (1 equiv), CH_2Cl_2, Et_3N, DMAP, -78°C to -40°C; (**vi**) NaH, DMF, MeI, 0°C, (51%, 6 steps); (**vii**) LiOH, THF, H_2O;(**viii**) PCC-Al_2O_3 (80%, 2 steps); (**ix**) MeMgCl, THF, -78°C, (83%); (**x**) PCC-Al_2O_3, (79%); (**xi**) $H_2C{=}CH(CH_2)_2MgBr$, THF, -78°C; (**xii**) O_3, CH_2Cl_2, Py, then Ph_3P; (**xiii**) PCC, Al_2O_3; (**xiv**) H_2, Pd-black, EtOH, (81%, 4 steps); (**xv**) PCC-Al_2O_3, CH_2Cl_2, 40°C, 12 h, (85%);(**xvi**) LDA, THF, -78°C, (75-85%); (**xvii**) n-Bu_4NF, THF, -30°C; (**xviii**) $(COCl)_2$, Me_2SO, -78°C, Et_3N,-78°C to -15°C, (65%); (**xix**) KOBu-t, $(MeO)_2P(O)CH_2CO_2Me$, THF, 22°C, 48 h, (70%).

Scheme 26

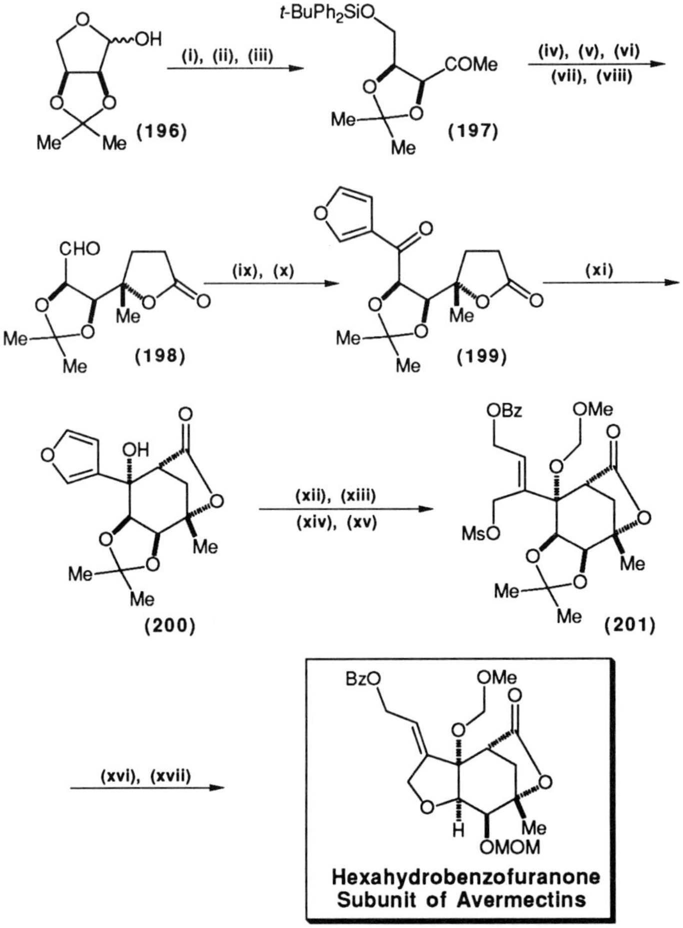

(196) —(i), (ii), (iii)→ (197) —(iv), (v), (vi), (vii), (viii)→ (198) —(ix), (x)→ (199) —(xi)→ (200) —(xii), (xiii), (xiv), (xv)→ (201) —(xvi), (xvii)→

Hexahydrobenzofuranone Subunit of Avermectins

Reagents: (**i**) MeMgCl, THF, -78°C, (88%); (**ii**) *t*-$BuPh_2SiCl$, Et_3N, DMAP, CH_2Cl_2, -30°C, (99%); (**iii**) PCC-Al_2O_3, CH_2Cl_2, (95%); (**iv**) $H_2C{=}CH(CH_2)_2MgBr$, THF, -78°C, (95%); (**v**) O_3, CH_2Cl_2, Py, -78°C, then Ph_3P, (90%); (**vi**) PCC, Al_2O_3, CH_2Cl_2, (89%); (**vii**) *n*-Bu_4NF, THF,0°C, (98%); (**viii**) PCC-Al_2O_3, NaOAc, CH_2Cl_2, (88%); (**ix**) 3-Bromofuran, *sec*-BuLi, Et_2O, -78°C, then $MgBr_2.Et_2O$ (0.9 equiv), -78°C to -40°C, (87%); (**x**) $(COCl)_2$, Me_2SO, CH_2Cl_2, -78°C, then Et_3N, -78°C to -20°C, (79%); (**xi**) LDA, THF, -78°C, (67%); (**xii**) MOM-Cl, DMF, 0°C, then add NaH, (82%); (**xiii**) *N*-Bromosuccinimide, H_2O, THF, (1:4), 0°C, then $NaBH_4$, (84%); (**xiv**) BzCl, (1equiv), CH_2Cl_2, Et_3N, DMAP, -78°C to -40°C, (82%); (**xv**) MsCl, Et_3N, CH_2Cl_2, 0°C, (90%); (**xvi**) Dowex 50W-X8 (H^+) resin, THF, H_2O, 70°C, (82%); (**xvii**) MOM-Cl (5 equiv), DMF, 0°C, add NaH, (81%).

participated readily in an intramolecular Claisen condensation to produce (**195**) as a single isomer. The selectivity observed in this deprotonation is particularly meritorious, given the considerable potential of this system for β-elimination and α–epimerisation. Deprotection of (**195**), followed by Swern oxidation and phosphonate anion condensation completed this interesting route to the southern hemisphere of the avermectins and milbemycins.

D.R. Williams and F.D. Klingler (*J. Org. Chem.*, 1988, 53, 2134) have also reported a variant of this approach (Scheme 26), in which the key intramolecular Claisen step was executed on (**199**). In this case, a 4:1 mixture of tertiary alcohols was obtained, and the major product (**200**) isolated in 67% yield. This was subsequently taken through a series of reactions that included ring-opening of the furan to the diol, selective benzoylation, and mesylation. Cleavage of the acetonide ring in the resulting product (**201**) then proved sufficient to induce heterocyclisation, and reprotection of the tertiary alcohol finally yielded the hexahydrobenzofuranone subunit.

(iii) Intramolecular Michael reactions

(-)-Cryptosporin

A powerful class of monosaccharide Michael acceptors are the 1-nitroglycals, and an excellent demonstration of their utility as chiral intermediates can be found in the synthesis of (-)-cryptosporin by W. Brade and A. Vasella (*Helv. Chim. Acta.*, 1989, 72, 1649). Here the nitro group of 1-nitro-L-fucal played a dual role in that it served to activate the C(2) position towards nucleophilic attack, while also acting as a good leaving for a subsequent elimination reaction. Scheme 27 illustrates these ideas more fully. 3,4-*O*-Isopropylidene-L-fucose (**202**) (J. Barbat, J. Gelas, D. Horton, *Carbohydr. Res.*, 1983, 116, 312) was transformed into 1,6-dideoxy-3,4-*O*-isopropylidene-1-nitro-α,β-L-galactopyranose (**203**) in the standard way, by formation of the C(1)-oxime, conversion to a mixture of nitrones with *p*-nitrobenzaldehyde, and ozonolysis. Acetylation and β-elimination with base afforded (**204**). The 1-nitroglycal reacted with the anion derived from phenylsulphonyl lactone (**205**) at -70°C to give (**206**) in 65% yield. This was then deprotected with aqueous acid to afford the opposite enantiomer of natural cryptosporin, which was in agreement with other work by R.B. Gupta and R.W. Franck (*loc. cit.*).

(-)-Fuscol

The synthesis of (-)-fuscol by M. Iwashima *et al.* (*Tetrahedron Lett.*, 1992, 33, 81) provides a very good illustration of the effectiveness of the double Michael reaction on monosaccharide derivatives for chiral carbocycle construction (Scheme 28). Thus, when

Scheme 27

HO HO O Me O O (202) —(i), (ii)→ O₂N HO O Me O O (203) —(iii)→ O₂N O Me O O (204)

+

MEMO O O SO₂Ph (205) —(iv)→ MEMO O O O Me O O (206)

(206) —(v)→

OH O O Me OH O OH

(-)-Cryptosporin

Reagents: (**i**) NaOMe, MeOH, NH_2OH. HCl, 50^oC, 4h, (66%); (**ii**) *p*-$NO_2C_6H_4CHO$, *p*-TsOH, $MgSO_4$, CH_2Cl_2, 4h; O_3, -78^oC, 45 min., (69%); (**iii**) Ac_2O, Py, CH_2Cl_2, DMAP, < 5^oC, 17 h; Amberlite IRA 93 (OH^-), $CHCl_3$, overnight, (84%); (**iv**) LDA, THF, -70^oC, 4A sieves, warmed to r. temp. and exposed to ultrasound for 20 h; -78^oC quench, aq. NH_4Cl, (65%); (**v**) 1M HCl, THF, 5 h, 60-64^oC, (62%).

Scheme 28

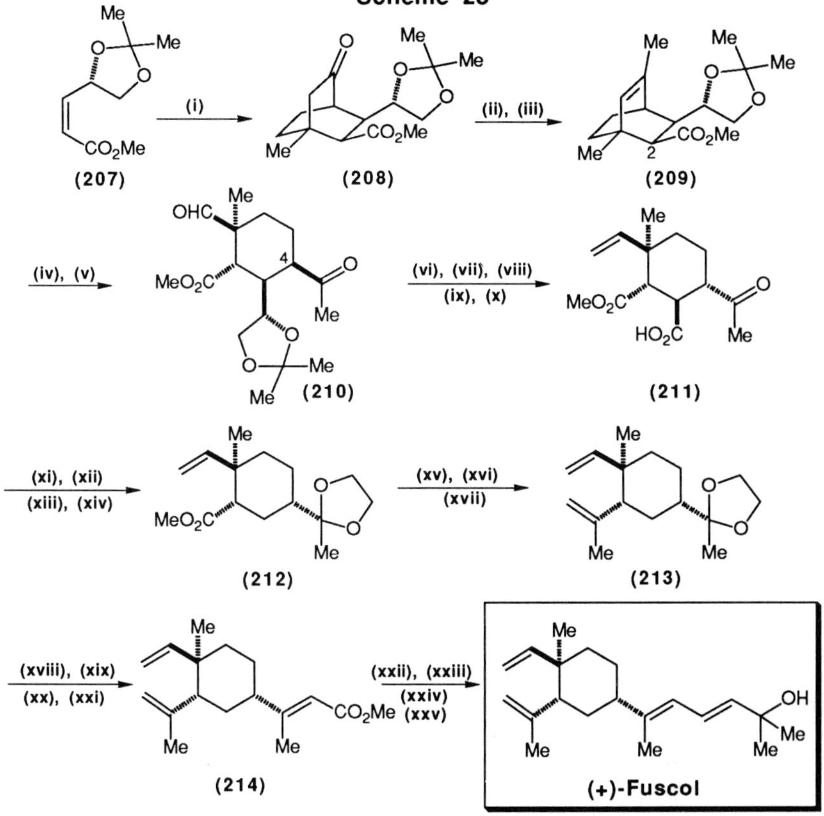

Reagents: (i) 3-Methyl-cyclohex-2-enone, LDA, THF, -78°C to -40°C, (93%); (**ii**) LDA, THF, -78°C, $(EtO)_2POCl$, warm to -20°C, (98%); (**iii**) MeMgI, $Ni(acac)_2$, THF, 0°C, (73%); (**iv**) KOBu-*t*, DMSO-THF, 23°C, (99%); (**v**) O_3, Py (0.2 equiv), MeOH-CH_2Cl_2, (2:1), -78°C then Me_2S, (96%); (**vi**) CH_2I_2, Zn, Me_3Al, THF, 20°C, (79%); (**vii**) NaOMe, MeOH, 50°C, (83%); (**viii**) 80% AcOH, 23°C; (**ix**) $NaIO_4$, silica gel, H_2O-CH_2Cl_2, (78%, 2 Steps); (**x**) $NaClO_2$, NaH_2PO_4, MeCH=CMe_2, *t*-BuOH-H_2O, 23°C, (78%); (**xi**) $(COCl)_2$, Py, C_6H_6, 5°C; (**xii**) *N*-Hydroxypyridine-2-thione sodium salt, DMAP, C_6H_6, 5°C to 25°C; (**xiii**) Bu_3SnH, AIBN, 50°C, (71%, 3 steps); (**xiv**) $(CH_2OH)_2$, *p*-TsOH, C_6H_6, 80°C, (89%); (**xv**) 20% KOH, DMSO, 40°C, (93%); (**xvi**) MeLi, THF, 0°C to 24°C, (98%); (**xvii**) CH_2Br_2, Zn, $TiCl_4$, THF, CH_2Cl_2, 25°C, (84%); (**xviii**) 80% AcOH, 26°C, (93%); (**xix**) $BrCH_2CO_2Me$, Zn, 1,3-dioxane, 40°C, ultrasound; (**xx**) AcCl, $PhNMe_2$, $CHCl_3$, 60°C, (60%, 2 steps); (**xxi**) DBU, C_6H_6, 80°C, (93%) (*E*:*Z* = 4.6:1); (**xxii**) DIBAL-H, CH_2Cl_2, -78°C; (**xxiii**) PDC, CH_2Cl_2, 4A MS, 26°C, (92%, 2 steps); (**xxiv**) KOBu-*t*, *t*-BuOH, $(i\text{-}PrO)_2P(O)CHCO_2Me$, THF, -78°C to 0°C, (95%); (**xxv**) MeLi, Et_2O, -30°C, (86%).

the kinetic enolate of 3-methyl-2-cyclohexenone was treated with α,β-unsaturated ester (**207**), prepared from D-mannitol, it participated in a double Michael sequence that led to bicyclo[2,2,0]octane (**208**) in high yield. The keto group of (**208**) was then converted into the trisubstituted olefin found in (**209**) by conversion to the enol phosphate and a nickel (0) catalysed cross coupling reaction with methylmagnesium bromide (T. Hayashi *et al.*, *Synthesis*, 1981, 1001). The next step in the synthesis was epimerisation at C(2) of methyl ester (**209**), which was accomplished with potassium *tert*-butoxide. Ozonolysis of the double bond provided keto-aldehyde (**210**), which was selectively methylenated with diiodomethane, zinc, and a catalytic amount of trimethylaluminium. At this point, the C(4) atom bearing the keto substituent was epimerised and the protected diol oxidatively degraded to the carboxylic acid. This was decarboxylated by the *O*-acyl thiohydroxamate method of D.H.R. Barton, D. Crich, and W.B. Motherwell (*Tetrahedron*, 1985, 41, 3901). The result, after ketalisation, was (**212**). The isopropenyl group in (**213**) was introduced by a three-step protocol that involved saponification, addition of methyllithium to the acid, and methylenation (L. Lombardo, *Tetrahedron Lett.*, 1982, 23, 4293). Subsequent removal of the ketal from (**213**) unveiled the ketone which participated in an ultrasound-induced Reformatsky reaction that led to (**214**) after β-elimination (*E:Z* = 4.6:1). Further manipulation of (**214**) furnished the α,β-unsaturated enal, which underwent a Horner-Emmons reaction, and addition of methyllithium to furnish (-)-fuscol.

(iv) Intramolecular Horner-Wadsworth-Emmons reactions

(-)-Shikimic acid

The intramolecular Wadsworth-Horner-Emmons reaction has now emerged as a very useful method for carbocyclic ring closure when applied on monosaccharide derivatives. Two outstanding examples of this type of cyclisation can be found in the shikimic acid syntheses of G.W.J. Fleet and T.K. Shing (*J. Chem. Soc., Chem. Comm.*, 1983, 849; G.W. Fleet, T.K.M. Shing, and S.M. Warr, *J. Chem. Soc. Perkin Trans. 1*, 1984, 905), and S. Mirza and A. Vasella, (*Helv. Chim. Acta*, 1984, 67, 1562). In the Fleet route (Scheme 29), which began from D-mannose, the key phosphonate (**218**) required for carbocylisation was formed by a nucleophilic displacement of the primary triflate (**216**), with the sodium salt of trimethylphosphonoacetate in *N,N*-dimethylformamide, containing 18-crown-6. After palladium-catalysed hydrogenolysis the mixture of lactols (**218**) underwent ring-closure to methyl *O,O*-isopropylidenshikimate, after treatment with base. Deacetonation subsequently furnished pure methyl shikimate, which was converted into shikimic acid by the action of alkali.

Scheme 29

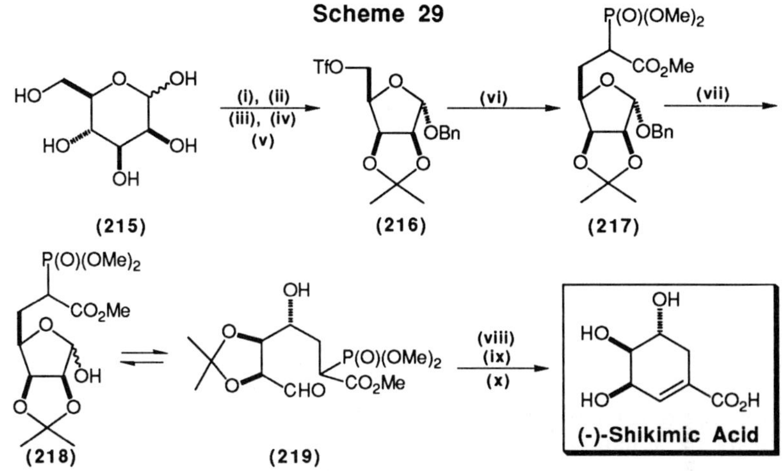

(215)
(216)
(217)
(218)
(219)
(i), (ii)
(iii), (iv)
(v)
(vi)
(vii)
(viii)
(ix)
(x)
(-)-Shikimic Acid

Reagents: (**i**) Acetone, H_2SO_4, 3-4 h, (92%); (**ii**) NaH, DMF, BnCl, (86%); (**iii**) HCl, MeOH, (90%); (**iv**) $NaIO_4$, H_2O, MeOH, 1.5 h; $NaBH_4$, EtOH, (89%); (**v**) $(CF_3SO_2)_2O$ (1.3 equiv), Py (2 equiv), CH_2Cl_2, -30°C, 30 min.; (**vi**) $[(MeO)_2P(O)CHCO_2Me]^-Na^+$, (1.5 equiv), 18-crown-6, DMF, 50°C, 4 h, (74%, 2 steps); (**vii**) Pd-C, MeOH, H_2, 10 h; (**viii**) NaOMe (3 equiv), MeOH, 2 h; (**ix**) Dowex 50 W X-8 H^+ resin, MeOH, (62%, 3 steps); (**x**) Alkaline Hydrolysis (80%).

Scheme 30

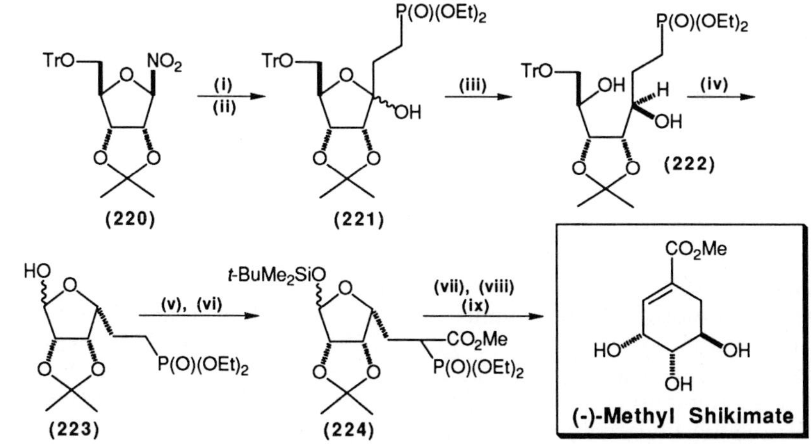

(220)
(221)
(222)
(223)
(224)
(i)
(ii)
(iii)
(iv)
(v), (vi)
(vii), (viii)
(ix)
(-)-Methyl Shikimate

Reagents: (**i**) n-Bu_4NF, $H_2C{=}CHP(O)(OEt)_2$,THF, 0°C,1h; (**ii**) Formamide, $NaHCO_3$, 60°C, 24h, (87%); (**iii**) $NaBH_4$, MeOH, 0°C to r. temp., (97%); (**iv**) Anh. $ZnBr_2$, CH_2Cl_2, 30 min.; $NaIO_4$, MeOH, H_2O, 0°C, (85%); (**v**) t-$BuMe_2SiCl$, Imidazole, CH_2Cl_2, (93%); (**vi**) n-BuLi, THF, 15 min.; MeOCOCl, -78°C, 3h, (94%); (**vii**) n-Bu_4NF, THF, 1h; (**viii**) NaOMe, MeOH, 1h, r. temp., (86%, 2 steps); (**ix**) Dowex (50 W)-H^+ Resin, MeOH, (97%).

In the Vasella strategy to (-)-methyl shikimate (Scheme 30), the 1-deoxy-1-nitro-D-ribose derivative (**220**) was initially chain extended at C(1) through a base-catalysed Michael addition of the glycosyl nitronate to diethyl vinylphosphonate. This gave ketose (**221**) after heating in wet formamide. The latter was reduced to (**222**), the trityl ether removed, and periodate cleavage performed to give the *lyxo*-phosphonates (**223**) as a 4.9:1 mixture of anomers. These were silylated with *tert*-butyldimethylsilyl chloride and the individual anomers separated by chromatography. After *C*-alkylation adjacent to the phosphonate with *n*-butyllithium and methyl chloroformate, each anomer provided a 1:1 mixture of diastereomers (**224**) which were deprotected with fluoride ion, and cyclised to methyl shikimate with base.

Other interesting examples of the Wadsworth-Horner-Emmons reaction being used for carbocyclic ring closure can be found in the synthesis of a naturally-occurring glyoxalase I inhibitor from D-mannose (S. Mirza, L.-P. Molleyres, and A. Vasella, *Helv. Chim Acta.*, 1985, 68, 988) and in the route to the nucleoside antibiotic (-)-neoplanocin by H.J. Haltenbach *et al.* (*Tetrahedron Lett.*, 1985, 26, 5295).

(v) Intramolecular nucleophilic alkylation

(+)-Lineatin

A number of monosaccharide-based natural product syntheses have employed nucleophilic substitution reactions to bring about carbocyclisation. One noteworthy case is the nucleophilic alkylation used for constructing the highly strained cyclobutane system in (+)-lineatin (A.A. Kandil and K.N. Slessor, *J. Org. Chem.*, 1985, 50, 5649). In this strategy (Scheme 31), D-ribonolactone was initially transformed into aldehyde (**225**), and this intermediate homologated by a phosphonate condensation. Hydrogenation of the resulting alkene gave (**226**) and deprotection of the SEM ether with acid furnished a 3:2 mixture of hemiacetals (**227**), which were selectively blocked at the anomeric position with *tert*-butyldimethylsilyl chloride. The β-silyl ethers, epimeric at the cyano centre, were then reacted with methanesulphonyl chloride to give (**228**). Hemiketal (**229**) was obtained after reaction with aqueous hydrofluoric acid, and this subjected to intramolecular acetalisation under acid catalysis. This led to (**230**), which underwent carbanionic ring closure in 69% yield after exposure to potassium amide in boiling tetrahydrofuran. Reduction of the nitrile present in (**231**) with diisobutylaluminium hydride led to aldehyde (**232**), which was deoxygenated by a Wolff-Kishner reaction.

Scheme 31

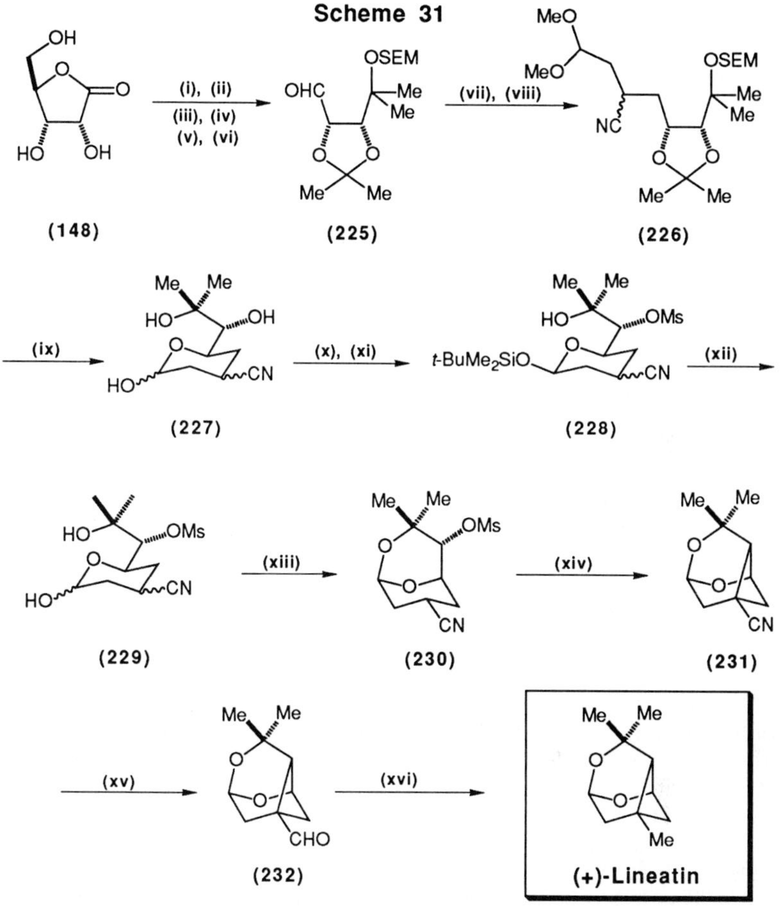

(148) → (i), (ii), (iii), (iv), (v), (vi) → (225) → (vii), (viii) → (226) → (ix) → (227) → (x), (xi) → (228) → (xii) → (229) → (xiii) → (230) → (xiv) → (231) → (xv) → (232) → (xvi) → **(+)-Lineatin**

Reagents: (**i**) $Me_2C{=}O$, H_2SO_4, 0°C to r. temp., (85%); (**ii**) MeMgI, Et_2O, THF, r. temp., 1 h, then Δ for 5 h, (75%); (**iii**) Ac_2O, Py, 12h, (75%); (**iv**) *i*-Pr_2NEt, THF, $Me_3SiO(CH_2)_2OCH_2Cl$, Δ, 3h, (88%); (**v**) NaOH, aq. MeOH, 12 h, (100%); (**vi**) $NaIO_4$, H_2O, (95%); (**vii**) NaH, THF, $(MeO)_2CH_2CH(CN)P(O)(OEt)_2$, 5°C, 5 h, (90%, *E*:*Z* = 1:2); (**viii**) H_2, Pd-C, EtOH, (97%); (**ix**) 1% aq. H_2SO_4, MeOH, r. temp., 36-48 h, (90%); (**x**) *t*-$BuMe_2SiCl$, Imidazole, DMF, r. temp, 24 h, (84%); (**xi**) MsCl, Py, 0°C to r. temp., 24 h, (75%); (**xii**) 49% HF, MeCN, r. temp., 24 h, (89%); (**xiii**) *p*-TsOH, CH_2Cl_2, C_6H_6, Δ, 3 h, (33%); (**xiv**) K, liq. NH_3, THF, -78°C then warm to r. temp.; Add (xx), THF, Δ, 3 h, (69%); (**xv**) DIBAL-H, THF, -78°C to r. temp., 5h, (75%); (**xvi**) NH_2NH_2-H_2O, KOH, Triethylene Glycol, 135°C, 1h; then 210°C, 3h, (73%).

(d) *Cationic cyclisation strategies*

Trichothecene skeleton

A novel approach to an optically active trichothecene intermediate has been devised by R. Tsang and B. Fraser-Reid (*J. Org. Chem.*, 1985, 50, 4659) which employed a carbocationic cylisation on allylic ether (**236**) for assembly of the A-ring (Scheme 32). This unusual reaction delivered intermediate (**237**) which was subsequently transformed into allylic alcohol (**239**) in six steps. The latter substrate readily participated in an Eschenmoser-Claisen rearrangement (A. Eschenmoser *et al.*, *Helv. Chim. Acta.*, 1964, 47, 2425) to furnish (**240**), which was deprotected and selectively sulphonylated. Intramolecular carbanionic alkylation of sulphonate (**241**) proceeded smoothly to complete the annulation of ring-C. The last two steps were simply benzylation and chemoselective reaction with methyllithium.

2. *Use of monosaccharides for the assembly of molecules with heterocyclic rings*

Clavalanine

Clavalanine is a recently discovered antibiotic obtained from *Streptomyces clavuligerus*, which exerts its antibacterial effects through interference with methionine biosynthesis (D.L. Pruess and M. Kellett, *J. Antibiot.*, 1983, 36, 208; R.H. Evans *et al.*, *ibid.*, 1983, 36, 213). M. Weigele *et al.* (*J. Org. Chem.*, 1985, 50, 3457) have announced a synthesis of this molecule from D-xylose (Scheme 33). The initial stages of this pathway involved deoxygenation of the thiocarbonylimidazolide (**243**) with tri-*n*-butyltin hydride, hydrolysis of the 1,2-*O*-isopropylidene acetal, and regioselective sulphonylation of the 1,2-*O*-stannylene derivative with *p*-chlorophenylsulphonyl chloride (D. Wagner, J.P.H. Verheyden, and J.G. Moffatt, *J. Org. Chem.*, 1974, 39, 24). This led to hemiacetal (**244**) which was oxidised and subjected to azide displacement. Hydrogenation resulted in the amine, which was acylated with benzyloxycarbonyl chloride and deacetylated. Compound (**245**) underwent facile ring-opening to the hydroxy-acid, and after esterification with diphenyldiazomethane, selective sulphonylation afforded (**246**). This underwent coupling to racemic 4-acetoxy 2-azetidinone in the presence of catalytic palladium acetate and triethylamine. After lithium bromide displacement, a 2.5:1 mixture of the *4S:4R* isomers (**248**) was isolated. This mixture cyclised to the desired bicyclic β-lactams in *N,N*-dimethylformamide after treatment with silver 2,2-dimethyl-6,6,7,7,8,8,8-heptafluoro-3,5-octanedionate, and at this point the diastereoisomers were separated. Hydrogenation of the *4S*-isomer removed the benzylic protecting groups to give clavalanine.

Scheme 32

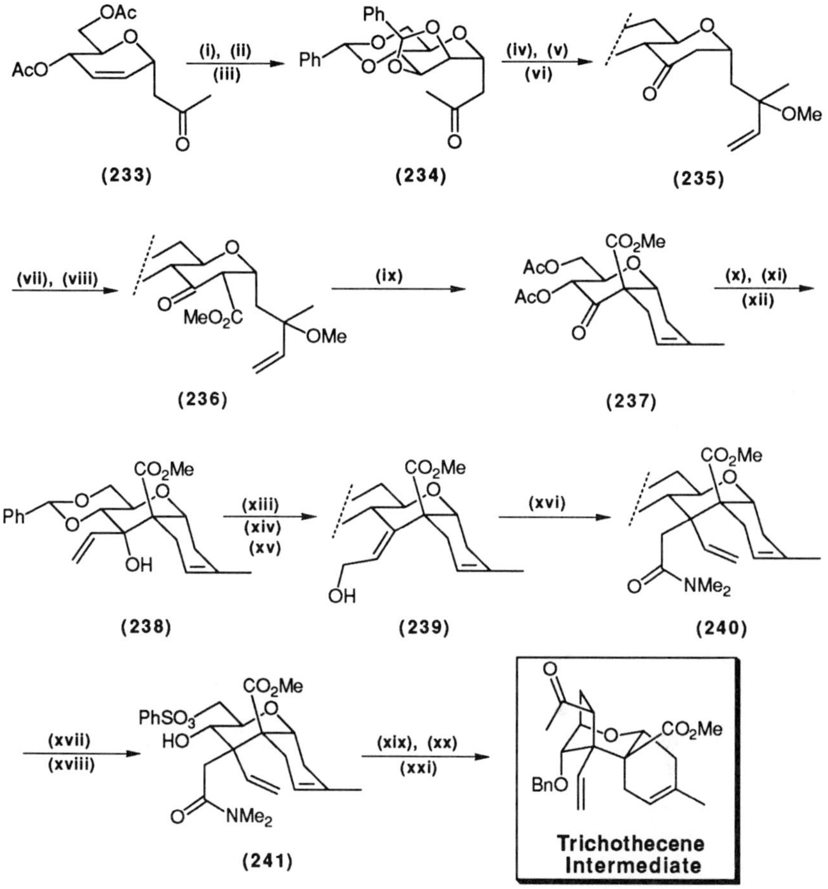

(233) (i), (ii) (iii) → (234) (iv), (v) (vi) → (235)

(vii), (viii) → (236) (ix) → (237) (x), (xi) (xii) →

(238) (xiii) (xiv) (xv) → (239) (xvi) → (240)

(xvii) (xviii) → (241) (xix), (xx) (xxi) → Trichothecene Intermediate

Reagents: (**i**) $KMnO_4$, aq. EtOH; (**ii**) Et_3N, MeOH, H_2O; (**iii**) $PhCH(OMe)_2$, CSA; (**iv**) $H_2C{=}CHMgBr$; (**v**) NaH, MeI; (**vi**) *n*-BuLi, THF, -40°C; (**vii**) $MeOMgCO_2Me$, MeOH; HCl; (**viii**) CH_2N_2 (65%, 8 steps); (**ix**) $SnCl_4$, Ac_2O, CH_2Cl_2; (**x**) NaOMe, MeOH; (**xi**) $PhCH(OMe)_2$, CSA, (55%, 3 steps); (**xii**) $H_2C{=}CHMgBr$, THF, (95%); (**xiii**) $SOCl_2$, Py, THF; (**xiv**) KOAc, DMF; (**xv**) NaOMe, MeOH, (90%, 3 steps); (**xvi**) $MeC(OMe)_2NMe_2$, PhMe, Δ, (88%); (**xvii**) CSA, MeOH; (**xviii**) $PhSO_2Cl$, Py, (83%, 2 steps); (**xix**) $KN(SiMe_3)_2$, THF, -78°C to 25°C; (**xx**) $PhCH_2Br$, (75%, 2 steps); (**xxi**) MeLi, Et_2O, -40°C, (78%).

Scheme 33

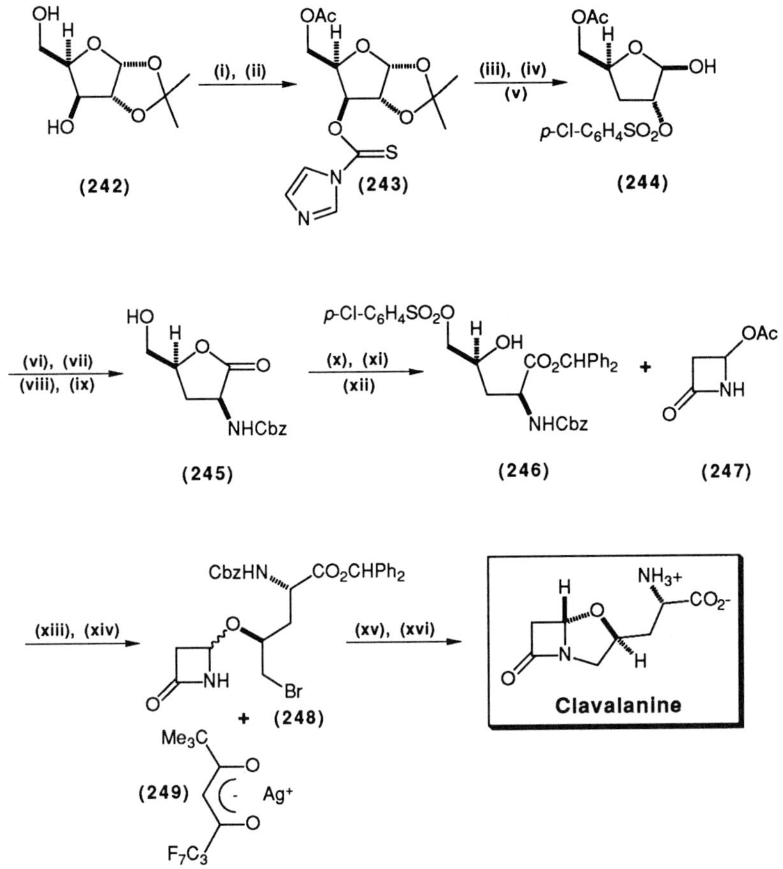

(242) —(i), (ii)→ (243) —(iii), (iv), (v)→ (244) —(vi), (vii), (viii), (ix)→ (245) —(x), (xi), (xii)→ (246) + (247) —(xiii), (xiv)→ (248) + (249) —(xv), (xvi)→ **Clavalanine**

Reagents: (**i**) Ac_2O, Py, CH_2Cl_2, -10°C to r. temp., (85%); (**ii**) 1,1'-Thiocarbonyldiimidazole, $ClCH_2CH_2Cl$, Δ, 1 h, (97%); (**iii**) *n*-Bu_3SnH, AIBN, PhMe, Δ, (95%); (**iv**) 50% TFA, 5°C, (78%); (**v**) Bu_2SnO, MeOH, Δ, 1.5 h, then cooled to 4°C, *p*-$ClC_6H_4SO_2Cl$, DME, 1 h, (61%); (**vi**) RuO_2, $NaIO_4$, $ClCH_2CH_2Cl$, CCl_4, (98%); (**vii**) LiN_3, DMF, -10°C, (90%); (**viii**) H_2, Pd-C, DME, H_2O, HCl; aq. $NaHCO_3$, $PhCH_2OCOCl$, (70%); (**ix**) Dowex AG 50W-X4 (H^+) 100-200 Mesh, 1,4-Dioxane/H_2O (1:1), (93%); (**x**) 0.1 N KOH, acidify; (**xi**) $Me_2C{=}O$, Ph_2CHN_2, r. temp, 4 h, (78%); (**xii**) *p*-$ClC_6H_4SO_2Cl$, $ClCH_2CH_2Cl$, Py, -10°C to r. temp, (63%); (**xiii**) C_6H_6, $Pd(OAc)_2$, Et_3N, (52%); (**xiv**) THF, LiBr, (95%); (**xv**) DMF, 70°C, 10 h, (50%); (**xvi**) H_2, Pd-C, MeOH, (97%).

(-)-Bulgecinine

(-)-Bulgecinine is one of the amino acid constituents of the bulgecins (A. Imada *et al.*, *J. Antibiot.*, 1982, 35, 1400), SQ-28504 and SQ-28546 (R. Cooper and S. Unger, *J. Org. Chem.*, 1986, 51, 3942). These are naturally-occurring glycosides that induce bulges in the cell walls of gram-negative bacteria when co-administered with β-lactam antibiotics. After bulge-formation has occurred, the microorganisms are rendered more susceptible to the action of the antibiotic agent. T. Shiba *et al.* (*Tetrahedron Lett.*, 1985, 26, 4759) have carried out a synthesis of (-)-bulgecinine from D-glucose. Their route began from 3-deoxy-D-glucose derivative (**250**) (E. Vis, P. Karrer, *Helv. Chim. Acta*, 1954, 37, 373) and is shown in Scheme 34. Tosylation followed by azide displacement led to (**251**). This was transformed into benzyloxyurethane (**252**) by reduction and protection. Benzylation of the 4,6-diol system, glycoside hydrolysis, and hemiacetal oxidation provided lactone (**253**). It underwent ring-opening to methyl ester (**254**) when heated in methanol. The hydroxyl group was then substituted by chloride, since this leaving group could withstand the hydrogenolysis conditions used for removing the benzyloxy groups in the following step. Deprotection was accompanied by formation of lactone (**256**), which was converted into (-)-bulgecinine by treatment with saturated barium hydroxide solution.

Another synthesis of (-)-bulgecinine has been described from D-glucuronic acid by B.P. Bashyal, H.-F. Chow, and G.W.J. Fleet (*Tetrahedron Lett.*, 1986, 27, 3205; *Tetrahedron*, 1987, 43, 423).

2S,3R,4R,5S-Trihydroxypipecolic acid

G.W.J. Fleet *et al.* (*Tetrahedron Lett.*, 1986, 27, 3205; *Tetrahedron,* 1987, 43, 415) have reported a concise synthesis of *2S, 3R, 4R, 5S*-trihydroxypipecolic acid that starts from D-glucuronic acid derivative (**257**). The key features of their strategy (Scheme 35) were the stereocontrolled introduction of the protected amine substituent at C(5), the clean removal of the 1,2-*O*-isopropylidene group, and the reductive amination step used to deliver the natural product.

(-)-Rosmarinecine

K. Tatsuta *et al.* (*J. Am. Chem. Soc.*, 1983, 105, 4096) have described an approach to (-)-rosmarinecine which highlights the utility of 2-amino-D-xylofuranose derivative (**260**) (K. Tatsuta *et al.*, *Bull. Chem. Soc. Jpn.*, 1982, 55, 3254) as a chiral starting material for alkaloid synthesis. Their route (Scheme 36) commenced with silylation of the primary hydroxyl groups in (**260**). This was followed by chelation-controlled addition of allylmagnesium bromide to produce amido-alcohol (**261**), as a single isomer. Oxidative degradation of the allyl group with potassium permanganate and sodium periodate led to

Scheme 34

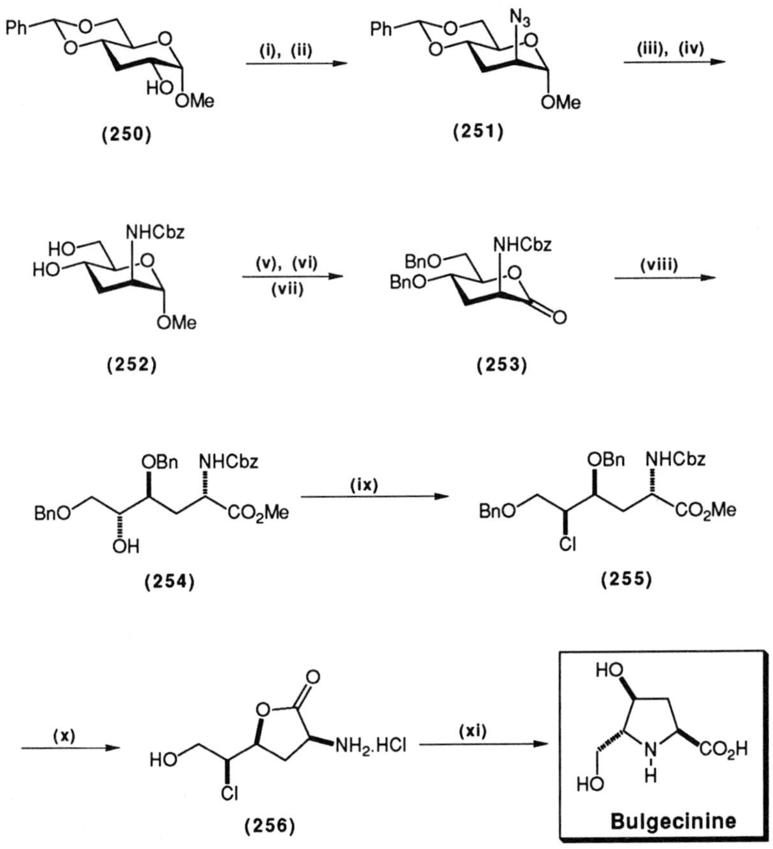

(250) —(i), (ii)→ (251) —(iii), (iv)→

(252) —(v), (vi), (vii)→ (253) —(viii)→

(254) —(ix)→ (255)

—(x)→ (256) —(xi)→ Bulgecinine

Reagents: (**i**) TsCl, Py, (88%); (**ii**) NaN_3, DMF, (73%); (**iii**) H_2, Pd Black, MeOH, HCl (1.0 equiv), (100%); (**iv**) *N*-Benzyloxycarbonyloxysuccinimide, CH_2Cl_2, (92%); (**v**) BnBr, NaOH, DMF, (61%); (**vi**) conc. HCl, AcOH, (66%); (**vii**) PDC, CH_2Cl_2, (59%); (**viii**) MeOH, Δ, (100%); (**ix**) Ph_3P, CCl_4, (43%); (**x**) H_2, Pd-Black, MeOH, conc. HCl, (100%); (**xi**) Sat. $Ba(OH)_2$, pH 9.0, (75%).

Scheme 35

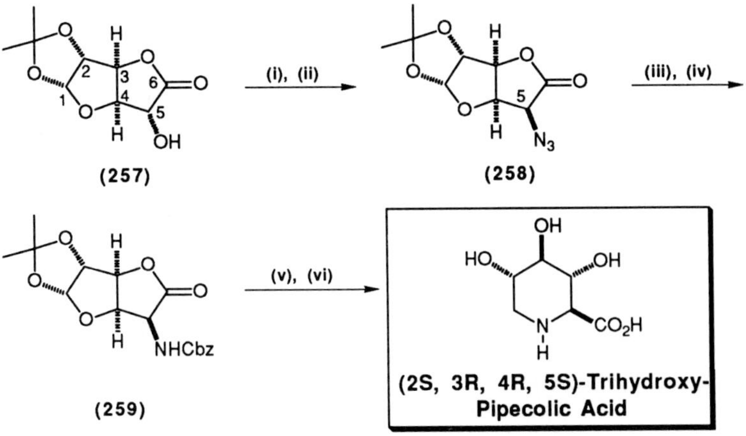
(i), (ii)
(iii), (iv)
(257)
(258)
N3
(v), (vi)
(259)
NHCbz
OH
HO
OH
CO2H
(2S, 3R, 4R, 5S)-Trihydroxy-Pipecolic Acid

Reagents: (**i**) $(CF_3SO_2)_2O$, Py, CH_2Cl_2, -20°C; (**ii**) NaN_3, DMF, -20°C; (**iii**) H_2, Pd-C, EtOAc; (**iv**) $PhCH_2OCOCl$, $NaHCO_3$, EtOAc, H_2O, (44%, 4 steps); (**v**) CF_3CO_2H, H_2O, r. temp.; (**vi**) H_2, Pd-Black, H_2O, AcOH (9:1), 4 days, (60%, 2 steps).

Scheme 36

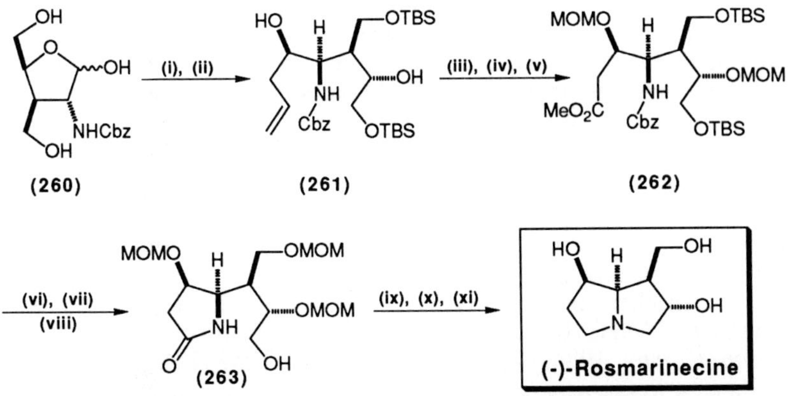
NHCbz
(i), (ii)
OTBS
Cbz
(iii), (iv), (v)
MOMO
OMOM
MeO2C
(260)
(261)
(262)
(vi), (vii)
(viii)
(263)
(ix), (x), (xi)
(-)-Rosmarinecine

Reagents: (**i**) *t*-$BuMe_2SiCl$, Pyridine, 15 h, (87%); (**ii**) $H_2C{=}CHCH_2MgBr$, Et_2O, 5°C, 30 min., 20°C, 30 min., Δ, 3 h, (92%); (**iii**) $KMnO_4$ (1.5 equiv), $NaIO_4$ (5.0 equiv), aq. *t*-BuOH, then 5% K_2CO_3,15 h; (**iv**) CH_2N_2, Et_2O, 30 min.; (**v**) MOM-Cl, *i*-Pr_2NEt, $CHCl_3$, 60°C, (92%, 3 steps); (**vi**) H_2 (3 atm), Pd-C, THF, AcOH, 3 h; (**vii**) *n*-Bu_4NF, THF, 5°C, 30 min., (93%, 2 steps); (**viii**) MOM-Cl (1.5 equiv), *i*-Pr_2NEt, THF, 6 h, (51%); (**ix**) MsCl, py, 1h, (93%); (**x**) BH_3-Me_2S, THF, 60°C; (**xi**) 0.5 N HCl, Dioxane, 80°C, 6 h, (54%, 2 steps).

the carboxylic acid which was esterified with diazomethane. After methoxymethylation, methyl ester (**262**) was obtained in 74% overall yield. Removal of the *N*-protecting group led to γ-lactam formation, and subsequent desilylation, followed by selective methoxymethylation generated intermediate (**263**). Mesylation, and reduction of the lactam with borane-methyl sulphide gave the pyrrolidine, and subsequent intramolecular S_N2 cyclisation completed the pyrrolizidine skeleton. Deprotection with acid yielded (-)-rosemarinecine.

(-)-Swainsonine

(-)-Swainsonine is an indolizidine alkaloid isolated from a number of sources (S.M. Colgate, P.R. Dorling, and C.R. Huxtable, *Aust. J. Chem.*, 1979, 32, 2257). It is of medical importance on account of its pronounced immunostimulatory and anticancer properties (G.W.J. Fleet *et al.*, *Tetrahedron*, 1988, 44, 2649). A stereospecific 13-step synthesis of (-)-swainsonine has been developed from D-mannose by G.W.J. Fleet, M.J. Gough, and P.W. Smith (*Tetrahedron Lett.*, 1984, 25, 1853; B.P. Bashyal *et al.*, *Tetrahedron*, 1987, 43, 3083). In their approach (Scheme 37), D-mannose was initially transformed into alcohol (**265**), and this oxidised to (**266**) with pyridinium chlorochromate. A stereospecific reduction from the less hindered face afforded the D-talose derivative (**267**), which was converted into triflate (**268**). Displacement of the sulphonate ester with azide enabled the nitrogen functionality to be installed at C(4) with the correct stereochemistry for manipulation into swainsonine. The next steps involved desilylation, oxidation and homologation of the C(6) position to give α,β-unsaturated aldehyde (**270**). Hydrogenation reduced the enal double bond, transformed the azide to the amine, and effected an intramolecular reductive amination to furnish (**271**). More vigorous hydrogenation also cleaved the benzyl glycoside allowing a second reductive amination to proceed. Deprotection of (**272**) with trifluoroacetic acid gave (-)-swainsonine in 16% overall yield.

Another conceptually interesting synthesis of (-)-swainsonine was that of M.H. Ali, L. Hough, and A.C. Richardson (*J. Chem. Soc., Chem. Comm.*, 1984, 447; *Carbohydr. Res.*, 1985, 136, 225) (Scheme 38) which began from methyl 3-amino-3-deoxy-α-D-mannopyranoside hydrochloride (**273**). This unusual starting material is a highly crystalline compound that is readily available on large scale from methyl α-D-glucopyranoside by application of the nitromethane dialdehyde cyclisation protocol (A.C. Richardson, *J. Chem. Soc.*, 1962, 373). At first, it was found necessary to temporarily protect the amino group in (**273**) in order to allow selective tosylation of the C(6) hydroxyl group.

Scheme 37

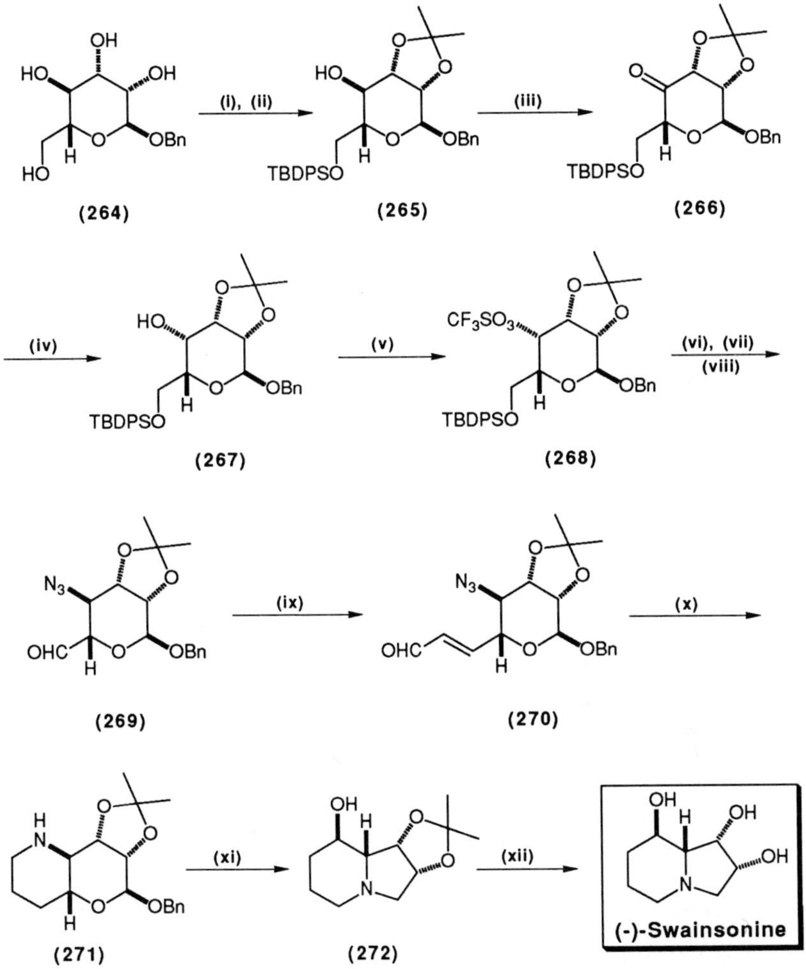

(264) **(265)** **(266)** **(267)** **(268)** **(269)** **(270)** **(271)** **(272)** **(-)-Swainsonine**

Reagents: (**i**) *t*-$BuPh_2SiCl$, Imidazole, DMF, (89%); (**ii**) $Me_2C{=}O$, $Me_2C(OMe)_2$, CSA, (97%); (**iii**) PCC, CH_2Cl_2, 4 A MS; (**iv**) $NaBH_4$, EtOH, (81%, 2 steps); (**v**) $(CF_3SO_2)_2O$ (3.0 equiv), Py; (**vi**) NaN_3, DMF, 20°C, (67%, 2 steps); (**vii**) *n*-Bu_4NF, THF, (97%); (**viii**) PCC, (2.0 equiv), CH_2Cl_2, 4 A MS; (**ix**) $Ph_3P{=}CHCHO$, CH_2Cl_2, (68%, 2 steps); (**x**) H_2, Pd-C, MeOH, r. temp.; (**xi**) H_2, AcOH, 3 days, (87%, 2 steps); (**xii**) CF_3CO_2H, H_2O, (74%).

Scheme 38

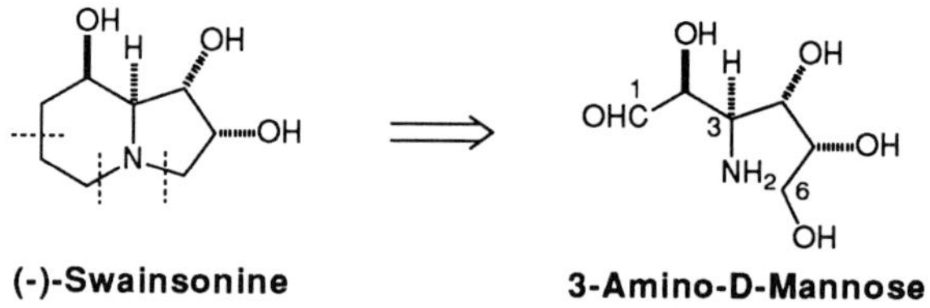

(-)-Swainsonine ⟹ 3-Amino-D-Mannose

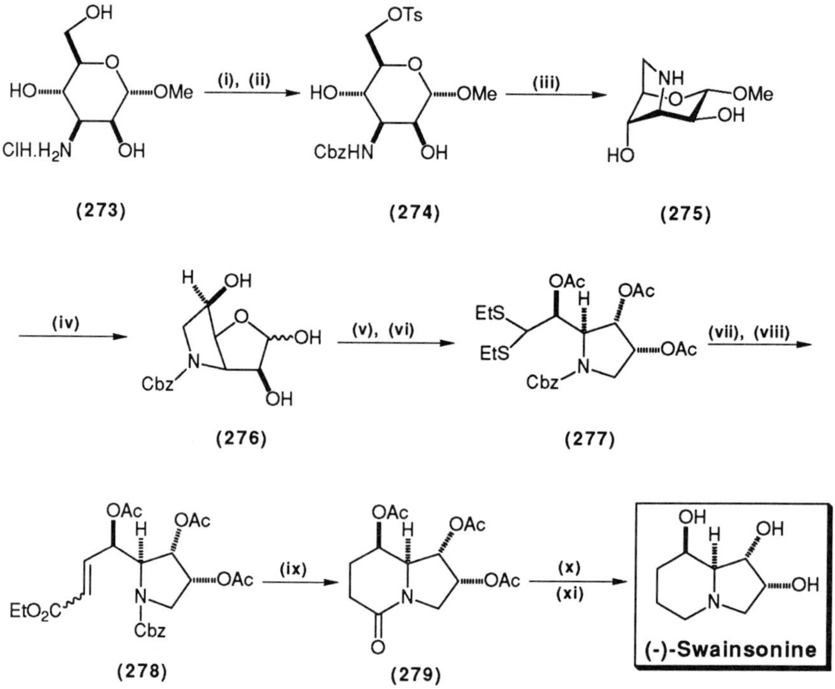

(273) →(i), (ii)→ (274) →(iii)→ (275) →(iv)→ (276) →(v), (vi)→ (277) →(vii), (viii)→ (278) →(ix)→ (279) →(x), (xi)→ (-)-Swainsonine

Reagents: (**i**) $NaHCO_3$, 50% aq. EtOH, $PhCH_2OCOCl$, 2 h, r. temp.; (**ii**) TsCl, Py, 36 h, r. temp., (82%, 2 steps); (**iii**) Pd-C, EtOH, 60 p.s.i., 18 h, (73%); (**iv**) 0.4 M HCl, 95-100°C, 16 h, (52%); (**v**) EtSH, conc. HCl, 15 min., (74%); (**vi**) Ac_2O, Py, (73%); (**vii**) $HgCl_2$, $CdCO_3$, 50% aq. $Me_2C{=}O$, Δ, 30 min., (96%); (**viii**) $EtO_2CCH{=}PPh_3$, MeCN, Δ, 15 min., (86%); (**ix**) H_2, Pd-C, 60 p.s.i., 2 h, (25%); (**x**) BH_3-Me_2S, THF, r. temp., 8 h, (71%); (**xi**) NaOMe, MeOH, (100%).

Hydrogenation then liberated the amine for cyclisation to the pyrrolidine ring system. Acid hydrolysis of (**275**) led to (**276**), which was thioketalised and acetylated to produce (**277**). After conversion to the aldehyde, Wittig olefination delivered (**278**) as a mixture of *E*- and *Z*-geometrical isomers. This stereochemical outcome was unimportant, however, since the next step involved hydrogenation and lactamisation to produce (**279**). Unfortunately, the hydrogenation step was plagued by a significant amount of *O*- to *N*-acyl migration, which lowered the yield of the cyclisation product to 25%. Even so, reduction of the lactam proceeded readily with borane-methyl sulphide complex, as did *O*-deacetylation. Together, they supplied (-)-swainsonine in 71% yield for the last two steps.

Other novel monosaccharide based approaches to (-)-swainsonine have been reported (T. Suami, K. Tadano, and Y. Imura, *Chem. Lett.*, 1984, 513; *Carbohydr. Res.*, 1985, 136, 67; N. Yasuda, H. Tsutsumi, and T. Takaya, *Chem. Lett.*, 1984, 1201; H. Setoi, H. Takeno, and M. Hashimoto, *J. Org. Chem.*, 1985, 50, 3948; R.B. Bennett, III, *et al.*, *J. Am. Chem. Soc.*, 1989, 111, 2580).

(+)-Castanospermine

(+)-Castanospermine is a potent inhibitor of α- and β-glucosidases, whose synthesis was first achieved by R.C. Bernotas and B. Ganem (*Tetrahedron Lett.*, 1984, 25, 165). These workers recognised that the absolute configurations of the hydroxyl groups at C(6), C(7) and C(8) in (+)-castanospermine were the same as those at C(2), C(3) and C(4) respectively in D-glucose. As a result, they prepared amino-epoxide (**285**) from D-glucose, and carried out an intramolecular nucleophilic cyclisation via a sodium borohydride mediated reduction of the *N*-trifluoroacetyl group (Scheme 39). Surprisingly, the desired product (**286**) was the minor component of this reaction, the azepine (**287**) tending to predominate. Nevertheless, the synthesis continued with oxidation of (**286**) to aldehyde (**288**), and reaction with lithio *tert*-butyl acetate. This gave a 1:1 mixture of diastereoisomers (**289**) which were separated and carried forward to (+)-castanospermine and 1-epicastanospermine in three steps. These were hydrogenolytic debenzylation, lactamisation with acid, and diisobutylaluminium hydride reduction to the indolizidine.

Neooxazolomycin

Neoxazolomycin is a novel polyene antitumour antibiotic that contains a densely functionalised bicyclic lactone-lactam ring system. An enantioselective total synthesis has recently been developed by A.S. Kende, K. Kawamura, and R.J. DeVita (*J. Am. Chem. Soc.*, 1990, 112, 4070) which made extensive use of monosaccharide chemistry for construction of the saturated bicyclic heterocycle.

Scheme 39

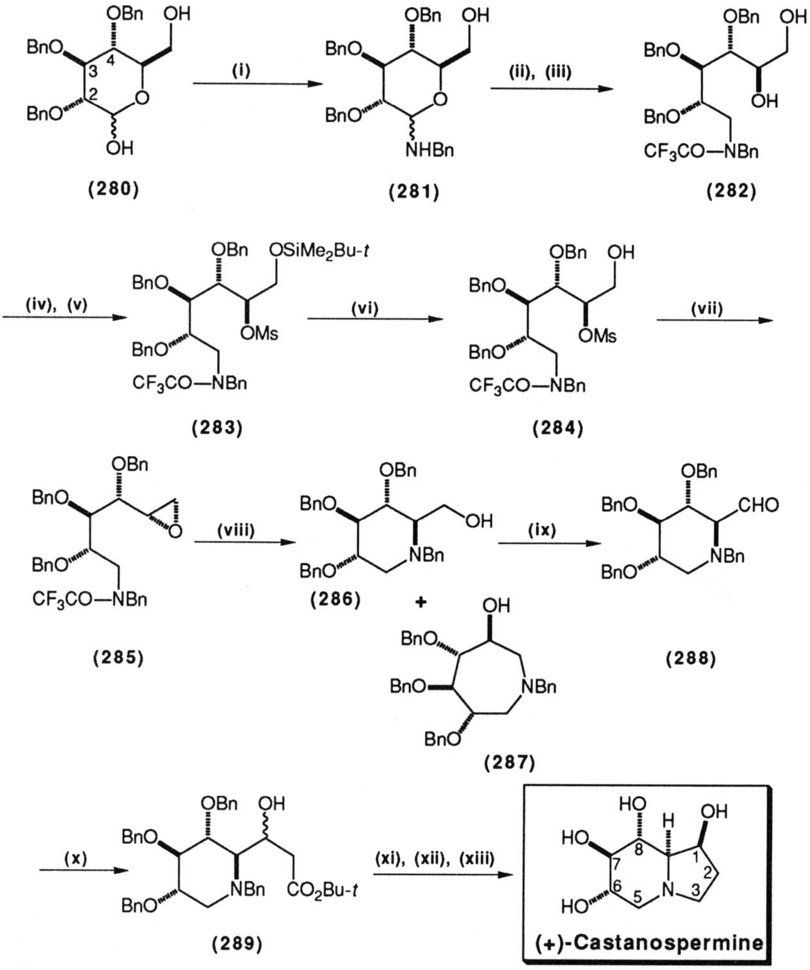

Reagents: (**i**) $PhCH_2NH_2$, $CHCl_3$, (77%); (**ii**) $LiAlH_4$, THF, Δ, 5 h; (**iii**) $(CF_3CO)_2O$, (78%, 2 steps); (**iv**) *t*-$BuMe_2SiCl$, Imidazole; (**v**) MsCl, Base; (**vi**) *n*-Bu_4NF, THF; (**vii**) NaOMe, MeOH, (75%, 4 steps); (**viii**) $NaBH_4$, EtOH, 40°C, (45% piperidine, 55% azepine); (**ix**) Me_2SO, $(COCl)_2$, CH_2Cl_2, Et_3N, -78°C, (90%); (**x**) *t*-BuO_2CCH_2Li; (**xi**) H_2, Pd-C; (**xii**) CF_3CO_2H-H_2O, 60°C, 3 h; (**xiii**) DIBAL-H.

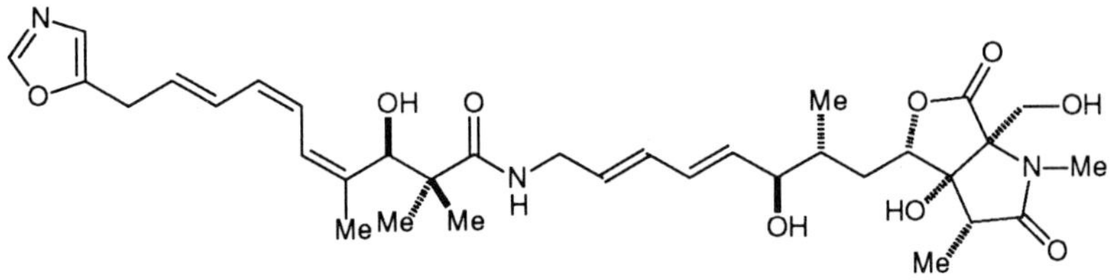

Neooxazolomycin

The synthesis of the left hand side of the antibiotic commenced from (Z)-3-bromo-2-methyl-2-propenol (**290**) (Scheme 40). Initially, this was converted to aldehyde (**292**) by a four-step sequence that involved *O*-silylation, palladium (0) catalysed cross-coupling with (trimethylsilyl)acetylene, selective *O*-desilylation with aqueous acetic acid, and oxidation with manganese dioxide. An asymmetric Reformatsky-type reaction was then executed between (**292**) and the tin (II) enolate derived from oxazolidinone (**293**) (A.S. Kende, K. Kawamura, M.J. Orwat, *Tetrahedron Lett.*, 1989, 30, 5821). This gave (**294**) as the sole product in 95% yield. Treatment of (**294**) with lithium hydroperoxide, followed by lithium hydroxide, caused removal of both the chiral auxiliary and the *C*-silyl group. The resulting acid was esterified with diazomethane, the β-hydroxy ester reacted with *tert*-butyldimethylsilyl triflate, and the terminal alkyne chemoselectively deprotonated with *n*-butyllithium to produce (**295**), after quenching with iodine. Diimide reduction cleanly afforded the *(Z,Z)*-dienyl iodide (**296**), which underwent palladium (0) catalysed cross-coupling with *(E)*-vinyl stannane (**297**) without olefin isomerisation (J.K. Stille and B.L. Groh, *J. Am. Chem. Soc.*, 1987, 109, 813). Desilylation followed by saponification, acetylation, and hydrolysis of the mixed anhydride eventually produced acid (**298**), which was converted into activated anhydride (**300**) by reaction with *N,N*-bis-(2-oxo-3-oxazolidinyl)phosphorodiamidic chloride (**299**) and triethylamine (J. Cabre and A.L. Palomo, *Synthesis*, 1984, 413).

In order to construct the right half of neooxazolomycin, methyl 2,3-di-*O*-benzyl-α-D-glucopyranoside (**301**) was chosen as the starting material (Scheme 41). It was transformed into epoxygalactoside (**303**) in five steps according to the procedure set out by J.G. Buchanan and R. Fletcher (*J. Chem. Soc.*, 1965, 6313). Oxirane (**303**) underwent *trans*-diaxial ring-opening with excess dimethylmagnesium (K.A. Parker and R.E. Babine, *Tetrahedron Lett.*, 1982, 23, 1763) to give (**304**) in 95% yield. Selective silylation of the equatorial hydroxyl group in (**304**) permitted efficient deoxygenation of the remaining hydroxy via the Barton protocol to give (**305**). *O*-Desilylation followed by acetonation and hydrogenolysis led to alcohol (**306**), which was

Scheme 40

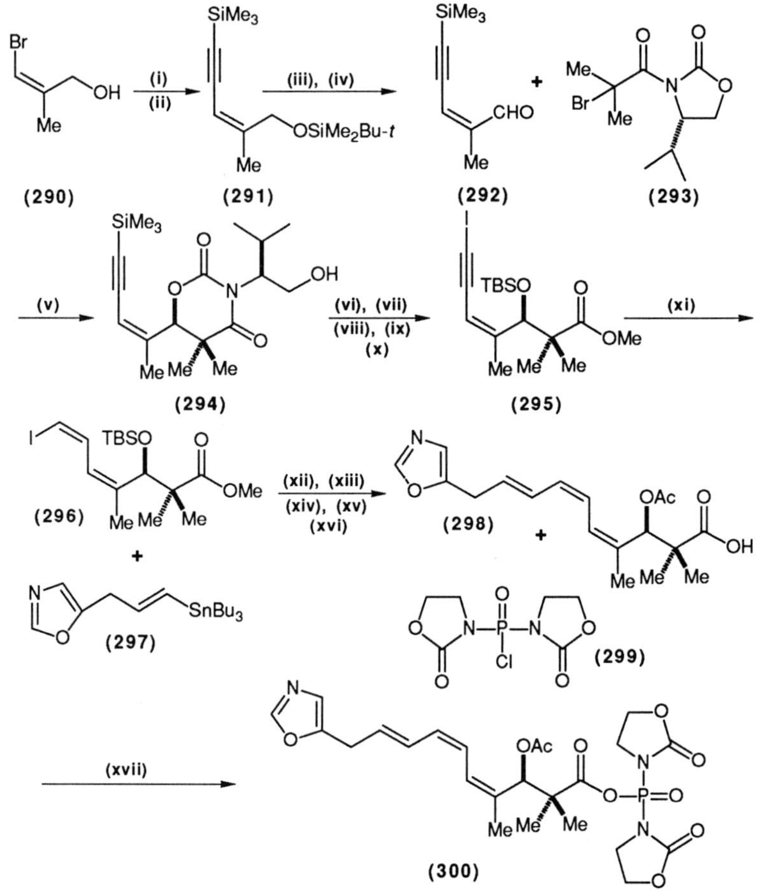

(290) (291) (292) (293) (294) (295) (296) (297) (298) (299) (300)

Reagents: (**i**) *t*-$BuMe_2SiCl$, Imidazole, DMF, 0°C, 5 min.; (**ii**) Me_3SiCCH, $Pd(Ph_3P)_4$, CuI, *n*-$BuNH_2$, C_6H_6, 23°C, 18 h; (**iii**) AcOH, THF, H_2O, 23°C, 18 h; (**iv**) Basic MnO_2, Hexane, CH_2Cl_2, 23°C, 2 h, (84%, 4 steps); (**v**) $SnCl_2$, $LiAlH_4$, THF, 20°C, 20 min. then add aldehyde in THF, 20°C, 16 h, (95%); (**vi**) 30% H_2O_2, LiOH, THF, H_2O, 23°C, 24 h; (**vii**) LiOH (3 equiv), THF, MeOH, 20°C, 24 h, (86%, 2 steps); (**viii**) CH_2N_2, Et_2O, 0°C, (81%); (**ix**) *t*-$BuMe_2SiOTf$, 2,6-Lutidine, CH_2Cl_2; (**x**) *n*-BuLi, THF, -78°C, 1h, then add I_2, THF, 1h; (**xi**) NH_2NH_2, $CuSO_4$-5 H_2O (0.1 equiv), 95% EtOH, 23°C, air bubbled, (72%, 3 steps); (**xii**) $PdCl_2(CH_3CN)_2$, DMF, 23°C, 91 h, (79%); (**xiii**) 50% HF, MeCN, (92%); (**xiv**) LiOH, THF, MeOH, H_2O, 24 h, (94%); (**xv**) Ac_2O, Py, 20 h; (**xvi**) $NaHCO_3$, MeOH, H_2O, 1h, (99%, 2 steps); (**xvii**) Et_3N (2.2 equiv), CH_2Cl_2.

Scheme 41

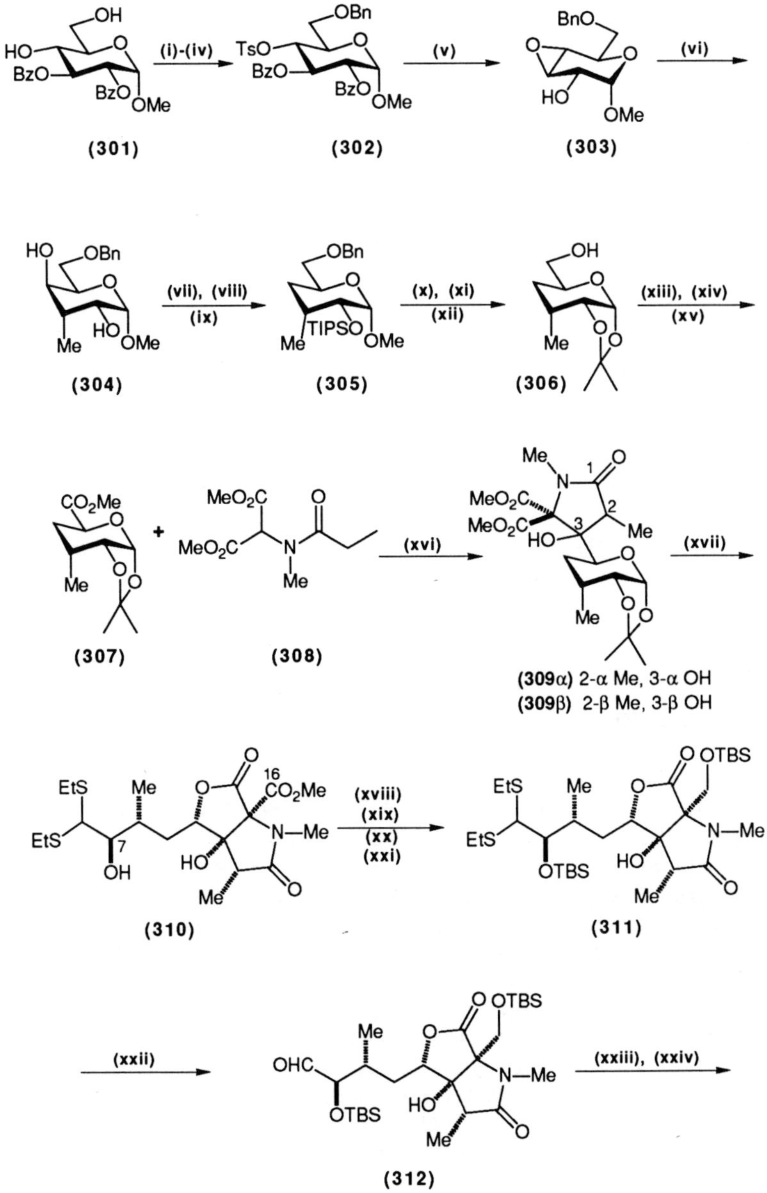

(301) —(i)-(iv)→ (302) —(v)→ (303) —(vi)→

(304) —(vii), (viii), (ix)→ (305) —(x), (xi), (xii)→ (306) —(xiii), (xiv), (xv)→

(307) + (308) —(xvi)→ (309) —(xvii)→

(309α) 2-α Me, 3-α OH
(309β) 2-β Me, 3-β OH

(310) —(xviii), (xix), (xx), (xxi)→ (311)

—(xxii)→ (312) —(xxiii), (xxiv)→

Scheme 41

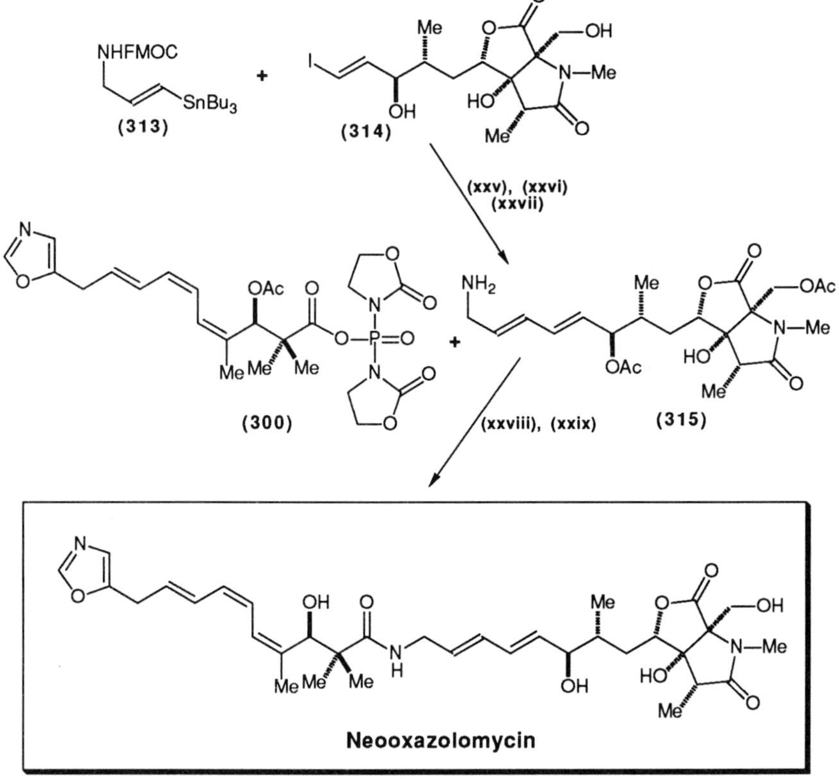

(313) + (314) → (xxv), (xxvi), (xxvii) → (315)

(300) + (315) → (xxviii), (xxix) → Neooxazolomycin

Neooxazolomycin

Reagents: (**i**) Ph_3CCl, Py, r. temp.; (**ii**) *p*-TsCl, Py, (28%, 2 steps); (**iii**) Cat. conc. HCl, EtOH, Δ, 1h, (90%); (**iv**) $AgClO_4$, Ag_2O, BnBr, $CHCl_3$, (81%); (**v**) NaOMe, MeOH, (86%); (**vi**) MeLi, MeMgCl, THF, Et_2O, 23°C, 16 h, (95%); (**vii**) *i*-Pr_3SiOTf, 2,6-Lutidine, THF, -78°C to 23°C, 16 h; (**viii**) $(Im)_2C{=}S$, CH_2Cl_2, Δ; (**ix**) Bu_3SnH, AIBN, Xylene, Δ, 6 h; (**x**) *n*-Bu_4NF, THF, 23°C; (**xi**) $Me_2C{=}O$, $FeCl_3$, (64%, 5 steps); (**xii**) H_2, $Pd(OH)_2$, MeOH, 4 h; (**xiii**) $(COCl)_2$, Me_2SO, CH_2Cl_2, Et_3N, -78°C, 20 min., then 20°C for 1h; (**xiv**) $KMnO_4$, *t*-BuOH, 5% NaH_2PO_4, 23°C; (**xv**) CH_2N_2, Et_2O, (70%, 4 steps); (**xvi**) *t*-BuLi (2.12 equiv), TMEDA (2.12 equiv), -78°C, then add ester at -78°C (82%, based on recovered ester [49%]); (**xvii**) EtSH, cat. conc. HCl, (100%); (**xviii**) *t*-$BuMe_2SiOTf$, 2,6-Lutidine, CH_2Cl_2, 23°C; (**xix**) LiOH, THF, H_2O, 23°C, 1 h, then 1N HCl; (**xx**) $[Me_2N{=}CHCl]^+Cl^-$, MeCN, THF, 0°C, 1 h; $NaBH_4$, DMF, -78°C, to 23°C, 16 h; (**xxi**) *t*-$BuMe_2SiOTf$, 2,6-Lutidine, CH_2Cl_2, 23°C, (59%, 4 steps); (**xxii**) $HgCl_2$, $CaCO_3$, MeCN, H_2O, 23°C, 1 h, (99%); (**xxiii**) CHI_3, $CrCl_2$, THF, 23°C, 1.5 h, (70%); (**xxiv**) *n*-Bu_4NF, THF, (100%); (**xxv**) $PdCl_2(MeCN)_2$, (5 mol %), DMF, 23°C, 24 h, (84%); (**xxvi**) Ac_2O, Py, 20 h, (96%); (**xxvii**) DBU (2 equiv), CH_2Cl_2, 2 h, (60%); (**xxviii**) CH_2Cl_2, Et_3N, (60%); (**xxix**) LiOH, THF, H_2O, 23°C, 1 h, then 1N HCl, (67%).

converted into methyl ester (**307**) in three steps. A cyclocondensation with the dianion derived from amidomalonate (**308**) was next performed at low temperature (A.S. Kende and R.J. DeVita, *Tetrahedron Lett.*, 1988, 29, 2521). It furnished a separable mixture of isomers (**309**α) and (**309**β) (1:1.4 ratio) in 82% yield based on recovered ester (49%). The desired lactam underwent rearrangement to (**310**) upon exposure to ethanethiol and concentrated acid. This was then *O*-silylated at C(7) and the methyl ester at C(16) hydrolysed to the acid. The latter underwent chemoselective reduction with *N,N*-dimethylchloromethylenimium chloride and sodium borohydride, according to the protocol developed by T. Fujisawa, T. Mori, and T. Sato (*Chem. Lett.*, 1983, 835). After selective silylation, the entire four-step sequence provided (**311**) in 59% yield. Dethioketalisation gave aldehyde (**312**), which was olefinated with iodoform and chromium dichloride (K. Takai, K. Nitta, and K. Utimoto, *J. Am. Chem. Soc.*, 1986, 108, 7408) to give the *(E)*-vinyl iodide (**314**) after desilylation. Coupling with vinyl stannane (**313**) under palladium (0) catalysis proceeded with retention of olefin geometry, and supplied the *(E,E)*-dienyl amine (**315**), after acetylation and deprotection of the *N*-fluorenyloxycarbonyl group. Union of (**315**) with (**300**) was complete within one hour, and careful hydrolysis of the *O*-acetate protecting groups with lithium hydroxide gave neoxazolomycin in 67% yield.

The BCD Framework of Brevetoxin A

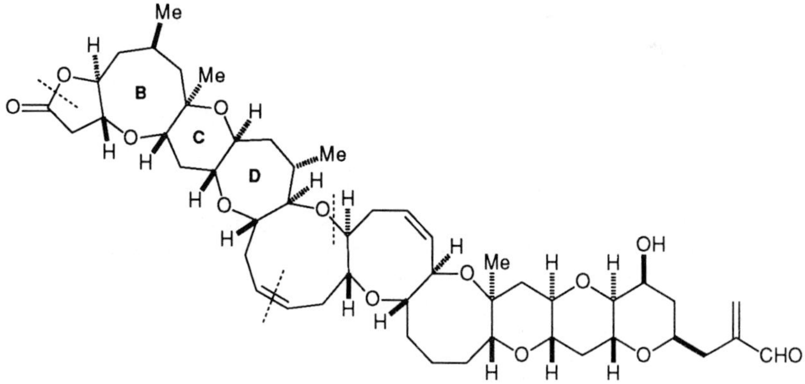

Brevetoxin A

Brevetoxin A is an extremely potent polyether toxin produced by the "red tide" dinoflagellate *Gymnodinium breve*. It is one of the causative agents of massive fish kills and food poisoning in the Gulf of

Mexico (Y. Shimizu *et al.*, *J. Am. Chem. Soc.*, 1986, 108, 514). A fascinating approach to its BCD ring system has been elaborated from 1,2:5,6-di-*O*-isopropylidene-α-D-glucofuranose (**34**) by K.C. Nicolaou, D.G. McGarry, and P.K. Somers (*J. Am. Chem. Soc.*, 1990, 112, 3696). They initially engaged this region of the molecule (Scheme 42) by deoxygenating the C(3) position of (**34**) and selectively hydrolysing the 5,6-*O*-isopropylidene group. The resulting diol was oxidatively cleaved to the carboxylic acid and this esterified with methyl iodide and potassium carbonate in *N,N*-dimethylformamide. This supplied methyl ester (**316**), which underwent condensation with the α-phenylsulphonyl anion derived from (**317**) in 79% yield. Reduction of the β-keto sulphone with aluminium amalgam produced ketone (**318**) which was subjected to a chelation-controlled Grignard addition to give a 13:1 mixture of tertiary methyl carbinols. Thioketalisation of the major product (**319**) and selective *O*-benzylation afforded (**320**). This was hydrolysed to the aldehyde and after treatment with acetylmethylene triphenylphosphorane produced (**321**). Michael addition and oxidative cleavage of the terminal olefin subsequently delivered (**322**), and from this point onwards, both carbonyls were functionalised using almost identical chemistry. Thus, (**322**) reacted smoothly with two equivalents of silyl ketene acetal (**323**) to give a diastereomeric mixture of (**324**) in 81% yield. Hydrogenolytic removal of both benzyl ethers led to bis(hydroxy acid) (**325**), which underwent double lactonisation via the thiopyridyl ester method. This step was crucial for establishing the BCD framework of brevetoxin A and was the preamble to successful installation of the stereocentres that would eventually form part of the A and E rings. The first stage of this operation was selective deprotection of the silyl ether present in the eight-membered lactone, followed by β-elimination of the hydroxyl group using the Martin sulphurane (R.J. Arhart and J.C. Martin, *J. Am. Chem. Soc.*, 1972, 94, 4997; *ibid.*, 1972, 94, 5003). Repetition of this sequence on the seven-membered β-silyloxy lactone provided (**327**). The stereochemistry of the methyl group in the seven-membered lactone was then set by hydrogenation of the double bonds in (**327**). The product was thionated using the reagent developed by S. Sheibye, E.S. Pederson, and S.O. Lawesson, (*Bull. Soc. Chim. Belg.*, 1978, 87, 229) to afford dithionolactone (**328**) in 63% yield. A double alkylation was then performed by adding tri-*n*-butylstannyllithium to the thiocarbonyl groups and trapping the two thiolates with excess iodomethane. This capitalised on previous work by K.C. Nicolaou *et al.* (*J. Am. Chem. Soc.*, 1987, 109, 2504) which demonstrated unequivocally that nucleophilic additions to medium-sized thionolactones yield stable, nucleophilic, tetrahedral intermediates which can be readily alkylated with electrophiles. These observations proved

Scheme 42

(34) —(i), (ii)→ (1) —(iii), (iv), (v)→ (316) + (317)

Me, SO_2Ph, MeO

—(vi), (vii)→ (318) —(viii)→ (319)

—(ix), (x)→ (320) —(xi), (xii)→ (321)

OBn, SEt, SEt, OBn, COMe, OH, BnO

—(xiii), (xiv), (xv)→ (322) + (323) —(xvi)→ (324)

$OSiMe_2Bu$-*t*, OBn, TBSO, OTBS, BnO

—(xvii)→ (325) —(xviii)→ (326)

HO_2C, CO_2H, HO, OH, TBSO, OTBS

Scheme 42

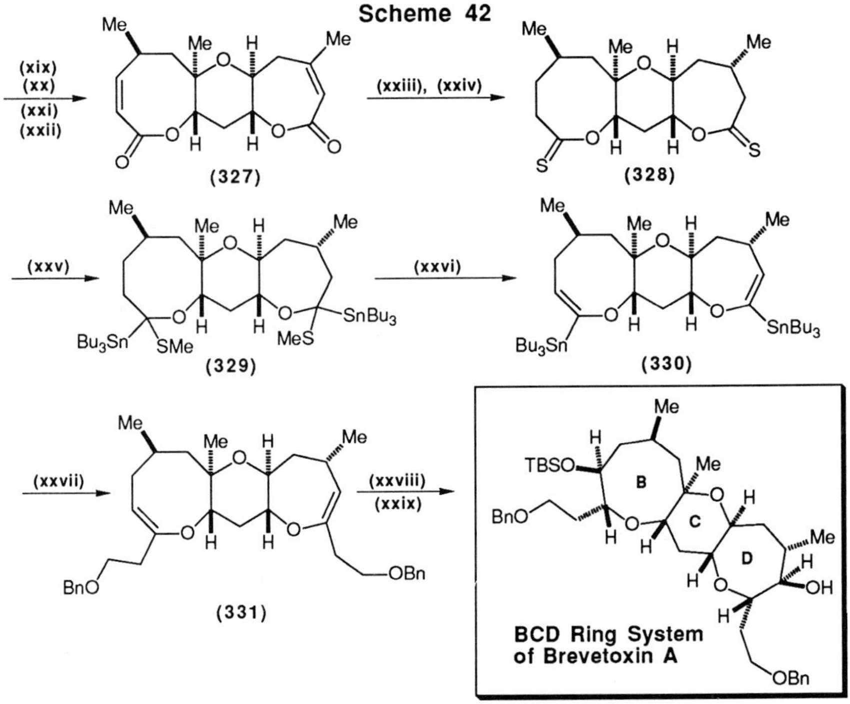

BCD Ring System of Brevetoxin A

Reagents: (i) NaH, THF, Cat. Imidazole, CS_2, then MeI; (**ii**) *n*-Bu_3SnH, AIBN, PhMe, Δ, (60%, 2 steps); (**iii**) H_2SO_4, MeOH, (55%); (**iv**) $NaIO_4$, $RuCl_3$, MeCN, H_2O, CCl_4 (3:2:2); (**v**) K_2CO_3, MeI, DMF, (63%, 2 steps); (**vi**) *n*-BuLi, THF, -78°C; AcOH quench at -78°C, (79%); (**vii**) Al/Hg, THF, H_2O, Δ, (86%); (**viii**) MeMgCl, DME, -78°C to r. temp., (95%, stereoselectivity 13:1); (**ix**) EtSH, CH_2Cl_2, $ZnCl_2$, (92%); (**x**) NaH, THF, BnBr, (86%); (**xi**) *N*-Chlorosuccinimide, 2,6-Lutidine, MeCN, H_2O, (80%); (**xii**) Ph_3P=CHCOMe, PhMe, 4 h, Δ; (**xiii**) NaH, THF, overnight, r. temp., (73%, 2 steps); (**xiv**) OsO_4, *N*-Methyl Morpholine *N*-Oxide, THF-H_2O (20:1); (**xv**) $NaIO_4$, THF-H_2O (10:1), (93%, 2 steps); (**xvi**) $ZnBr_2$ (1.0 equiv), Et_2O, -78°C, then add Silyl Ketene Acetal (3.0 equiv), 30 min., (81%); (**xvii**) H_2, $Pd(OH)_2$, THF, 25°C, 3 h, (100%); (**xviii**) PyS-SPy (2.5 equiv), Ph_3P (2.5 equiv), $AgClO_4$ (2.2 equiv), PhMe, Δ, 4 h, (76%); (**xix**) HF-Py (1 equiv), THF, 0°C, 3h, (85%); (**xx**) Martin Sulphurane, CH_2Cl_2, 0°C, 30 min., (87%); (**xxi**) HF-Py (1 equiv), THF, 0°C, 4 h, (92%); (**xxii**) Martin Sulphurane, CH_2Cl_2, 0°C, 30 min., (92%); (**xxiii**) H_2, Pd-C, THF, 4 h, (100%); (**xxiv**) Lawesson's Reagent (3.0 equiv), 1,1,3,3-Tetramethylthiourea, Xylenes, 115°C, 3 h, (63%); (**xxv**) *n*-Bu_3SnLi (3.0 equiv), THF, -78°C, 10 min., MeI (6.0 equiv), -78°C, 15 min., (86%); (**xxvi**) $(CuOTf)_2$ (4.0 equiv), C_6H_6, Pentamethylpiperidine (4.05 equiv), 25°C, (45%); (**xxvii**) *n*-BuLi (3.0 equiv), THF, -78°C, 5 min., then add $BnOCH_2CH_2OTf$ (5.0 equiv), HMPA, Et_3N, THF, warm to r. temp, 45 min, (65%); (**xxviii**) Thexylborane (4.0 equiv), THF, 0°C, 5 h, NaOH-H_2O_2 (20.0 equiv), 0°C, 2 h, (73%); (**xxix**) *t*-$BuPh_2SiCl$ (1.5 equiv), Imidazole (3.0 equiv), DMF, 24 h, r. temp., (82%).

critical to the success of this approach, since they provided a pathway for obtaining the bis(vinyl stannane) (**330**) via copper (I) triflate mediated elimination of the thiomethyl groups in (**329**). Transmetallation then generated a stable dianion which was alkylated with (benzyloxy)ethyl triflate to append the two-carbon side chains found in the BCD-ring system. Hydroboration of (**331**) with thexylborane proceeded stereoselectively from the less hindered β-face, to yield the target molecule after selective silylation.

Okadaic Acid

The total synthesis of okadaic acid (M. Isobe *et al.*, *Tetrahedron Lett.*, 1984, 25, 5049; *ibid*, 1985, 26, 5199, 5203; *ibid.*, 1986, 27, 963; *Tetrahedron*, 1987, 43, 4737, 4749, 4759, 4767) provides a perfect illustration of the utility of pyranoid glycals for complex polyether construction. Schemes 43, 44 and 45 show the retrosynthetic thinking that was used to guide the assembly of okadaic acid. A Julia olefination reaction was planned for unification of fragments (**332**) and (**333**). An acid-catalysed transketalisation of dihydroxy ketone (**334**) was envisaged for building the spiroketal system in (**332**). Michael addition of a methyl carbanion to (**335**) would secure trisubstituted enone (**334**), and coupling of the acetylenic anion from (**336**) with lactone (**337**) was contemplated for obtaining (**335**). Tri-*O*-acetyl-D-glucal (**102**) and 2-acetoxy glucal (**338**) were picked as starting materials for the preparation of (**336**) and (**337**) respectively. For the C(15)-C(38) segment (**333**), the C(27)-C(28) bond was the site chosen for disconnection. This led to aldehyde (**339**) and phenylsulphone (**340**) which were to be generated from (**102**) and (**338**) respectively.

The synthesis of lactone (**337**) is shown in Scheme 46. Compound (**338**) underwent Ferrier rearrangement with isopropanol and boron trifluoride etherate (R.J. Ferrier and N. Prasad, *J. Chem. Soc. (C)*, 1969, 570) to give (**348**) after stereoselective reduction of the enol acetate with lithium aluminium hydride. Hydrogenation of the olefin and protecting group manipulation enabled the secondary alcohol to be benzylated. This allowed oxidation of the primary alcohol to the aldehyde and subsequent Wittig reaction to give (**349**). Transketalisation with chloroethanol led to a mixture of anomeric chloroethyl glycosides, which on treatment with diisobutylaluminium hydride provided the allylic alcohol. This participated in a stereoselective oxymercuration reaction to afford acetonide (**350**) after reductive work up with sodium borohydride, and acetal exchange. A novel deprotection strategy was adopted for unravelling the hemiacetal. It involved heating the chloroethyl glycoside in *N,N*-dimethylformamide with potassium iodide and sodium benzenesulphinate. This produced the (phenylsulphonyl)ethyl glycoside which underwent a facile β-

Scheme 43

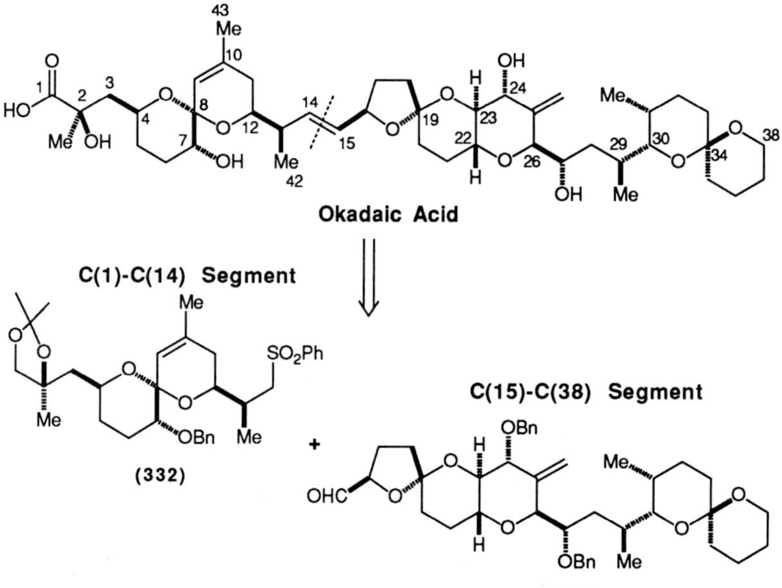

Okadaic Acid

C(1)-C(14) Segment

(332)

C(15)-C(38) Segment

(333)

Scheme 44

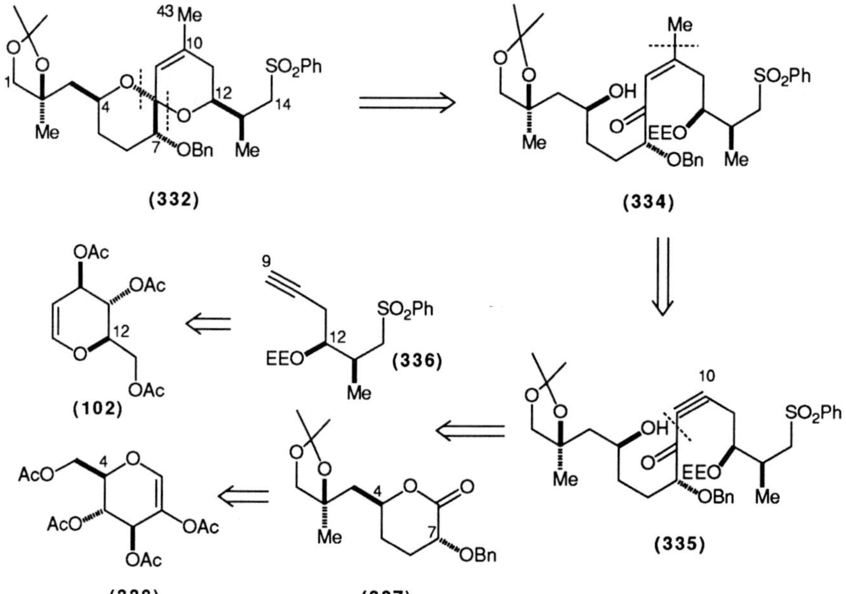

(332)

(334)

(336)

(102)

(335)

(337)

(338)

Scheme 45

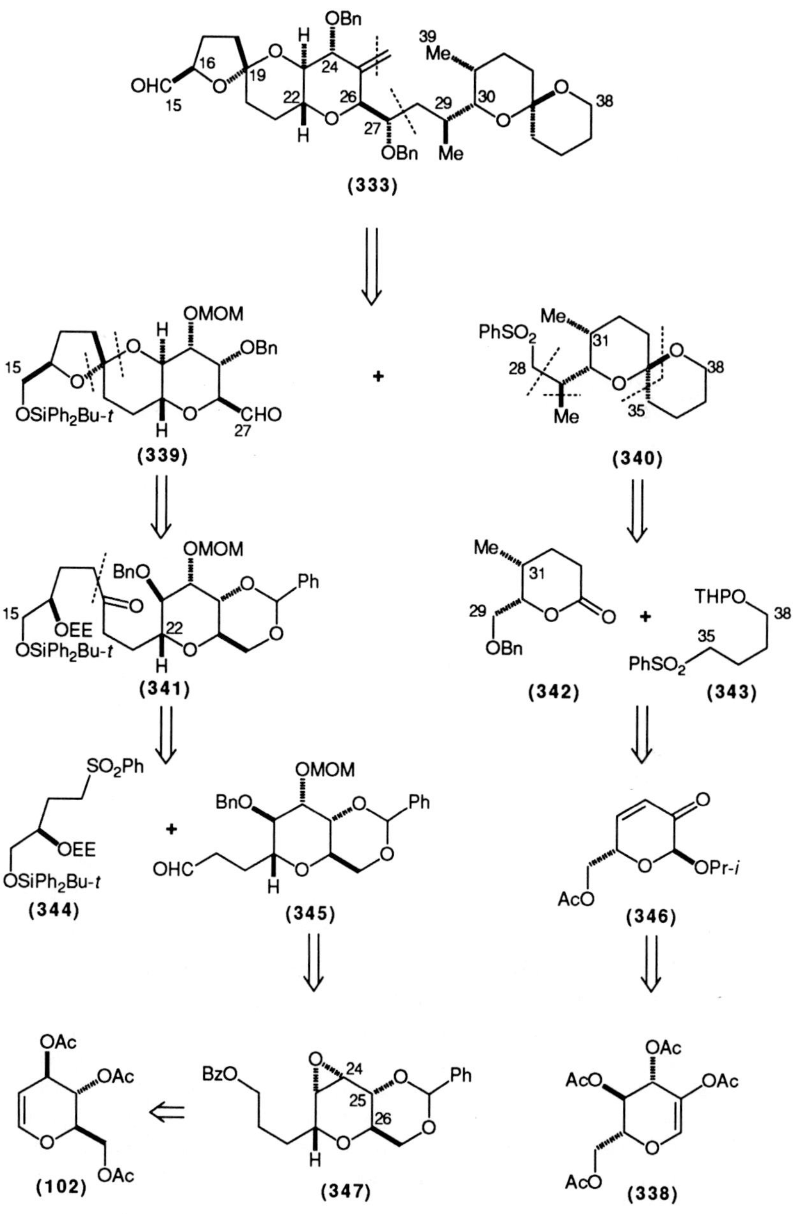

OBn
H
39
Me
OHC
15
16
19
24
22
26
27
29
30
38
H
OBn
Me
(333)
OMOM
H
OBn
15
OSiPh2Bu-t
H
CHO
27
(339)
+
Me
PhSO2
31
28
38
35
Me
(340)
OMOM
BnO
Ph
15
OEE
OSiPh2Bu-t
22
H
(341)
Me
31
29
OBn
(342)
+
THPO
38
35
PhSO2
(343)
SO2Ph
OEE
OSiPh2Bu-t
(344)
+
OMOM
BnO
Ph
OHC
H
(345)
O
OPr-i
AcO
(346)
OAc
OAc
OAc
(102)
BzO
24
25
26
Ph
H
(347)
OAc
AcO
OAc
AcO
(338)

Scheme 46

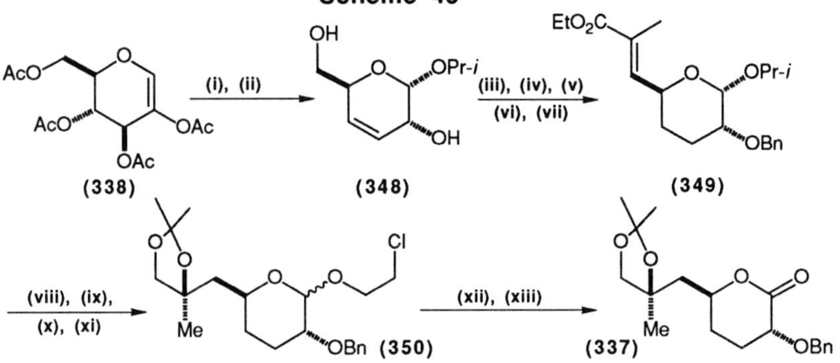

Reagents: (**i**) *i*-PrOH, BF_3-Et_2O, (63%); (**ii**) $LiAlH_4$, THF, 0°C, (100%); (**iii**) H_2, Pd-C, EtOAc, (96%); (**iv**) *t*-$BuMe_2SiCl$, Imidazole, DMF, 0°C; (**v**) NaH, THF, $PhCH_2Br$; (**vi**) *n*-Bu_4NF, THF, (80%, 3 steps); (**vii**) $(COCl)_2$, Me_2SO, CH_2Cl_2, -78°C, Et_3N; Ph_3P=C(Me)CO_2Et, -20°C to r. temp., (83%); (**viii**) $ClCH_2CH_2OH$, CSA, 50°C, (79%); (**ix**) DIBAL-H, CH_2Cl_2, -78°C, (100%); (**x**) $Hg(OAc)_2$, THF, H_2O, 0°C, 1.5 days; $NaBH_4$, r. temp., 1 h, (81%); (**xi**) $Me_2C(OMe)_2$, CH_2Cl_2, cat. PPTS, (91%); (**xii**) $NaSO_2Ph$, KI, DMF, 100°C, 10 h, (72%); (**xiii**) DMF, AcOH, AcONa, Br_2 (8.7 equiv), 30 min., 0°C, (54%).

Scheme 47

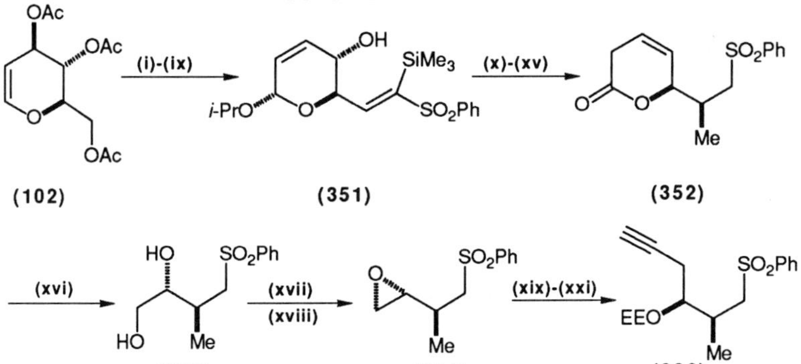

Reagents: (i) *i*-PrOH, BF_3-Et_2O, 0.5 h; (**ii**) Et_3N, MeOH; ; (**iii**) BzCl, Py, 0°C to r. temp.; (**iv**) EtOCH=CH_2, PPTS, CH_2Cl_2; (**v**) KOH, MeOH; 1 N HCl; (**vi**) Me_2SO, $(COCl)_2$, CH_2Cl_2, -78°C, Et_3N, (61%, 6 steps); (**vii**) $PhS(Me_3Si)_2CLi$, THF; (**viii**) *m*-CPBA, CH_2Cl_2; (**ix**) PPTS, 2-Propanol, CH_2Cl_2, Δ, 55 min.; (**x**) MeMgBr, Hexane, THF, -20°C, (80%, 91:9); (**xi**) KF, MeOH; (**xii**) Ac_2O, Py, (63%, 3 steps); (**xiii**) AcOH, H_2O, 40°C, 12 h; (**xiv**) PDC, CH_2Cl_2, (63%, 2 steps); (**xv**) Zn, $CuSO_4$-5 H_2O, AcONa, AcOH, H_2O, 0°C to r. temp., 5h, (64%); (**xvi**) O_3, CH_2Cl_2, MeOH, -78°C; $NaBH_4$, r. temp.; (**xvii**) *p*-TsCl, Et_3N, CH_2Cl_2, (41%, 2 steps); (**xviii**) *t*-BuOK, *t*-BuOH, (98%); (**xix**) Me_3SiCC-Li, BF_3-Et_2O, THF, -78°C, (95%); (**xx**) *n*-Bu_4NF, THF, (100%); (**xxi**) PPTS, EtOCH=CH_2,CH_2Cl_2, (100%).

Scheme 48

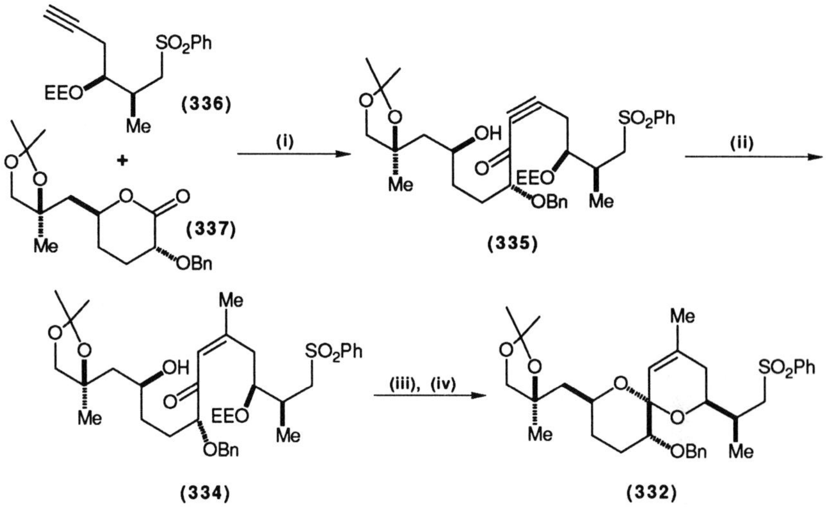

Reagents: (**i**) Alkyne, *n*-BuLi, THF, -78°C, 15 min., then lactone added, stirred 15 min., (80%); (**ii**) Me_2CuLi, Et_2O, -78°C, (87%); (**iii**) PPTS, MeOH, r. temp., 2.5 h; (**iv**) CH_2Cl_2, $Me_2C(OMe)_2$, PPTS, r. temp., overnight, (29% 2 steps).

Scheme 49

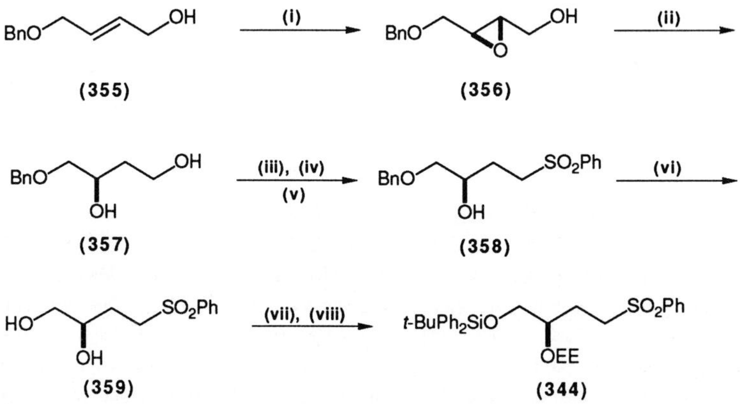

Reagents: (**i**) L-(+)-Diethyl Tartrate, *t*-BuO_2H, Ti(OPr-*i*)$_4$, CH_2Cl_2, -20°C, (85%); (**ii**) REDAL, THF, 22°C, 3 h, (98%); (**iii**) TsCl, Et_3N, 5°C, 48 h; (**iv**) PhSNa, THF, DMF, r. temp., 1h; (**v**) *m*-CPBA, CH_2Cl_2, 1 h, 0°C, (70%, 3 steps); (**vi**) AcOH, EtOH, Pd-C, H_2, r. temp, 5 h, (100%); (**vii**) *t*-$BuPh_2SiCl$, DMF, Imidazole, r. temp.; (**viii**) PPTS, CH_2Cl_2, EtOCH=CH_2, r. temp., ovenight, (90%, 2 steps).

Scheme 50

(102) →[(i), (ii), (iii)] (360) →[(iv), (v), (vi), (vii)] (347)

→[(viii), (ix), (x), (xi)] (361) →[(xii), (xiii)] (345) + (344)

→[(xiv), (xv)] (362) →[(xvi), (xvii)] (363)

→[(xviii), (xix)] (364) →[(xx), (xxi), (xxii)] (339)

Reagents: (i) H_2C=$CHCH_2SiMe_3$, BF_3-Et_2O, CH_2Cl_2, -50°C to r. temp. (α/β = 16:1); (**ii**) Et_3N, MeOH,(89%, 2 steps); (**iii**) $PhCH(OMe)_2$, CSA, CH_2Cl_2, (80%); (**iv**) B_2H_6, THF, -25°C, 12 h; H_2O_2, 2 N NaOH; (**v**) BzCl, Py; (**vi**) Dowex 50W (H^+) resin, MeOH, 9.5 h; (**vii**) *m*-CPBA, $CHCl_3$, 0°C-5°C, overnight, then in same pot $PhCH(OMe)_2$, CSA; (**viii**) NaOMe, THF, MeOH, 5°C, (65%, 6 steps); (**ix**) $(COCl)_2$, DMSO, CH_2Cl_2, -78°C, Et_3N; (**x**) PPTS $(MeO)_3CH$, CH_2Cl_2; (**xi**) NaH, DMF, $PhCH_2OH$, 70°C, (51%, 3 steps); (**xii**) KH, DMF, MOM-Cl, (100%); (**xiii**) 0.5 N HCl, THF, r. temp., 3 h, (98%); (**xiv**) *n*-BuLi, THF, -78°C, 30 min, then add aldehyde, stir at -78°C for 30 min., (86%); (**xv**) $(COCl)_2$, Me_2SO, CH_2Cl_2, -78°C, Et_3N; (**xvi**) Al/Hg, THF, H_2O, 60°C, (75%, 2 steps); (**xvii**) AcOH, EtOH, Δ, 72 h, (60%); (**xviii**) Pd-C, EtOH, AcOH, 15 h, r. temp; (**xix**) TrCl, Py, 65°C, (79%, 2 steps); (**xx**) NaH, THF, DMF, $PhCH_2Br$; (**xxi**) Et_2AlCl, CH_2Cl_2, -78°C, 20 min.,(83%, 2 steps); (**xxii**) $(COCl)_2$, Me_2SO, CH_2Cl_2, -78°C, Et_3N.

elimination reaction. Bromine-water oxidation of the lactol finally delivered the target lactone (**337**).

Scheme 47 depicts the route used for obtaining (**336**). A pivotal step was the stereospecific hydroxyl-directed addition of methyl magnesium bromide to (**351**). Hydrolysis of the isopropyl group then liberated the lactol which was oxidised to the α,β-unsaturated lactone and reductively deconjugated with zinc-copper couple. Ozonolysis of (**352**) and reductive work up furnished diol (**353**) which was selectively tosylated and treated with base to form the epoxide. Opening of (**354**) with lithium (trimethylsilyl)acetylide was followed by desilylation and ethoxyethylation and led to sulphone (**336**) in 2.1% overall yield.

Treatment of (**336**) with *n*-butyllithium converted it to the lithium acetylide which coupled readily with lactone (**337**) at low temperature to afford adduct (**335**) in 80% yield (Scheme 48). Lithium dimethylcuprate added to the alkynone to produce the *(Z)*-trisubstituted olefin (**334**) as a single isomer in 87% yield. Spiroketalisation was accomplished by exposing (**334**) to pyridinium *p*-toluenesulphonate in methanol. This caused partial deacetalation which was rectified by reaction with 2,2-dimethoxypropane to give (**332**).

In order to build the aldehyde component (**339**) needed for the preparation the C(15)-C(38) sub-unit, a route needed to be developed for obtaining sulphone (**344**) in homochiral form. The sequence that was devised is shown in Scheme 49. It included an asymmetric epoxidation on allylic alcohol (**355**) (T. Katsuki and K.B. Sharpless, *J. Am. Chem. Soc.*, 1980, 102, 5974; Y. Gao *et al.*, *ibid.*, 1987, 109, 5765) and a regioselective REDAL reduction of the resulting 2,3-epoxy alcohol (P. Ma *et al.*, *J. Org. Chem.*, 1982, 47, 1378).

The approach to (**339**) commenced with a highly stereoselective Lewis acid mediated allylation reaction on tri-*O*-acetyl-D-glucal with allyltrimethylsilane (A. Hosomi and H. Sakurai, *Tetrahedron Lett.*, 1976, 1295; S. Danishefsky and J. Kerwin, *J. Org. Chem.*, 1982, 3803, 5428) (Scheme 50). This afforded (**360**) after *O*-deacetylation and *O*-benzylidenation. Hydroboration and oxidative work-up of the terminal olefin in (**360**) gave a primary alcohol which was immediately benzoylated. The benzylidene acetal was temporarily removed so that the allylic hydroxyl at C(4) could be used to control the stereochemical outcome of the olefin epoxidation reaction, but the group was reinstated immediately afterwards. Treatment of (**347**) with base removed the benzoate ester to provide an alcohol that was oxidised and converted to its dimethyl acetal. The epoxide underwent exclusive *trans*-diaxial ring opening with sodium benzyloxide to set the stereochemistry at C(23) in the target molecule. Standard chemical operations on (**361**) supplied aldehyde (**345**) which reacted with the anion derived from (**344**) to give

Scheme 51

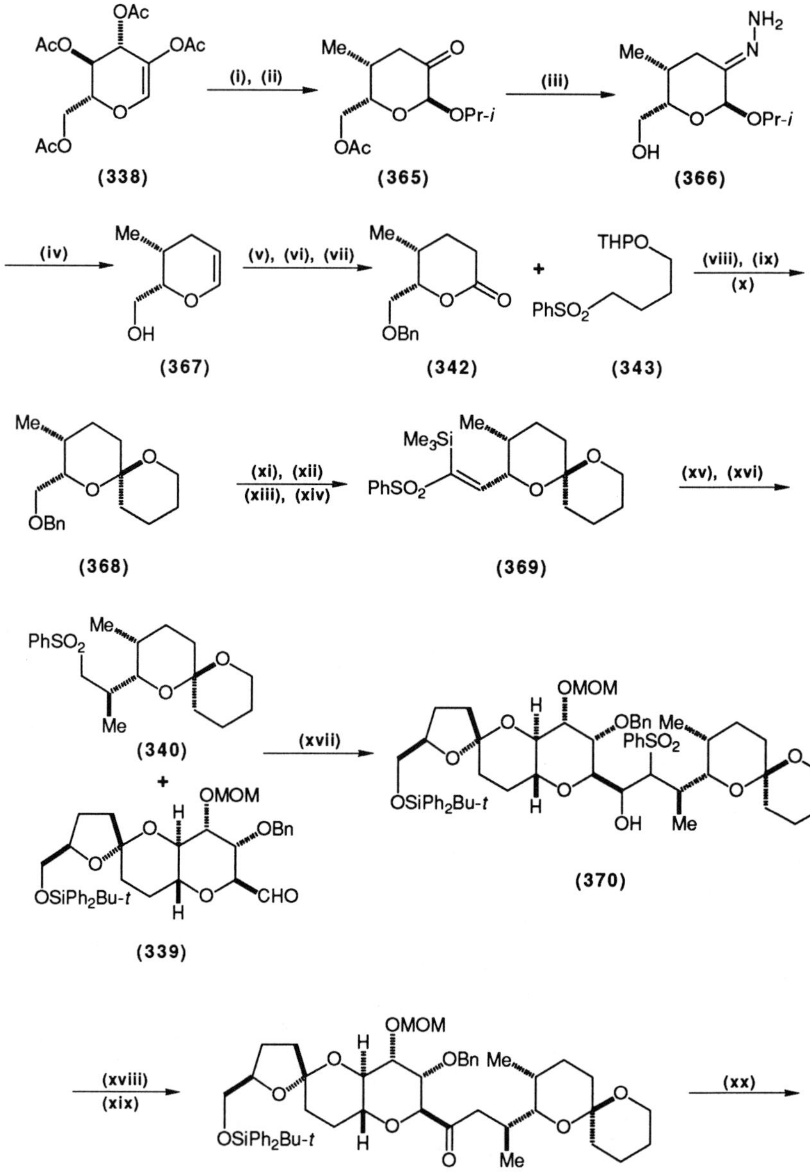

(338) —(i), (ii)→ (365) —(iii)→ (366)

—(iv)→ (367) —(v), (vi), (vii)→ (342) + (343) —(viii), (ix), (x)→

(368) —(xi), (xii), (xiii), (xiv)→ (369) —(xv), (xvi)→

(340) + (339) —(xvii)→ (370)

—(xviii), (xix)→ (371) —(xx)→

Scheme 51

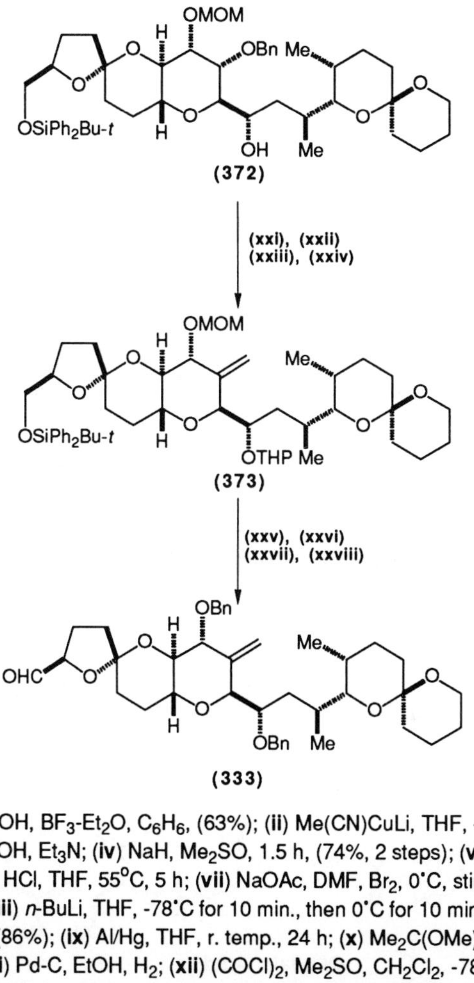

(372)

(xxi), (xxii)
(xxiii), (xxiv)

(373)

(xxv), (xxvi)
(xxvii), (xxviii)

(333)

Reagents: (**i**) *i*-PrOH, BF_3-Et_2O, C_6H_6, (63%); (**ii**) Me(CN)CuLi, THF, -20°C, 3 h, (79%); (**iii**) N_2H_4-H_2O, EtOH, Et_3N; (**iv**) NaH, Me_2SO, 1.5 h, (74%, 2 steps); (**v**) NaH,THF, DMF, $PhCH_2Br$; (**vi**) 1 N HCl, THF, 55°C, 5 h; (**vii**) NaOAc, DMF, Br_2, 0°C, stir at r.temp., 3 h, (52%, 3 steps); (**viii**) *n*-BuLi, THF, -78°C for 10 min., then 0°C for 10 min.; add lactone at -78°C stir for 1 h, (86%); (**ix**) Al/Hg, THF, r. temp., 24 h; (**x**) $Me_2C(OMe)_2$, PPTS, EtOH, (81%, 2 steps); (**xi**) Pd-C, EtOH, H_2; (**xii**) $(COCl)_2$, Me_2SO, CH_2Cl_2, -78°C, Et_3N; (**xiii**) $PhS(Me_3Si)_2CLi$, THF, - 40°C to 0°C, (56%, 3 steps); (**xiv**) *m*-CPBA, CH_2Cl_2,0°C to r. temp., (100%); (**xv**) MeLi, THF, -78°C to 0°C; (**xvi**) KF, MeOH, (89%, 2 steps); (**xvii**) *n*-BuLi, Et_2O, 0°C, 10 min.; cool to - 42°C add aldehyde, 10 min., (92%); (**xviii**) CrO_3-2Py, CH_2Cl_2, r. temp., 30 min.; (**xix**) Al/Hg, THF, H_2O, r. temp.; (**xx**) $NaBH_4$, EtOH, 0°C, 1 h, (57%, 3 steps); (**xxi**) PPTS, CH_2Cl_2, Dihydropyran; (**xxii**) $Pd(OH)_2$, EtOAc, H_2, 6 h, (69%, 2 steps); (**xxiii**) $(COCl)_2$, Me_2SO, CH_2Cl_2, -78°C, Et_3N; (**xxiv**) *n*-BuLi, THF, Ph_3PCH_3Br, Δ, 3 h, (79%, 2 steps); (**xxv**) Me_3SiBr, CH_2Cl_2, -78°C; (**xxvi**) NaH, THF, DMF, $PhCH_2Br$; (**xxvii**) n-Bu_4NF, THF, (50%, 3 steps); (**xxviii**) $(COCl)_2$, Me_2SO, -78°C, CH_2Cl_2, Et_3N, (81%).

Scheme 52

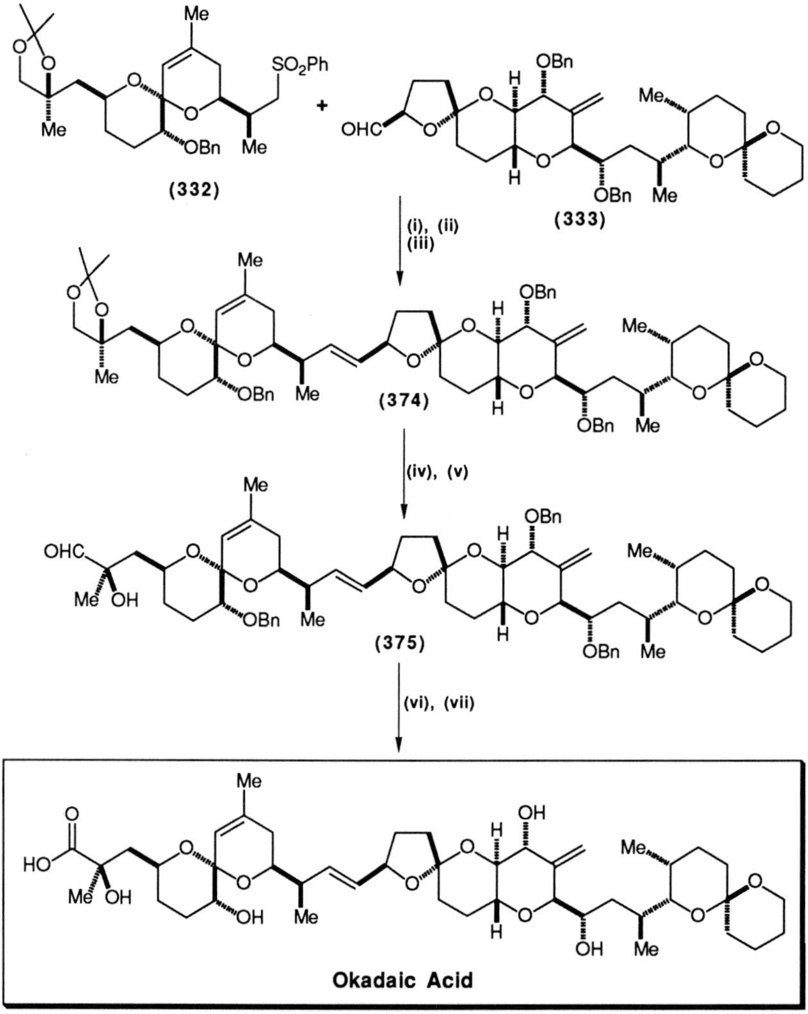

Reagents: (**i**) *sec*-BuLi, THF, -78°C, 10 min; add aldehyde stir 20 min.; (**ii**) Ac_2O, Py; (**iii**) 5% Na/Hg, MeOH, EtOAc, (32%, 4 steps); (**iv**) AcOH, THF, H_2O, 55°C, 36 h, (63%); (**v**) SO_3-Py, Me_2SO, Et_3N, 30 min.; (**vi**) $NaClO_2$,*t*-BuOH, H_2O, Na_2PO_4, 2-Methyl-2-Butene, r. temp., 2.5 h, (66%, 2 steps); (**vii**) trace Li, liq. NH_3, EtOH, -78°C, 30 min., (87%).

β-keto sulphone (**362**) after oxidation. Aluminium amalgam cleanly desulphonylated (**362**) and acid treatment gave triol (**363**). Hydrogenolysis under acidic conditions brought about spiroketalisation, affording (**364**) after tritylation. Benzylation, detritylation, and Swern oxidation completed this route to aldehyde (**339**).

The preparation of sulphone (**340**) is illustrated in Scheme 51. After Ferrier rearrangement of (**338**), a important finding was that treatment of the product enol acetate with Me(CN)CuLi led to a single product (**365**) in one step. Clearly, the reaction must have proceeded by conjugate addition to the intermediate enone. An eliminative Wolff-Kishner reaction was then used to obtain glycal (**367**), and this benzylated, hydrolysed, and oxidised to lactone (**342**). Reaction with the α-phenylsulphonyl anion derived from (**343**) afforded a β-keto sulphone mixture which was desulphonylated, and spiroketalised. Debenzylation of (**368**), Swern oxidation, Peterson olefination, and oxidation provided the spiro heteroolefin (**369**). This entered into a conjugate addition reaction with methyllithium, the stereochemistry of attack being governed by the reagent chelating with the pyranoid ring oxygen. Upon desilylation, sulphone (**340**) was isolated in high yield.

The aldehydic segment (**339**) combined with the sulphone carbanion obtained from (**340**) in 92% yield. The diasteromeric mixture of β-hydroxysulphones was then oxidised and desulphonylated to supply ketone (**371**). This was reduced with sodium borohydride to give a single alcohol (**372**). After protection as a tetrahydropyranyl ether, debenzylation and oxidation installed a keto group at C(25), which was subjected to Wittig methylenation. Removal of both acetal protecting groups was then accomplished selectively with trimethylsilyl bromide. The remaining steps used for completing (**333**) were benzylation, desilylation and Swern oxidation.

The best conditions for metallating phenylsulphone (**332**) employed *sec*-butyllithium as the base in a tetrahydrofuran-hexane solvent system at low temperature. The resulting carbanion added readily to aldehyde (**333**) (Scheme 52). After acetylation, and reductive elimination with sodium amalgam, the *(E)*-olefin (**374**) was obtained as a single isomer in 32% yield for the three steps. After removal of the acetonide at C(1)-C(2) with mild acid, a two-step oxidation procedure led to the C(1) carboxylic acid. The final hurdle was deprotection of the *O*-benzyl ethers to obtain okadaic acid, which was accomplished in 87% yield by dissolving metal reduction.

(+)-Streptolic acid

(+)-Streptolic acid is a degradation product obtained from cleavage of streptolydigin, the most potent member of the naturally-occurring 3-acyltetramic acid family of antibiotics (K.L. Rinehart *et al.*,

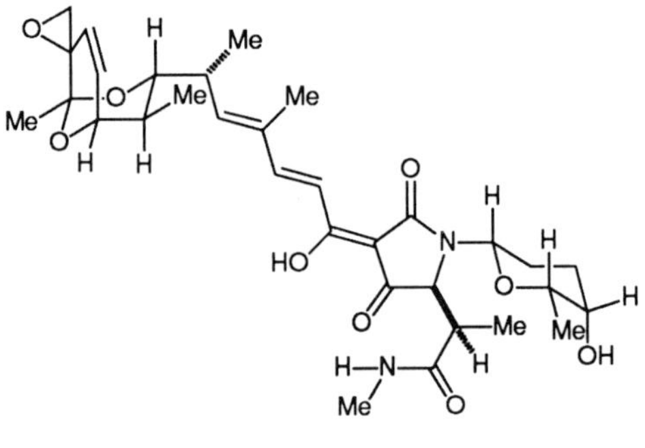

Streptolydigin

J. Am. Chem. Soc., 1973, 95, 4077; R.A. DiCioccio and B.I.S. Srivastava, *Biochem. Biophys. Res. Commun.*, 1976, 72, 1343; F. Reusser, *J. Bacteriol.*, 1969, 99, 151; *ibid*, 1969, 100, 1335). An efficient synthesis of (+)-streptolic acid has been described by R.E. Ireland and M.G. Smith (*J. Am. Chem. Soc.*, 1988, 110, 854) beginning from tri-*O*-acetyl-D-glucal (Scheme 53). One of the first objectives in their strategy was to prepare glycal (**376**). This was done by a procedure developed during the synthesis of tirandamycic acid by R.E. Ireland, P.G.M. Wuts, and B. Ernst (*J. Am. Chem. Soc.*, 1981, 103, 3205). Desilylation of (**376**), followed by iodination at C(6) and elimination, led to the stable bis(enol ether) (**377**). Application of the Ireland ester enolate Claisen rearrangement on (**377**) furnished an 86:14 mixture of methyl epimers (**378**) and (**379**), which could not be separated until much later in the synthesis. Ketalisation of this mixture in acidic methanol provided a readily separable 4:1 mixture of glycosides in which the *(R)*-anomer predominated, and further routine manipulations eventually afforded allylic ether (**380**). The latter was debenzylated in almost quantitative yield with lithium di-*tert*-butylbiphenyl radical anion (P.K. Freeman and L.L. Hutchinson, *J. Org. Chem.*, 1980, 45, 1924; R.E. Ireland *et al.*, *J. Am. Chem. Soc.*, 1985, 107, 3285). Oxidation provided α,β-enone (**381**), which underwent stereospecific axial attack with [(benzyloxy)methyl]lithium at -78°C (W.C. Still, *J. Am. Chem. Soc.*, 1978, 100, 1481). Cleavage of the silyl ether and regioselective Sharpless asymmetric epoxidation then gave epoxy alcohol (**382**), which opened regioselectively with lithium dimethylcuprate (Y. Kishi and H. Nagaoka, *Tetrahedron*, 1981, 37, 3873). Cyclisation of (**383**) proceeded readily in chloroform containing a trace amount of *p*-toluenesulphonic acid. It led to bicyclic ketals (**384**) and (**385**), which were separated by careful chromatography. After oxidation of alcohol (**384**) to the aldehyde, stabilised Wittig reaction with (carboxyethylidene)triphenylphosphorane delivered (**386**) in 60%

Scheme 53

OAc OAc OAc (102) —(i), (ii), (iii) (iv), (v), (vi)→ O OBn OSiMe$_2$Bu-*t* (376) —(vii), (viii) (ix), (x)→

(377) OBn —(xi), (xii) (xiii), (xiv)→ OBn H Me CO_2Me + OBn H Me CO_2Me —(xv), (xvi) (xvii), (xviii) (xix), (xx)→

(378) : (379)
86 : 14
Inseparable mixture at this stage

(380) OBn H Me OMe Me OSiMe$_2$Bu-*t* —(xxi), (xxii)→ (381) O H Me OMe Me OSiMe$_2$Bu-*t* —(xxiii) (xxiv), (xxv)→ (382) OH OBn H Me OMe Me O OH

—(xxvi)→ (383) OH OBn H Me OMe Me OH Me OH —(xxvii)→ (384) OH OBn H Me O Me O Me OH + (385) OH OBn H Me O Me O Me OH

(384) : (385)
86 : 14
Isomers separable at this stage

Scheme 53

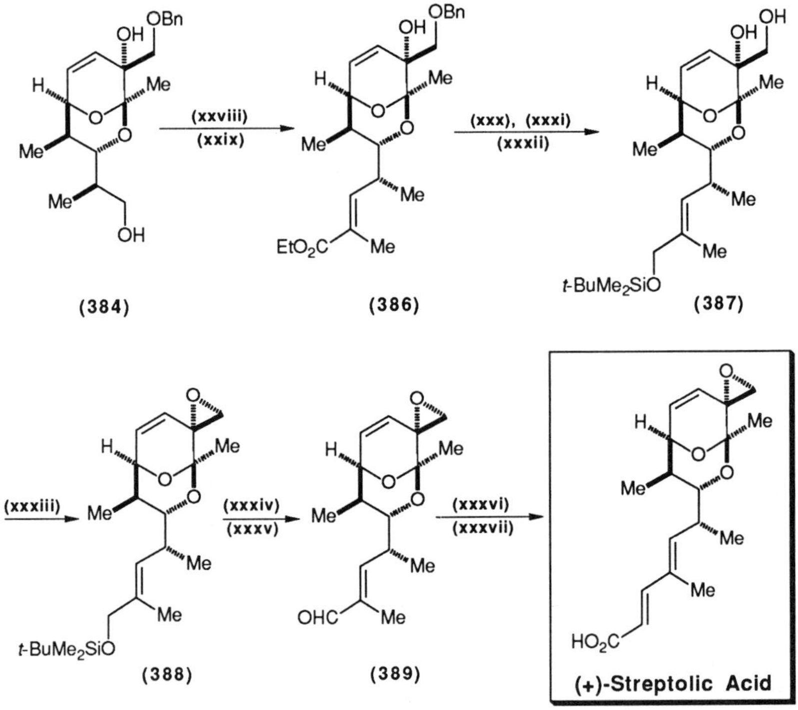

Reagents: (**i**) NaOMe, MeOH; (**ii**) *t*-$BuMe_2SiCl$, Py; (**iii**) BzCl, Py, -35°C to 0°C; (**iv**) KH, BnBr, THF, 0°C; (**v**) NaOH, EtOH; (**vi**) EtCOCl, Py, r. temp., (68%, 6 steps); (**vii**) *n*-Bu_4NF, THF, (98%); (**viii**) TsCl, Py, CH_2Cl_2, (89%); (**ix**) NaI, MeCOEt, Δ, 4 h, (87%); (**x**) DBU, C_6H_6, Δ, 1 h, (99%); (**xi**) $(Me_3Si)_2NLi$, THF, HMPA, *t*-$BuMe_2SiCl$, -78°C; (**xii**) C_6H_6, Δ; (**xiii**) KF. $2H_2O$, $KHCO_3$, HMPA; (**xiv**) MeI, HMPA, (85%, 4 steps); (**xv**) MeOH, *p*-TsOH, (99%); (**xvi**) DIBAL-H, Et_2O, -78°C; (**xvii**) $(COCl)_2$, Me_2SO, CH_2Cl_2, -78°C, *i*-Pr_2NEt; (**xviii**) $EtO_2CCH{=}PPh_3$, CH_2Cl_2, (87%, 3 steps); (**xix**) DIBAL-H, Et_2O, -78°C, (96%); (**xx**) *t*-$BuMe_2SiCl$, Imidazole, DMF, (97%); (**xxi**) Lithium Di-*t*-butylbiphenyl radical anion, THF, -78°C, (97%); (**xxii**) $(COCl)_2$, Me_2SO, *i*-Pr_2NEt, CH_2Cl_2, -78°C, (93%); (**xxiii**) THF, $BnOCH_2Li$, -78°C, (89%); (**xxiv**) *n*-Bu_4NF, THF, (93%); (**xxv**) (+)-DIPT, Ti(OPr-*i*)$_4$, *t*-BuO_2H, CH_2Cl_2, -20°C, (90%); (**xxvi**) Me_2CuLi, Et_2O, 0°C; (**xxvii**) cat. *p*-TsOH, $CHCl_3$, (68%, 2 steps); (**xxviii**) PCC, CH_2Cl_2; (**xxix**) $EtO_2CC(Me){=}PPh_3$, C_6H_6, 80°C, 12 h, (61%, 2 steps); (**xxx**) DIBAL-H, Et_2O, -78°C, (95%); (**xxxi**) *t*-$BuMe_2SiCl$, Imidazole, DMF, (93%); (**xxxii**) Lithium Di-*t*-butylbiphenyl radical anion, THF, -78°C, (94%); (**xxxiii**) NaH, THF, 0°C, (Tolylsulphonyl)imidazolide, (96%); (**xxxiv**) *n*-Bu_4NF, THF, (97%); (**xxxv**) $(COCl)_2$, Me_2SO, CH_2Cl_2, -78°C, Et_3N; (**xxxvi**) $EtO_2CCH{=}PPh_3$, CH_2Cl_2, (74%,2 steps); (**xxxvii**) 10% NaOH, MeOH, 10% HCl quench, (86%).

yield, along with 5% of a product arising from chromate ester rearrangement of the tertiary allylic alcohol (W.G. Dauben and D.M. Michno, *J. Org. Chem.*, 1977, 42, 682). Reduction of ester (**386**), silylation, and debenzylation provided vicinal diol (**387**), which was converted to the epoxide with tosylimidazole and sodium hydride (R.E. Ireland, L. Courtney, B.F. Fitzsimmons, *J. Org. Chem.*, 1983, 48, 5186). Desilylation of (**388**) and Swern oxidation afforded aldehyde (**389**) which was transformed into (+)-streptolic acid by Wittig olefination and saponification.

(+)-Anamarine

A very elegant synthesis of (+)-anamarine has been completed by K. Lorenz and F. Lichtenhaler (*Tetrahedron Lett.*, 1987, 28, 6437) from D-gulonolactone and *(R,R)*-diethyl tartrate. It exploited the seminal observations of J.A. Secrist, III and S.R. Wu (*J. Org. Chem.*, 1977, 42, 4084; *ibid.*, 1979, 44, 1434) on the Wittig olefination reactions of glycosyl 6-phosphoranes with protected aldoses. Initially, D-gulonolactone was transformed into (**391**) by *O*-silylation, diisobutylaluminium hydride reduction, and thioketalisation (Scheme 54). Selective pivaloylation of the primary hydroxyl in (**391**) was followed by acetalation of the remaining hydroxyl groups and provided (**392**). This was converted to aldehyde (**393**) by desulphurisation with Raney nickel, depivaloylation with base, and oxidation with pyridinium chlorochromate. Aldehyde (**393**) readily participated in a low temperature Wittig reaction with the phosphorane obtained from deprotonation of (**394**). The latter was prepared from *(R,R)*-diethyl tartrate. The Wittig process generated a 9:1 mixture of (**395**) and its C(5) epimer in 60% yield, and in both cases, the *(Z)*-olefin was formed exclusively. Transketalisation of this mixture unmasked the anomeric hydroxyl while keeping the isopropylidene groups intact, and oxidation subsequently afforded enelactone (**396**). After separation from the minor component, photochemical isomerisation to the *(E)*-olefin was accomplished with phenyldisulphide in benzene. Deacetonation and acetylation supplied (+)-anamarine.

F.W. Lichtenhaler. K. Lorenz, and W. Ma (*Tetrahedron Lett.*, 1987, 28, 47) have reported a variant of this strategy for obtaining the (-)-enantiomer from D-glucose.

(-)-Isoavenaciolide

(-)-Isoavenaciolide is an antifungal agent with an intriguing α-methylene bis(butyrolactone) ring system in which there are three asymmetric centres. An interesting approach to its construction has been devised by C.E. McDonald and R.W. Dugger (*Tetrahedron Lett.*, 1988, 2413) (Scheme 55). The key steps entailed introduction of the *n*-octyl side chain by S_N2' cuprate displacement on allylic carbonate (**400**), and

Scheme 54

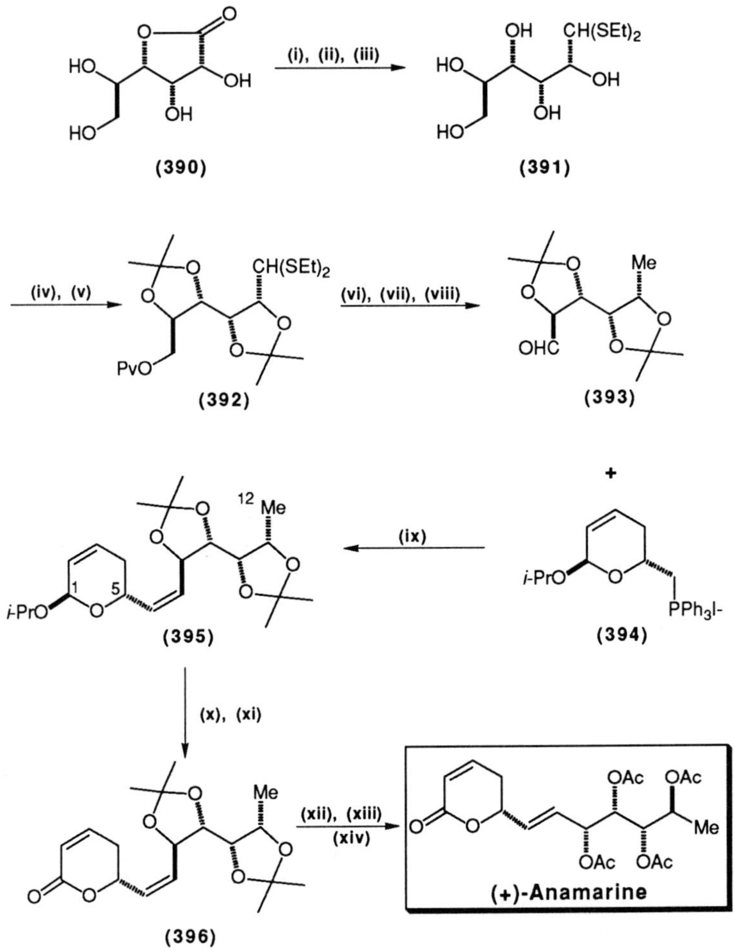

Reagents: (**i**) Me_3SiCl, Py, Imidazole; (**ii**) DIBAL-H, PhMe, -65°C; (**iii**) EtSH, HCl, 0°C, (86%, 3 steps); (**iv**) $(Me)_3CCOCl$, Py, -10°C to 40°C, (81%);(**v**) P_2O_5, Acetone, 6 h, 25°C, (84%); (**vi**) Raney Ni, EtOH, Δ, 1 h, (90%); (**vii**) KOBu-*t*, H_2O, Et_2O, 2 h, 0°C to 25°C, (92%); (**viii**) PCC, Al_2O_3, NaOAc, 5 h, 25°C, (75%); (**ix**) *n*-BuLi, THF, HMPA, 2:1, -78°C to 0°C, 1 h, (60%); (**x**) 0.1 M HCl, Me_2CO, (1:2); (**xi**) PCC, Al_2O_3, NaOAc, CH_2Cl_2, (64%, 2 steps); (**xii**) Ph_2S_2, *hv*, C_6H_6, 5 h, 25°C, (73%); (**xiii**) CF_3CO_2H, 25°C, 10 min; (**xiv**) Ac_2O, Py, (67%, 2 steps).

tri-*n*-butyltin hydride mediated free-radical cyclisation of α-bromo acetal (**401**) (G. Stork *et al.*, *J. Am. Chem. Soc.*, 1983, 105, 3741). The stereochemical outcome of the intramolecular radical addition process was controlled by the chirality at C(2), and the tendency for the tri-*n*-butyltin hydride to approach the intermediary secondary radical from the least hindered side of the bicycle. Deprotection of both acetals followed by oxidation provided (**403**), which had previously been converted to the natural product by W.L. Parker and F. Johnson, (*J. Org. Chem.*, 1973, 38, 2489). An alternative approach to (-)-isoavenaciolide from D-glucose has been disclosed by R.C. Anderson and B. Fraser-Reid, (*J. Org. Chem.*, 1985, 50, 4781).

C(1)-C(9) Segment of (+)-asteltoxin

Asteltoxin is a potent inhibitor of *E. coli* BF_1-ATPase (M. Satre, Biochem. *Biophys. Res. Commun.*, 1981, 100, 267). It is obtained from toxic maize cultures of *Aspergillus stellatus* (G.J. Kruger *et al.*, *J. Chem. Soc., Chem. Commun.*, 1979, 441). Its absolute stereochemistry was recently assigned by S.L. Shreiber and K. Satake (*Tetrahedron Lett.*, 1986, 27, 2575) after synthesis of a degradation product corresponding to C(1)-C(9) segment from R-isopropylidene glyceraldehyde. The cornerstone of their synthetic strategy was the Paterno-Buchi [2+2] photocycloaddition between (**405**) and 3,4-dimethylfuran (Scheme 56). This gave rise to a 1.2:1.0 mixture of two separable diastereomeric adducts in which (**406**) predominated. Treatment of (**406**) with *m*-chloroperbenzoic acid in water resulted in hemiacetal formation via opening of the glycal epoxide. After conversion to the γ-hydroxy hydrazone, sequential exposure to ethyl magnesium bromide and acidic methanol provided the target bistetrahydrofuran in 40% overall yield for the three steps. Unfortunately, the synthetic material only had an enantiomeric excess of 54%.

(-)-Bissetone

A very good demonstration of the utility of monosaccharide pyrones as enantiomerically pure building blocks for tetrahydropyranoid natural product synthesis is provided by the route to (-)-*(S,S)*-bissetone by F.W. Lichtenhaler *et al.* (*Angew. Chem. Int. Ed. Engl.*, 1987, 26, 1271). 2-Hydroxy glucal perbenzoate (**408**) was prepared according to the procedure of R.J. Ferrier and G.H. Sankey (*J. Chem. Soc. C*, 1966, 2339). The enol ester was then selectively hydrolysed with hydroxylamine, the oxime converted to ketone (**409**), and β-elimination induced by mild base. Enol benzoate (**410**) reacted readily with the lithium enolate of acetone to give (-)-bissetone after Zemplen deacetylation .

Scheme 55

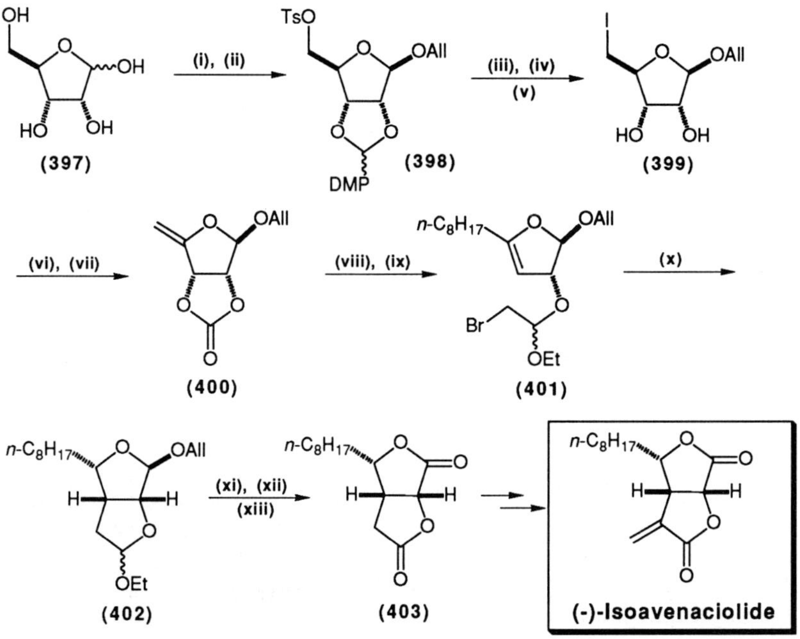

Reagents: (**i**) $H_2C{=}CHCH_2OH$, 2,4-Dimethoxybenzaldehyde, *p*-TsOH, C_6H_6, Δ, (81%); (**ii**) TsCl, Py; (**iii**) DDQ, H_2O, MeCN; (**iv**) aq. KOH, MeOH, (69%, 3 steps); (**v**) NaI, DMF, 80°C; (**vi**) DBU, DMF, 80°C; (**vii**) $(Imid)_2C{=}O$, DMF, 80°C, (65%, 3 steps); (**viii**) Et_2O, $(n\text{-}C_7H_{15})CuCN$, -23°C to 0°C, (85%); (**ix**) $EtOCH(Br)CH_2Br$, Et_3N, CH_2Cl_2, -78°C; (**x**) Bu_3SnH, AIBN, C_6H_6, D, (76%, 2 steps); (**xi**) $[COD]Ir(Ph_2MeP)_2PF_6$; (**xii**) aq. H_2SO_4; (**xiii**) $CrO_3.2$ Py, (78%, 3 steps).

Scheme 56

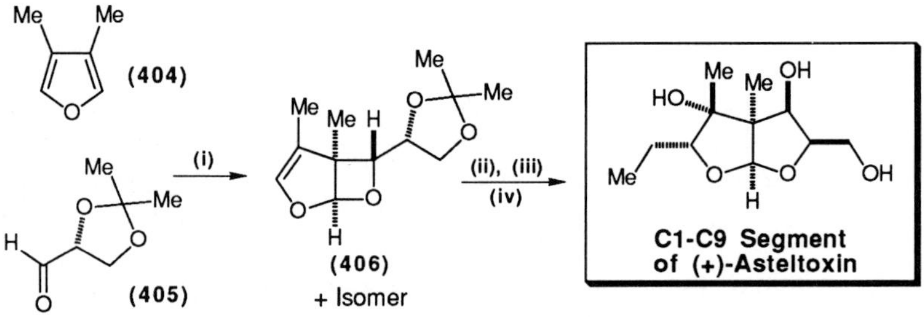

Reagents: (**i**) *hv*, 450 W Hanovia lamp, Vicor Filter, Et_2O, (35%); (**ii**) *m*-CPBA, H_2O, THF; then add Me_2NHNH_2; (**iii**) EtMgBr; (**iv**) Amberlyst 15 (H^+) resin, MeOH, (40%, 3 steps).

Scheme 57

(407) (408) (409) (410) (-)-Bissetone

Reagents: (i) NaI, Acetone, then Et_2NH, (70%); (ii) NH_2OH-HCl, Py, 70°C, 14 h; (iii) aq. $NaHCO_3$, (84%, 3 steps); (iv) MeC(OLi)=CH_2, THF, -78°C, 10 min., then H_2O, -20°C, 2h, (60%, +5% C(5)-epimer); (v) NaOMe, MeOH, (92%).

3. *Use of monosaccharides in the total synthesis of naturally-occurring macrolides*

FK-506

In the mid-1980's, H. Tanaka *et al.* (*J. Am. Chem. Soc.*, 1987, 109, 5031) isolated an unusual macrolide from culture filtrates of *Streptomyces tsukubaensis* which was termed FK-506. Its structure was assigned through a combination of spectrocopic techniques and X-ray crystallography. FK-506 has elicited considerable medical and synthetic attention mainly because of its powerful immunosuppressant properties (T. Ochiai *et al.*, *Transplantation*, 1987, 44, 729, 734; V. Warty *et al.*, *ibid.*, 1988, 46, 453; A.W. Thompson, *Immunol. Today*, 1988, 10, 6). S. Danishefsky *et al.* (*J. Org. Chem.*, 1990, 55, 2771, 2776, 2786) have recently completed a very elegant formal total synthesis of FK-506 that utilised methyl β-D-galactopyranoside and tri-*O*-acetyl-D-galactal as two key starting materials. Their route intersected with an advanced intermediate prepared by T.K. Jones *et al.* (*J. Am. Chem. Soc.*, 1990, 112, 2998) in their total synthesis of FK-506. The precise retrosynthetic planning is shown in Schemes 58, 59 and 60. It called for a macrolactamisation reaction to close the 21-membered ring, and a Julia olefination between (**413**) and (**414**) to create the *(E)*-olefin present at C(19)-C(20). Intermediate (**413**), which corresponded to the C(20)-C(34) segment, was itself disconnected across the C(27)-C(28) olefin

Scheme 58

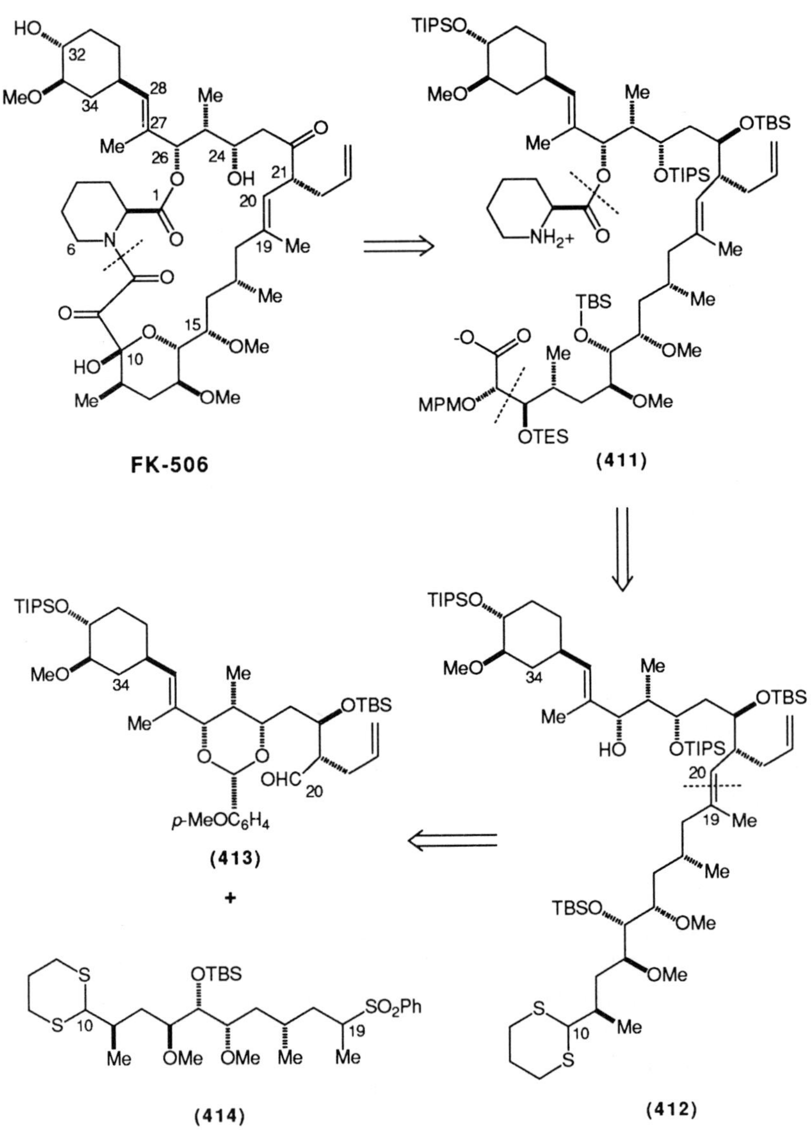

FK-506

(411)

(412)

(413)

+

(414)

Scheme 59

TIPSO 32 28 MeO 34 Me 23 OTBS Me 27 22 21 20 OHC p-MeOC$_6$H$_4$

(413)

TBSO MeO 34 28 SO$_2$Ph + Me OHC 27 23 22 p-MeOC$_6$H$_4$ OAc OAc O OAc CO$_2$H

(415) **(416)** **(417)** **(418)**

Scheme 60

S S 10 11 13 OTBS 14 15 17 SO$_2$Ph 19 Me OMe OMe Me Me

(414)

OH MeO$_2$C 10 12 16 11 17 Me OMe OMe Me OBz

(419)

O O OBn CO$_2$Me OMe OMe Me

(420)

OH HO O OMe OH OH

OH MeO O OBn OH OMe

O O OBn 4 3 2 CHO OMe OMe

Sugar Numbering

(423) **(422)** **(421)**

into sulphone (**416**) and aldehyde (**417**). The former was considered accessible from (R)-cyclohexenecarboxylic acid and the latter from tri-*O*-acetyl-D-galactal (**418**). For the preparation of dithiane (**414**), a hydroxyl-directed hydrogenation on (**419**) was envisaged for setting the C(11) and C(17) stereocentres. Since the relative and absolute stereochemistry at C(13), C(14), and C(15) in (**419**) corresponded to that at C(4), C(3) and C(2) (sugar numbering) in the protected D-galactose derivative (**421**), methyl β-D-galactopyranoside (**423**) was selected as its precursor.

Scheme 61 illustrates the route that was used to obtain the C(10)-C(19) segment. Methyl β-D-galactopyranoside was selectively silylated at the primary position, and then benzylated at O(3) via the *O*-stannylene. Methylation of the remaining hydroxyl groups furnished (**424**). Heating with acid cleaved both the silyl ether and the glycoside to afford lactol (**422**). Reduction of (**422**) produced a triol, which was acetonated and oxidised to aldehyde (**421**). Wittig olefination produced (**420**) with complete stereocontrol. The ester was then reduced to the allylic alcohol, this converted to the chloride, and displacement effected with sodium cyanide. This delivered (**425**) in 69% yield along with the α,β-unsaturated nitrile in 22% yield. Multiple reduction of the cyano group in (**425**) afforded alcohol (**426**), which was transformed into the corresponding tetraol by debenzylation and acid hydrolysis of the acetonide. Periodate cleavage of the terminal diol led to the aldehyde, which was olefinated and benzoylated to give enoate (**419**). A two-directional hydroxyl directed hydrogenation was then carried out at 1000 p.s.i. in the presence of a cationic rhodium-phosphine catalyst (D. A. Evans, M.M. Morrissey, R.C. Dow, *Tetrahedron Lett.*, 1985, 26, 6005; D.A. Evans, M.M. Morrissey, *J. Am. Chem. Soc.*, 1984, 106, 3866; J.M. Brown, R.G. Naik, *J. Chem. Soc., Chem. Comm.*, 1982, 348). This gave rise to a 19:2.2:1 mixture of three diastereoisomers in which the desired product (**427**) was present in excess. Treatment of (**427**) with mild acid was sufficient to bring about lactonisation. Selective reduction of (**428**) to the lactol, was followed by ring-opening to the hydroxythioketal, silylation and debenzoylation. The resulting primary alcohol was then converted to the phenylsulphone, and this alkylated with methyl iodide to give (**414**).

The preparation of the C(20)-C(34) sector is detailed in Scheme 62. Initially, it involved Ferrier rearrangement of (**418**) with stannic chloride and methanol, *O*-deacetylation, selective silylation of the primary alcohol, and hydroxyl-directed epoxidation. This supplied epoxy alcohol (**429**), which underwent *trans*-diaxial opening with Li_2Me_2CuCN (B.H. Lipsutz, J. Kozlowski, and R.S. Wilhelm, *J. Am. Chem. Soc.*, 1982, 104, 2305). Compound (**430**) was protected at C(2) and C(4) by a *p*-methoxybenzylidene acetal. Desilylation and

Scheme 61

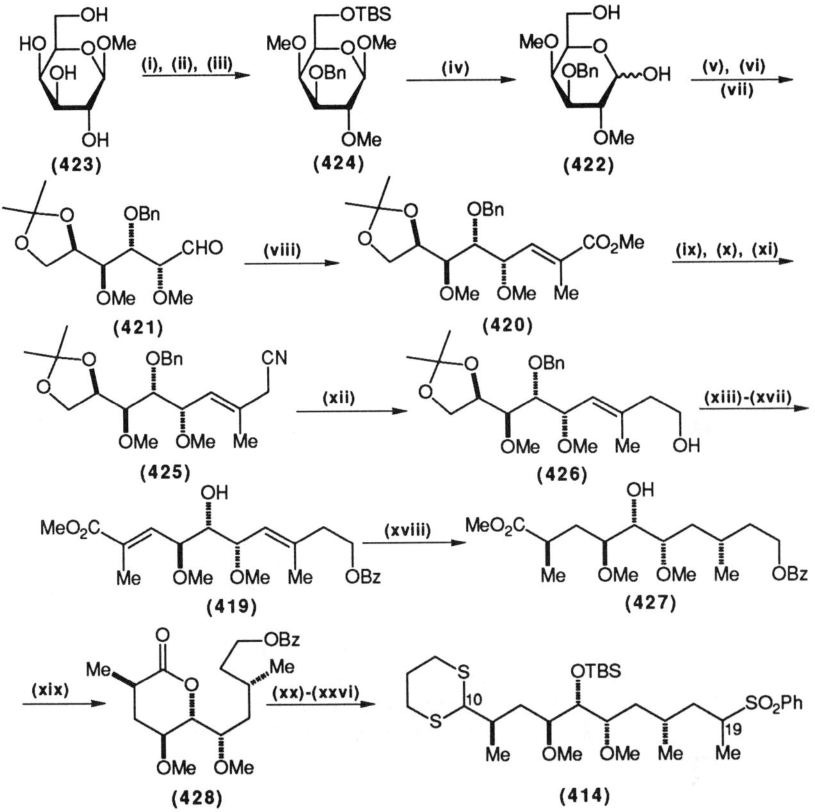

Reagents: (i) *t*-BuMe$_2$SiCl, Et$_3$N, DMAP, CH$_2$Cl$_2$, (72%); (**ii**) (*n*-Bu$_3$Sn)$_2$O, PhMe, Δ, 6h; PhCH$_2$Br, *n*-Bu$_4$NBr, 80°C, (98%); (**iii**) NaH, THF, MeI, (95%); (**iv**) 3M HCl, THF, Δ, 21 h, (73%); (**v**) NaBH$_4$, EtOH, (90%); (**vi**) *p*-TsOH, Acetone, r. temp, 4h, (89%); (**vii**) Me$_2$SO, (COCl)$_2$, CH$_2$Cl$_2$, -78°C, Et$_3$N; (**viii**) Ph$_3$P=C(Me)CO$_2$Me, CH$_2$Cl$_2$, 12 h, (90%, 2 steps); (**ix**) Li(Et)$_3$BH, THF, -78°C, then -20°C for 1.5 h, then r. temp., (100%); (**x**) LiCl, DMF, *s*-Collidine, 0°C, CF$_3$SO$_2$Cl, (67%); (**xi**) NaCN, DMF, (69%); (**xii**) DIBAL-H, CH$_2$Cl$_2$, -78°C; NaBH$_4$, EtOH, (63%, 2 steps); (**xiii**) Na, liq. NH$_3$, -78°C, THF, (90%); (**xiv**) AcOH, H$_2$O, THF, Δ, 2h; (**xv**) NaIO$_4$, THF:H$_2$O, (4:1), 2h, r. temp.; (**xvi**) CH$_2$Cl$_2$, Ph$_3$P=C(Me)CO$_2$Me, 12 h; (**xvii**) BzCl, Py,THF, 11 h, (55%, 4 steps); (**xviii**) CH$_2$Cl$_2$, [Ph$_2$P(CH$_2$)$_4$PPh$_2$]Rh(NBD), 1000 p.s.i., H$_2$, 5 h, (89%); (**xix**) *p*-TsOH, CH$_2$Cl$_2$, 4 A MS 3 h, r. temp.; (**xx**) Li(Et)$_3$BH, THF, -78°C, (87%, 2 steps); (**xxi**) 1,3-Propanedithiol, BF$_3$-Et$_2$O, -78°C to 0°C, (85%); (**xxii**) *t*-BuMe$_2$SiOTf, 2,6-Lutidine, CH$_2$Cl$_2$, (97%); (**xxiii**) K$_2$CO$_3$, MeOH, (76%); (**xxiv**) Ph$_3$P, I$_2$, Py, C$_6$H$_6$, Δ, (90%); (**xxv**) NaSO$_2$Ph, DMF, 20 h, r. temp., (81%); (**xxvi**) *n*-BuLi, THF,-78°C, 10-15 min, add MeI at -78°C, warm to r. temp, 45 min., (93%).

Scheme 62

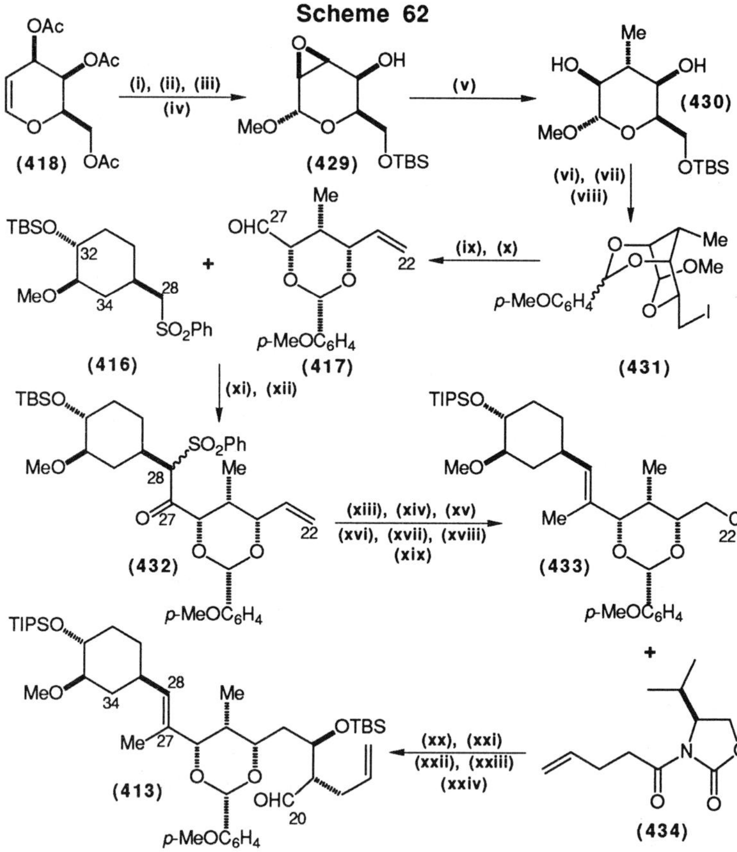

Reagents: (**i**) $SnCl_4$ (0.3 equiv), MeOH, $ClCH_2CH_2Cl$, 30 min; (**ii**) NaOMe, MeOH, 45 min.; (**iii**) *t*-$BuMe_2SiCl$, Et_3N, CH_2Cl_2, (72%, 3 steps); (**iv**) *m*-CPBA, CH_2Cl_2, 4,4'-thio-bis(2-*t*-butyl-6-methylphenol), Δ, 38 h, (73%, *Syn:Anti* = 97:3); (**v**) Li_2Me_2CuCN, Et_2O, -78°C to 0°C, then stir 40 min.; (**vi**) *p*-(MeO)$C_6H_4CH(OMe)_2$, CSA, C_6H_6, Δ, 1h; (**vii**) *n*-Bu_4NF, THF, (56%, 3 steps); (**viii**) Ph_3P, I_2, Py, C_6H_6, 45°C, 45 h (87%); (**ix**) Zn, Py, 95% EtOH, Δ; (**x**) PPTS, CH_2Cl_2, (93%, 2 steps); (**xi**) *n*-BuLi, THF, -78°C to r. temp.; (**xii**) Dess-Martin Periodinane, CH_2Cl_2, 2h; (**xiii**) Lithium Napthalenide, THF, -78°C, ultrasound, (58%, 3 steps); (**xiv**) MeMgBr, THF, 0°C; (**xv**) $MeO_2CN^-SO_2^+NEt_3$, C_6H_6, (76%, 2 steps); (**xvi**) *n*-Bu_4NF, THF; (**xvii**) *i*-Pr_3SiOTf, 2,6-Lutidine, CH_2Cl_2, DMAP, (95%, 2 steps); (**xviii**) 9-BBN, THF, 0°C; 1 N NaOH, 30% H_2O_2, (73%); (**xix**) CH_2Cl_2, Py, Dess-Martin Periodinane, (86%); (**xx**) *n*-Bu_2BOTf, CH_2Cl_2, *i*-Pr_2NEt, -78°C to r. temp; 30% H_2O_2; (**xxi**) *t*-$BuMe_2SiOTf$, 2,6-Lutidine, CH_2Cl_2, (90%, 2 steps); (**xxii**) BnOLi, THF, 0°C, (87%); (**xxiii**) DIBAL-H, PhMe, -78°C; (**xxiv**) $(COCl)_2$, Me_2SO, CH_2Cl_2, -78°C, Et_3N, (88%, 2 steps).

iodination then afforded (**431**) which was reductively eliminated and subjected to acetal equilibration with mild acid to give aldehyde (**417**). After metallation of sulphone (**416**) with *n*-butyllithium, the resulting anion was added onto aldehyde (**417**) to provide β-keto sulphone (**432**), after Dess-Martin oxidation. This was desulphonylated with lithium napthalenide, and the product ketone treated with methyl magnesium bromide to give a mixture of tertiary methyl carbinols. These were dehydrated with the Burgess reagent to afford a 6:1.5:1 mixture of three olefins enriched in the desired C(27)-C(28) *E*-isomer (Burgess, E.M. and H.R. Penton, *Org. Synth.*, 1977, 56, 40; E.M. Burgess, H.R. Penton, and E.A. Taylor, *J. Org. Chem.*, 1973, 38, 26). After protecting group interchange in the cyclohexyl ring, hydroboration and oxidation furnished the pure *(E)*-aldehyde (**433**). The latter participated in a *syn*-selective aldol reaction with chiral oxazolidinone (**434**), mediated by di-*n*-butylboron triflate (D.A. Evans, J. Bartroli, and T.L. Shih, *J. Am. Chem. Soc.*, 1981, 103, 2127). This afforded aldehyde (**413**) after removal of the auxiliary and oxidation.

The key coupling between the anion derived from (**414**) and (**413**) (Scheme 63) proceeded in 88% yield. The resulting mixture of β-hydroxy sulphones were trifluoroacetylated to give (**435**), and reductive elimination performed with lithium napthalenide to generate a 2.5:1 mixture of *E:Z* olefins (**436**). Difficulties were encountered in the selective removal of the acetal unit from (**436**) using pyridinium *p*-toluenesulphonate in propanol and acetonitrile. Apparently, the silyl ethers were also being cleaved. The desired C(24),C(26) diol was isolated in 33% yield, along with 28% yield of recovered starting material. The diol was selectively monosilylated at C(24) with excess triisopropylsilyl triflate to give (**412**). This operation allowed the protected pipecolic acid moiety to be introduced at C(26). After the 1,3-dithiane had been transketalised with methanol, treatment with pyridinium *p*-toluenesulphonate in dichloromethane furnished the Merck aldehyde (**437**).

T.K. Jones *et al.* (*loc cit.*) have converted (**437**) to FK-506 by the series of reactions shown in Scheme 64. The first of these was an Evans aldol addition to append the necessary two-carbon unit at C(8) and C(9). After hydrolysis of the chiral auxilliary with lithium hydroperoxide, the *t*-butylcarbamate was exposed to triethylsilyl triflate and then silica gel. This provided amino acid (**411**) which underwent macrolactamisation via the 2-acyloxypyridinium salt (T. Mukaiyama, M.Usui, and K. Saigo, *Chem. Lett.*, 1976, 49). Deprotection of the *p*-methoxybenzyl group with DDQ, and hydrolysis of the triethylsilyl ether with aqueous trifluoroacetic acid supplied the C(8),C(9)-diol which was oxidised to the diketoamide. This was reacted with aqueous hydrofluoric acid in acetonitrile to remove all the silyl protecting groups to give 22-

Scheme 63

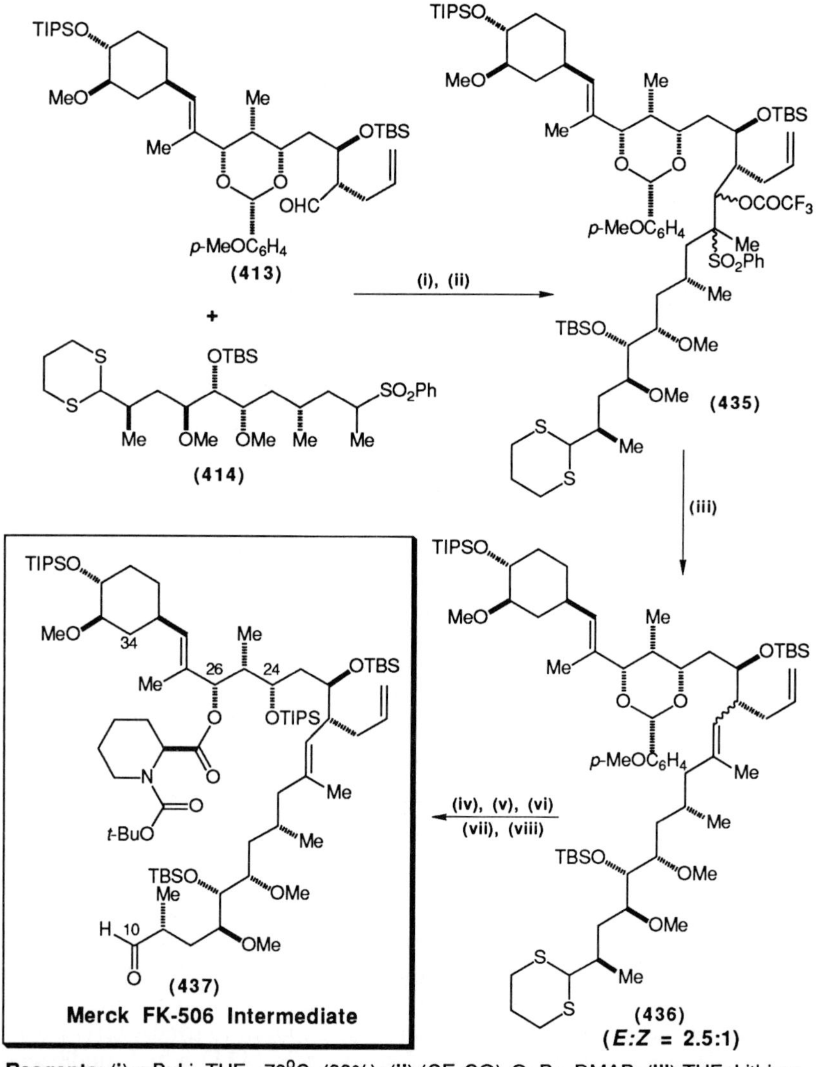

Reagents: (i) n-BuLi, THF, -78°C, (88%); (ii) $(CF_3CO)_2O$, Py, DMAP; (iii) THF, Lithium Napthalenide, -20°C, ultrasound, (68%, 2 steps); (iv) PPTS, 2-Propanol, MeCN, 70°C, 26 h, (33%, + 28% recovered starting material); (v) i-Pr_3SiOTf (2.5 equiv), 2,6-Lutidine, CH_2Cl_2, r. temp., 20 min., (75%); (vi) t-BOC-Pipecolic Acid, CH_2Cl_2, DCC, DMAP, -20°C, 12 h, (81%); (vii) N-Chlorosuccinimide, $AgNO_3$, 2,6-Lutidine, THF, MeOH, r. temp.; (viii) PPTS, CH_2Cl_2, 2 h, (75%).

Scheme 64

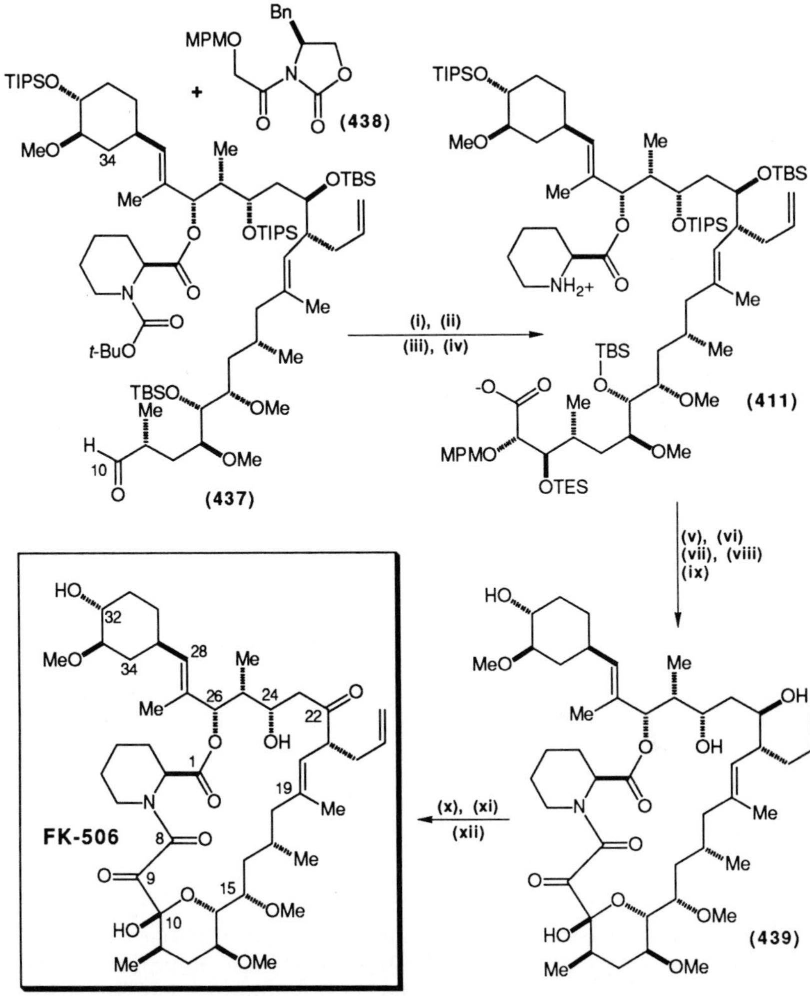

Reagents: (**i**) n-Bu_2BOTf, Et_3N, -50°C to -30°C, (88%); (**ii**) LiOH, 30% H_2O_2, THF, H_2O, 0°C; (**iii**) Et_3SiOTf (4.5 equiv), 2,6-Lutidine (6 equiv), CH_2Cl_2, 0°C; (**iv**) SiO_2, CH_2Cl_2, 80 min., (80%, 3 steps); (**v**) 2-Chloro-1-methylpyridinium iodide, Et_3N, CH_2Cl_2, 20°C, (81%); (**vi**) DDQ, H_2O, CH_2Cl_2; (**vii**) CF_3CO_2H, H_2O, THF, r. temp., (74%, 2 steps); (**viii**) $(COCl)_2$, Me_2SO, CH_2Cl_2, -60°C, Et_3N; after work-up Swern Oxidation repeated, (80%); (**ix**) HF, MeCN, 20°C, (81%); (**x**) Et_3SiCl, Py, 0°C for 2h, then -30°C for 12 h (70%); (**xi**) Dess-Martin Periodinane, CH_2Cl_2, Py, 1.5 h, r. temp., (61%); (**xii**) MeCN: 48% aq. HF: H_2O (85:15:5), CH_2Cl_2, 5 min., (81%).

dihydro-FK-506 (**439**). The latter was selectively silylated at the C(24) and C(32) hydroxyls with triethylsilyl chloride, and the C(22) alcohol converted to the ketone with the Dess-Martin reagent. Deprotection with aqueous hydrofluoric acid in acetonitrile afforded (-)-FK-506 in 81% yield.

Interestingly, H. Tanaka *et al.* (*J. Antibiotics*, 1988, 41, 1592) have isolated two naturally-occurring macrolides known as FR-900520 and FR-900523, which differ from FK-506 in that they have an ethyl and a methyl substituent replacing the allyl unit at C(21). These compounds show much lower immunosuppressant activity than FK-506, being of similar potency to cyclosporin A. The great differences in biological activity led A.B. Smith, III and K.J. Hale (*Tetrahedron Lett.*, 1989, 30, 1037) to pursue a strategy to a C(10)-C(23) intermediate that would allow the synthesis all three natural products, as well as many analogues, for biological studies into the mechanism of immunosuppression. Their route utilised a monosaccharide starting material for building the C(10)-C(15) sub-unit. It commenced from exocyclic olefin (**442**), which was first prepared in six steps from methyl α-D-glucopyranoside by T.D. Inch and G.J. Lewis (*Carbohydrate Res.*, 1972, 22, 91). Stereoselective hydrogenation of (**442**) with palladium on charcoal in ethanol led to a 3.5:1 mixture of (**444**):(**443**) from which (**444**) was isolated pure in 67% yield after recrystallisation. A second hydrogenation in the presence of palladium hydroxide cleanly removed the benzylidene acetal from (**444**). The resulting diol was selectively esterified at the primary position in 84% yield by the Mitsunobu reaction (S. Bottle and I.D. Jenkins, *J. Chem. Soc., Chem. Comm.*, 1984, 385), and the remaining hydroxyl silylated with *tert*-butyldimethylsilyl chloride. The benzoate was removed with diisobutylaluminium hydride to prevent silyl migration, and the primary alcohol (**446**) oxidised to the carboxylic acid (K.B. Sharpless *et al.*, *J. Org. Chem.*, 1981, 46, 3936). Esterification occurred with methyl iodide and potassium carbonate in *N,N*-dimethylformamide. Ester (**447**) reacted with two equivalents of the anion derived from sulphone (**451**) to give a diastereomeric mixture of β-ketosulphones (**452**) in 74% yield (Scheme 66). Difficulties were encountered in the desulphonylation of (**452**) to the corresponding ketone using conventional methodology. After considerable experimentation, it was eventually discovered that tri-*n*-butyltin hydride could accomplish the desired transformation cleanly and rapidly in 84% yield (A.B. Smith, III, K.J. Hale, and J. P. McCauley, Jr., *Tetrahedron Lett.*, 1989, 30, 5579). Desilylation of the product with hydrogen fluoride-pyridine complex led to β-hydroxy ketone (**453**), which was subjected to hydroxyl-directed reduction with tetra-*n*-methylammonium triacetoxy-borohydride (D.A. Evans, K.T. Chapman, E.M. Carreira, *J. Am.*

Scheme 65

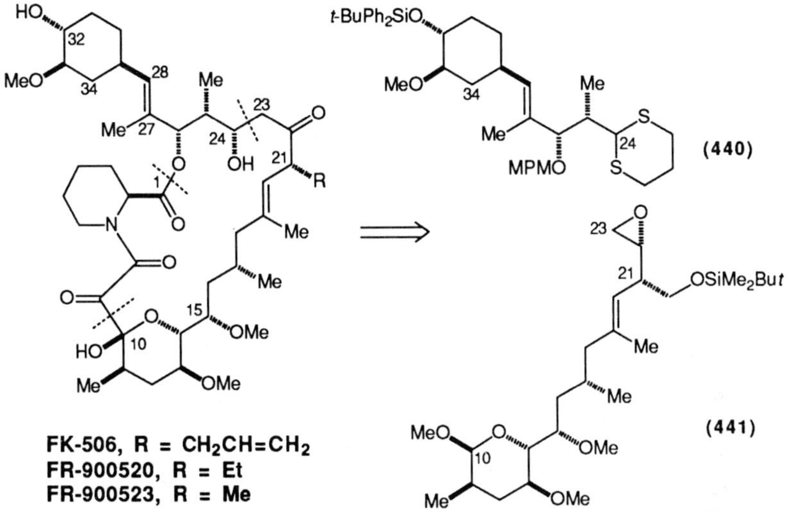

FK-506, R = $CH_2CH=CH_2$
FR-900520, R = Et
FR-900523, R = Me

(440)

(441)

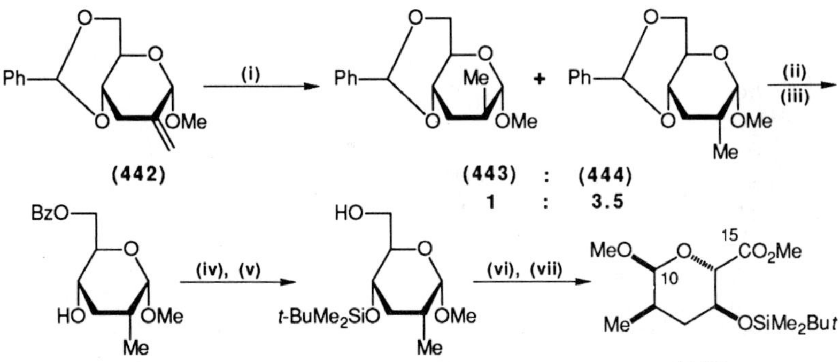

(442)

(443) : (444)
1 : 3.5

(445)

(446)

(447)

Reagents: (**i**) H_2, 10% Pd-C, EtOH, r. temp., (67%); (**ii**) H_2, $Pd(OH)_2$, EtOH, r. temp., (92%); (**iii**) Ph_3P, *i*-$PrO_2CN{=}NCO_2Pr$-*i*, $PhCO_2H$, THF, (84%); (**iv**) *t*-$BuMe_2SiCl$, DMF, Imidazole, 70°C, (89%); (**v**) DIBAL-H, CH_2Cl_2, -78°C,(96%); (**vi**) $RuCl_3$-xH_2O, $NaIO_4$, $MeCN:CCl_4:H_2O$, 3:2:2; (**vii**) K_2CO_3, MeI, DMF, (75%, 2 steps).

Scheme 66

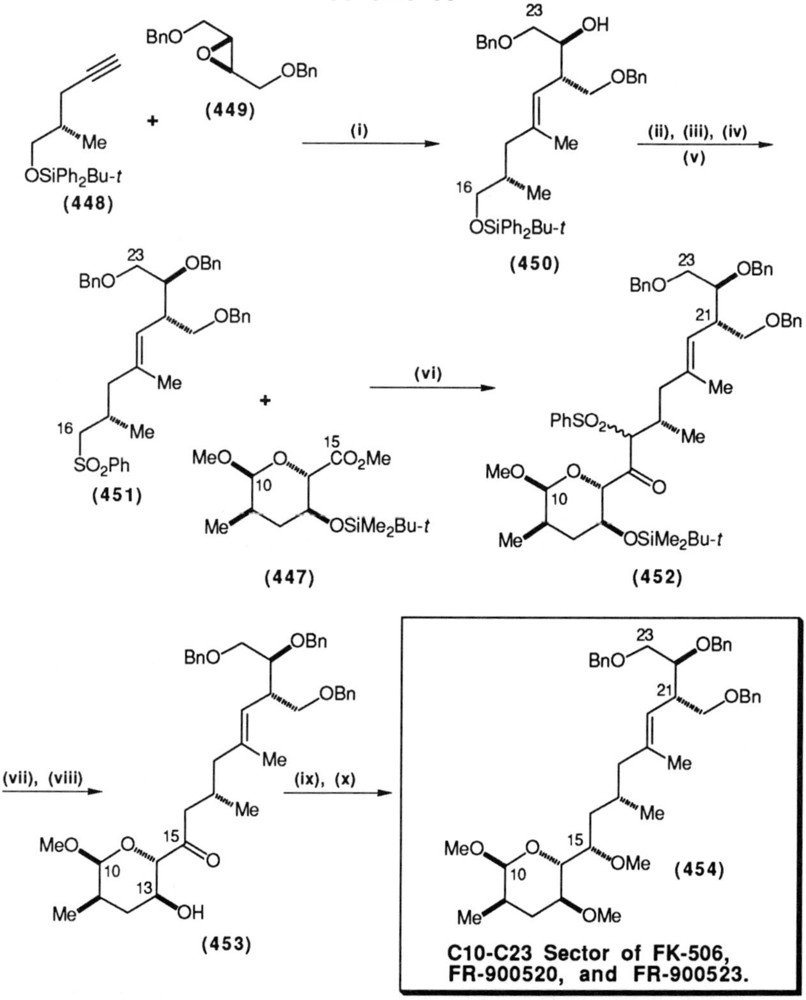

Reagents: (i) Me_3Al, Cp_2ZrCl_2, 1,2-dichloroethane, 2 h, r. temp. then distill solvents and Me_3Al; *n*-BuLi, Hexanes, 0°C, 15 min., then add epoxide (**449**) in C_6H_6 and warm to r. temp for 1h, (66%) ; (**ii**) 60% NaH, THF, *n*-Bu_4NI, BnBr, (71%); (**iii**) *n*-Bu_4NF, THF, (89%); (**iv**) $(PhS)_2$, Bu_3P, DMF, (86%);(**v**) Oxone ®, THF, MeOH, H_2O, (91%); (**vi**) Sulphone (2.0 equiv), *n*-BuLi, -78°C, 10 min., then add ester (1.0 equiv), warm to 0°C, stir 1 h, (74%); (**vii**) *n*-Bu_3SnH (4.0 equiv), PhMe, Δ, add AIBN (2.0 equiv) in portions over 30 min., (84%); (**viii**) HF-Py, THF, r. temp., (92%); (**ix**) $Me_4NBH(OAc)_3$, AcOH, MeCN, -40°C, 18 h (84%, 20:1, 1,3-*Anti:Syn*); (**x**) NaH, MeI, DMF, (76%).

Chem. Soc., 1988, 110, 3561). This afforded a 20:1 mixture of *anti:syn* 1,3-diols which were separated by chromatography. Methylation of the major *anti*-diol completed the synthesis of (**454**), which was suitably functionalised for further elaboration towards the natural products by a variety of different strategies, one of which is shown in Scheme 65.

Avermectin A_{1a}

The synthesis of the potent anthelmintic agent avermectin A_{1a} by S.J. Danishefsky *et al* .(*J. Am. Chem. Soc.*, 1987, 109, 8117, 8119; *ibid.*, 1989, 111, 2967) represents an important milestone in the synthesis of complex macrolides from monosaccharide precursors. The Danishefsky retrosynthetic analysis (Scheme 67) split avermectin A_{1a} into glycal (**455**) and aglycone (**456**). A Mukaiyama lactonisation was planned for closure of the macrolide ring via (**457**). The oxahydrindane unit in (**458**) was to be established through selective heteroconjugate addition onto the enal in (**459**), followed by intramolecular aldol ring closure of the resulting enolate, and β-elimination of the heterosubstituent. Further dissection at the C(8)-C(9) olefin in (**459**) suggested (**460**) and (**461**) as intermediates. A hetero Diels-Alder reaction between (**468**) and (**469**) was to be employed for setting the C(19) stereocentre. This cycloadduct was then to be converted into (**467**) by acid hydrolysis, reduction, and silylation. After ring-opening of the glycal and selective protection, an oxidative spiroketalisation on (**466**) would be the key step in the preparation of aldehyde (**463**). It was envisaged that Wittig and crotylboronate chemistry could be applied on (**463**) for the preparation of (**460**). An intermolecular Sakurai-Ferrier reaction on (**471**) with a *(Z)*-crotylsilane, followed by selective hydrogenation was planned for obtaining (**470**). This material would be transposed into (**468**) by allylic displacement with a suitable methyl carbanion. D-Ribosyl aldehyde (**464**) appeared a good starting material for elaboration into (**461**), since a Lewis acid mediated addition of crotylstannane (**465**) could be used to set the C(5)-stereocentre. Oxidative cleavage of the resulting olefin would then provide aldehyde (**462**) which would be appropriately functionalised for conversion into (**461**).

The synthesis of spiroketal aldehyde (**460**) is contained in Scheme 68. It began by reaction of tripivalate (**471**) with *(Z)*-crotyltriphenylsilane to afford a 4.5:1 mixture of C_{26}*syn*:C_{26}*anti* isomers which were separable after chromatography on silver nitrate impregnated silica gel. The major component was selectively hydrogenated across the terminal olefin to install the ethyl group at C(26). A novel S_N2'-type *anti*-displacement of the allylic pivaloyloxy group in (**470**) was then used to introduce the C(24) methyl group with the correct stereochemistry (H.L. Goering and V.D. Singleton, Jr, *J. Am. Chem. Soc.*, 1976, 98, 7854). Depivaloylation, formation of the

Scheme 67

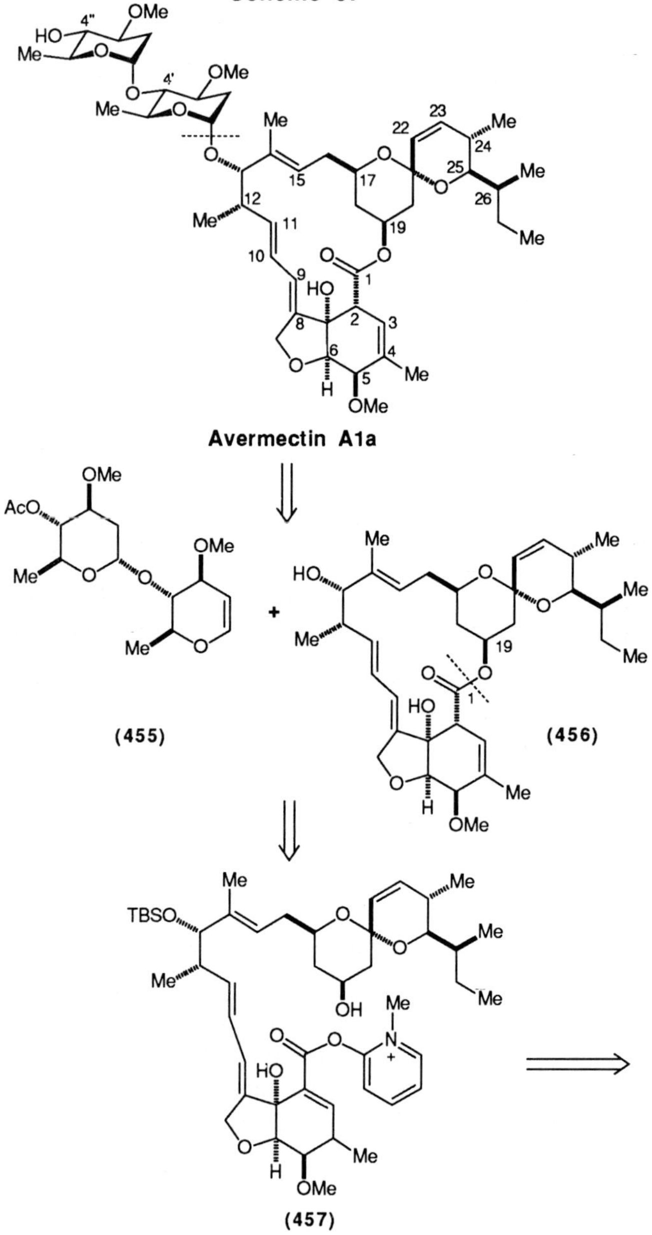

HO 4" OMe Me O O 4' OMe Me O Me 23 22 Me O 24 25 Me 15 17 O 26 12 Me 11 19 Me 10 O O 9 1 HO 8 2 3 6 4 O H 5 Me OMe

Avermectin A1a

⟹

OMe AcO OMe Me O O Me O + Me HO Me O Me O Me Me 19 Me O O 1 HO O H Me OMe

(455) **(456)**

⟹

Me Me TBSO O Me O Me Me OH Me Me O O N + HO O H Me OMe

(457)

⟹

Scheme 67

primary triflate and nucleophilic substitution with sodium cyanide led to (**472**). After reduction with diisobutylaluminium hydride, aldehyde (**468**) participated in a low-temperature magnesium bromide-catalysed cyclocondensation reaction (S. Danishefsky *et al., J. Am. Chem. Soc.*, 1985, 107, 1246; M.D. Bednarski and J.P. Lyssikatos, *Comprehensive Organic Synthesis*, Eds. B.M. Trost, I. Fleming, Pergamon Press, 1991, Vol. 2, 661) to give an inseparable 4:1 mixture of two adducts in which (**473**) predominated. After stereospecific ketone reduction, separation of the isomers was possible and the major product silylated, converted to the bromohydrin, and the bromide reduced under free radical conditions to provide (**474**). Reductive ring-opening of the lactol, pivaloylation and desilylation then furnished (**466**). This underwent oxidative spiroketalisation in the presence of iodine and mercuric oxide (I.T. Kay and D. Bartholomew, *Tetrahedron Lett.*, 1984, 25, 2035; I.T. Kay and E.G. Williams, *ibid.*, 1983, 24, 5915; F.E. Wincott and S. Danishefsky, *ibid.*, 1987, 28, 49) to give aldehyde (**463**), after selective reduction of the primary pivaloate and Swern oxidation. Wittig olefination provided enal (**475**), which reacted with chiral allyl boronate (**476**) (W.R. Roush and R.L. Halterman, *J. Am. Chem. Soc.*, 1986, 108, 294) to afford an inseparable 4:1 mixture of *anti*-alcohols epimeric at C(12) and C(13). After silylation and oxidative cleavage of the double bond, the major product (**477**) was separated in pure form. Wittig olefination, ester reduction, and oxidation concluded this pathway to aldehyde (**460**).

The preparation of ketone (**462**) and its conversion to avermectin A_{1a} are shown in Scheme 69. Ribosyl aldehyde (**464**) was prepared according to the method of G.H. Jones and J.G. Moffatt (*Methods in Carbohydrate Chemistry*, Academic Press: New York, 1972, Vol VI, 315). It reacted readily with *(E)*-tributylcrotylstannane in the presence boron trifluoride etherate to afford a 10:1 mixture of homoallylic alcohols (G.E. Keck *et al., Tetrahedron Lett.*, 1984, 25, 3927), from which the major product (**478**) was isolated after methylation. Deacetonation with methanolic hydrogen chloride followed by ionic hydrogenation with triethylsilane removed the glycosidic methoxy. Subsequent treatment with α-acetoxypivaloyl bromide and base led to epoxide (**479**) in 72% overall yield. Reduction of (**479**) with lithium triethylborohydride was regiospecific giving alcohol (**480**) in 96% yield. Ozonolytic cleavage of the olefin, Wittig condensation, diisobutylaluminium hydride reduction and silylation with *tert*-butyldimethylsilyl chloride converted (**480**) into (**481**). This was then oxidised to ketone (**462**) and coupled with aldehyde (**460**) by a kinetically-controlled aldol reaction. The *(E,E)*-diene spanning the C(8)-C(11) sector was installed by β-elimination of the β-mesyloxy ketone. After selective desilylation of the primary silyl ether, the alcohol was

Scheme 68

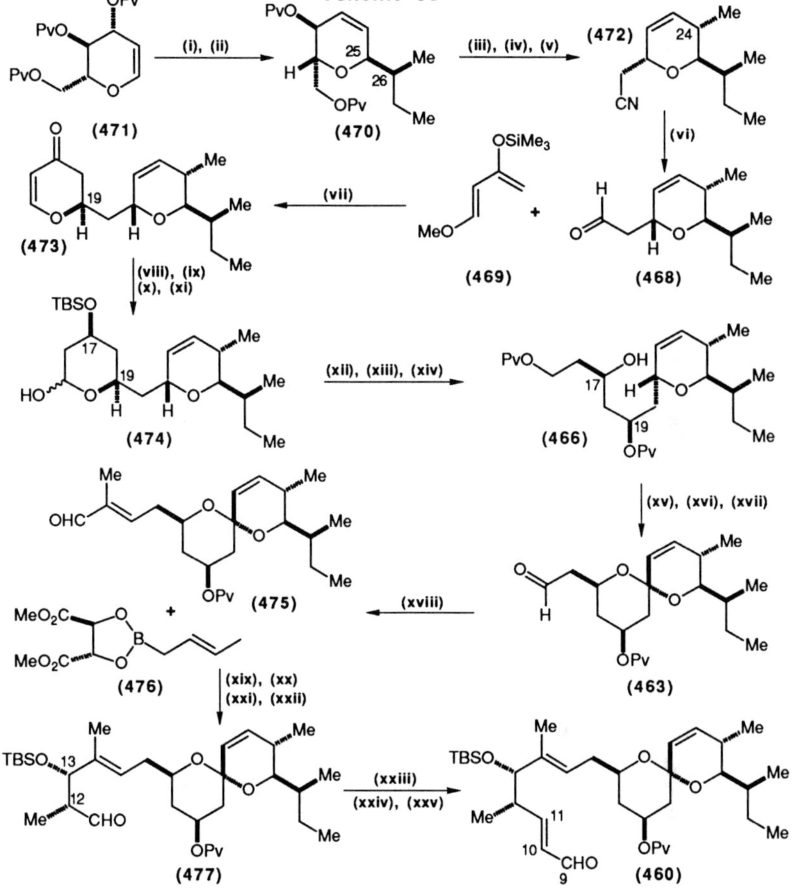

Reagents: (**i**) *(Z)*-$Ph_3SiCH_2CH{=}CHMe$, BF_3-Et_2O, CH_3CH_2CN, -30°C, (73%); (**ii**) H_2, Pd-C, MeOH, 45 min., (90%); (**iii**) MeLi, CuI, Et_2O, -40°C, (73%); (**iv**) LiOH, THF, MeOH, H_2O, (82%); (**v**) $(CF_3SO_2)_2O$, Py, CH_2Cl_2, -25°C to 0°C; NaCN, DMF, r. temp., (82%); (**vi**) DIBAL-H, Et_2O, -25°C, 1h, (90%); (**vii**) $MgBr_2$-Et_2O, CH_2Cl_2, -78°C, 15 min, then add diene dropwise, stir 30 min.; CCl_4, TFA, 15 min., (77%); (**viii**) $CeCl_3$-$7H_2O$, EtOH, 15 min., cooled to -78°C, then $NaBH_4$, added over 1 h, (63%); (**ix**) *t*-$BuMe_2SiOTf$, CH_2Cl_2, 2,6-Lutidine, -78°C, (87%); (**x**) THF, H_2O, $NaHCO_3$, *N*-Bromosuccinimide, (76%); (**xi**) Ph_3SnH, AIBN, PhMe, Δ, (93%); (**xii**) $LiBH_4$, THF, Δ, (93%); (**xiii**) PvCl, CH_2Cl_2, Et_3N, DMAP, 24h, (89%); (**xiv**) HF, MeCN, (87%); (**xv**) HgO, I_2, CCl_4, 275 W light, 1.5 h, (53%); (**xvi**) Super-Hydride® (3 equiv), THF, -78°C, 1 h, (78%); (**xvii**) $(COCl)_2$, Me_2SO, CH_2Cl_2, -78°C, Et_3N, (91%); (**xviii**) $Ph_3P{=}C(Me)CHO$, C_6H_6, Δ, 36 h, (94%); (**xix**) 4 A MS, PhMe, -78°C, 1.5 h, (92%); (**xx**) *t*-$BuMe_2SiOTf$, CH_2Cl_2, 2,6-Lutidine, (94%); (**xxi**) OsO_4, THF, Py; $NaHSO_3$, EtOAc, H_2O; (**xxii**) $Pb(OAc)_4$, CH_2Cl_2, (66%, 2 steps); (**xxiii**) $Ph_3P{=}CHCO_2Me$, CH_2Cl_2, 24 h, r. temp., (95%, all *E*); (**xxiv**) DIBAL-H, CH_2Cl_2, -78°C, 2 h, (81%); (**xxv**) $(COCl)_2$, Me_2SO, CH_2Cl_2, -78°C, Et_3N, (95%).

Scheme 69

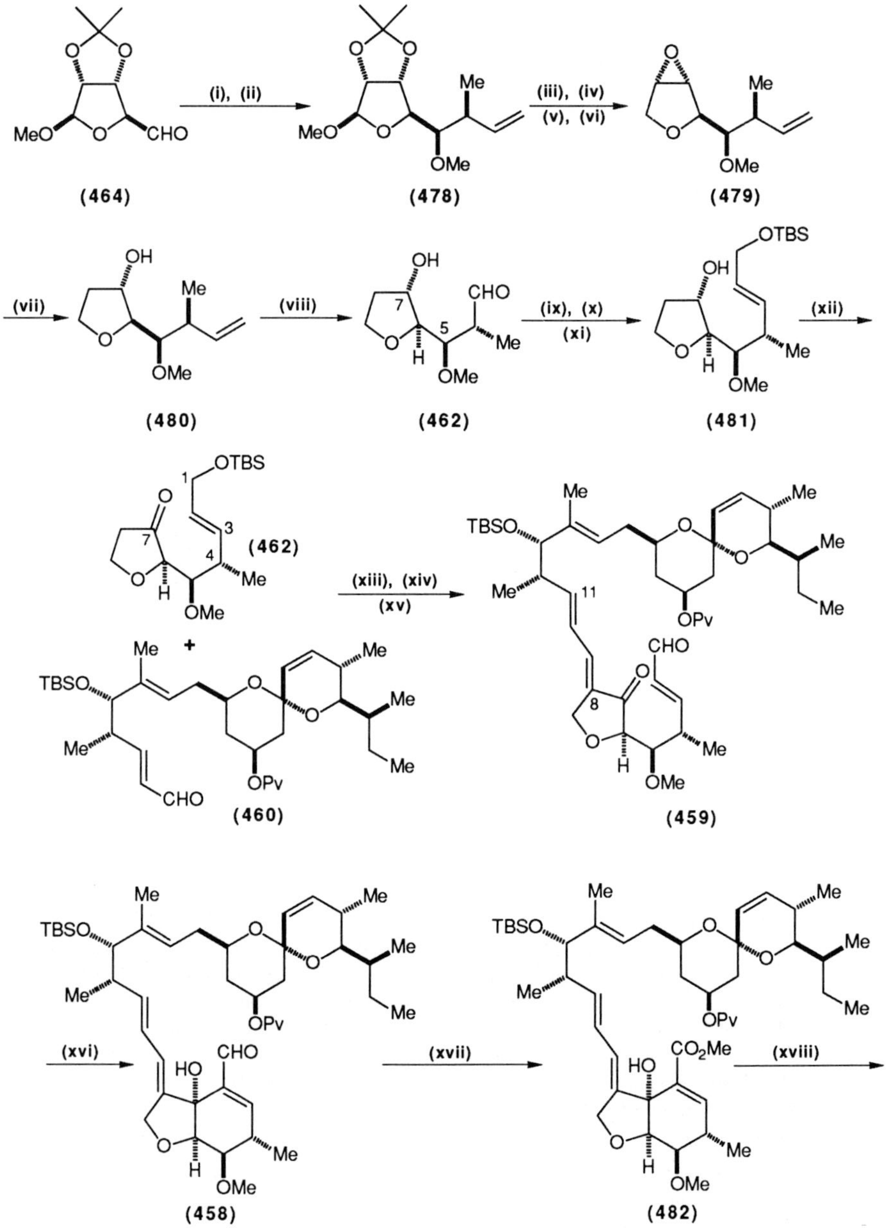

(464) —(i), (ii)→ (478) —(iii), (iv), (v), (vi)→ (479) —(vii)→ (480) —(viii)→ (462) —(ix), (x), (xi)→ (481) —(xii)→ (462)

(462) + (460) —(xiii), (xiv), (xv)→ (459) —(xvi)→ (458) —(xvii)→ (482) —(xviii)→

Scheme 69

(483) → (xix), (xx), (xxi) → (484) 2-*epi*-Avermectin Aglycone → (xxii), (xxiii), (xxiv), (xxv) → Avermectin A1a

Reagents: (**i**) *(E)*-MeCH=$CHCH_2SnBu_3$, BF_3-Et_2O, CH_2Cl_2, -78°C, (79%); (**ii**) NaH, THF, MeI, 0°C to r. temp., (86%); (**iii**) 1N HCl, MeOH, Δ, 5 h; (**iv**) Et_3SiH, BF_3-Et_2O, CH_2Cl_2, 0°C to r. temp., (79%, 2 steps); (**v**) 2-Acetoxyisobutyryl Bromide, CH_2Cl_2, 1h; (**vi**) Amberlite IR-400 (OH^-) resin, MeOH, (92%, 2 steps); (**vii**) $LiEt_3BH$, THF, 0°C, 6 h, (96%); (**viii**) O_3, CH_2Cl_2, -78°C; Zn, AcOH, work up; (**ix**) Ph_3P=$CHCO_2Me$, CH_2Cl_2, 12 h, (84%, 2 steps); (**x**) DIBAL-H, Et_2O, 2h, 0°C, (97%); (**xi**) *t*-$BuMe_2SiCl$, Imidazole, DMAP, CH_2Cl_2, (97%); (**xii**) PCC, NaOAc, CH_2Cl_2, (89%); (**xiii**) $LiN(SiMe_3)_2$, THF, -78°C, 5 min, then 0°C for 30 min.; add aldehyde at -78°C, stir 1 h, then add MsCl, Et_3N, (67%); (**xiv**) 40% aq. HF, MeCN (5:95), -20°C, (90%); (**xv**) PCC, NaOAc, CH_2Cl_2, (88%); (**xvi**) Me_3Al, LiSPh, THF, 0°C, 10 min.; *m*-CPBA, CH_2Cl_2, -20°C, 2 h; PhMe, Δ, 30 min., (76%); (**xvii**) 80% $NaClO_2$, NaH_2PO_4, H_2O, *t*-BuOH, 2-methyl-2-butene, 4 h; CH_2N_2, Et_2O, (79%); (**xviii**) LiOH, MeOH, H_2O, (95%); (**xix**) Et_3N, 2-Chloro-*N*-Methyl-Pyridinium Iodide, MeCN, Δ, (62%); (**xx**) *n*-Bu_4NF, THF, (87%); (**xxi**) LDA, -78°C, THF, (71%); (**xxii**) Imidazole, C_6H_6, 1.5 h, (32%); (**xxiii**) *N*-Iodosuccinimide, MeCN, (**455**), 1 h, (64%); (**xxiv**) *n*-Bu_3SnH, AIBN, PhMe, Δ, (78%); (**xxv**) $LiEt_3BH$, THF, -78°C, (97%).

oxidised to aldehyde (**459**). Ring-closure to the oxahydrindene was realised through selective conjugate addition of thiophenoxide to the enal and intramolecular aldol cyclisation (H. Nozaki *et al.*, *Tetrahedron Lett.*, 1980, 21, 361; *Bull. Chem. Soc. Jpn.*, 1981, 54, 274). Initially, this generated a cyclised β-thiophenoxy aldehyde which underwent elimination to enal (**458**) after oxidation with *m*-chloroperoxybenzoic acid. The enal was then driven to methyl ester (**482**) by sodium chlorite oxidation and reaction with diazomethane. Seco-acid (**483**) was obtained by base hydrolysis. It underwent macrolactonisation, desilylation, and deconjugation with lithium diisopropylamide to give *2-epi*-avermectin A_{1a} aglycone (**484**). Epimerisation at C(2) was achieved with imidazole in benzene according to the method of S. Hanessian, D. Dube and P.J. Hodges (*J. Am. Chem. Soc.*, 1987, 109, 7063) to give the avermectin A_{1a} aglycone. Coupling with disaccharide (**455**) was achieved in the presence of *N*-iodosuccinimide (J. Thiem, H. Karl, and J. Schwentner, *Synthesis,* 1978, 696). It afforded avermectin A_{1a} as the sole product after deiodination with tri-*n*-butyltin hydride, and selective reduction of the acetate at C(4").

Halichondrin B

Undoubtedly, the most spectacular achievement to date in the area of macrolide synthesis from monosaccharides has been the total synthesis of halichondrin B by Y. Kishi *et al.* (*J. Am. Chem. Soc.*, 1992, 114, 3162; *Tetrahedron Lett.*, 1992, 33, 1549, 1553; *ibid.*, 1987, 28, 3463). Halichondrin B belongs to an architecturally novel class of polyether macrolides of marine origin that exhibit powerful antitumour properties *in vivo* (Y. Hirata *et al.*, *J. Am. Chem. Soc.*, 1985, 107, 4796; Y. Hirata and D. Uemura, *Pure Appl. Chem.*, 1986, 58, 701). Kishi's retrosynthetic analysis of the target molecule is contained in Schemes 70 and 71. An initial incision was made at the C(38)-C(39) and C(38)-O(41) bonds, to provide vinyl iodide (**485**) and aldehyde (**486**). Vinyl iodide (**485**) was further disconnected at the C(43)-C(44) linkage into aldehyde (**487**) and bromide (**488**). Halogen-metal exchange on bromide (**488**) was anticipated to lead to an organometallic reagent that could be added to the aldehyde to give the requisite C(44) ketone after oxidation. The tetrahydrofuran ring of (**487**) was to be constructed by an intramolecular attack of the C(47) hydroxyl on the epoxide obtained from C(48)-hydroxyl-directed epoxidation of homo-allylic alcohol (**490**). This intermediate was to be prepared from chiral butenolide (**492**), readily available from L-ascorbic acid (G. Chittenden *et al.*, *J. Org. Chem.*, 1988, 53, 627). A suitable genesis of the C(38)-C(1) fragment (**486**) was considered to be a Ni(II)/Cr(II) mediated coupling between (**489**) and (**491**), followed by macrolactonisation between the C(30) and C(1) termini. The C(14)-C(38) unit was to be assembled by a Ni(II)/Cr(II) mediated coupling between (**493**) and (**494**) followed by

Scheme 70

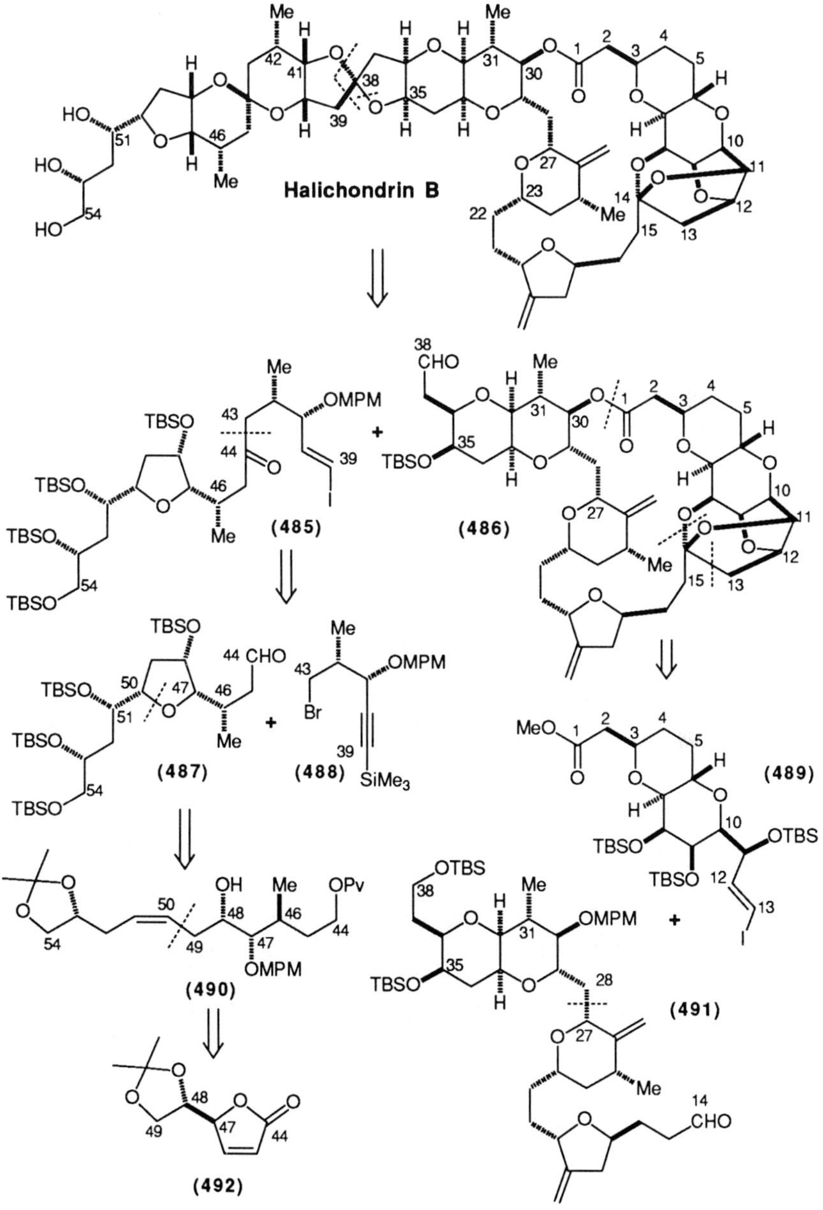

Halichondrin B

(485) (486) (487) (488) (489) (490) (491) (492)

Scheme 71

intramolecular nucleophilic displacement on the C(23) mesylate by the C(27) hydroxyl that would result. An intramolecular Michael reaction on (**497**) was planned for forming the bicyclic pyran system of (**493**), while the C(31) stereocentre in (**500**) was to be established by an Ireland ester enolate Claisen rearrangement on a suitable derivative of D-galactal (**502**). Wittig condensation between (**498**) and (**501**) would be utilised for C(21)-C(22) bond construction, and after 1,4-enone reduction, stereospecific ketone reduction and mesylation would give (**494**). The C(1)-C(13) sector was to be synthesised from 1,2:5,6-di-*O*-isopropylidene-α-D-glucofuranose (**34**).

The main steps involved in the preparation of the C(1)-C(13) fragment are contained in Scheme 72. D-Glucose diacetonide was converted to (**499**) by an undisclosed route. After benzoylation, the *p*-anisylidene acetal was removed by transketalisation to give (**503**). The two hydroxyl groups were then silylated, the benzoate ester saponified, and Swern oxidation performed. The resulting aldehyde (**495**) underwent a Ni(II)/Cr(II) mediated coupling (K. Takai *et al.*, *Tetrahedron Lett.*, 1985, 26, 5585; *J. Am. Chem. Soc.*, 1986, 108, 6048; Y. Kishi *et al., ibid.*, 1986, 108, 5644) with iodoalkyne (**496**), which was followed by a selective *C*-desilylation to provide alcohol (**504**) as the major product of a 9:1 mixture of hydroxyl epimers at C(11). Hydrostannation led to the *(E)*-vinyl stannane (**505**) which underwent iodination and *O*-silylation to furnish vinyl iodide (**489**) in a spectacular 50-55% overall yield for these ten steps.

Assembly of the C(14)-C(38) fragment (**491**) began with preparation of the C(14)-C(26) sub-unit (**494**) (Scheme 73). The synthesis commenced from 2-deoxy-L-arabinose diethylthioacetal 4,5-acetonide (**506**) (M.Y.H. Wong, G.R. Gray, *J. Am. Chem. Soc.*, 1978, 100, 3548). The isopropylidene group was hydrolysed with aqueous acetic acid to give (**507**), and this selectively silylated on the primary alcohol. Treatment with aqueous iodine and sodium bicarbonate removed the thioketal, and acetylation of the resulting hemiacetal provided (**508**). *C*-Glycoside formation was accomplished with allylsilane and boron trifluoride etherate. A hydroboration/oxidation sequence effected hydration of the terminal olefin, and the primary alcohol was protected as a *p*-methoxyphenyldiphenylmethyl ether (MMTrOR). Deprotection of the acetate from this product followed by oxidation led to ketone (**509**). Olefination was performed according to the method of F.N. Tebbe, G.W. Parshall, and G.S. Reddy (*J. Am. Chem. Soc.*, 1978, 100, 3611; L.F. Cannizzo and R.H. Grubbs, *J. Org. Chem.*, 1985, 50, 2386). After removal of the *p*-methoxyphenyl-diphenylmethyl ether with mild acid in methanol, pivaloylation, desilylation, and Dess-Martin oxidation supplied aldehyde (**501**) in 47% overall yield for the 14 steps. In order to synthesise the required coupling partner (**498**), corresponding to the C(22)-C(26) region,

Scheme 72

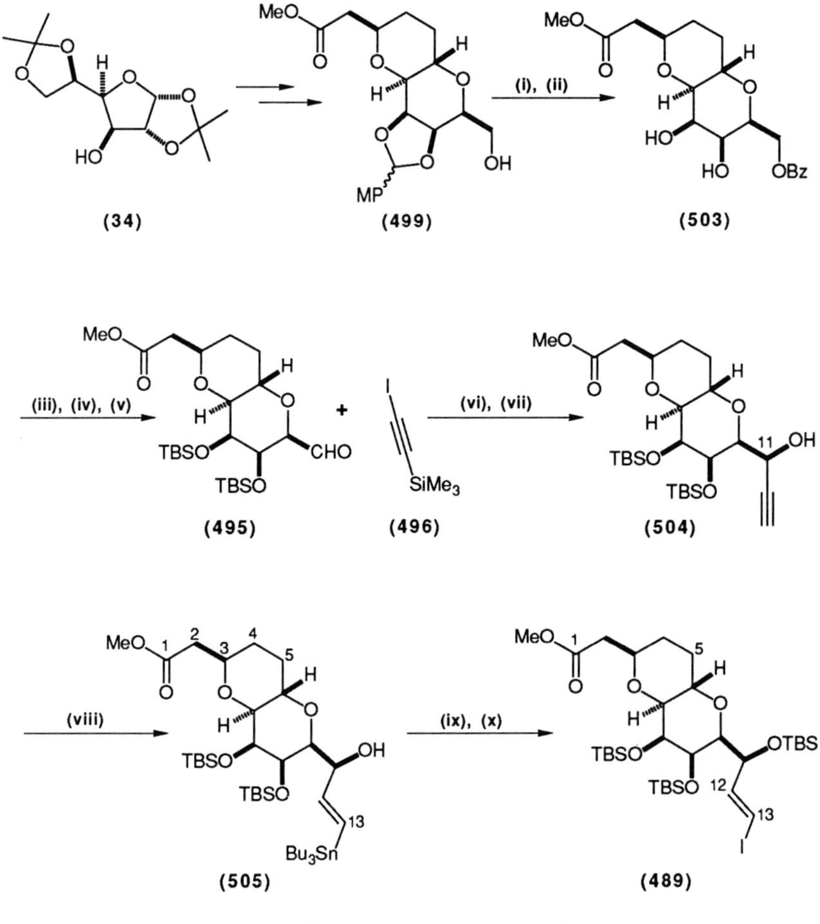

(34) → → (499) —(i), (ii)→ (503)

—(iii), (iv), (v)→ (495) + (496) —(vi), (vii)→ (504)

—(viii)→ (505) —(ix), (x)→ (489)

Reagents: (**i**) PhCOCl, Py; (**ii**) *p*-TsOH, MeOH, r. temp.; (**iii**) *t*-$BuMe_2SiOTf$, Et_3N, CH_2Cl_2; (**iv**) MeONa, MeOH; (**v**) Me_2SO, $(COCl)_2$, Et_3N, CH_2Cl_2, -78°C; (**vi**) $NiCl_2$ (0.01%)-$CrCl_2$, THF, r. temp.,(stereoselectivity = 8-9:1); (**vii**) $AgNO_3$, H_2O-EtOH (1:4); (**viii**) *n*-Bu_3SnH, AIBN, PhMe, 80°C; (**ix**) I_2, CH_2Cl_2, 0°C; (**x**) *t*-$BuMe_2SiOTf$, Et_3N, CH_2Cl_2, r. temp.

Scheme 73

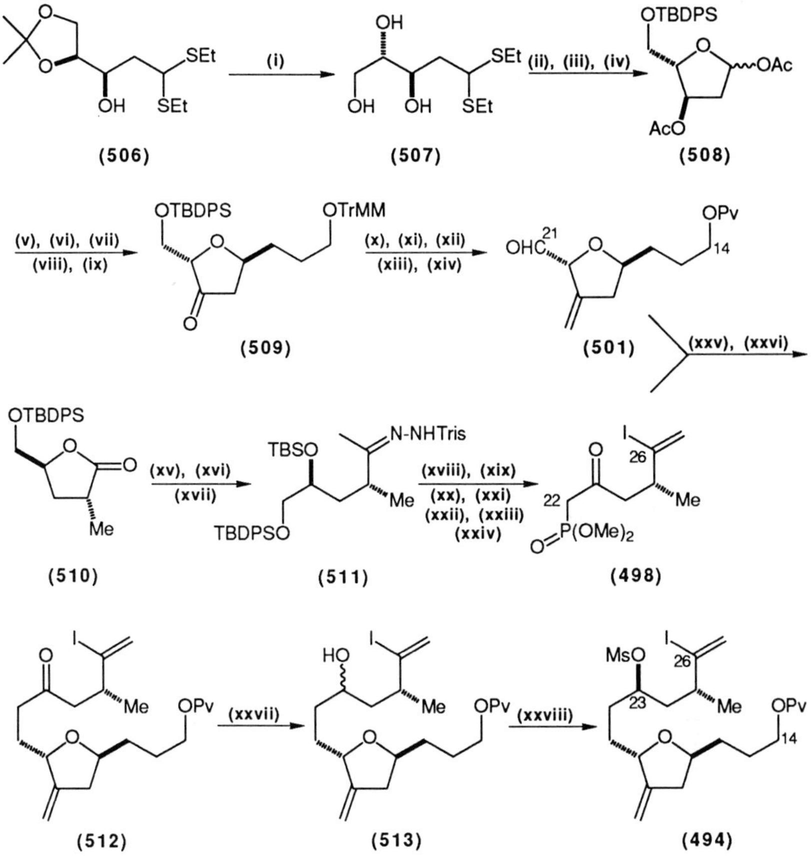

(506) (507) (508) (509) (501) (510) (511) (498) (512) (513) (494)

Reagents: (**i**) AcOH, H_2O; (**ii**) *t*-$BuPh_2SiCl$, Imidazole; (**iii**) I_2, $NaHCO_3$, H_2O; (**iv**) Ac_2O, Py; (**v**) $H_2C{=}CHCH_2SiMe_3$, BF_3-Et_2O, MeCN, 0°C; (**vi**) 9-BBN; H_2O_2; (**vii**) MMTrCl, Et_3N, CH_2Cl_2; (**viii**) K_2CO_3, MeOH; (**ix**) Me_2SO, $(COCl)_2$,CH_2Cl_2, Et_3N, -78°C; (**x**) Me_3Al, Cp_2TiCl_2; (**xi**) MeOH, PPTS; (**xii**) PvCl, Py; (**xiii**) *n*-Bu_4NF; (**xiv**) Dess-Martin Periodinane, CH_2Cl_2; (**xv**) MeLi, Et_2O; (**xvi**) *t*-$BuMe_2SiCl$, Imidazole, CH_2Cl_2; (**xvii**) THF, $TrisNHNH_2$, cat. HCl; (**xviii**) *n*-BuLi, THF, *n*-Bu_3SnCl; (**xix**) I_2, CH_2Cl_2; (**xx**) HF-Py, MeCN; (**xxi**) $NaIO_4$, THF-H_2O (2:1); (**xxii**) Jones Reagent; (**xxiii**) $PhCH_2OH$, DCC; (**xxiv**) $(MeO)_2P(O)CH_2Li$; (**xxv**) NaH, THF, 0°C, followed by treatment with the aldehyde at 0°C; (**xxvi**) $[(Ph_3P)CuH]_6$, C_6H_6, H_2O, r. temp; (**xxvii**) $NaBH_4$, MeOH, 0°C; (**xxviii**) Ms_2O, Et_3N, 0°C.

Scheme 74

Reagents: (**i**) *t*-$BuMe_2SiCl$, Imidazole, DMF; (**ii**) NaH, BnBr, THF, DMF (10:1), 0^oC to r. temp.; (**iii**) *n*-Bu_4NF, THF; (**iv**) $(EtCO)_2O$, Et_3N, CH_2Cl_2; (**v**) $LiN(SiMe_3)_2$, *t*-$BuMe_2SiCl$, HMPA, THF, (1:9), -78^oC to 0^oC; (**vi**) C_6H_6, Δ; (**vii**) 1N NaOH, H_2O, THF, (1:1), r. temp.; (**viii**) I_2, KI, Sat. $NaHCO_3$,; (**ix**) *n*-Bu_3SnH, AIBN, C_6H_6, Δ; (**x**) DIBAL-H, THF, -78^oC; (**xi**) *p*-TsOH, MeOH; (**xii**) Tf_2O, Py, CH_2Cl_2; (**xiii**) NaCN, DMF, r. temp.; (**xiv**) DIBAL-H, CH_2Cl_2, -78^oC; (**xv**) $NaBH_4$, MeOH, 0^oC; (**xvi**) H_2, $Pd(OH)_2$, MeOH; (**xvii**) EtSH, BF_3-Et_2O, CH_2Cl_2; (**xviii**) *t*-$BuMe_2SiOTf$, Et_3N, CH_2Cl_2, r. temp.; (**xix**) I_2, $NaHCO_3$, Me_2CO, H_2O; (**xx**) $NiCl_2$(1.0%)-$CrCl_2$, THF, r. temp.; (**xxi**) MPMOC(=NH)CCl_3, CH_2Cl_2, BF_3-Et_2O, 0^oC; (**xxii**) HF, Py, MeCN, r. temp.; (**xxiii**) $Me_2C(OMe)_2$, PPTS, CH_2Cl_2,; (**xxiv**) *n*-Bu_4NF, THF; (**xxv**) PPTS, MeOH; (**xxvi**) *t*-$BuMe_2SiOTf$, Et_3N, CH_2Cl_2; (**xxvii**) $LiAlH_4$, Et_2O, 0^oC; (**xxviii**) Dess-Martin Periodinane, CH_2Cl_2.

(**510**) was chosen as the starting material (K. Tomooka *et al.*, *Tetrahedron Lett.*, 1989, 30, 1563). Addition of methyllithium led to the open-chain methyl ketone, which was silylated with *tert*-butyldimethylsilyl chloride, and transformed into the tripsylhydrazone. Treatment of (**511**) with *n*-butyllithium led to the vinyl anion, which was converted into the iodoalkene via the vinyl stannane. After desilylation, periodate cleavage of the 1,2-diol gave the aldehyde which was transformed into the benzyl ester. The latter reacted with the carbanion derived from dimethyl methylphosphonate to afford β-keto phosphonate (**498**). These ten steps proceeded in 60% overall yield. Condensation of (**498**) with aldehyde (**501**) produced an enone which underwent chemoselective conjugate reduction with (triphenylphosphine)copper hydride hexamer (W.S. Mahoney, D.M. Brestensky, and J.M. Stryker, *J. Am. Chem. Soc.*, 1988, 110, 291) to give (**512**). Reduction of the keto group with sodium borohydride furnished a 1:1 mixture of both possible diastereomers (**513**) which were separated at this stage. The undesired isomer was recycled back into the synthesis by Mitsunobu inversion and treatment of the ester with base. The desired alcohol was then mesylated with methanesulphonic anhydride to produce (**494**).

The C(27)-C(38) section (**493**) of halichondrin B was built from D-galactal (**502**) (Scheme 74). The first steps were silylation of the primary alcohol with *tert*-butyldimethylsilyl chloride, regioselective benzylation of the axial hydroxyl group, desilylation, and propionylation. Formation of the bis(silylketene)acetal from (**514**), followed by Ireland-Claisen rearrangement gave (**515**) as the major product of an 8:1 mixture, after ester hydrolysis, iodolactonisation, and tri-*n*-butyltin hydride reduction. Reduction of the lactone and Fischer glycosidation paved the way for reduction of nitrile (**516**) to the aldehyde with diisobutylaluminium hydride. This was further reduced to the alcohol, and the furanoside ring opened up to (**517**) by thioketalisation. Exhaustive silylation and thioketal hydrolysis afforded (**518**). Ni(II)/Cr(II) mediated addition of vinyl iodide (**519**) onto aldehyde (**518**) provided a 2:1 mixture of alcohols (**520**) in favour of the desired C(30)R-isomer. A Mitsunobu inversion recycling sequence was again applied on the minor diastereomer to optimise the yield of desired product. *p*-Methoxybenzylation of C(30)R-(**520**) was accomplished with the benzyl imidate under Lewis acid catalysis, and afforded (**497**) after selective desilylation with hydrogen fluoride-pyridine complex, and acetal exchange. Unmasking of the C(33) hydroxyl with tetra-*n*-butylammonium fluoride then brought about Michael ring closure. All that remained for completing the synthesis of aldehyde (**493**) were acetal hydrolysis, *O*-silylation, reduction of the methyl ester, and Dess-Martin oxidation.

The route developed to access the C(39)-C(54) unit is shown in

Scheme 75

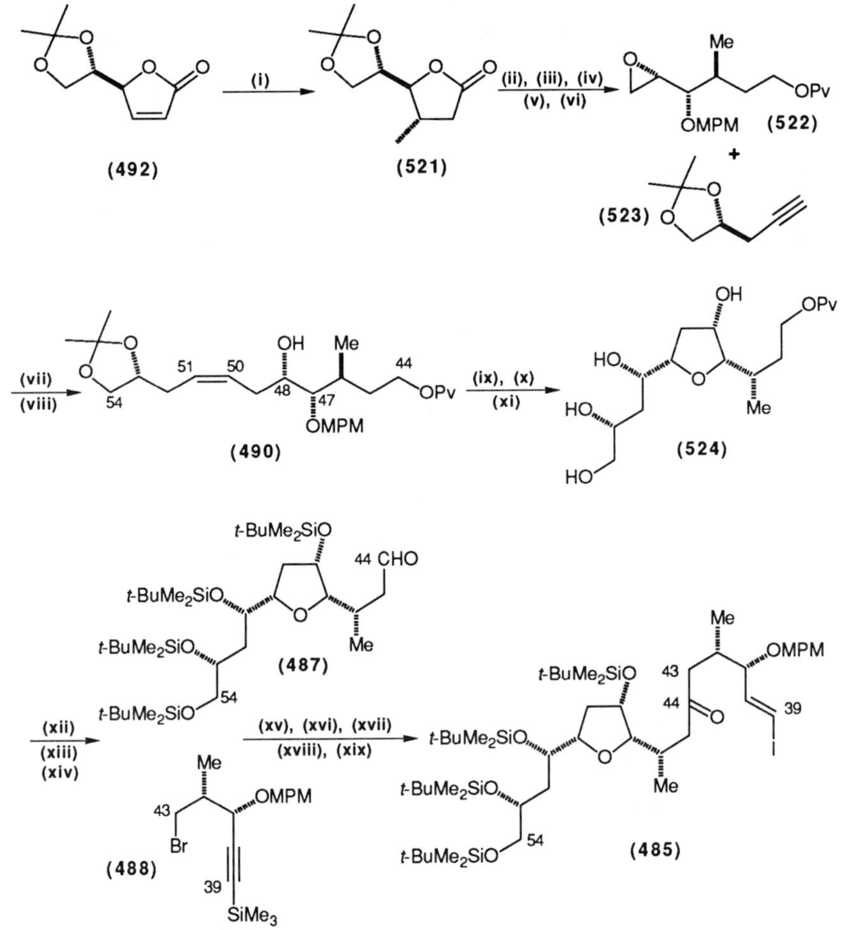

(492) (521) (522) (523) (490) (524) (487) (488) (485)

(i) (ii), (iii), (iv), (v), (vi) (vii), (viii) (ix), (x), (xi) (xii), (xiii), (xiv) (xv), (xvi), (xvii), (xviii), (xix)

Reagents: (**i**) Me_2CuLi, Me_3SiCl, THF, -78°C to r. temp.; (**ii**) $LiAlH_4$, Et_2O, 0°C; (**iii**) PvCl, Py, CH_2Cl_2; (**iv**) *p*-$MeOC_6H_4CH_2Br$, KH, THF, r. temp.; (**v**) AcOH, H_2O, (4:1), r. temp.; (**vi**) NaH, THF, r. temp., followed by *p*-TsImd, THF, r. temp.; (**vii**) *n*-BuLi, THF, -78°C followed by BF_3-Et_2O, then RCHO; (**viii**) H_2, Lindar Catalyst, Quinoline, Hexanes, r. temp.; (**ix**) *t*-BuO_2H, $V(O)(acac)_2$, C_6H_6, r. temp.; (**x**) CF_3CO_2H, CH_2Cl_2; (**xi**) AcOH, H_2O (4:1), r. temp.; (**xii**) *t*-$BuMe_2SiOTf$, Et_3N, CH_2Cl_2; (**xiii**) $LiAlH_4$, Et_2O, 0°C; (**xiv**) Dess-Martin Periodinane, CH_2Cl_2, r. temp; (**xv**) *t*-BuLi, Et_2O, -78°C then RCHO; (**xvi**) $AgNO_3$ (6 equiv), $(Me_3Si)_2NH$ (7 equiv), H_2O-EtOH (1:4); (**xvii**) *n*-Bu_3SnH, AIBN, PhMe, 80°C; (**xviii**) I_2, CH_2Cl_2, r. temp; (**xix**) Dess-Martin Periodinane, CH_2Cl_2, r. temp.

Scheme 75. Conjugate addition of lithium dimethyl cuprate in the presence of chlorotrimethylsilane (E.J. Corey and N.W. Boaz, *Tetrahedron Lett.*, 1985, 26, 6019) proceeded from the less hindered face of (**492**) to produce (**521**). A five step protocol involving reduction, selective pivaloylation, *p*-methoxybenzylation, acetonide cleavage, and terminal epoxide formation led to (**522**). This underwent ring-opening with the alkynyl lithium obtained from (**523**) provided boron trifluoride etherate was present. Lindlar hydrogenation furnished the *cis*-homoallylic alcohol (**490**). This was epoxidised under chelation controlled conditions (K.B. Sharpless and R.C. Michaelson, *J. Am. Chem. Soc.*, 1973, 95, 6136), and after acid treatment the desired tetrahydrofuran was isolated as the major component of a 7-8:1 mixture, in 61% overall yield. After removal of the acetonide with acetic acid, the remaining hydroxyls in (**524**) were exhaustively silylated, the pivaloate reduced to the alcohol, and oxidation performed with the Dess-Martin periodinane. Aldehyde (**487**) combined with the organolithium reagent derived from (**488**), to afford ketone (**485**) after *C*-desilylation, hydrostannation of the alkyne, iododestannylation, and oxidation.

The key Ni(II)/Cr(II) mediated coupling used for joining (**493**) with (**494**) is shown in Scheme 76. It resulted in a 6:1 mixture of allylic alcohols. The major product cyclised upon treatment with base to give (**491**), after reduction and oxidation. Another Ni(II)/Cr(II)-mediated coupling/Dess-Martin oxidation was employed for the preparation of (**525**). These two steps generated the *trans*-enone in 77% overall yield. Buffered DDQ removed the *p*-methoxybenzyl group from (**525**), and hydrolysis of the methyl ester supplied the seco-acid, which was macrolactonised using the 2,4,6-trichlorobenzoic anhydride protocol of M. Yamaguchi *et al.* (*Bull. Chem. Soc. Jpn.*, 1979, 52, 1989). Michael addition of the C(9) hydroxyl to C(12) occurred under the conditions used for desilylation, the stereoselectivity being 5-6:1 in favour of the desired isomer. Treatment with pyridinium *p*-toluenesulphonate subsequently completed the polycyclic ring system around C(8)-C(14). Four steps were necessary for preparing aldehyde (**486**). These involved selective esterification of the primary alcohol at C(38), silylation of the remaining hydroxyl at C(35), debenzoylation, and Dess-Martin oxidation. Another Ni(II)/Cr(II) promoted coupling was then utilised for coupling the right half aldehyde (**486**) with the left half vinyl iodide (**485**). Again the *trans*-enone was obtained exclusively after oxidation, in 60% overall yield. The final stages of the synthesis are contained in Scheme 77. The first step was desilylation of enone (**526**) with tetra-*n*-butylammonium fluoride. This induced hemiketal formation between the C(48)-hydroxyl and the C(44)-ketone, and Michael addition of the C(44)-hemiketal hydroxyl at C(40). At the same time, the C(35)-hydroxyl attacked the C(38)-ketone to give (**527**). After removal of the *p*-methoxybenzyl group from C(41) of (**527**), acid treatment resulted in

Scheme 76

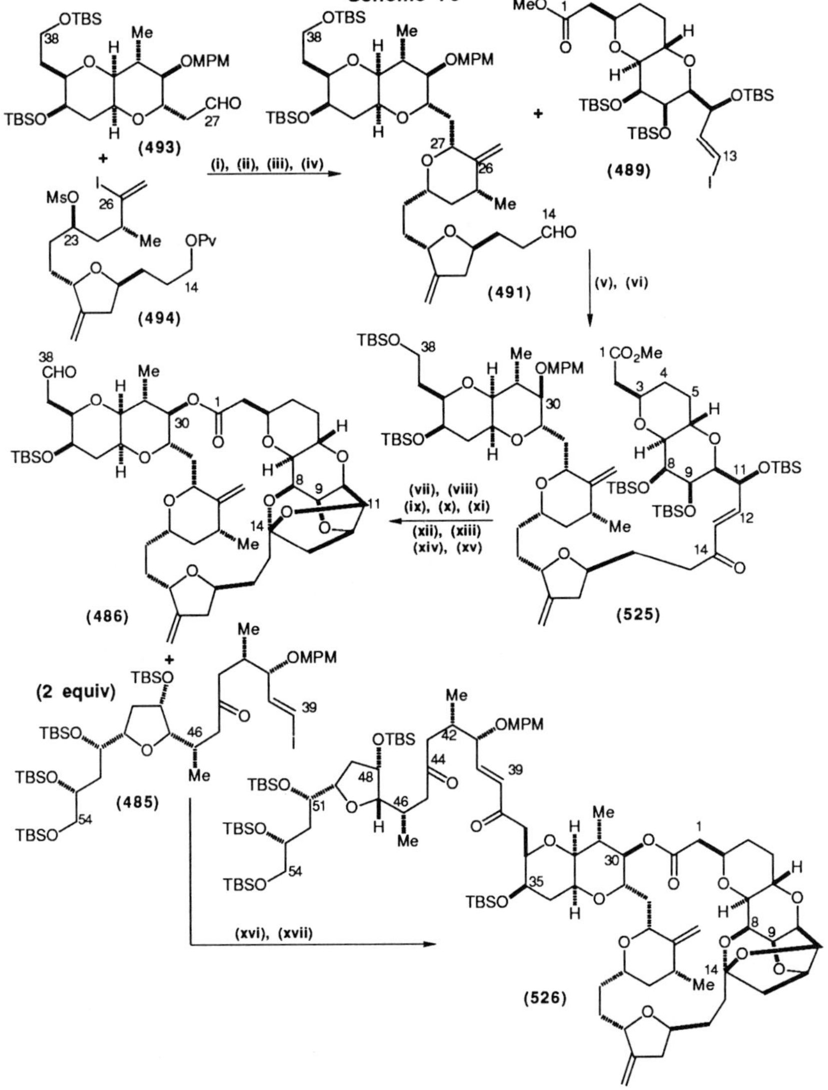

Reagents: (**i**) $NiCl_2$ (0.5%)-$CrCl_2$, DMF, THF (1:5), r. temp.; (**ii**) KH, DME, 80°C; (**iii**) $LiAlH_4$, Et_2O, 0°C; (**iv**) Dess-Martin Periodinane, CH_2Cl_2; (**v**) $NiCl_2$ (0.1%)-$CrCl_2$, DMF, r. temp; (**vi**) Dess-Martin Periodinane, CH_2Cl_2; (**vii**) DDQ, pH 7 phosphate buffer, *t*-BuOH, CH_2Cl_2 (10:1:100), r. temp; (**viii**) LiOH, H_2O, THF (1:3), r. temp; (**ix**) Yamaguchi Lactonisation; (**x**) *n*-Bu_4NF, THF; (**xi**) PPTS, CH_2Cl_2; (**xii**) *p*-$NO_2C_6H_4COCl$, Py, r. temp.; (**xiii**) *t*-$BuMe_2SiOTf$, Et_3N, CH_2Cl_2; (**xiv**) K_2CO_3, MeOH; (**xv**) Dess-Martin Periodinane, CH_2Cl_2; (**xvi**) $NiCl_2$ (0.1%)-$CrCl_2$, DMF, r. temp.; (**xvii**) Dess-Martin Periodinane, CH_2Cl_2.

Scheme 77

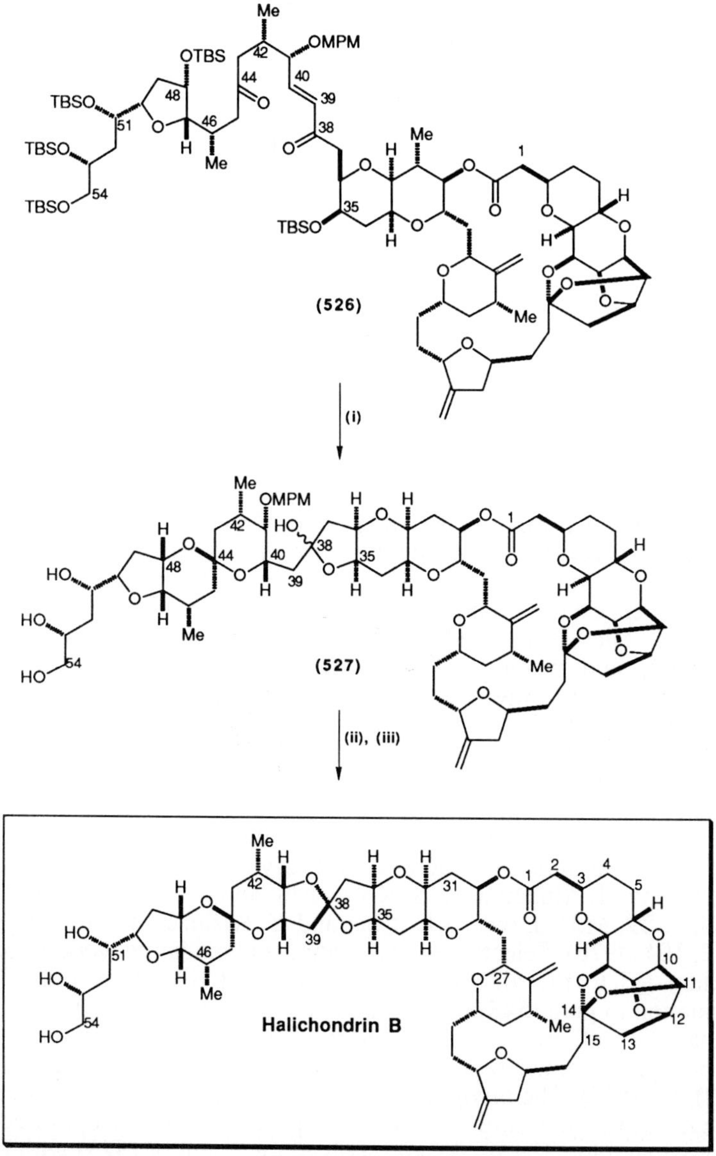

(526)

(i)

(527)

(ii), (iii)

Halichondrin B

Reagents: **(i)** n-Bu_4NF, DMF, r. temp; **(ii)** DDQ, pH 7 phosphate buffer, t-BuOH, CH_2Cl_2 (10:1:100); **(iii)** CSA, CH_2Cl_2, r. temp.

5,5-spiroketalisation, to deliver halichondrin B in 50-60% overall yield for the last three steps.

Another interesting synthesis of the C(1)-C(15) region of halichondrin B has been reported from D-ribose, by A.J. Cooper and R.G. Salomon (*Tetrahedron Lett.*, 1990, 27, 3813). Their retrosynthetic analysis, shown in Scheme 78, made disconnections at the C(3)-O(7) bond and at the C(14) acetal of (**528**) to give (**529**). A Michael addition was planned for closure of the pyran ring, while an acid-catalysed ketalisation on the C(14) keto group was envisaged for assembly of the polycyclic acetal. Further clearance of the C(1)-C(5) chain, and retro-Michael cleavage at O(9)-C(12), suggested (**531**) as a possible intermediate. After dissection at the C(10)-C(11) site, the relationship between the B-ring and D-ribose was immediately apparent.

The actual route to (**528**), depicted in Schemes 79 and 80, focussed in its early stages on elaborating the C(10)-C(15) region. Thus, methyl 2,3-0-isopropylidene-β-D-ribofuranose was oxidised, olefinated, and subjected to asymmetric osmylation with osmium tetroxide and dihydroquinidine *p*-chlorobenzoate (K.B. Sharpless *et al.*, *J. Am. Chem. Soc.*, 1988, 110, 1968; *ibid.*, 1989, 11, 1123). This afforded a 2.5:1 mixture of (**535**) and (**536**). It was necessary to selectively invert the C(11) stereocentre (halichondrin numbering) in (**535**). This was accomplished by a regioselective nucleophilic displacement on the cyclic sulphate (**537**) with tetra-*n*-butylammonium benzoate. The product (**538**) was debenzoylated, the resulting diol ketalised, and a chelation-controlled reduction achieved with diisobutylaluminium hydride to deliver aldehyde (**539**). Wittig olefination followed by acid hydrolysis led initially to (**541**) and then to (**531**) which spontaneously cyclised to (**542**) under the reaction conditions. It should be noted that this product had the opposite stereochemistry to that found at C(12) in halichondrin B. After acetylation, peracetate (**530**) participated in a chemoselective intermolecular Sakurai reaction with allyltrimethylsilane and trityl perchlorate (T. Mukaiyama, S. Kobayashi, and S, Shoda, *Chem. Lett.*, 1984, 1529; M.D. Lewis, J.K. Cha, and Y. Kishi, *J. Am. Chem. Soc.*, 1982, 104, 4976). Selective hydroboration and oxidative work up of the product led to the alcohol which underwent oxidation and Wittig olefination to give (**543**). At this point deacetylation was effected with methanolic sodium methoxide, and this simultaneously brought about the necessary epimerisation at C(12) to give a 2:1 mixture of epimers (**529**) and (**544**). The mixture was ketalised directly with pyridinium *p*-toluenesulphonate and the desired product (**545**) separated by chromatography. The final step involved treatment of (**545**) with base. This delivered the C(1)-C(15) segment of halichondrin B in 73% yield.

Scheme 78

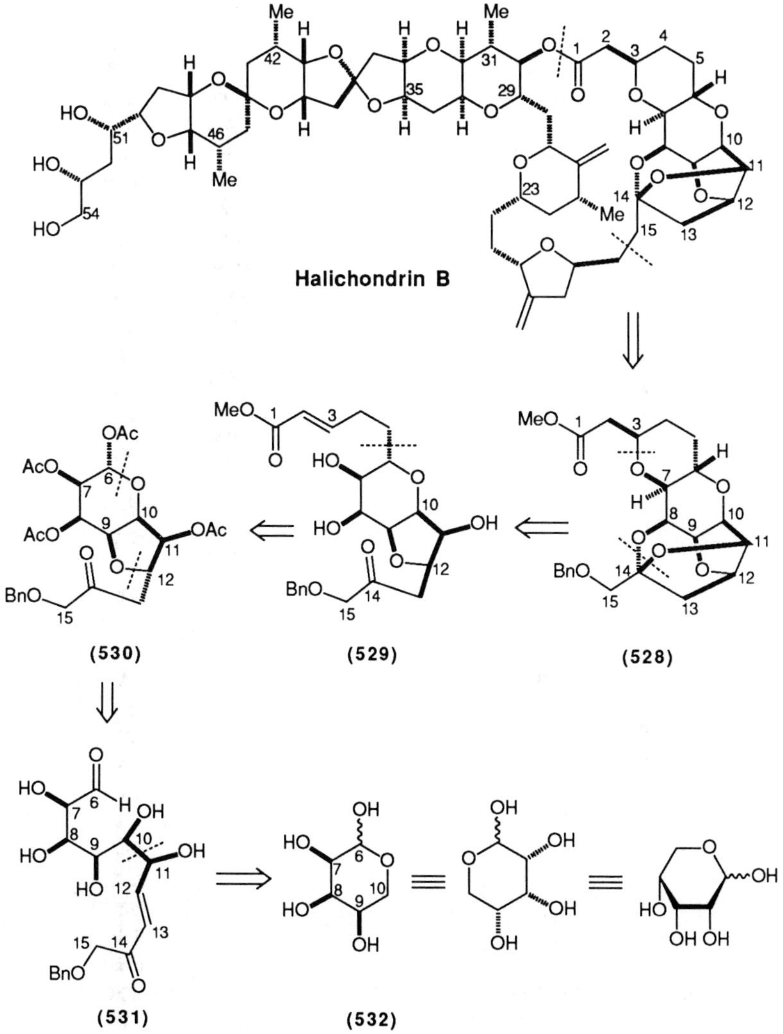

Halichondrin B

(530) (529) (528)

(531) (532)

Scheme 79

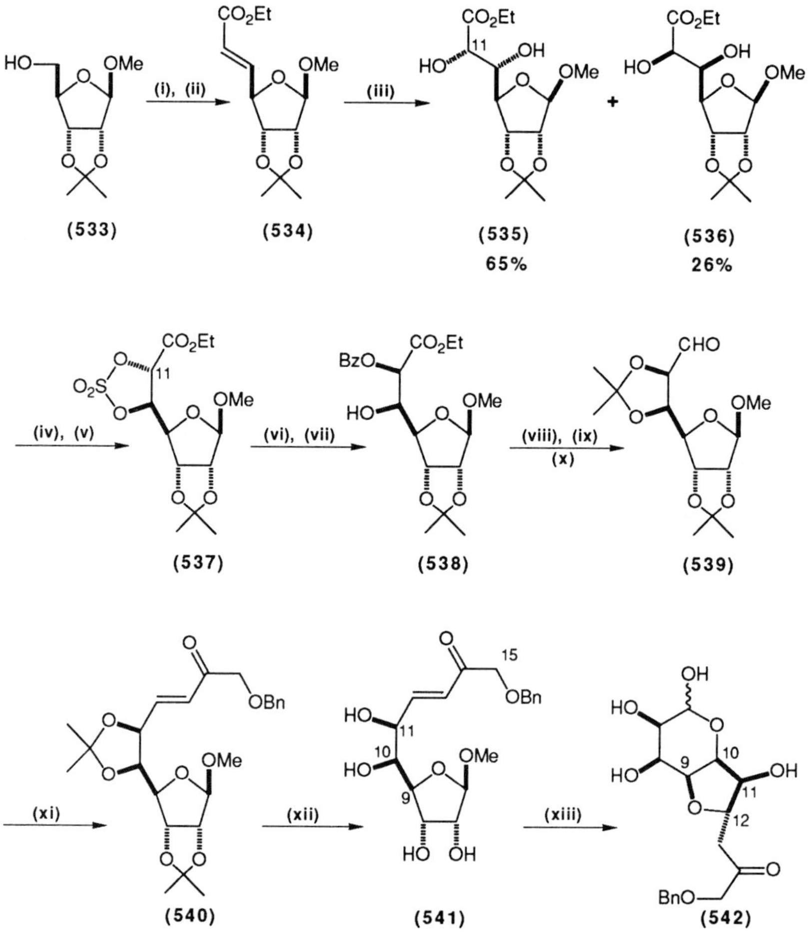

Reagents: (**i**) Me_2SO, $(CF_3CO)_2O$, -60°C, CH_2Cl_2, Et_3N; (**ii**) $Ph_3P{=}CHCO_2Et$, (85%, 2 steps); (**iii**) OsO_4, *N*-Methyl Morpholine Oxide, Dihydroquinidine *p*-Chlorobenzoate, Me_2CO , H_2O; (**iv**) $SOCl_2$, Et_3N, CH_2Cl_2; (**v**) $RuCl_3$, $NaIO_4$, MeCN, H_2O,(90%, 2 steps); (**vi**) *n*-Bu_4NOBz, C_6H_6; (**vii**) H_2SO_4, H_2O, THF, (88%, 2 steps); (**viii**) BaOMe, MeOH; (**ix**) PPTS, Me_2CO, $Me_2C(OMe)_2$; (**x**) DIBAL-H, CH_2Cl_2, -78°C, (75%, 3 steps); (**xi**) $Ph_3P{=}CHCO(CH_2OBn)$, (82%); (**xii**) CF_3CO_2H, H_2O, (10:1); (**xiii**) H_2O, Dowex 50W (H^+) Resin;

Scheme 80

(542) (543) (544) (529) (545)

1 : 2

(528) C1-C15 Segment of Halichondrin B

Reagents: (i) Ac_2O, Py, DMAP, (26%, 3 steps); (**ii**) H_2C=$CHCH_2SiMe_3$, $PhCClO_4$, -5 to 0°C, (79%); (**iii**) $[(Me)_2CHCH(Me)]_2BH$, THF; NaOH, H_2O_2, (69%); (**iv**) SO_3 Py complex, Me_2SO, Et_3N, CH_2Cl_2; (**v**) Ph_3P=$CHCO_2Me$, (57%, 2 steps); (**vi**) MeOH-MeOAc (10:1), NaOMe, (100%); (**vii**) PPTS, CH_2Cl_2, (100%); (**viii**) $Bn(Me)_3NOMe$, MeOH, MeOAc (10:1), (73%).

Second Supplements to the 2nd Edition of Rodd's Chemistry of Carbon Compounds, Vol. IE/F/G, edited by M. Sainsbury

Chapter 24

DISACCHARIDES AND OLIGOSACCHARIDES

R. DARCY AND K. MC CARTHY

Introduction

With increasing understanding of the importance of complex carbohydrates in biological information transfer and molecular recognition, there are widening possibilities for the design and synthesis of artificial molecular systems based on these phenomena (R.U. Lemieux, in "Frontiers of Chemistry", ed. K.J. Laidler, Pergamon, Oxford, 1982, p. 3; J.F. Kennedy and C.A. White, "Bioactive Carbohydrates," Ellis Horwood, Chichester, 1983; R.C. Hughes, "Glycoproteins", Chapman and Hall, London, 1983; J.N. Kanfer and S. Hakomori, "Handbook of Lipid Research", vol. 3 "Sphingolipid Biochemistry", Plenum, New York, 1983; "Biology of Carbohydrates," vol. 2, eds. V. Ginsburg and P.W. Robbins, Wiley, New York, 1984; "The Receptors," vol. 2, ed. P.M. Conn, Academic Press, New York, 1985; T. Feizi and R.A. Childs, Trends Biochem. Sci., 1985, *19*, 24; B.K. Brandley and R.L. Schnaar, J. Leuk. Biol., 1986, *40*, 97; T.W. Rademacher, R.B. Parekh and R.A. Dwek, Ann. Rev. Biochem., 1988, *57*, 785; N. Sharon and H. Lis, Science, 1989, *246*, 227; Chem. Britain, 1990, *26*, 679; "Carbohydrate Recognition in Cellular Function", CIBA Foundation Symp. 145, Wiley, New York, 1989).

These factors, and the complexity of carbohydrates, have made stereoselective glycosylation probably the most challenging reaction in synthetic organic chemistry. Since the last review in this series (R. Khan, J.K. Wold and B.S. Paulsen, in Rodd's Chemistry of Carbon Compounds, Suppl. vol. I FG, ed. M.F. Ansell, Elsevier, 1983, p.231) a number of methods have become more generally used (H. Paulsen, Angew. Chem. Int. Ed. Engl., 1982, *21*, 155; *ibid*, 1984, *21*, 155; Proc. IUPAC Symp. 5th Organic, 1984, 317; Chem. Soc. Rev., 1984, *13*, 15; Studies in Org. Chem., 1986, *25*, 243; GBF Monograph Ser., 1987, *8*, 115;

Angew.Chem. Int. Ed.Engl., 1990, *29*, 823; R.R. Schmidt, Angew. Chem. Int. Ed. Engl.,1986, *25,* 212; P.J. Garegg, Chem. Britain, 1990, *26*, 669; P.J. Garegg and A.A. Lindberg, in "Carbohydrate Chemistry", ed. J.F. Kennedy, Oxford University Press, Oxford, 1988, p.500; H.M. Flowers, Methods in Enzymology, 1987, *50,* 93; R.W. Binkley, J. Carbohydr. Chem., 1988, *7,* vii; RSC Specialist Periodical Reports on Carbohydr. Chem., 1983-1991, *14-23*.

This review deals with glycosylation methods rather than with strategies for oligosaccharide synthesis, which have been covered elsewhere (H. Paulsen, *loc. cit.*; P.J. Garegg and A.A. Lindberg, *loc. cit.*). The use of enzymes in oligosaccharide synthesis is also reviewed, as well as some methods of preparation by polysaccharide degradation. Neighbouring-group effects often enter into the reactions discussed here, however protection-group strategies, and fuller discussion of influences of protecting groups such as steric effects, may be found in the above mentioned reviews (and in particular P.J. Garegg, *loc. cit.*; A.H. Haines, Adv. Carbohydr. Chem. Biochem., 1981, *39,* 13; U. Zehavi, *ibid*, 1988, *46,* 179; N.M. Spijker and C.A.A. van Boeckel, Angew.Chem. 1991, *103*, 179).

The attachment of oligosaccharides to proteins and other macromolecular carriers is necessary to form glycoconjugates (glycoproteins, glycolipids) which are used for such purposes as isolation of enzymes and antibodies (P.J. Garegg and A.A. Lindberg, *loc. cit.;* P.J. Garegg, *loc. cit.*; H. Paulsen, *loc. cit.*; H. Kunz, Angew. Chem. Int. Ed. Engl., 1987, *26*, 294).

The molecular recognition effects associated with complex carbohydrates require analysis of three-dimensional structure, and this is emphasised in the review of physical methods, particularly of NMR spectroscopy.

For nomenclature and structural conventions, see J.F. Kennedy and C.A. White, "Carbohydrate Chemistry", Oxford Univ. Press. 1988; N. Sharon, Eur. J. Biochem., 1986, *159*, 1.

1. Di- and Oligosaccharides

(a) Synthesis

Methods of synthesis have been classified in this review as follows (Figure 1, showing D-configurations):

(i) Neighbouring-group assisted procedures for the synthesis of β-glycosidic linkages (**3**);

(ii) Halide-assisted glycosylation methods, for α-glycosidic linkages (**1**) and (**2**);

(iii) Synthesis of β-glycosidic linkages, without neighbouring-group participation, (**4**) and (**6**);

(iv) The imidate procedure for synthesis of (**1**) and (**3**);

(v) The *n*-pentenyl methodology: synthesis of (**1**), (**3**) and (**2**);

(vi) The 1,2-halonium ion and related methods, synthesis of (**5**), (**6**);

(vii) Synthesis of linkages, (**7**), (**8**), in the oct-2-ulosonic and non-2-ulosonic (KDO and neuraminic) acid series;

(viii) Synthesis of thioglycosides;

(ix) Preparation by degradation of polysaccharides;

(x) Enzyme-catalysed synthesis.

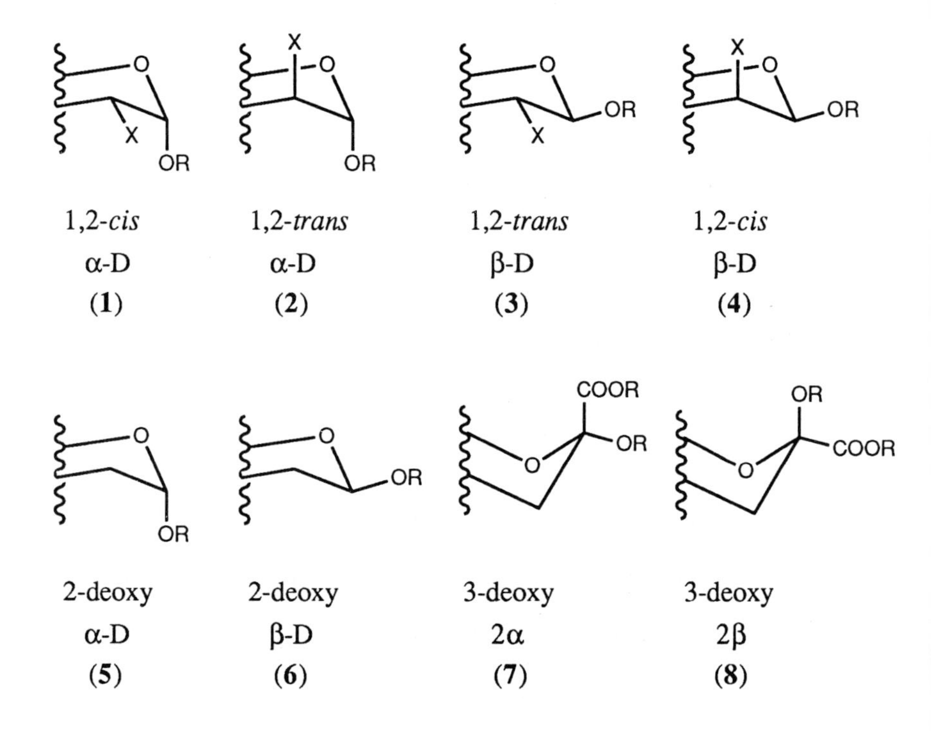

Figure 1

(i) *Neighbouring-group assisted procedures for the synthesis of β-glycosidic linkages* **(3)**

The most widely used method for creating a 1,2-*trans*-β-glycosidic linkage in the D-gluco and D-galacto series is that involving a neighbouring 2-substituent to stabilise the transition state leading to a dioxolonium ion pair intermediate. There is stereoselective attack by nucleophile from the face opposite to the neighbouring group as represented in a simplified manner in Scheme 1.

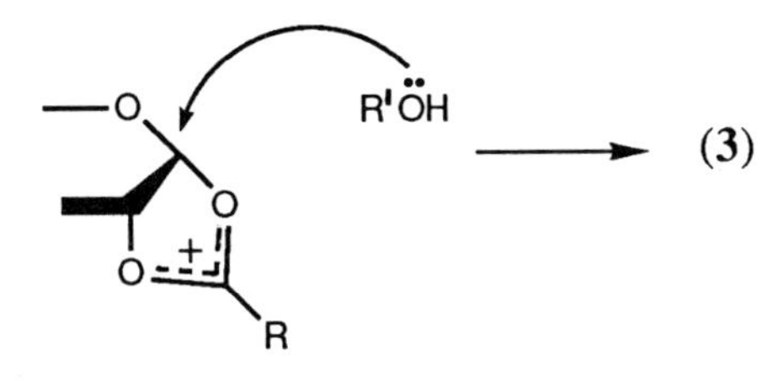

Scheme 1

Benzoate as 2-substituent in a glycosyl bromide, and the soluble halophilic promoter silver triflate, have proved to be an efficient combination for 1,2-*trans*-glycosylation with oxygen at C-2 (S. Hanessian and J. Banoub, Carbohydr. Res., 1977, *53*, C13; O. Hindsgaul *et al*, Carbohydr. Res., 1982, *109*, 109.

The use of base as proton-scavenger is to be avoided as it may result in orthoester products, and is unnecessary as long as 4A molecular sieves are present. Other promoters used to improve on this classical Koenigs-Knorr reaction have been soluble mercury(II) promoters such as the bromide or cyanide (H. Paulsen, *loc. cit.*), however here the steroselectivity is sometimes not very good, particularly in the galacto series. A competing reaction may be the formation of products with a free 2-hydroxyl group which can also give rise to (1 → 2)-linked glycoside. This is less of a problem with a 2-benzoate (P.J. Garegg *et al*, Acta Chem. Scand., 1985, *B39*, 569).

Glycosylation of methyl 4,6-*O*-benzylidene-β-D-glucopyranoside with tetra-*O*-acetyl-α-D-glucopyranosyl bromide in the presence of silver triflate and tetramethylurea gave the 2-linked β-disaccharide in 28% yield (K. Takeo, Carbohydr. Res., 1980, *87*, 147). Silver imidazolate and zinc chloride afforded the β-linked disaccharides from acetobromoglucose with primary and secondary sugar alcohols, the yields being 90% and 40% respectively (P.J. Garegg, R. Johansson and B. Samuelsson, Acta Chem. Scand. Ser. B, 1982, *36*, 249). Tin triflate catalysed the condensation of the acetobromoglucose with other monosaccharides giving the β-configuration exclusively (30-60% yield) (A. Rubineau and A. Malkeron, Tet. Lett., 1985, *26*, 1713).

Formation of 1,2-*trans* glycosidic linkages using trimethysilyl trifluoromethanesulphonate as catalyst, and 1,2-*trans*-diacetate as glycosylating agent, dispenses with the need for glycosyl halide as intermediate (T. Ogawa, K. Beppu and S. Nakabayashi, Carbohydr. Res., 1981, *93*, C6; S. Nakabayashi and C.D. Warren, Carbohydr. Res., 1986, *150*, C7).

This catalyst was also successfully employed in reactions involving orthoesters and oxazolines (see below). The coupling reaction of *O*-trimethylsilyl glycosides and 6-*O*-(*t*-Bu Ph_2Si)- protected galactosides in the presence of trimethylsilyl triflate led to β-(1→6) linkages (E.M. Nashed *et al.*, J. Org. Chem., 1989, *54*, 6116).

In a scheme of oligosaccharide synthesis, the preparation of a glycosyl chloride or bromide as glycosyl donor for extension of the oligosaccharide structure is often not possible at a particular stage. Thioglycoside can be used as a protected glycosyl donor, convertible into the fluoride for further 1,2-*trans*-glycosylation (K.C. Nicolau *et al.* J. Am. Chem. Soc., 1984, *106*, 4189; K.C. Nicolau, Chem. Britain, 1985, *21*, 813). More recently, thioglycosides have been used directly by activation with methyl triflate and other promoters. This produces a sulphonium intermediate which in the presence of a participating 2-substituent leads to 1,2-*trans*-glycosylation with high stereoselectivity (H. Lönn, Carbohydr. Res., 1985, *139*, 105, 115; P. Fügedi *et al*, Glycoconjugate J., 1987, *4*, 97; F. Dasgupta and P.J. Garegg, Carbohydr. Res., 1988, *177*, C13; V. Pozgay and H.J. Jennings, *ibid*, 1988, *179*, 81; J.F.G. Vliegenthart *et al.*, *ibid*, 1989, *195*, 75). Improved promoters are being developed by these workers (F. Andersson *et al.*, in "Trends in Synthetic Carbohydrate Chemistry", Amer. Chem. Soc. Symp. series, *386*, 1989, p.117), as well as new sulphur-based glycosyl donors.

The β-1-thioglycosides have been used in the block synthesis of an oligosaccharide involving the formation of β-(1→6) and β-(1→3)glucopyranosyl glucopyranoside linkages (F. Andersson *et al.*, Tetrahedron Lett., 1986, *27*, 3917).

In 1,2-*trans*-glycosylations of 2-amino-sugars, the presence of an *N*-phthalimido protecting group has been shown to give high selectivity with silver triflate as promoter (R.U. Lemieux, T. Takeda and B.Y. Chung, in "Synthetic Methods for Carbohydrates", ed. S. El Khadem, Amer. Chem. Soc. Symp. series, 1976, *39*, p.90) and probably functions as neighbouring group in the reaction in the same manner as, for example, benzoate. Methods in this series which avoid the preparation of glycosyl halide uses 1-acetates, and ferric chloride as promoter (M. Kiso and L. Anderson, Carbohydr. Res., 1985, *136*, 309), or 1-thioglycoside and ferric chloride (F. Dasgupta and P.J. Garegg, Acta. Chem. Scand., 1989, *43*, 471).

Aminosugar-derived 1,2-oxazolines have been used in Lewis acid catalysed reactions which are formally related to the Koenigs-Knorr method (M.A. Nashed, *et al.*, Carbohydr. Res., 1980, *82*, 237; R. Gigg and R. Conant, Carbohydr. Res., 1982, *100*, C1).

The oxazolines were preceded in use by the 1,2-orthoesters of Kochetkov and co-workers, (P.A. Gorin, E.M. Barreto-Bergter and F.S. Da Cruz, Carbohydr. Res., 1981, *88,* 177) and to these have been added the 1,2-thioorthoesters (L.V. Backinowsky *et al.*, Carbohydr. Res., 1980, *85,* 209) and 1,2-cyanoorthoesters (A.F. Bochkov and N.K. Kochetkov, Carbohydr. Res., 1975, *39*, 355; V.I. Betaneli *et al.*, *ibid*, 1981, *94,* C1). Preformed cyanoorthoesters can be glycosylated under Helferich (mercuric cyanide catalysed) conditions without affecting the cyanoorthoester groups (V.I. Betaneli *et al.,* Carbohydr. Res., 1983, *113,* C1). Reaction of this group is with the trityl ether of the hydroxyl function to be glycosylated (Scheme 2), typically at room temperature in dichloromethane with 10% perchlorate.

$TrClO_4$

-TrCN

R ÖTr

CN

Me

Me

Scheme 2

The cyanoorthoester method is stereospecific in the majority of cases. Where there is some formation of 1,2-*cis* glycosidic linkage, it has been shown that this can be corrected under high pressure (~10 Kbar) (N.K. Kochetkov *et al.*, Carbohydr. Res., 1987, *164*, 241; *ibid*, 1987, *167*, C8). An examination of 3-glucosylation of 3-*O*-trityl-D-glucopyranose tetraacetate for example showed that increase in pressure changed poor selectivity to specificity.

If the reaction is carried out with a derivative bearing the cyanoorthoester and *O*-trityl groups it results in the formation of a chain of monomeric units linked by 1,2-*trans* glycosidic linkages. Starting from a hetero-oligosaccharide, a regular heteropolysaccharide is produced, built up of repeating units, and there has been much success with this method in the synthesis of microbial polysaccharides (N.K. Kochetkov, Chem. Soc. Rev., 1990, *19*, 29).

(ii) Halide-ion catalysed synthesis of glycosides **(1)** *and* **(2)**

In this *in situ* anomerisation procedure, the more stable α-D-glycopyranosyl halide with a non-active substituent at C-2 is used. A catalyst such as tetraalkyl ammonium halide causes equilibration of the α-D-glycopyranosyl and β-D-glycopyranosyl halides (Scheme 3) (H. Paulsen, *loc. cit.*). Reaction of the more unstable β-D-halide would be the faster and thus the reaction is considered to proceed under kinetic control, the small concentration of the β-D-glycopyranosyl halide allowing selective formation of the desired α-*O*-glycopyranoside.

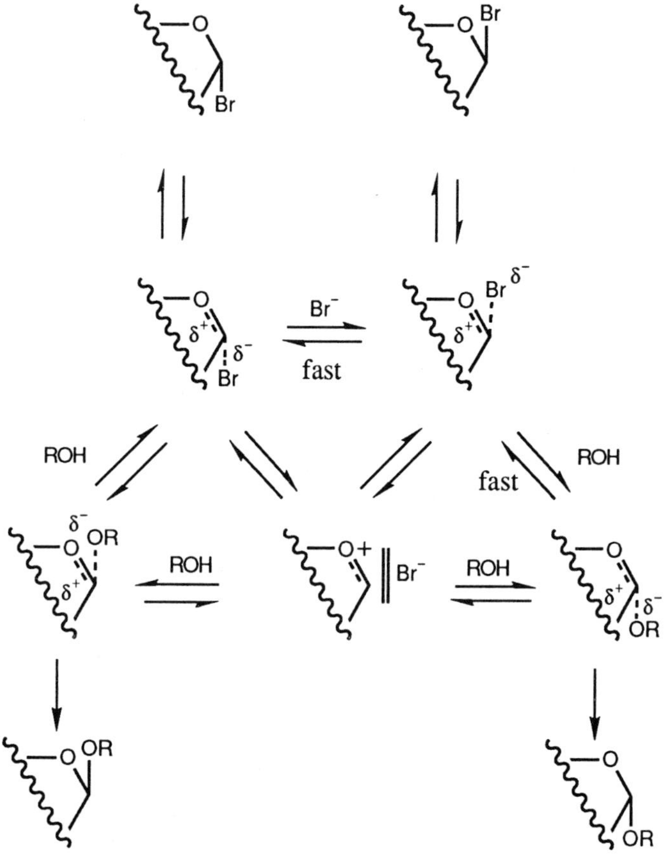

Scheme 3

The most important parameters determining selectivity and yield are the reactivities of the halide, catalyst and alcohol. The less stable bromides are more reactive than chlorides. For a given halogen, benzyl derivatives are more reactive than acetylated or benzoylated compounds, and the positions of the benzyl groups are more important than their number. Trichloroacetyl groups decrease, and deoxy groups increase reactivity.

The activity of the catalyst increases in the order, tetraalkylammonium halide, mercury salts (cyanide, bromide), silver salts (perchlorate, triflate). Secondary hydroxyl groups in pyranoses are of moderate reactivity and are matched by reactive halide and catalyst in solvent of low dielectric constant (dichloromethane or mixture) with 4Å molecular sieves at room temperature. With more reactive primary hydroxyl groups the stereoselectivity decreases unless a milder catalyst is used or the pure β-halide is used directly. Better selectivity is obtained in the galacto series (including fucose) than in the gluco series with soluble silver salts (P.J. Garegg, H. Hultberg and C. Lindberg, Carbohydr. Res., 1980, *83*, 157; P.J. Garegg and H. Hultberg, *ibid,* 1982, *110,* 261). Tetra-*O*-benzyl-α-D-glucopyranosyl chloride used in the presence of silver imidazolate and mercury (II) chloride with 1,2:3,4-di-*O*-isopropylidene-α-D-galactopyranose and 1,2:5,6-di-*O*-isopropylidene-α-D-glucofuranose gave the α-linked disaccharides in high yield (P.J. Garegg, C. Ortega and B. Samuelsson, Acta. Chem. Scand., Ser. B, 1981, *35,* 631)

Glycosyl fluorides may be prepared using (diethylamino) sulphur trifluoride (DAST) (W. Rosenbrook, D.A. Riley and P.A. Lartey, Tetrahedron Lett., 1985, *26*, 3; G.H. Posner and S.R. Haines, *ibid*, 1985, *26*, 5) and have been used for α-glycosylations (T. Mukaiyama, Y. Murai and S. Shoda Chem. Lett., 1981, 431; T. Mukaiyama, Y. Hashimoto and S. Shoda, Chem. Lett., 1983, 935; S. Hashimoto, M. Hayashi and R. Noyori, Tetrahedron Lett., 1984, *25*, 1379; H. Kunz and W. Sager, Helv. Chim. *Acta,* 1985, *68*, 283).

Thioglycosides, which are stable to most carbohydrate protection group strategies, are convertible to bromides with *N*-bromo-succinimide (K.C. Nicolau, S.P. Seitz and D.P. Papahatjis, J. Am. Chem. Soc., 1983, *105,* 2430) and to fluorides by treatment with NBS-DAST or NBS/HF-pyridine complex (K.C. Nicolau *et al.,* J. Chem. Soc., Chem. Commun., 1984, 1155; K.C. Nicolau *et al.*, J. Am. Chem. Soc., 1984, *106*, 4189 M. Kreuzer and J. Thiem, Carbohydr. Res., 1986, *149*, 347) and used in glycosylations. The use of tetraalkylammonium bromide with thioglycoside and a non-participating 2-substituent converts the reaction to a halide-assisted one. References to conditions for the uses of thioglycosides are given in sections (i) and (vi). It has been shown that *in situ* activation of thioglycosides with bromine provides another glycosylation method (J.O. Kihlberg *et al.*, J. Org. Chem., 1990, *55*, 2860). A mild method for the synthesis of α-linked disaccharides involves reaction of the glycosyl thiopyridyl with a sugar alcohol in the presence of iodomethane (V.G. Reddy *et al.*, Tetrahedron Lett., 1989, *30*, 4283).

(iii) Synthesis of β-glycosidic linkages **(4)** *and* **(6)** *without neighbouring-group participation*

These configurations present the greatest difficulties, since participation from a C-2 group is not available to offset preferred α-glycosylation of a cationic intermediate. Displacement reaction via a bimolecular-type transition state, starting from an α-oriented leaving group, is possible with certain promoters (P.J. Garegg and A.A. Lindberg in "Carbohydrate Chemistry," ed. J.F. Kennedy, Oxford Univ. Press, 1988, p.500). The insoluble silver zeolite yields β-D-mannopyranosides from 6-*O*-acetyl-2,3,4-tri-*O*-benzyl-α-D-mannopyranosyl bromide (P.J. Garegg and P. Ossowski, Acta Chem. Scand., 1983, *B37*, 249; P.J. Garegg *et al.*, Carbohydr. Res., 1983, *119*, 95) and from other mannose and glucose derivatives (C.A.A. van Boeckel, T. Beetz and S.F. van Aelst, Tetrahedron Lett. 1984, *40,* 4097). Acyl groups in the 4-position, or the use of galactosyl bromide, favours β-linkage (C.A. van Boeckel and T.A. Beetz, Rec. Trav. Chim. Pay-Bas, 1986, *104*, 171).

Glyculosyl bromides (**9**) react with sugar alcohols in the presence of silver carbonate to give β-glycosuloses, and this method has been used to synthesise glycosyl-(1→4)-β-D-mannose and glycosyl-(1→4)-β-D-mannosamine units (F.W. Lichtenthaler, E. Cuny and S. Weprek, Angew. Chem. Int. Edn. Eng., 1983, *22*, 891; F.W. Lichtenthaler, E. Kaji and S. Weprek, J. Org. Chem., 1985, *50*, 3505).

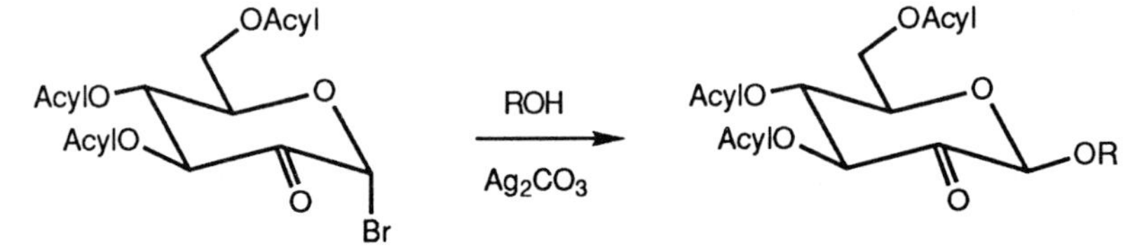

(**9**)

Scheme 4

The stable crystalline benzoylated gluculosyl bromide is prepared in four steps, but in high overall yield, and will undergo heterogeneous exchange with iodide (E. Kaji *et al.*, Bull. Chem. Soc. Japan, 1988, *61*, 1291; F.W. Lichtenthaler, S. Swidetsky and K. Nakamura, Tetrahedron Lett. 1990, *31,* 71).

Insoluble halophilic promoters such as silver silicate (H. Paulsen and O. Lockhoff, Chem. Ber., 1981, *114*, 3102; H. Paulsen *et al.*, Carbohydr.

Res., 1982, *103*, C7) or silver zeolite (P.J. Garegg and P. Ossowski, *loc. cit.*) have been used with acylated 2-deoxyglycosyl halides to improve yields.

(iv) The imidate procedure: synthesis of **(1)** *and* **(3)**

Glycosylation of amides with glycosyl halides having non-participating 2-substituents, in the presence of silver salts, produces imidates by *O*-glycosylation (Scheme 5) (P. Sinaÿ, Pure Appl. Chem., 1978, *50*, 1437; J.R. Pougny *et al.*, Nouv. J. Chim., 1978, *2*, 389). These have been used in oligosaccharide synthesis to give yields and stereoselectivities comparable with halide-assisted glycosylation, even in the α-D-galactosylation of unreactive hydroxy compounds (M.L. Milat and P. Sinaÿ, Carbohydr. Res., 1982, *100*, 263). Useful selectivities have been found in partial glycosylations involving different secondary hydroxyl groups (P.J. Garegg and I. Kvarnström, Carbohydr. Res., 1981, *90*, 61).

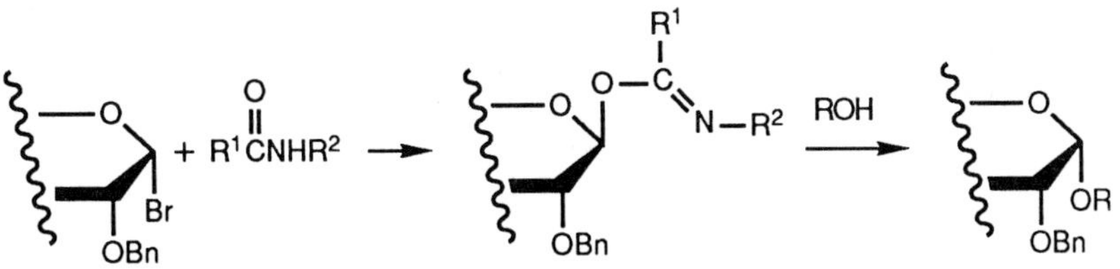

Scheme 5

The imidate method has been extended to trichlorimidates (R^1=CCl_3, R^2=H) (R.R. Schmidt, J. Michel and M. Roos, Liebigs Ann. Chem., 1984, 1343). The β-imidate is formed first under kinetic control, in this case starting from trichloroacetonitrile, and may be equilibrated to the α-imidate under strongly basic reaction conditions. Reactions of both types of imidate have been carried out with a range of hydroxy compounds, including protected pyranosides with one free hydroxyl group. Reactions give predominant inversion of configuration, thus the procedure, with a non-participating 2-substituent, affords the choice of 1,2-*cis*-α-D or 1,2-*trans* - configuration (R.R. Schmidt, Angew. Chem. Int. Ed. Engl., 1986, *25*, 212).

Convenient preparations of α- and β-trichloroacetimidates of 3,4,6-protected 2-azidogalactosyl, 2-azido-glucosyl and 2-azido-lactosyl derivatives have been reported and their use in glycosylation reactions explored (R.R. Schmidt and G. Grundler, Angew. Chem. Suppl. 1982,

1707; R.R. Schmidt, J. Michel and M. Roos, *loc. cit.*; R.R. Schmidt, *loc. cit.*).

The influences of catalyst, solvent and temperature on the reactions of *O*-(tetra-*O*-benzyl-α-D-glucopyranoside) trichloroacetimidate with primary and secondary sugar hydroxyl groups were examined (R.R. Schmidt and J. Michel, J. Carbohydr. Chem., 1985, *4*, 141). With BF_3 in dichloromethane at between -18° and -40°, anomeric ratios of about 1:4 (α:β) were obtained for secondary hydroxyl groups.

(v) The n-pentenyl methodology: synthesis of **(1)**, **(3)** *and* **(2)**

The glycosyl donors, *n*-pentenyl glycosides, can be prepared directly from an aldose by modified Fischer glycosidation methods, are stable to diverse chemical conditions, and are readily converted into glycosyl halides for Koenigs-Knorr reactions (D.R. Mootoo, V. Date and B. Fraser-Reid, J. Am. Chem. Soc., 1988, *110*, 2662; D.R. Mootoo *et al.*, *ibid*, 1988, *110*, 5582; D.R. Mootoo, P. Konradsson and B. Fraser-Reid, *ibid*, 1989, *111*, 8540; B. Fraser-Reid *et al.*, J. Org. Chem. 1990, *55*, 6068; A.J. Ratcliffe, P. Konradsson and B. Fraser-Reid, J. Am. Chem. Soc., 1990, *112,* 5665). Activation by halonium ion addition, with *N*-iodosuccinimide and triflic acid for example as source, is mild and chemospecific (Scheme 6).

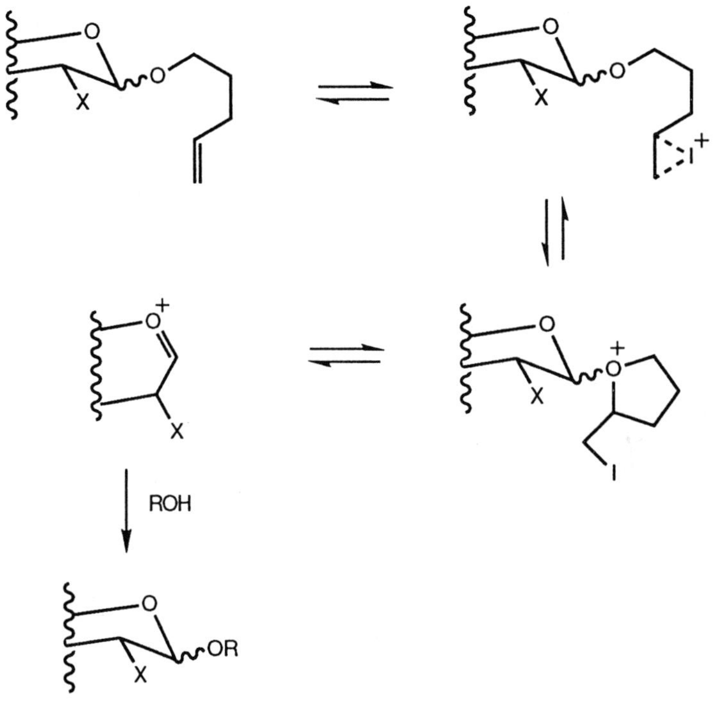

Scheme 6

The α,β-selectivity can be controlled by choice of solvent and by other methods (B. Frazer-Reid *et al.,* J. Chem. Soc. Chem. Commun., 1988, 823; D.R. Mootoo and B. Fraser-Reid, Tetrahedron Lett., 1989, *30*, 2363). The 2-*N*-substituted glycosyl donors behave similarly to the corresponding *N*-substituted glycosyl halides, but with contrasting sturdiness, and under a milder activation procedure. The *n*-pentenyl glycosides, and similar glycosyl donors, may be "disarmed" by an adjacent substituent (for example OCOR). Presence of these electron-withdrawing groups means that the positively-charged intermediate is less favoured than when there is an adjacent alkoxy group.

(vi) 1,2-halonium ion and related methods: synthesis of **(5)**, **(6)**

Most stereoselective syntheses of these structures have developed from the discovery that an acetylated glycal can be converted to such a glycoside *via* an iodonium compound (Scheme 7) (R.U. Lemieux and A.R. Morgan, Canad.

J. Chem., 1965, *43*, 2190). Improvements have been the use of N-bromosuccinimide, N-iodosuccinimide (J. Thiem and B. Meyer, Chem. Ber., 1980, *113,* 3058; J. Thiem and S. Köpper, Angew. Chem. Int. Edn. Engl., 1982, *21*, 779) or silver imidazolate-zinc chloride, or -mercury(II) chloride and iodine (P.J. Garegg and B. Samuelsson, Carbohydr. Res., 1980, *84*, C1; P.J. Garegg *et al. ibid* 1981, *92*, 157). That using iodosuccinimide has been applied extensively (S.J. Danishefsky *et al.*, J. Am. Chem. Soc., 1987, *109*, 8119; J. Thiem, in "Trends in Synthetic Carbohydr. Chem.," ed. D. Horton, L.D. Hawkins and G.J. McGarvey, ACS Symposium Series No. 386, 1989, p. 131).

A related method uses glycosyloxyselenation-deselenation (G. Jaurand *et al.,* J. Chem. Soc. Chem. Commun., 1981, 572; *ibid*, 1982, 701).

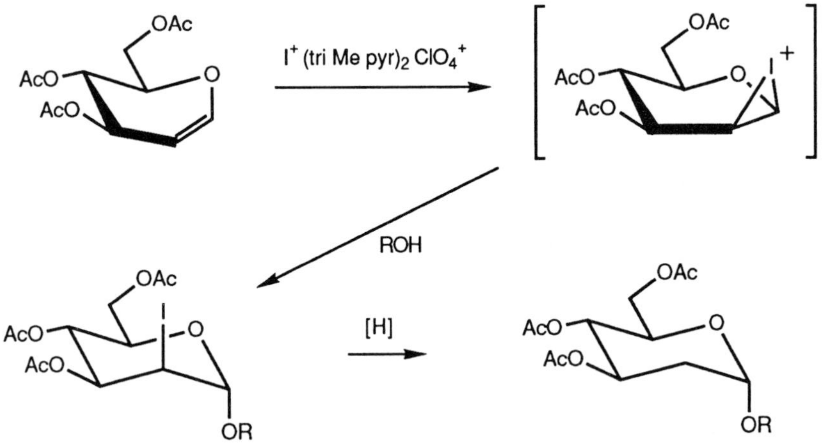

Scheme 7

Participation by the 2-bromosubstituent has been used in Koenigs-Knorr type glycosylations of 3,4-di-*O*-acetyl-2,6-dibromo-2,6-dideoxy-α-D-manno- and -glucopyranosyl bromides (K. Bock *et al.*, Carbohydr. Res., 1984, *130,* 125). The α-D-glycosides (from the manno starting material) were formed in high yields, but stereoselectivity for β-glycosylation (from the gluco starting material) was lower. The products were reduced to the corresponding deoxy glycosides.

The addition of sugar sulphenate esters to glycals catalysed by Lewis acids (trimethylsilyl triflate the most efficient) has been investigated (Y. Ito and T. Ogawa, Tetrahedron Lett., 1987, *28,* 2723) (Scheme 8).

Scheme 8

Acid-labile trityl and silyl ethers survived the reaction conditions. For the 3,4,6-tri-*O*-benzyl glucal and galactal, 1,2-diequatorial addition was a major reaction course, suggesting that the stereochemical outcome is mainly governed by steric rather than coordinate interaction between incoming phenyl-sulphenyl group and benzyloxy group. This result is in contrast to the two-step α-selective glycosyl selenation (G. Jaurand *et al., loc. cit.*) which may be explained as diastereofacial selective formation of episelenonium ion and subsequent diaxial ring opening.

Stereospecific 1,2-migrations (Scheme 9) make possible deoxygenation at C-2 as well as stereocontrolled synthesis of α-and β-glycosides of the resulting deoxysugars (K.C. Nicolau *et al.*, J. Am. Chem. Soc., 1986, *108*, 2466).

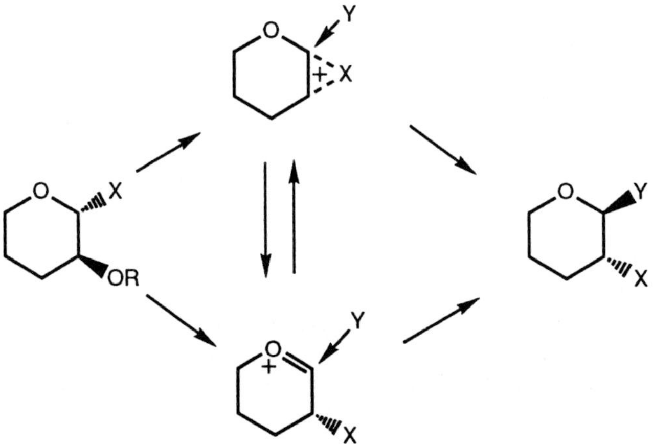

Scheme 9

Reaction of a thiophenyl α-D-glycoside for example with (diethylamino) sulphur trifluoride (DAST) gave the 1-fluoro-2-thiophenyl sugar derivative (**10**).

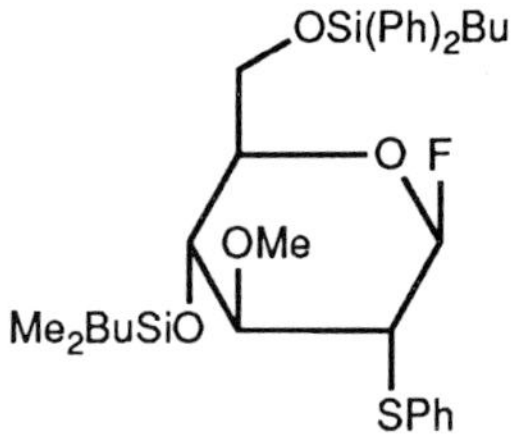

(**10**)

This was then used in α- or β-directed glycosylation by neighbouring group and/or solvent participation. Finally desulphurisation afforded the 2-deoxy-α-glycoside or its β-isomer. Coupling was achieved using stannous chloride, and in the presence of a tin-complexing solvent or reagent (Et_2O or Me_2S) the SPh group remained free to direct the stereochemical course of glycosylation almost exclusively to the otherwise disfavoured isomer.

The increased scope for the use of phenyl thioglycosides by their conversion into glycals has been described by the group of P. Sinaÿ (A. Fernandez-Mayorales *et al.,* Carbohydr. Res., 1989, *188*, 91).

(vii) Synthesis of linkages, (**7**),(**8**), *in the octulosonic and nonulosonic acid series*

The synthesis of α-linked *N*-acetyl neuraminic acid (NeuAc) and KDO (3-deoxy-D-*manno*-2-octulosonic acid) glycosides is a special problem due to the impossibility of neighbouring-group participation, as well as the fact that anomeric effect favours formation of a β-linkage (H.Paulsen *et al.,*Liebigs Ann. Chem., 1984, 1270; T. Ogawa and M. Sugimoto, Carbohydr. Res., 1985, *135*, C5-C9). A glycosyloxyselenation procedure carried out on the KDO derivative (**11**) has yielded the α-2,6-disaccharide in 40% yield before deselenation, no β-anomer being formed (K.Ikeda *et al.,* Carbohydr. Res., 1989, *189,* C1-C4).

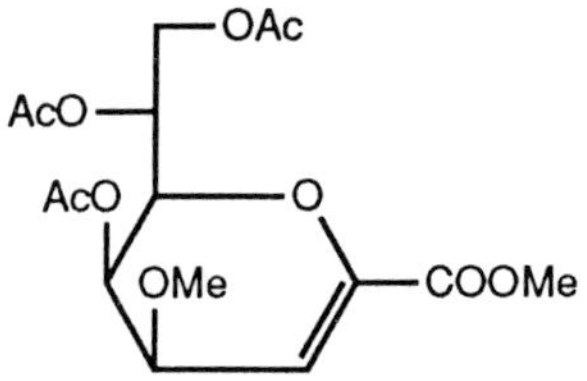

(**11**)

Glycosylations of 2β-chloro and 2β-bromo-3β-hydroxy-*N*-acetyl neuraminic acid derivatives, prepared from 2-deoxy-2,3-dehydro-NeuAc methyl ester *via* the 2,3-epoxides, with various acceptors such as 6-unprotected glucose, 3-unprotected galactose and 3'-unprotected lactose derivatives were carried out in the presence of silver triflate to give the 2α-neuraminyl glycoside in preference to the corresponding β-glycosides, except for the lactoside. These results were obtained in toluene at -15°, and in benzene at room temperature the β-glycoside predominated. The 3β-hydroxy group in the products could be removed by tri-*n*-butyl SnH reduction of the thiobenzoate (K. Okamoto *et al.*,Tetrahedron Lett., 1986, *27*, 5233; Tetrahedron, 1987, *43*, 5919; *ibid*, 1988, *44*, 1291). Use of 2-halo-2-deoxy-3β-phenylthio derivatives with neighbouring-group participation of the 3β-substituent has been developed as a more direct method (Y. Ito and T. Ogawa, Tetrahedron Lett., 1987, *28*, 6221; T. Kondo, H. Abe and T. Goto, Chem. Lett., 1988, 1657; T. Ogawa *et al.*, Carbohydr. Res. 1988, *172*, 183; *ibid*, 1989, *186*, 95; Y. Ito and T. Ogawa, Tetrahedron, 1990, *46*, 89).

(viii) Synthesis of thioglycosides

Thiosugars have been used to prepare analogues of oligosaccharide substrates for glycosidases, using the concept of enhancement of thiolate as nucleophile in polar aprotic solvents (more recently, 1,3-dimethyl-2-oxohexahydropyrimidine in preference to the toxic hexamethylphosphoric triamide) (M. Apparu *et al.*, Canad. J. Chem., 1981, *59*, 314; M. Blanc-Muesser, J. Defaye and H. Driguez, J. Chem. Soc. Perkin Trans I, 1982, 15; M. Blanc-Muesser *et al.*, *ibid*, 1984, 1885). The (1→4)-thio-disaccharides and trisaccharides were prepared by displacement of sulphonate with glycose 1-thiolates.

The branched thiocyclomalto-oligosaccharides, 6-*S*-α- and 6-*S*-β-D-glucopyranosyl-6-thiocyclomaltoheptaose, have been prepared similarly from 6-*O*-*p*-tolylsulphonylcyclomaltoheptaose (J. Defaye *et al.*, Carbohydr. Res., 1989, *192*, 25).

From the sodium salt of protected 2-thio-*N*-acetyl-neuraminic acid, a series of (2→6)-linked and (2→3)-linked α-thioglycoside disaccharides have been prepared (A. Hasegawa, J. Nakamura and M. Kiso, J. Carbohydr. Chem., 1986, *5*, 21; O. Kanie *et al.*, *ibid*, 1987, *6*, 117).

(ix) Preparation by degradation of polysaccharides

Apart from the elucidation of polysaccharide structure by degradation to oligosaccharide fragments, solvolysis with hydrogen fluoride or hydrogen fluoride in methanol has been used to prepare unique oligosaccharides (Y.A. Knirel and E.V. Vinogradov, Adv. Carbohydr. Chem Biochem., 1989, *47,* 167). Labilities of glycosidic linkages to solvolysis varies greatly, and amidic bonds, including *N*-acetyl groups, are unaffected. These also stabilise adjacent glycosidic linkages. Thus, partial solvolysis at -30° of an *E. coli O*-antigen polysaccharide cleaved the glycosidic linkages of the D-mannosyl residues but not those of *N*-acetylglucosamine, to give the trisaccharide GlcNAc β(1→4)GlcNAc β(1→4)Man in high yield (P.-E. Jansson *et al., ibid*, 1987, *165*, 87).

The β(1→6)-linked oligosaccharides of 2-acetamido-2-deoxy-D-glucose (d.p. range up to six) were obtained from HF treatment of chitin (J. Defaye, A. Gadelle and C. Pedersen, Carbohydr. Res., 1989, *186*, 177).

Controlled acid hydrolysis of poly-α(2→8)-linked homopolymers of *N*-acetylneuraminic acid (colominic acid) and of its homologous poly-*N*-glycolylneuraminic acid afforded the corresponding disaccharides (R. Roy and R.A. Pon, Glycoconj. J., 1990, *7,* 3).

(x) Enzyme-catalysed synthesis

The problems of glycosyl activation, selectivity, complexity and aqueous solubility associated with chemical synthesis of oligosaccharides are potentially soluble more directly by the use of enzymes. Two groups of enzymes are involved in the synthesis of oligosaccharides *in vivo.* The largest group, those of the Leloir pathway, transfer sugars activated as sugar nucleoside phosphates (NDP-sugars) to the growing oligosaccharide chain (L.F. Leloir, Science, 1971, *172,* 1299; N. Sharon, "Complex Carbohydrates", Addison-Wesley, Reading, MA, 1975). Non-Leloir pathway enzymes transfer carbohydrate units activated simply as sugar phosphates.

The subject of *in vitro* synthesis of oligosaccharides using enzymes has been comprehensively reviewed (E.J. Toone *et al.,* Tetrahedron, 1989, *45,* 5365). Two strategies are available. In the first, glycosyl transferases (EC 2.4) are used which *in vivo* elaborate oligo-and polysaccharides. The mechanism of this catalysis has been reviewed (M.L. Sinnot, RSC Chem. Rev., 1990, *90,* 1171). The second strategy uses glycosidases or glycosyl hydrolases (EC 3.2) which *in vivo* have a catabolic function, and cleave glycosidic linkages to form mono- or oligosaccharides. They can be used to form glycosidic linkages in the presence of a suitable nucleophile other than water. The nature of the intermediate involved is still under debate (C.B. Post and M. Karplus, J. Am. Chem. Soc., 1986, *108,* 1317; S.G. Withers and I.P. Street, *ibid*, 1988, *110,* 8551.

Only a small number of glycosyltransferases are presently available, and genetic engineering (cloning or modification) will have an important influence. Several transferases have already been cloned, including glucosyl-, fucosyl- and sialyltransferases (E.S. Creeger and L.I. Rothfield, Methods Enzymol., 1982, *83*, 326; J. Weinstein *et al.*, J. Biol. Chem. 1987, *262,* 17735; J.C. Paulson *et al.,* Biochem. Soc. Trans., 1987, *15,* 618; L.K. Ernst *et al.,* J. Biol. Chem., 1989, *264*, 3436; V.P. Rajan *et al.*, J. Biol. Chem., 1989, *264*, 11158). A number of syntheses on the 10-100 mg scale have been carried out, in particular using the galactosyl and sialyltransferases (E.J. Toone *et al., loc. cit.*)

A transferase from bovine milk catalysed the preparation of β-D-galactopyranosyl disaccharides using UDP-galactose and an acceptor. In this way Gal β(1→4)GlcNAc N-acetyllactosamine has been prepared on a ten-gram scale (H.A. Nunez and R. Barker, Biochemistry, 1980, *19*, 489; C.-H. Wong, S.H. Haynie and G.M. Whitesides, J. Org. Chem., 1982, *47* 5416). However, preparation of the trisaccharide sialyl *N*-acetyllactosamine by addition of NeuAc to *N*-acetyllactosamine (Scheme 10) as well as preparation of other sialylated oligosaccharides, was carried out on milligram quantities due to limited availability of the 2,6-sialyltransferase (S. Salesan, and J.C. Paulson, J. Am. Chem. Soc., 1986, *108,* 2068; E.S. Simon, M.D. Bednarski, and G.M. Whitesides, J. Am. Chem. Soc., 1988, *110,* 7159; J. Thiem and W. Treder, Angew. Chem. Int. Ed. Engl., 1986, *25,* 1096).

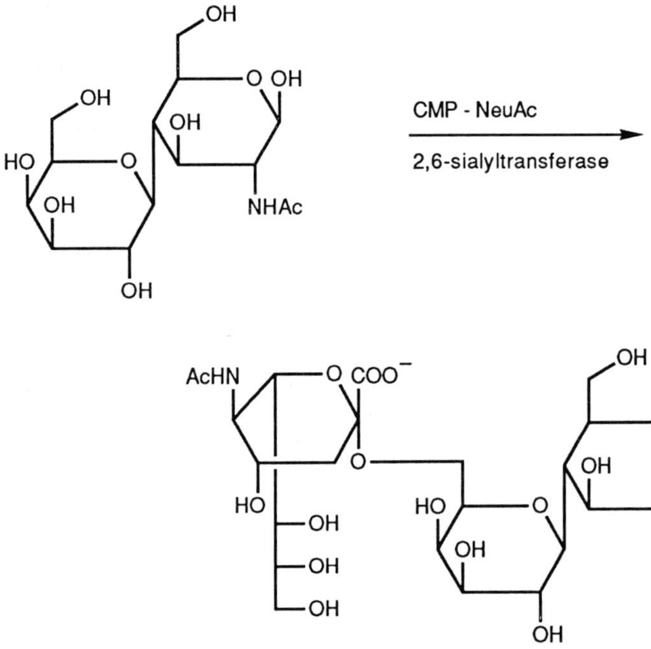

CMP - NeuAc

2,6-sialyltransferase

Scheme 10

Techniques of solid-phase synthesis have been applied using galactosyl transferases (U. Zehavi, S. Sadeh and M. Herchman, Carbohydr. Res., 1983, *124,* 23; U. Zehavi and M. Herchman, *ibid*, 1984, *128,* 160; U. Zehavi and M. Herchman, *ibid*, 1984, *133*, 339). The donor was UDP-Gal, and acceptor polymer-bound, on a scale of less than 10mg.

Sucrose synthetase has been used to transfer glucose from UDP-glucose to 1-deoxy-1-fluorofructose, to generate 1'-deoxy-1'-fluorosucrose (P.J. Card and W.D. Hitz, J. Am. Chem. Soc., 1984, *106,* 5348; P.J. Card, W.D. Hitz and K.G. Ripp, J. Am. Chem. Soc., 1986, *108,* 158).

UDP-Gal transferase acts on a wide range of acceptor substrates (M.M. Palcic, O.P. Srivastava and O.Hindsgaul, Carbohydr. Res., 1987, *159,* 315; C. Augé, *et al.,* Tetrahedron Lett., 1984, *25,* 1467). The specificity of other glycosyltransferases for acceptor sugars, NDP-sugars and their analogues is being evaluated (E.J. Noone *et al., loc. cit.).*

The glycosidases show high specificity for both the glycosyl moiety and the glycosidic linkage, but little if any specificity for the aglycon nucleophile which reacts with the intermediate enzyme complex. Greater availability and stability of glycosidases, and their non-requirement for activated sugar nucleosides, must be considered along with low yields and formation of mixtures (E.J. Noone *et al., loc. cit.*; C. Bucke and R.A. Rastall, Chem. Britain, 1990, *26*, 675). They can be used synthetically either starting from the glycoside, or from the specific free monosaccharide.. The first mode, known as "transglycosylation", uses aryl glycosides, glycosyl fluorides and disaccharides for example as donors. By derivatising the substrate with superior leaving groups, the kinetics of binding are altered - the kinetic approach. This approach has been studied sporadically since the 1950's (S. Chiba *et al.,* Agric. Biol. Chem., 1962, *26,* 787; D.H. Hutson, Biochem. J., 1964, *92,* 142; K.W. Knox, *ibid*, 1965, *14,* 534; Y.-T.Li, J. Biol. Chem., 1967, *242*, 5474; D.J. Manners *et al.,* Carbohydr. Res., 1968, *7,* 291; M. Takanashi, T. Shimomura and S. Chiba, Agric. Biol. Chem., 1969, *33*, 1399; S. Chiba and T. Shimomura, *ibid*, 1971, *35*, 1292). Recently, Nilsson has investigated ways of controlling the specificity. The nature of the glycosyl acceptor affects the regiospecificity (K.G.I. Nilsson, Carbohydr. Res., 1987, *167*, 95). For example, when α-galactosidase acts on *p*-nitrophenyl-α-galactoside as donor, and methyl α-galactoside as acceptor, predominantly 1→3 linkage is formed, together with 1→6 linkage. When *o*-nitrophenyl-α-galactoside is used as acceptor, predominantly 1→2 linkage is formed with a trace of 1→3.

Significant work has been done in recent years on the use of glycosidases in the second mode mentioned above, starting from the monosaccharide rather than the glycoside (A.M. Stephen, S. Kirkwood and F. Smith, Canad. J. Chem., 1962, *40*, 151; K. Ajisaka, M. Nishida and M. Fujimoto, Biotechnol. Lett., 1987, *9,* 243; A. Ajisaka, M. Fujimoto and M. Nishida, Carbohydr. Res., 1988, *180*, 35; E. Johansson *et al.,* Enzyme Microbiol. Technol., 1989, *11,* 347; A. Ajisaka and H. Fujimoto, *ibid*, 1989, *185,* 139; Z.L. Nikolov, M.M. Meagher and P.J. Reilly, Biotechnol. Bioeng., 1989, *34* ,694). Conditions of low water activity favour "reverse hydrolysis" - the thermodynamic or equilibrium approach. In a two-phase system, for example of dextran and polyethylene glycol, the enzymes tend to partition into one phase, making it possible to separate them from oligosaccharide products. Typically a 50:50 mixture of sugar substrate and acceptor, up to a total concentration of over 45% by weight, is used (C. Bucke and R.A. Rastall, *loc. cit.*). Under these conditions, with yields usually 5-10%, α-mannosidase from jack bean has been shown to produce a wide range of hetero-oligosaccharides.

Branched cyclodextrins (cyclomalto-oligosaccharide) have been prepared from cyclodextrins, and maltose or maltotriose or glucosyl fluoride, with pullulanase or a bacterial isoamylase. (Y. Sakano, M. Sano and T. Kobayashi, Agric. Biol. Chem., 1985, *49*, 3391; J.-I. Abe *et al.,*

Carbohydr. Res. 1986, *154,* 81; S. Kitahata, Y. Yoshimura and S. Okada, *ibid,* 1987, *159* 303; *ibid,* 1987, *168*, 285; Agric. Biol. Chem., 1988, *52,* 1655; S. Hizukuri *et al.,* J. Jpn. Soc. Starch Sci., 1986, *33,* 119).

(b) Physical Methods

(i) Nuclear Magnetic Resonance Spectroscopy

References to published NMR investigations on many di- and oligosaccharides have been tabulated (RSC Specialist Periodical Reports on Nuclear Magnetic Resonance, *10-20*, ed. G.A. Webb, Chem. Soc., London, 1981-1991).

High-field NMR spectroscopy presents the possibility of determining the complete structures of complex carbohydrates (S.W. Homans, Progr. in NMR Spectros., 1990, *22*, 55; C.A. Bush, Bull. Magn. Reson., 1988, *10,* 73) if necessary in conjunction with other analytical methods such as methylation analysis and FAB-mass spectrometry. The oligosaccharide is released from the proteoglycan or glycoprotein by chemical or enzymatic means and purified by gel filtration and HPLC (H. Rauvala *et al.*, Adv. Carbohydr. Chem. Biochem., 1981, *38*, 389; E.F. Hounsell in "HPLC of Small Molecules", ed. C.K. Lim, IRL Press, Oxford, 1986, p.49; E.F. Hounsell, Chem. Soc. Rev., 1987, *16,* 161; Chem. Britain, 1990, *26,* 684; K.B. Hicks, Adv. Carbohydr. Chem. Biochem., 1988, *46,* 17; J.K. Welply, Trends Biotechnol., 1989, *7*, 5).

Sequential exoglycosidase digestion and high-resolution gel permeation chromatography have been used successfully (K. Yamashita, T. Mizuochi and A. Kobata, Methods Enzymol., 1982, *83,* 105) but the determination of novel branched structures by chemical degradation is difficult.

^{1}H and ^{13}C spectroscopy, including homo- and hetero-nuclear 2D NMR, are used to glean information on the anomeric form, ring structure, linkage between units, branching and sequence. Often a conformational analysis is important for understanding biological activity. Many oligosaccharides are not amenable to crystallisation, and for solution conformations, NMR remains the only high resolution method. With 50-500μg of sample, ^{1}H-NMR may determine complete structure; for 2D and 3D ^{1}H-NMR, and for natural abundance ^{13}C-NMR, more of the sample is required.

Even with relatively large glycoprotein-derived oligosaccharides, resonances of certain structural reporter groups can be discerned (J.F.G. Vliegenthart, L. Dorland and H. van Halbeek, Adv. Carbohydr.

Chem. Biochem., 1983, *41*, 209). The chemical shift and coupling constant of anomeric (H1) protons provide information on the kind of sugar residue, as well as on the type and configuration of its glycosidic linkage. A $J_{1,2}$ value of 2-4 Hz indicates α-D-Gal, α-D-GlcNAc or α-L-Fuc, while a $J_{1,2}$ value of 7-9 Hz indicates the corresponding β-anomer. For mannose, with its equatorial H1, a $J_{1,2}$ value of 1.6 Hz is found for the α-anomer, and $J_{1,2}$ of 0.8Hz for the β-anomer. Taken with chemical shift, the resonances may be assigned to particular residues. Mannose H2 and H3 atoms indicate the type of substitution of the common mannotriose branching core. Other protons which are sensitive to the structural environment are: the neuraminic acid H3, fucose H5 and CH_3 protons, galactose H3 and H4, and amino sugar *N*-acetyl CH_3 protons.

By 500 MHz ^{1}H-NMR the complete primary structures of the oligosaccharides $Man_{2\text{-}9}$GlcNAc were revealed (H. Van Halbeek, *et al.*, *FEBS Lett.*, 1980, *121*, 65; *ibid*, *121*, 71). The proposed structures were mainly based upon interpretation of the sets of chemical shift values of H1s and H2s of mannose residues.

From these 1D methods have evolved a series of 2D NMR techniques, including ^{1}H-^{1}H correlated spectroscopy (COSY) (S.W. Homans, *loc. cit.*; S.W. Homans, R.A. Dwek and T.W. Rademacher, Biochem., 1987, *26*, 6571) which trace through-bond coupling networks starting from readily identifiable anomeric protons. The absence of coupling between the anomeric and H5 protons gives a unidirectional trace around the ring. Resonance overlap usually however remains a problem unless COSY is supplemented by other 2D techniques. The 2D J-coupled COSY spectra of arabinose, xylose, sorbose, cellobiose, lactose and melibiose have been presented (B.E. Wilson, W. Lindberg and B.R. Kowalski, J. Am. Chem. Soc., 1989, *111*, 3797).

A promising approach for the study of large molecules is 3D NMR spectroscopy (S.W. Fesik, R.T. Gampe and E.R.P. Zuiderweg, J. Am. Chem. Soc., 1989, *111*, 770; S.W. Fesik and E.R.P. Zuiderweg, Quart. Rev. Biophys., 1990, *23*, 97) in which a homonuclear 2D NMR experiment and a heteronuclear shift correlation are combined so that the 2D spectra are edited by the heteronuclear chemical shifts.

Comparison of the assignments of an oligosaccharide (derived by one or more of the above techniques) with those that are available for constituent monosaccharides in their free form give chemical shift differences which can be interpreted in terms of linkage position. In principle, the protons assigned from COSY and related 2D spectra can be included with Vliegenthart's reporter groups, and large resonance shifts (>0.05 ppm) sought as evidence for nearby monosaccharide substitution or deletion, the effects of which are approximately additive. This fingerprint method is effective for a glycan showing homology to a database such as that for N-linked oligosaccharides (J.P. Carver and A.A. Grey, Biochem., 1981, *20,* 6607; E.F. Hounsell *et al*., Biochem. J., 1984, *223,* 129), or for saccharides containing the Gal β(1→4)Glc sequence (N. Platzer *et al*., Carbohydr. Res., 1989, *191,* 191).

Nuclear Overhauser effect spectroscopy (NOESY) (L. Muller and R.R. Ernst, Mol. Phys., 1980, *38,* 963; D.J. States, R.A. Haberkorn and D.J. Ruben, J. Magn. Reson., 1982, *48*, 286) is based on a short-range (<4Å) through-space interaction betwen nuclei. Intra-residue NOEs confirm assignments made by the above methods, and an interresidue effect from an anomeric proton to proton(s) in another residue define a linkage. There is a strong interaction for example between the anomeric proton and the aglycon proton (that on the carbon to which the glycosidic linkage is attached) in a -Man α(1→2)Man α- fragment. This approach was consistent with GC-MS methylation analysis for a novel class of glycan (M.A.J. Ferguson *et al*., Science, 1988, *239,* 753).

Solution conformations are important for the biological activity of oligosaccharides (K. Bock, Pure and Appl. Chem., 1983, *55*, 605; J.P. Carver and J.-R. Brisson, in "Biology of Carbohydrates", vol. 2, ed. V. Ginsburg and P.W. Robbins, Wiley, N.Y., 1984, p.289; S.W. Homans, *loc. cit.*; E.F. Hounsell, *loc. cit.*; S.W. Homans, Progr. NMR Spectroscopy, 1990, *22,* 55; H. Kunz and C. Unverzagt, Angew. Chem. Int. Ed. Engl., 1988, *27*, 1697; B. Meyer, in "Topics in Current Chemistry 154, Carbohydrate Chemistry", ed.J.Thiem, Springer, Berlin, 1990; H. Kessler, M.Gehrke and C. Griesinger, Angew. Chem. Int. Ed. Engl., 1988, *27*, 490; H. Paulsen, *ibid,* 1990, *29*, 823). COSY confirms an expected ring state, which remains essentially fixed on the NMR timescale, at least for pyranosides. For conformations about 1→6 linkages stereospecific assignments for the C6 protons are required. These have been assigned for gluco- and galactopyranosides by the development of a general synthesis of sugars with a chirally deuterated hydroxymethyl group (Y. Nishida, H. Ohrui and H. Meguro, Tetrahedron Lett., 1984, *25*, 1575; H. Ohrui *et al*., *ibid*, 1985, *26*, 3251; H. Ohrui *et al*., Canad. J. Chem., 1987, *65*, 1145. The C6 protons of galactose both resonate at around 3.7 ppm unless *O*-substituted (A. de Bruyn, M. Anteunis and G. Verhegge, Acta. Ciencia Indica, 1975, *1,* 83). Otherwise, for degenerate protons, spectral simulation employs the multiplicity of the C5 proton (Homans *et al.,* Biochem., 1986, *25,* 6342).

By observing the time-development of NOE between nuclei, and using intraresidue NOEs for calibration, short interresidue distances can be measured and the solution conformation of the oligosaccharide determined (G.M. Clore and A.M. Gronenborn, J. Magn. Reson., 1985, *61,* 158;

S.W. Homans, R.A. Dwek and T.W. Rademacher, Biochem., 1987, *26*, 6553). Since the NOE is proportional to the inverse sixth power of internuclear distance, it is particularly sensitive to small conformational changes, but also artificially weights conformers with close approach of the two given nuclei in situations where more than one conformer exists. Theoretical potential energy surfaces provide information on such internal motion and on its possible NMR results (S.W. Homans, *loc. cit.*) Molecular dynamics simulation has been used to compare the magnitude of torsional oscillations in oligomannose-branched oligosaccharides (S.W. Homans *et al.*, Biochem., 1987, *26,* 6649). The small torsional oscillations ($\pm 20^{o}$) of the Man $\alpha(1\rightarrow3)$ Man $\alpha(\beta)$ linkages validated the NOE-derived average structure. For $1\rightarrow6$ linkages however, there could be large torsional fluctuations, and NOE measurements were considered to reflect the true average structure only when conformers had similar probability of close approach of the aglyconic H1 atom to H6 or H6'.

Conformations derived from ^{1}H-NMR may be supported by HSEA calculations which can be performed on small computers. These take into account Hard Sphere (van der Waals) interactions and Exo-Anomeric effect (R.U. Lemieux *et al.,* Canad. J. Chem., 1980, *58,* 631). This effect causes the aglyconic carbon to assume an orientation with $\phi \sim +60^{o}$ in β-D-glycosides and $\phi \sim -60^{o}$ in α-D-glycosides. Coulombic and hydrogen-bond interactions are ignored because of solvent water. A number of oligosaccharides have been conformationally analysed in this way (J.P. Carver and J.-R. Brisson, *loc. cit.;* H. Paulsen *et al.,* Carbohydr. Res., 1986, *156*, 87; *ibid*, 1987, *165,* 251; Angew. Chem. Int. Ed. Engl., 1990, *29,* 823). For example, HSEA calculations have produced conformations in which strongly deshielded protons are close to oxygen atoms in other sugar units (H. Thφgersen *et al.*, Canad. J. Chem., 1982, *60,* 44; K. Bock, Pure Appl. Chem., 1983, *55*, 605).

A study has been made of chemical shift, relaxation (T_1) and NOE data for these model compounds of antennary glycans in relation to solution conformations: GlcNAc $\beta(1\rightarrow6)$Man αOMe, GlcNAc $\beta(1\rightarrow2)$Man αOMe, GlcNAc $\beta(1\rightarrow2)$[GlcNAc $\beta(1\rightarrow6)$]Man αOMe, Man $\alpha(1\rightarrow6)$Man βOMe, GlcNAc $\beta(1\rightarrow2)$[GlcNAc $\beta(1\rightarrow4)$]Man αOMe (D.A. Cumming *et al.*, Biochem., 1987, *26,* 6655; D.A. Cumming and J.P. Carver, *ibid*, 1987, *26*, 6664; ibid, 1987, *26,* 6676). In this study, the HSEA algorithm, and related potential energy algorithms which additionally take into account electrostatic, hydrogen bonding and Hassel-Ottar effect (O. Hassel and B. Ottar, Acta. Chem. Scand., 1947, *1,* 929) were assessed in view of the fact that they deal superficially with solvent effects.

The proton-proton dipolar contributions to proton spin-lattice relaxation rates (T_1) are, to a reasonable degree of approximation, pairwise additive and a measure of interproton distance. The method has been compared with the above more widely used approaches to carbohydrate structure and conformation (P. Dais and A.S. Perlin, Adv. Carbohydr. Chem. Biochem., 1987, *45,* 125). Spin-lattice relaxation and ^{13}C-NMR studies on cellobiose in the presence of β-glucosidase showed a fast interaction between enzyme and disaccharide which was shown from T_1 data to have a nearly spherical shape in solution (N. Marchettini *et al.*, Spectros. Lett., 1987, *20,* 81).

Unlike proton resonances, where the chemical shift reflects multiple short- and long-range effects, ^{13}C resonance shifts depend largely on short-range glycosylation and substitution. Inductive effects, which strongly affect chemical shifts, are less important for glycosylation. More important is the presence or absence of inter-unit non-bonded proton-proton interactions which polarise CH bonds and change the paramagnetic component of the shielding of the carbon nuclei (N.K. Kochetkov, Chem. Rev., 1990, *19,* 29).

For complete assignment, the ^{13}C-spectra of suitable model compounds are usually required. Data for a large number of such compounds are available (K. Bock, C. Pedersen and H. Pedersen, Adv. Carbohydr. Chem. Biochem., 1984, *42,* 193; K. Dill, E. Berman and A. Pavia, *ibid,* 1985, *43,* 1; G.W. Small and M.K. McIntyre, Anal. Chem., 1989, *61,* 666). Useful rules have been established (N.K. Kochetkov, O.S. Chizhov and A.S. Shaskov, Carbohydr. Res., 1984, *133,* 173; R.C. Beier, B.P. Mundy and G.A. Strobel, Canad. J. Chem., 1980, *58,* 2800; S. Seo *et al.*, J. Am. Chem. Soc., 1978, *100*, 3331; K. Mizutani *et al.,* Carbohydr. Res., 1984, *126,* 177; J.H. Bradbury and G.A. Jenkins, *ibid*, 1984, *126*, 125; N.K. Kochetkov, *loc. cit.;* A.S. Shashkov *et al.*, Magn. Reson. Chem., 1988, *26,* 735).

NMR data (^1H and ^{13}C) and HSEA calculations have been reported for a range of disaccharides (P. E. Jansson *et al.,* J. Chem. Soc. Perkin Trans. I, 1990, 591 and refs. therein; H. Baumann, P.-E. Jansson and L. Kenne, *ibid*, 1988, 209; I. Backman *et al., ibid,* 1988, 889; M. Forsgren, P.-E. Jansson and L. Keane, *ibid,* 1985, 2383), acetylated disaccharides (W.J. Goux, Carbohydr. Res., 1988, *184,* 47), and trisaccharides with a 2,3 or 3,4-branched galactose residue (G.M. Lipkind *et al.*, Carbohydr. Res., 1989, *195,* 11; *ibid*, 1989, *195,* 27) and a (1→6) branch (Y. Nishida *et al.*, Chem. Abstr., 1989, *111*, 23826). Other multinuclear investigations of di-and oligosaccharides have been reviewed (RSC Spec.Per. Report on Carbohydr. Chem., ed. R.J. Ferrier, vol. 22, 1990, p.232-236). Various methods have been used to calculate conformations representing energy minima for maltose, cellobiose and derivatives which have been correlated with experimental data such as NOE (G.M. Lipkind, V.E. Verovsky and N.K. Kochetkov, Carbohydr. Res., 1984, *133*, 1; A.S. Shashkov, G.M. Lipkind and N.K. Kochetkov, *ibid*, 1986, *147*, 175; N.K. Kochetkov, *loc. cit.;* R.U. Lemieux and K. Bock, Arch. Biochem. Biophys., 1983, *221*, 125; I. Tvaroska and S. Perez, Carbohydr. Res., 1986, *149,* 389). A correlation between ^{13}C-glycosylation shifts and one of the dihedral angles in the glycosidic linkage in the minimum energy conformation has been observed (K. Bock, A. Brignole and B.W. Sigurskjold, J. Chem. Soc., Perkin Trans. *2*, 1986, 1711).

Carbon NMR of polysaccharides and related oligosaccharides has been reviewed (P.A.J. Gorin, Adv. Carbohydr. Chem. Biochem., 1981, *38,* 13; H.J. Jennings, *ibid,* 1983, 41, 155); also NMR of carbohydrates linked to aminoacids and proteins (K. Dill, E. Berman and A.A. Pavia, Adv. Carbohydr. Chem. Biochem, 1985, *43*, 1; H. Paulsen, Angew. Chem. Int. Ed. Engl., 1990, *29*, 823).

(ii) Mass Spectrometry

Mass spectrometry of oligosaccharides has been reviewed (V.N. Reinhold in "Mass Spectrometry in Biomedical Research," ed. S. Gaskell, Wiley, 1986, p.181; A. Dell and M. Panico, *ibid*, p.149; A. Dell, Adv. Carbohydr. Chem. Biochem., 1987, *45*, 19; A.L. Burlingame *et al.*, Anal. Chem., 1988, *60,* 294R; "Analysis of Carbohydrates by GLC and MS", eds. C.J. Biermann and G.D. McGinnis, CRC Press, Boca Raton, Florida, 1989).

In general, small quantities of oligosaccharides are purified by HPLC before MS analysis. HPLC/MS and MS/MS interfacing is developing (V.N. Reinhold, *loc. cit.*). Separation by TLC has been used in direct TLC-MS, where strips from a silica-coated aluminium plate are placed on the probe tip (A.M. Lawson *et al.*, Carbohydr. Res., 1990, *200,* 47; T. Mizouchi *et al.,* J. Biol. Chem., 1989, *264*, 13834).

Analysis of oligosaccharide composition by permethylation, acid hydrolysis, hydride reduction and acetylation is an established method (G. Lindberg in "Methods in Enzymology", ed. V. Ginsburg, Academic Press, 1972, p. 178). The method does not differentiate 4-linked aldopyranosyl from 5-linked aldofuranosyl residues and certain moieties are destroyed in acid hydrolysis. There are advantages in methanolysis followed by acetylation (B. Fournet *et al.*, Anal. Biochem., 1981, *116*, 489) or in reductive cleavage with trimethylsilyltriflate and boron trifluoride etherate (D. Rolf and G.R. Gray, J. Am. Chem. Soc., 1982, *104,* 3539; J.-G. Jun and G.R. Gray, Carbohydr. Res., 1987, *163*, 247; P.R. Gruber and G.R. Gray, *ibid*, 1990, *203*, 79). The acetylated products are analysed by GC-MS. The products from the tetrasaccharide stachyose which has 1→6 and 1→2 linkages are fully methylated deoxyhexitols (terminal Gal and Fruct) and 6-*O*-acetyl methylated 1-deoxy-glucitol and -galactitol (internal sequence). The two stereoisomeric fructose derivatives have different GC retention times but identical mass spectra. Two components have been reported for fructosyl (V.N. Reinhold, *loc. cit.*; D. Rolf and G.R. Gray, Carbohydr. Res., 1984, *131*, 17) as well as for neuraminic acid residues (V.N. Reinhold, *loc. cit.*). Positional isomers, and therefore the linkage point are resolved chromatographically, and distinguished by their EI mass spectra. The GC and MS data for a series of non-, mono-, and di-acetylated methylated 1-deoxy hexitols have been published (A. Van Langenhove and V.N. Reinhold, Carbohydr. Res., 1985, *140*, 1).

In electron-impact MS of permethylated glycolipids (H. Egge and P. Hanfland, Arch. Biochem. Biophys., 1981, *210*, 396) sequence information can be obtained for analytes up to 3000 daltons, however at higher molecular weights the thermal desorption causes an excess of pyrolytic fragments.

Chemical ionisation MS in the presence of gases such as methane or ammonia has been shown to be a useful technique (R.J. Cotter, Anal.Chem., 1980, *52,* 1589; V.N. Reinhold, *loc. cit.*; V.N. Reinhold and S.A. Carr, Mass Spectrom. Rev., 1983, *2*, 153; K. Harade *et al.,* Biomed. Mass Spectrom., 1983, *10*, 5).

A number of attempts have been made to solve the general problem of sequencing underivatised oligosaccharides by MS (J.P. Kamerling *et al.*, Biomed. Mass Spectrom., 1983, *10*, 420; Y. Chen *et al.*, Biomed. Mass Spectrom., 1987, *14,* 9; Z. Lam *et al.*, Rapid Commun. Mass Spectrom., 1987, 1, 83; J.C. Prome *et al.*, Org. Mass Spectrom., 1987, *22*, 6; R.A. Laine *et al.*, J. Am. Chem. Soc., 1988, *110,* 6931; Z. Lam, M.B. Camisaron and G.G.S. Dutton, Anal. Chem., 1988, *60,* 2304; D.R. Muller, B. Doman and W.J. Richter, in "Advances in Mass Spectrometry", *11B*, ed. Longevialle, Heyden and Son, Chichester, 1989). Negative-ion fast atom bombardment (FAB) mass spectrometry with linked scans has been used to determine linkage position and sequence in a series of underivatised disaccharides and linear oligosaccharides (D. Garozzo *et al.*, Anal. Chem., 1990, *62*, 279). The method relies on selective deprotonation of the anomeric reducing-end hydroxyl group and variation in the subsequent fragmentation process.

High field magnets and soft ionisation procedures - fast atom bombardment (FAB) and liquid secondary ion (LSIMS) techniques (in which a sample in liquid matrix is bombarded with neutral atoms or with ions) have made MS a powerful method for analysing oligosaccharide structure using microgram quantities. Samples having molecular weights above 3800 are best examined as the permethylated or peracetylated derivatives, which H-bond less with the matrix and facilitate desorption. A naturally methylated glucose polysaccharide was found to give intense molecular ion signals (L.S. Forsberg *et al.*, J. Biol. Chem., 1982, *257*, 3555). The absence of background ions for derivatised samples aids analysis of mixtures (M.N. Fukuda *et al.,* J. Biol. Chem., 1984, *259*, 10925). Mass spectrometry of suitable native molecules is possible when acyl or other substituents make derivatisation undesirable. Negative ion FAB-MS with a glycerol matrix for a series of sulphated di- to decasaccharides gave the number of sulphate, hexose and N-acetylhexosamine residues. A pentasulphated molecule of sequence $(Hex\text{-}HexNAc)_3$ gave a spectrum showing sequential loss of sodium ions and of $NaSO_3$ (with addition of H^+) units from M-5H+4Na (P. Scudder *et al.*, Eur. J. Biochem., 1986, *157,* 136). A heptagalacturonide derived from plant cell wall, when applied to glycerol matrix in 5% acetic acid, showed negative ions from $[M\text{-}H]^-$ to $[M\text{+}6Na\text{-}7H]^-$; when applied in 0.1M HCl, only $[M\text{-}H]^-$ and $[M\text{+}Na\text{-}2H]^-$ were observed. Larger oligomers were converted to pentafluorobenzyloxime derivatives to improve solubility and desorption (K.R. Davis *et al.*, Plant Physiol., 1986, *80*, 568; E.A. Nothnagel *et al., ibid*, 1983, *71*, 916).

Derivatives must be prepared in order to obtain unambiguous sequence data for two reasons (A. Dell, *loc. cit.*). First, fragmentation with H-transfer can occur on either side of the glycosidic bond (Figure 2).

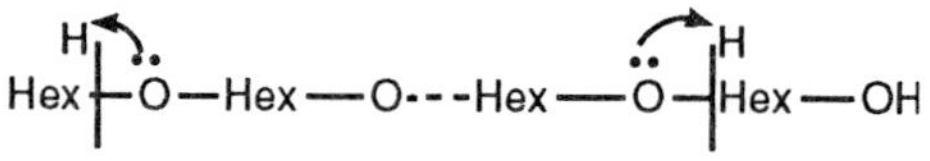

Hex—O—Hex—O- - -Hex—O—Hex—OH

(i) (ii)

- cleavage without charge involvement

non-reducing end

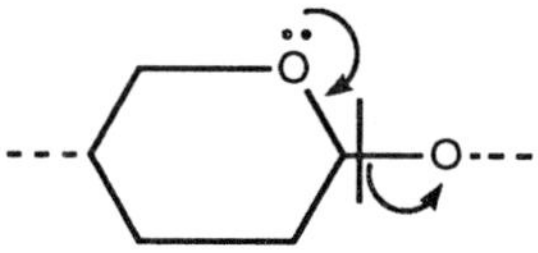

(iii)

- positive charge on the non-reducing fragment

Figure 2

Unless the sample is derivatised, it is not possible to tell whether the fragment ion is derived from the non-reducing or reducing end of the oligosaccharide. Secondly, double cleavages (A. Dell and C.E. Ballou, Carbohydr. Res., 1983, *120,* 95) are possible. For example, fragment ions frequently result from a combination of type - (i) cleavage with type - (iii) further along the chain. This gives apparent non-reducing end-sequence ions which are not, in fact, derived from the non-reducing terminus. Both these ambiguities are resolved with derivatisation (either by permethylation, or by peracetylation with trifluoroacetic anhydride-acetic acid) which limits fragmentation to well-defined pathways. In the positive mode, type - (iii) cleavage occurs as major fragmentation, from which sequence and branch residues can be assigned. If type - (i) cleavage occurs, the non-reducing terminus will have one fewer methyl or acetyl group than an original non-reducing terminus.

Linkage-site-specific fragmentation in FABMS is observed for permethylated molecules containing HexNAc residues (A. Dell, *loc. cit.*; H. Egge, J. Dabrowski and P. Hanfland, Pure Appl. Chem., 1984, *56,* 807; D. Abraham *et al.*, Biochem. J., 1984, *221*, 25; P. Hanfland *et al.* , Eur. J. Biochem., 1984, *145*, 531; M.N. Fukuda *et al.*, J. Biol. Chem., 1986, *261,* 5145). Such compounds give abundant ions from type - iii cleavage (see above), which are accompanied by related ions that provide information on the substitution pattern of the HexNAc residue.

Reactions such as acetolysis or methanolysis, which partially cleave the glycosidic bonds, have been used in analysis of glycoconjugate structures and in screening of glycoproteins for sugar type (A. Dell, *loc. cit.;* S. Naik *et al*., Biochem. Biophys. Res. Commun., 1985, *132*, 1). These salt-free reaction media are compatible with the FAB matrix.

(iii) X-Ray Crystallography

Bibliographies of crystal structures of carbohydrates have been published (G.A. Jeffrey and M. Sundaralingham, Adv. Carbohydr. Chem. Biochem., 1980, *37,* 373; ibid., 1981, *38*, 417; ibid., 1985, *43*, 203; P.R. Sundararajan and R.H. Marchessault, ibid., 1982, *40,* 381).

An analysis has been made of hydrogen-bonding in crystal structures of a number of disaccharides (G.A. Jeffrey and J. Mitra, Acta Crystallogr., Sect. B, 1983, *39,* 469). From this it was concluded that it is an oversimplification to assume that the crystal structure will be that which allows the maximum possible number of two-centre or linear hydrogen bonds. Two additional factors have to be considered. One is the importance of hydrogen-bond cooperativity. The second is that three-centre (bifurcated) hydrogen bonds are energetically comparable to two-centre bonds despite the longer H···O bond distances. The anomeric effect leads to the anomeric hydroxy groups behaving as a strong donor and weak acceptor.

The exo-anomeric effect is the preference of the aglycone carbon C (Fig.3) for the *sc* (synclinal or *gauche*) orientation, due to C1-O1 bond rotation, specified by the angle ϕ. Its most important outcome is the relative stability of relative orientations of neighbouring saccharide units in oligo- and polysaccharides. In these cases the exo-anomeric effect is more important than the anomeric effect (I. Tvaroska and T. Bleha, Adv. Carbohydr. Chem. Biochem., 1989, *47,* 45).

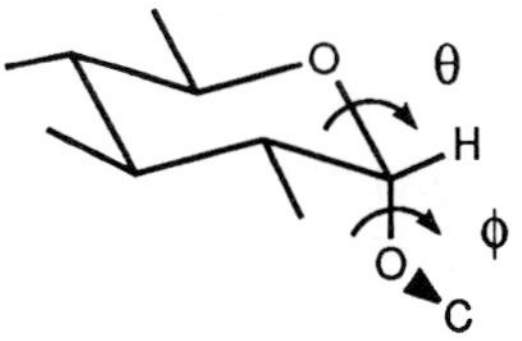

(+*sc*, +*sc*)

Figure 3

The conformation of α,α-trehalose provides a clear consequence of the exo-anomeric effect. In the solid anhydrous state this compound exhibits an approximate C_2 symmetry with two glycosidic-linkage torsion angles corresponding to the orientation (+*sc*, +*sc*) (G.A. Jeffrey and R. Nanni, Carbohydr. Res., 1985, *137*, 21). The preferred conformation in solution is very similar (K. Bock *et al.*, Eur. J. Chem., 1983, 131, 595). The evolution of molecular graphics from X-ray data for sucrose, isomaltulose, leucrose, glucosyl-α(1→1)-mannitol, and glucosyl-α(1→6)-sorbitol has been reviewed (F.W. Lichtenthaler, Starch/Stärke, 1991, *43*, 121).

X-ray crystallographic studies of the following structures have been published:

α-laminaribiose, β-D-glucopyranosyl-(1→3)-α-D-glucopyranosyl octa-acetate (D. Lamba *et al.*, Carbohydr. Res., 1986, *153*, 205);

β-maltose octaacetate (F. Brisse *et al.*, J. Am. Chem. Soc., 1982, *104*, 7470);

α-D-(1→4)-mannobiose (B. Shedrick, W. Mackie and D. Akrigg, Carbohydr. Res., 1984, *132*, 1;

lactose (Y.Nakai *et al.*, Chem. Pharm. Bull., 1982, *30,* 1811);

α-lactose monohydrate (a redetermination) (J.H. Noordik *et al.*, Z. Kristallogr., 1984, *168,* 59);

galabiose, α-D-galactopyranosyl-(1→4)-α,β-D-galactopyranose (G. Svensson *et al.*, Carbohydr. Res., 1986, *146,* 29);

4-*O*-β-D-galactopyranosyl-α-D-mannopyranose (C. Burden, W. Mackie and B. Sheldrick, Acta Crystallogr., 1986, *C42,* 177);

sucrose octaacetate (at 173K) (J.D. Oliver and L.C. Strickland, Acta Crystallogr., 1984, *C40,* 820);

2-acetamido-1,3,6-tri-*O*-acetyl-2-deoxy-4-*O*-(2,3,4,6-tetra-*O*-acetyl-α-L-idopyranosyl)-α-D-glucopyranose (A. Neuman *et al.*, Carbohydr. Res., 1985, *139,* 23);

methyl α-maltotrioside tetrahydrate (W. Pangborn, D. Langs and S. Perez, Int. J. Biol. Macromol., 1985, *7,* 363);

α-panose, α-D-glucopyranosyl-(1→6)-α-D-glucopyranosyl--(1→4)-α-D-glucopyranose (A. Imberty and S. Perez, Carbohydr. Res., 1988, *181,* 41);

melezitose, α-D-glucopyranosyl-(1→3)-β-D-fructofuranosyl α-D-glucopyranoside (J. Becquart, A. Neuman and H. Gillier-Pandraud, Carbohydr. Res., 1982, *111,* 9);

6-kestose, 6-*O*-fructofuranosyl sucrose monohydrate (V. Ferretti *et al.*, Acta Crystallogr., 1984, *C40,* 531);

2-acetamido-3-*O*-(2-acetamido-3,4,6-tri-*O*-acetyl-2-deoxy-α-D-galactopyranosyl)-1,4,6-tri-*O*-acetyl-2-deoxy-α-D-galactopyranose (P.Luger *et al.*, Carbohydr. Res., 1983, *117,* 23);

stachyose, α-D-galactopyranosyl-(1→6)-α-D-galactopyranosyl-(1→6)-α-D-glucopyranosyl-(1→2)-α-D-fructofuranose (R. Gilardi and J.L. Flippen-Anderson, Acta Crystallogr., 1987, *C43,* 806);

cyclomaltohexaose, α-cyclodextrin (B.Klar, B. Hingerty and W. Saenger, Acta Crystallogr. Sect. B., 1980, *36,* 1154);

cyclomaltoheptaose, β-cyclodextrin (W. Saenger *et al.*, Nature, London, 1982, *296,* 581; C. Betzel *et al.*, J. Am. Chem. Soc., 1984, *106,* 7545);

cyclomaltooctaose, γ-cyclodextrin (K. Harata, Chem. Lett. 1984, 641);

oligosaccharide chains from the core regions of glycoproteins (H. Paulsen, Angew. Chem. Int. Ed. Engl., 1990, *29,* 823);

maltoheptaose - phosphorylase A (E. Goldsmith, S. Sprang and R. Fletterick, J. Mol. Biol., 1982, *156,* 411).

(iv) Other Spectroscopic Methods

Vibrational spectroscopy, in particular the development of Fourier-transform infra-red (FTIR) and Laser Raman spectroscopy, has increased understanding of the structure of sugars in aqueous solutions and in hydrated form (M. Mathlouthi and J.L. Koenig, Adv. Carbohydr. Chem. Biochem., 1986, *44,* 7). The former allows analysis of species by spectrum-subtraction, while the latter represents an improvement on a method which has the primary advantage of low interference by liquid water. Ordering in sucrose as the quenched-melt, freeze-dried and saturated solution has been compared by FTIR (M. Mathlouthi, *loc. cit.*). The greater order in the last two states was related to results from X-ray diffraction (M. Mathlouthi, Carbohydr. Res., 1981, *91,* 113). Laser-Raman may be interpreted as showing the flexibility of the disaccharide in water (M. Mathlouthi *et al.*, Carbohydr. Res., 1980, *81,* 213); this spectroscopic method has also been applied to a study of the effects on the frequencies of the CH_2OH groups, which are sensitive to intra- and inter-molecular hydrogen bonding, of linking D-glucose and D-fructose to form sucrose (M. Mathlouthi and D.V. Luu, Carbohydr. Res., 1980, *81,* 203). IR spectroscopy has been used to determine the short-range order of water molecules in the immediate vicinity of carbohydrates, in the form of "hydration numbers" (J.L. Hollenberg and D.O. Hall, J. Phys. Chem., 1983, *87,* 695). However the procedure has been criticised as involving contradictory assumptions regarding the species of water of hydration (J. Jayne, J. Phys. Chem., 1983, *87,* 527).

2. Disaccharides and related compounds

A comprehensive survey of disaccharide and oligosaccharide structures lists those synthesised from 1960 to 1986, with derivatives, physical properties and methods of synthesis (A. Lipták *et al.*, "CRC Handbook of Oligosaccharides", CRC Press, Boca Raton, Florida, from 1991); and literature references are also listed annually (RSC Specialist Periodical Report, Carbohydrate Chemistry, ed. R.J. Ferrier).

(a) Tetrosylhexoses

4-*O*-(3-*C*-Hydroxymethyl-β-D-*glycero*-tetrofuranosyl)-D-glucose, 4-*O*-β-D-apiofuranosyl-D-glucose. Benzyl β-glycoside hexaacetate, $[\alpha]_D$ -80.7° (*c* 0.84, chloroform) (Y. Suzuki *et al.*, Agric. Biol. Chem., 1988, *52*, 1261).

6-*O*-α-D-Apiofuranosyl-D-glucose. Benzyl β-glycoside hexaacetate, preparation and NMR data (Y. Suzuki *et al., loc. cit.*).

6-*O*-β-D-Apiofuranosyl-D-glucose. Benzyl β-glycoside, $[\alpha]_D$ -95° (*c* 1.09, ethanol) (Y. Suzuki *et al., loc. cit.*).

(b) Pentosylpentoses

2-*O*-α-L-Arabinofuranosyl-L-arabinofuranose. Methyl α-glycoside pentaacetate, $[\alpha]_D$ -102° (*c* 1.5, chloroform) (L.V. Backinowsky *et al.*, Carbohydr. Res., 1985, *138*, 41).

3-*O*-α-L-Arabinofuranosyl-L-arabinofuranose. Methyl α-glycoside pentaacetate, m.p. 83.5-85° (from toluene-hexane), $[\alpha]_D$ -132° (*c* 1.2, chloroform) (L.V. Backinowsky *et al., loc. cit.*).

4-*O*-β-L-Arabinofuranosyl-L-arabinose. Methyl β-glycoside hexaacetate, $[\alpha]_D$ +90° (*c* 1.2, chloroform) is prepared by glycosylation, under pressure, of the 4-trityl ether of L-arabinose methyl β-glycoside diacetate with 1,2-*O*-cyanoalkylidene-L-arabinofuranose diacetate in the presence of ten molar percent trityl perchlorate in methylene chloride at 20° (S.A. Nepogod'ev, L. Backinowsky and N.K. Kochetkov, Bioorg.Khim., 1986, *12*, 1139).

4-*O*-β-L-Arabinofuranosyl-D-xylose. Methyl β-glycoside, $[\alpha]_D$ -107° (*c* 1, chloroform) (S.A. Nepogod'ev, L. Backinowsky and N.K. Kochetkov, *loc. cit.*)

2-*O*-β-D-Ribofuranosyl-D-ribose. Methyl β-glycoside, $[\alpha]_D$ -69° (*c* 0.85, methanol), $[\alpha]_D$ -54° (*c* 1.0, water) (J. Gass *et al.*, Carbohydr. Res., 1988, *180,* 243).

3-*O*-β-D-Ribofuranosyl-D-ribose. Methyl β-glycoside, m.p. 172-173°, $[\alpha]_D$ -67° (*c* 0.5, water) (J. Gass *et al., loc. cit.*).

5-*O*-β-D-Ribofuranosyl-D-ribose. Methyl β-glycoside, $[\alpha]_D$ -23.5° (*c* 1.5, water) (J.Gass *et al., loc. cit.*).

3-*O*-β-D-Xylopyranosyl-D-xylopyranose. Methyl β-glycoside pentaacetate, $[\alpha]_D$ -90.0° (*c* 2, chloroform) (L.V. Backinowsky, N.E. Nifant'ev and N.K. Kochetkov, Bioorg. Khim., 1984, *10*, 226).

(c) *Pentosylhexoses, Hexosylpentoses*

4-*O*-β-L-Arabinofuranosyl-D-glucose. β-Heptaacetate, $[\alpha]_D$ -40° (*c* 1, chloroform) is prepared by glycosylation of the 4-trityl ether of glucose tetraacetate with 1,2-*O*-cyanoalkylidene-L-arabinofuranose diacetate in the presence of ten molar percent trityl perchlorate in methylene chloride at 20° under pressure (S.A. Nepogod'ev, L.V. Backinowsky and N.K. Kochetkov, Bioorg. Khim., 1986, *12*, 1139).

2-*O*-β-D-Glucopyranosyl-α-L-arabinopyranose, m.p. 193-194°, $[\alpha]_D$ -36.8° (*c* 0.8, pyridine) is prepared from the methyl 3,4-*O*-isopropylidene-α-L-arabinoside by glycosylation with tetra-*O*-acetylglucosyl bromide and mercury(II) cyanide in dry toluene, followed by removal of protecting groups (M. Mizutani *et al.*, Carbohydr. Res., 1984, *126*, 177).

2-*O*-β-D-Glucopyranosyl-β-L-arabinopyranose, $[\alpha]_D$ +91.7 (*c* 0.69, pyridine) (M. Mizutani *et al.*, Carbohydr. Res., 1984, *126*, 177).

3-*O*-β-D-Glucopyranosyl-D-arabinofuranose. α-Heptaacetate, m.p. 127-128°, $[\alpha]_D$ -7.1° (*c* 0.4, chloroform) is prepared by reaction of octa-*O*-acetylcellobiononitrile with sodium methoxide and acetylation of the resulting anomeric mixture, followed by fractional recrystallisation (M.E. Gelpi and R.A. Cadenas, Carbohydr. Res., 1981, *88*, 277). β-Heptaacetate, m.p. 154-155°, $[\alpha]_D$ -36° (*c* 0.5, chloroform).

3-*O* β-D-Glucopyranosyl-D-arabinopyranose. α-Heptaacetate, m.p. 161-162° (rectangular plates, methanol) then 196-197° (needles), $[\alpha]_D$ -16.9° (*c* 0.98, chloroform) (M.E. Gelpi and R.A. Cadenas, *loc. cit.*; G. Zemplén, Ber., 1936, *59*, 1254). β-Heptaacetate, m.p. 139-140°, $[\alpha]_D$ -85.4° (*c* 0.14, chloroform) (M.E. Gelpi and R.A. Cadenas, *loc. cit.*).

(d) Deoxyhexosylpentoses

2-*O*-α-L-Rhamnopyranosyl-L-arabinopyranose, $[\alpha]_D$ -14° (*c* 1, ethanol) is prepared by glycosylation of methyl 3,4-di-*O*-benzyl-α-L-arabinopyranoside with 2,3,4-tri-*O*-acetyl-α-L-rhamnopyranosyl bromide in acetonitrile in the presence of mercury (II) cyanide and bromide, followed by deprotection (S.Kamiya, S. Esaki and R. Tamaka, Agric. Biol. Chem., 1984, *48*, 1353). Methyl α-glycoside, $[\alpha]_D$ -46° (*c* 1, ethanol) (S. Kamiya, S. Esaki and N. Shiba, *loc. cit.*).

2-*O*-(α-L-Rhamnopyranosyl)-L-arabinose. Methyl α-glycoside, $[\alpha]_D$ +8° (*c* 0.5, ethanol) (S. Kamiya, S. Esaki and N. Shiba, *loc. cit.*).

3-*O*-α-L-Rhamnopyranosyl-L-arabinopyranose. Benzyl β-glycoside, m.p. 220°, $[\alpha]_D$ +78° (*c* 1, 50% ethanol) is prepared by a standard glycosylation method (S. Kamiya *et al.*, Agric. Biol. Chem., 1986, *50*, 2147).

5-*O*-α-L-Rhamnopyranosyl-α-L-arabinofuranose. Methyl α-glycoside, $[\alpha]_D$ -103° (*c* 1, water) is prepared by a standard glycosylation method (S. Kamiya *et al., loc. cit.*).

(e) Deoxyhexosyldeoxyhexoses

3-*O*-(2,6-Dideoxy-α-D-*arabino*-hexopyranosyl)-2,6-dideoxy-α-D-*lyxo*-hexopyranose. Methyl α-glycoside, m.p. 140°, $[\alpha]_D$ +200° (acetone) is prepared by condensation of methyl 4-*O*-acetyl-2,6-dideoxy-α-D-*lyxo*-hexopyranoside with 3,4-di-*O*-acetyl-2,6-dideoxy-α-D-*arabino*-hexopyranosyl bromide in toluene-nitromethane (2:1) in presence of silver triflate, *sym*-collidine and 4A molecular sieves at -45°, followed by removal of acetate groups. The β-linked disaccharide was also formed. (J. Thiem and G. Schneider, Angew. Chem. Int. Ed. Engl., 1983, *22,* 58).

3-*O*-(2,6-Dideoxy-β-D-*arabino*-hexopyranosyl)-2,6-dideoxy-α-D-*lyxo*-hexopyranose. Methyl α-glycoside, $[\alpha]_D$ +38° (acetone) (J. Thiem and G. Schneider, *loc. cit.*)

4-O-(2,6-Dideoxy-α-D-*arabino*-hexopyranosyl)-2,6-dideoxy-α-D-*lyxo*-hexopyranose. Methyl α-glycoside, m.p. 134°, $[\alpha]_D$ +64° (acetone) is prepared by condensation of methyl 3-*O*-acetyl-2,6-dideoxy-α-D-*lyxo*-hexopyranoside with 3,4-di-*O*-acetyl-2,6-dideoxy-α-D-*arabino*-hexopyranosyl bromide in toluene-nitromethane (2:1) in presence of silver triflate, *sym*-collidine and 4A molecular sieves at -45°, followed by removal of acetate groups. (J. Thiem and G. Schneider, *loc. cit.*).

4-O-(2,6-Dideoxy-β-D-*arabino*-hexopyranosyl)-2,6-dideoxy-α-D-*lyxo*-hexopyranose. Methyl α-glycoside, syrup, $[\alpha]_D$ +37° (acetone) is formed in minor amount from methyl 3-*O*-acetyl-2,6-dideoxy-α-D-*lyxo*-hexopyranoside and 3,4-di-*O*-acetyl-2,6-dideoxy-α-D-*arabino*-hexopyranosyl bromide (J. Thiem and G. Schneider, *loc. cit.*).

2-O-β-L-Rhamnopyranosyl-L-rhamnopyranose. Methyl α-glycoside, $[\alpha]_D$ +37.4° (*c* 1.4, water) is prepared by reaction of 4-*O*-benzoyl-2,3-*O*-cyclohexylidene-α-L-rhamnopyranosyl bromide with methyl 3,4-di-*O*-benzyl-α-L-rhamnopyranoside in dichloromethane at -40° in the presence of silver carbonate and 4A molecular sieves, followed by removal of protecting groups. The α-linked disaccharide is also formed (T. Iversen and D.R. Bundle, J. Org. Chem., 1981, *46*, 5389; L.V. Backinowsky *et al.*, Carbohydr. Res., 1980, *84*, 225).

2-*O*-α-D-Rhamnopyranosyl-L-rhamnose, $[\alpha]_D$ +49° (*c* 0.6, ethanol) (A. Jaworska and A. Zamojski, Carbohydr. Res., 1984, *126*, 191).

3-*O*-β-L-Rhamnopyranosyl-L-rhamnose. Methyl α-glycoside, $[\alpha]_D$ +15° (*c* 1.6, water) is prepared from methyl 4-*O*-acetyl-2-*O*-benzoyl-α-L-rhamnopyranoside by glycosylation with 4-*O* benzoyl-2,3-*O*-cyclohexylidene-α-L-rhamnopyranosyl bromide in dichloromethane in the presence of silver carbonate and 4A molecular sieves at -40°, followed by removal of blocking groups (T. Iversen and D.R. Bundle, *loc. cit.*).

3-*O*-α-D-Rhamnopyranosyl-L-rhamnose, $[\alpha]_D$ +44° (*c* 0.35, ethanol) (A. Jaworska and A. Zamojski, *loc. cit.*).

4-O-β-L-Rhamnopyranosyl-L-rhamnose, $[\alpha]_D$ +4.3° (*c* 2.02, water). Methyl β-glycoside, $[\alpha]_D$ -16.5° (*c* 2.45, water) (L.V. Backinowsky *et al.*, Bioorg. Khim., 1980, *6*, 464). Methyl α-glycoside, $[\alpha]_D$ -12.1° (*c* 1.2, water) is prepared by reaction of 4-*O*-benzoyl-2,3-*O*-cyclohexylidene-α-L-rhamnopyranosyl bromide with methyl 2,3-*O*-isopropylidine-α-L-rhamnopyranoside in dichloromethane at -40° in the presence of silver carbonate and 4A molecular sieves, followed by removal of protecting groups (T. Iversen and D.R. Bundle, *loc. cit;* (L.V. Backinowsky *et al.,* Carbohydr. Res., 1980, *84*, 225).

(*f*) *Deoxyhexosylhexoses*

2-*O*-α-L-Fucopyranosyl-D-galactose, $[\alpha]_D$ -56° (*c* 1.2, water) has been synthesised *via* the trichloracetimidate method (B. Wegmann and R.R. Schmidt, Carbohydr. Res., 1988, *184*, 254). *p*-Nitrophenyl β-glycoside, $[\alpha]_D$ -108° (*c* 1, chloroform) (K.L. Matta, C.F. Piskorz and J.L. Barlow, Carbohydr. Chem., 1981, *90*, C1).

A series of D- and L-fucosylated galactoses, glucoses and mannoses has been synthesised, for NMR studies on conformation, using glycosyl halides (with silver triflate) or alkyl 1-thioglycosides (with methyl triflate) as glycosyl donors (H. Baumann, P.-E. Jansson and L. Kenna, J. Chem. Soc. Perkin Trans I, 1988, 209; I. Backman *et al., ibid,* 1988, 889; P.-E. Jansson, L. Kenne and E. Schweda, *ibid.,* 1988, 2729).

3-*O*-α-L-Fucopyranosyl-D-galactose. Methyl α-glycoside, $[\alpha]_D$ +22° (*c* 1.0, water) (H. Baumann, P.-E. Jansson and L. Kenna, *loc. cit.*).

3-*O*-α-D-Fucopyranosyl-D-galactose. Methyl α-glycoside, $[\alpha]_D$ +241° (*c* 1.0, water) (H. Baumann, P.-E. Jansson and L. Kenna, *loc. cit.*).

3-*O*-β-L-Fucopyranosyl-D-galactose. Methyl α-glycoside, $[\alpha]_D$ +153° (*c* 1.1, water) (H. Baumann *et al.*, J. Chem. Soc. Perkin I, 1989, 2145).

4-*O*-α-L-Fucopyranosyl-D-galactose. Methyl α-glycoside, $[\alpha]_D$ +18° (*c* 1.0, water) (H. Baumann *et al., loc. cit.*).

4-*O*-β-L-Fucopyranosyl-D-galactose. Methyl α-glycoside, $[\alpha]_D$ +127° (*c* 1.0, water) (H. Baumann *et al., loc. cit.*).

2-*O*-α-L-Fucopyranosyl-D-glucose. Methyl α-glycoside, $[\alpha]_D$ -16° (*c* 1.0, water). Methyl β-glycoside, $[\alpha]_D$ -99° (*c* 1.0, water) (P.-E. Jansson, L. Kenne, and E. Schweda, *loc. cit.*).

3-*O*-α-L-Fucopyranosyl-D-glucose, $[\alpha]_D$ -47° (*c* 1.47, methanol). Methyl α-glycoside, m.p. 197-199°, $[\alpha]_D$ -20° (*c* 0.7, methanol) (H.M. Flowers, Carbohydr. Res., 1983, *119*, 75). Methyl β-glycoside, $[\alpha]_D$ -138° (*c* 1.0, water) (H. Baumann, P.-E. Jansson and L. Kenne, J. Chem. Soc. Perkin Trans. I, 1988, 209).

3-*O*-α-D-Fucopyranosyl-D-glucose. Methyl α-glycoside, m.p. 192-195°, $[\alpha]_D$ +243° (*c* 1.0, water). Methyl β-glycoside, m.p. 91-92° (isopropanol), $[\alpha]_D$ +105° (*c* 1.1, water) (H. Baumann, P.-E. Jansson and L. Kenne, *loc. cit.*).

4-*O*-α-L-Fucopyranosyl-D-glucose. Methyl α-glycoside, $[\alpha]_D$ -17° (*c* 1.0, water) (I. Backman *et al., loc. cit.*).

4-*O*-β-L-Fucopyranosyl-D-glucose. Methyl α-glycoside, $[\alpha]_D$ +102° (*c* 1.0, water) (I. Backman *et al., loc. cit.*).

3-*O*-α-L-Fucopyranosyl-D-mannose. Methyl α-glycoside, $[\alpha]_D$ -65° (*c* 1.0, water) (H. Baumann, P.-E. Jansson and L. Kenne, *loc. cit.*).

3-*O*-α-D-Fucopyranosyl-D-mannose. Methyl α-glycoside, $[\alpha]_D$ +148° (*c* 1.0, water) (H. Baumann, P.-E. Jansson and L. Kenne, *loc. cit.*).

6-*O*-α-L-Rhamnopyranosyl-D-galactose, robinobiose, $[\alpha]_D$ 0 → +2.7° (H_2O). Heptaacetate m.p. 84-85° (113°), $[\alpha]_D$ -9.9° (chloroform), -19° (water) (L.V. Backinowsky *et al.,* Biorg. Khim., 1980, *6*, 464).

6-*O*-β-L-Rhamnopyranosyl-D-galactose, $[\alpha]_D$ +60.8° (*c* 2.4, water) (L.V. Backinowsky *et al.,* Carbohydr. Res., 1980, *84*, 225).

2-*O*-β-L-Rhamnopyranosyl-D-glucose. Methyl β-glycoside, $[\alpha]_D$ +37° (*c* 1.0, methanol) (S. Kamiya, S. Esaki and N. Shiba, Agric. Biol. Chem., 1987, *51*, 2207).

3-*O*-α-L-Rhamnopyranosyl-D-glucose. Methyl α-glycoside, $[\alpha]_D$ +37° (*c* 1.4, water). Methyl β-glycoside, $[\alpha]_D$ -61° (*c* 1, water) (K. Bock, J. Fernandez-Bolanos Guzman and R. Norrestam, *loc. cit.*; H. Baumann, P.-E. Jansson and L. Kenne, *loc. cit.*).

3-*O*-α-D-Rhamnopyranosyl-D-glucose. Methyl α-glycoside, $[\alpha]_D$ +137° (*c* 1.3, water). Methyl β-glycoside, $[\alpha]_D$ +45° (*c* 1.0, water) (K. Bock, J. Fernandez-Bolanos Guzman and R. Norrestam, Carbohydr. Res., 1988, *179*, 97; H. Baumann, P.-E. Jansson and L. Kenna, *loc. cit.*).

6-*O*-β-L-Rhamnopyranosyl-D-glucose, $[\alpha]_D$ +63.7° (*c* 0.8, water). (L.V. Backinowsky *et al., loc. cit.*).

3-*O*-α-L-Rhamnopyranosyl-D-talose. Methyl β-glycoside hexaacetate, $[\alpha]_D$ -58.6° (*c* 2.0, chloroform) (N.E. Nefant'ev *et al.,* Biorg. Khim., 1988, *14*, 187).

3-*O*-α-D-Rhamnopyranosyl-D-talose. Methyl β-glycoside, $[\alpha]_D$ -67.8° (*c* 2.8, water) (N.E. Nefant'ev *et al., loc. cit.).*

3-*O*-β-D-Rhamnopyranosyl-D-talose. Methyl β-glycoside, $[\alpha]_D$ +37.8° (*c* 3.5, water) (N.E. Nefant'ev *et. al., loc. cit.).*

(g) Hexosyldeoxyhexoses

4-*O*-β-D-Galactopyranosyl-L-fucose, m.p. 182-184°, $[\alpha]_D$ -39° (5 min) → -48° (18h) (*c* 1, water). Methyl β-glycoside, m.p.~150°, $[\alpha]_D$ -103° (*c* 1.5, methanol). α-Heptaacetate, $[\alpha]_D$ -84° (*c* 2, dichloromethane) (H.M. Flowers, Carbohydr. Res., 1983, *119*, 75).

2-*O*-α-D-Galactopyranosyl-L-rhamnose, $[\alpha]_D$ +94°, is prepared by methyl triflate-promoted glycosidation of benzyl 3,4-di-*O*-benzoyl-α-L-rhamnopyranoside with methyl 2,3,4,6-tetra-*O*-benzyl-1-thio-β-D-galactopyranoside followed by removal of protecting groups (T. Norberg, S. Oscarson and M. Szönyi, Carbohydr. Res., 1986, *152,* 301).

2-*O*-α-D-Glucopyranosyl-L-rhamnose. Methyl α-glycoside, $[\alpha]_D$ +90° (*c* 2, water) (S.S. Mamyan *et al.,* Bioorg. Khim., 1988, *14*, 205).

2-*O*-β-D-Glucopyranosyl-L-rhamnose. Methyl α-glycoside, $[\alpha]_D$ -12° (*c* 1, water) (S.S. Mamyan *et al., loc. cit.*).

3-*O*-α-D-Glucopyranosyl-L-rhamnose. Methyl α-glycoside, $[\alpha]_D$ +72.8° (*c* 1, methanol) (S.S. Mamyan *et al., loc. cit.*).

3-*O*-β-D-glucopyranosyl-L-rhamnose. Methyl α-glycoside, $[\alpha]_D$ -52° (*c* 1, water) (S.S. Mamyan *et al., loc. cit.*).

(h) Hexosylhexoses

(i) Galactosylgalactoses

6-*O*-β-D-Galactofuranosyl-D-galactofuranose. Methyl β-glycoside, $[\alpha]_D$ -90° (*c* 0.5, water) (C. Marino, O. Varela and R.M. de Lederkeremer, Carbohydr. Res., 1989, *190*, 65).

2-*O*-α-D-galactopyranosyl-D-galactose. *p*-Nitrophenyl β-glycoside, m.p. 282° (dec.), $[\alpha]_D$ +35.4° (*c* 0.3, water) (S.A. Abbas, J.J. Barlow and K.L. Matta, Carbohydr. Res., 1982, *106*, 59).

2-*O*-β-D-galactopyranosyl-D-galactose. Methyl β-glycoside, $[\alpha]_D$ +4.3° (*c* 1, water) is prepared by glycosylation of methyl 3,4,6-tri-*O*-benzyl-β-D-galactopyranoside with 2-*O*-benzoyl-3,4,6-tri-*O*-benzyl-1-*O*-tosyl-D-galactopyranose in acetonitrile, followed by removal of blocking groups (R. Ely and C. Schuerch, Carbohydr. Res., 1981, *92*, 149; H.F. Vernay *et al.,ibid*, 1980, *78,* 267). *p*-Nitrophenyl β-glycoside, m.p. 287° (dec.), $[\alpha]_D$ -58° (*c* 0.3, water) (S.A. Abbas, J.J. Barlow and K.L. Matta, *loc. cit.*).

3-*O*-α-D-Galactopyranosyl-D-galactose. *p*-Nitrophenyl β-glycoside, $[\alpha]_D$ +85.9° (*c* 0.69, water) (S.A. Abbas, J.J. Barlow and K.L. Matta, Carbohydr. Res., 1982, *101*, 231).

3-*O*-β-D-Galactopyranosyl-D-galactose. *p*-Nitrophenyl β-glycoside, m.p. 254-256° (dec.), $[\alpha]_D$ -35° (*c* 0.44, water) (S.A. Abbas, J.J. Barlow and K.L. Matta, *loc. cit.*).

4-*O*-α-D-galactopyranosyl-D-galactose. *p*-Nitrophenyl β-glycoside, amorphous, $[\alpha]_D$ +53° (*c* 1, chloroform) is prepared by glycosylation of D-galactose 1,2,3,6-tetra-*O*-benzoate with 2,3,4,6-tetra-*O*-benzyl-α-D-galactopyranosyl chloride using silver triflate as promoter, followed by hydrogenolysis, benzoylation, and *p*-nitrophenylation of the bromide before removal of protecting groups. Methyl β-glycoside, amorphous, $[\alpha]_D$ +88° (*c* 1.2, water) (P.J. Garegg and H. Hultberg, Carbohydr. Res., 1982, *110*, 261; M.-L. Milat, P.A. Zollo and P. Sinaÿ, Carbohydr. Res., 1982, *100*, 263.

5-*O*-β-D-Galactofuranosyl-β-D-galactofuranose, $[\alpha]_D$ -60° (*c* 0.1, water) is prepared by condensation between tetra-*O*-acetyl-β-D-galactofuranosyl chloride and allyl 2,3,6-tri-*O*-benzyl-β-D-galactofuranoside in dichloromethane in the presence of mercury (II) bromide and activated molecular sieves followed by removal of protecting groups (F. Sugawara, H. Nakayama and T. Ogawa, Agric. Biol. Chem., 1986, *50*, 1557; R.M. de Lederkremer, C. Marino and O. Varela, Carbohydr. Res., 1990, *200*, 227).

6-*O*-β-D-Galactofuranosyl-D-galactitol, $[\alpha]_D$ -41° (*c* 0.6, water) (C. Marino, O. Varela and R.M. de Lederkremer, Carbohydr. Res., 1989, *190*, 65).

(ii) Glucosylglucoses

2-*O*-α-D-Glucopyranosyl-D-glucose. Methyl α-glycoside, $[\alpha]_D$ +144.8° (*c* 2.5, water). Methyl α-glycoside heptaacetate, m.p. 158-159°, $[\alpha]_D$ +176.2° (*c* 3.0, chloroform) (A Temeriusz *et al.,* Polish J. Chem., 1982, *56*, 141).

4-*O*-α-D-Glucopyranosyl-D-glucose, maltose, chemistry has been reviewed (R. Khan, Adv. Carbohydr. Chem. Biochem, 1981, *39*, 214).

4-*O*-β-D-Glucopyranosyl-D-glucose, cellobiose, m.p. 224-225°, $[\alpha]_D$ +13.0 (3min) → +34.0° (*c* 5.0, water), is synthesised by condensation of 2,3,4,6-tetra-*O*-acetyl-α-D-glucopyranosyl bromide with benzyl 2,3,6-tri-*O* benzyl-β-D-glucopyranoside in benzene-nitromethane in the presence of mercuric cyanide at 50°, followed by removal of protecting groups

(K. Takeo *et al.,* Carbohydr. Res., 1983, *121*, 163). Methyl α-glycoside, m.p. 143-144° (ethanol), $[\alpha]_D$ +97.6° (*c* 2.0, water).

6-*O*-α-D-Glucopyranosyl-D-glucose. Methyl β-glycoside, anomeric purity 91%, $[\alpha]_D$ +51° (water) (M. Forsgren, P.-E. Jansson and L. Kenne, J. Chem. Soc. Perkin Trans I, 1985, 2383).

(iii) Mannosylmannoses

2-*O*-α-D-Mannopyranosyl-D-mannose. Methyl β-glycoside, $[\alpha]_D$ +8° (*c* 1.2, water) is prepared by mercury (II) cyanide catalysed glycosylation of methyl 3,4,6-tri-*O*-benzyl-β-D-mannopyranoside with 2,3,4,6-tetra-*O*-acetyl-α-D-mannopyranosyl bromide in benzene-nitromethane (S.H. Khan, R.K. Jain and K.L. Matta, Carbohydr. Res., 1990, *207*, 57). *p*-Nitrophenyl α-glycoside, m.p. 203-208°, $[\alpha]_D$ +99.1° (*c* 1.62, water) (G. Ekborg and C.P.J. Glaudemans, Carbohydr. Res., 1984, *134,* 83).

3-*O*-α-D-Mannopyranosyl-D-mannose. Methyl α-glycoside, $[\alpha]_D$ +93.5° (*c* 0.25, water) has been synthesised *via* alternative intermediates, by glycosylation of methyl 2-*O*-allyl-4,6-*O*-benzylidene-α-D-mannopyranoside with 2,3,4,6-tetra-*O*-acetyl-α-D-mannopyranosyl bromide in acetonitrile in the presence of mercury (II) bromide, cyanide and 4A molecular sieves (F.M. Winnik *et al.,* Carbohydr. Res., 1982, *103*, 15). *p*-Nitrophenyl α-glycoside, m.p. 155-160° (ethanol), $[\alpha]_D$ +185° (*c* 1.18, water) (G. Ekborg and C.P.J. Glaudemans, *loc. cit.*).

6-*O*-α-D-Mannopyranosyl-D-mannose. Methyl α-glycoside, $[\alpha]_D$ +94.5° (*c* 0.25, water) (F.M. Winnik *et al.,* Carbohydr. Res., 1982, *103*, 15). Methyl β-glycoside, amorphous, $[\alpha]_D$ +8° (*c* 0.4, methanol); 4-nitrophenyl β-glycoside, amorphous, $[\alpha]_D$ -62° (*c* 1.1, water) (S.H. Khan, R.K. Jain and K.L. Matta, *loc. cit.*).

(iv) Other hexosylhexoses

O-β-D-Galactopyranosyl-D-fructose. A mixture of isomers is prepared by reverse enzyme activity of β-D-galactosidase (K. Ajisaka, H. Fujimoto and H. Nishida, Carbohydr. Res. 1988, *180*, 35).

O-β-D-Galactopyranosyl-D-glucose. A mixture of isomers is prepared by reverse hydrolytic activity of β-D-galactosidases from *E. coli* and *A. oryzae*, by circulating a solution of monosaccharide through columns of immobilised enzyme and activated carbon in series, and eluting the disaccharides from the carbon column with 50% ethanol after 24 h (yield 6-11%) (K. Ajisaka, H. Fujimoto and H. Nishida, *loc. cit.).*

4-*O*-α-D-galactopyanosyl-D-glucose. Methyl α-glycoside heptaacetate, m.p. 70-72° (ethyl ether/ petroleum ether), $[\alpha]_D$ +138.9° (*c* 1, chloroform) is prepared from the β-trichloroacetimidate of 2,3,4,6-tetra-*O*-benzyl-D-galactose as donor in ether with trimethylsilyl triflate as catalyst (yield 65% of α/β 6/1 isolated) (B. Wegmann and R.R. Schmidt, J. Carbohydr. Chem., 1987, *6*, 357).

4-*O*-β-D-Galactopyranosyl-D-glucose, lactose. A review of the relative reactivity of the hydroxyl groups of lactose has been published (L.A.W. Thelwall, J. Dairy Res., 1982, *49*, 713).

3-*O*-β-D-galactofuranosyl-D-mannopyranose. Methyl α-glycoside, $[\alpha]_D$ -37° (*c* 0.6, water) is prepared from 3,5,6-tri-*O*-acetyl-α-D-galactofuranose 1,2-(methyl orthoacetate) and methyl 4,6-*O*-benzylidene-α-D-mannopyranoside, followed by removal of blocking groups (P.A.J. Gorin, E.M. Barreto-Bergter and F.S. da Cruz, Carbohydr. Res., 1981, *88*, 177).

6-*O*-β-D-galactofuranosyl-D-mannopyranose. Methyl α-glycoside, $[\alpha]_D$ -13° (*c* 1.2, water) is prepared from 3,5,6-tri-*O*-acetyl-α-D-galactofuranose 1,2-(methylorthoacetate) which is obtained from 2,3,5,6-tetra-*O*-acetyl-β-D-galactofuranosyl chloride. The orthoester is reacted with methyl 2,3,4,-tri-*O*-benzyl-α-D-mannopyranoside in nitromethane in the presence of mercury (II) bromide for 1h at 100°, then protecting groups are removed (P.A.J. Gorin, E.M. Barreto-Bergter and F.S. da Cruz, *loc. cit.*).

2-*O*-α-D-Galactopyranosyl-D-mannose, $[\alpha]_D$ +96° (*c* 0.5, water), is prepared by silver triflate-promoted glycosylation of benzyl 3,4,6-tri-*O*-benzyl-α-D-mannopyranoside with 2,3,4,6-tetra-*O*-benzyl-α-D-galacto-pyranosyl chloride in toluene in the presence of 2,4,6-trimethylpyridine and molecular sieves, followed by catalytic hydrogenation. (T. Norberg, S. Oscarson and M. Szönyi, Carbohydr. Res., 1986, *152*, 301).

3-*O*-β-D-Glucopyranosyl-D-fructose, $[\alpha]_D$ -43° (*c* 0.5, water) is prepared by condensation of tetra-*O*-acetyl-α-D-glucopyranosyl bromide with 1,2:4,5-di-*O*-isopropylidene-β-D-fructopyranose in 1:1 nitromethane-benzene in the presence of mercury (II) cyanide at 45°, followed by removal of protecting groups (A. Couto and R.M. de Lederkremer, Carbohydr. Res., 1984, *126*, 313).

2-*O*-α-D-Glucopyranosyl-D-galactose. Methyl β-glycoside, m.p. 188-189° (ethanol), $[\alpha]_D$ +91° (*c* 1, water) (A. Temeriusz *et al.*, Carbohydr. Res., 1985, *142*, 146).

2-*O*-β-D-Glucopyranosyl-D-galactose. Methyl β-glycoside, m.p. 112-115°, $[\alpha]_D$ -13° (*c* 1, water). α-Octaacetate, $[\alpha]_D$ +39° (*c* 1, chloroform) (A. Temeriusz *et al., loc. cit.*).

3-*O*-α-D-Glucopyranosyl-D-galactopyranose. Methyl α-glycoside, $[\alpha]_D$ 217° (*c* 1.0, water) is prepared using ethyl 2,3,4,6-tetra-*O*-benzyl-1-thio-β-D-glucopyranoside as glycosyl donor, and methyl 4-*O*-acetyl-2,6-di-*O*-benzyl-α-D-galactopyranoside, in diethyl ether with methyl triflate and ground molecular sieves (4A) at 0° then 20°, followed by removal of protecting groups (H. Baumann *et al.*, J. Chem. Soc. Perkin I, 1989, 2145). Methyl β-glycoside, m.p. 224-225°, $[\alpha]_D$ +104° (*c* 0.5, water) (A. Temeriusz *et al., loc. cit.*).

3-*O*-β-D-Glucopyranosyl-D-galactose. α-Octaacetate, m.p. 175-176°, $[\alpha]_D$ +67° (*c* 1, water) (E. Temeriusz *et al., loc. cit.*). Methyl α-glycoside, $[\alpha]_D$ +106° (*c* 1.0, water) (H. Baumann *et al., loc. cit.*).

4-*O*-α-D-Glucopyranosyl-D-galactopyranose. Methyl α-glycoside, $[\alpha]_D$ 210° (*c* 1.0, water) is synthesised using 2,3,4,6-tetra-*O*-benzyl-1-thioglucose ethyl β-glycoside as glucosyl donor, and methyl 3-*O*-benzoyl-2,6-di-*O*-benzyl-α-D-galactopyranoside, in diethyl ether with methyl triflate and ground molecular sieves (4A) at 0° then 20°. The β-linked product is also formed and separated by chromatography before deprotection. (H. Baumann *et al., loc. cit.*).

4-*O*-β-D-Glucopyranosyl-D-galactopyranose. Methyl α-glycoside, $[\alpha]_D$ 104° (*c* 0.9, water) (H. Baumann *et al., loc. cit.*).

6-*O*-α-D-Glucopyranosyl-D-galactose. Methyl α-glycoside, anomeric purity 92%, $[\alpha]_D$ +176° (water). Methyl β-glycoside, anomeric purity 92%, $[\alpha]_D$ +87° (water) (M. Forsgren, P.-E. Jansson and L. Kenne, *loc. cit.*).

6-*O*-β-D-Glucopyranosyl-D-galactose. Methyl α-glycoside, $[\alpha]_D$ +72° (water). Methyl β-glycoside, $[\alpha]_D$ -28° (water) (M.Forsgren, P.-E. Jansson and L. Kenne, *loc. cit.*). Also: methyl β-glycoside, m.p. 130°, $[\alpha]_D$ +6° (*c* 0.5, water); methyl β-glycoside heptaacetate, m.p. 120-122°, $[\alpha]_D$ -18° (*c* 1, chloroform) (A.Temeriusz *et al., loc. cit.).*

2-O-α-D-Mannopyranosyl-D-glucose, $[\alpha]_D$ +61.4° (c 2.2, water) is prepared by glycosylation of benzyl 3,4,6-tri-O-benzyl-β-D-glucopyranoside with mannose tetraacetyl α-bromide in acetonitrile in the presence of mercury (II) cyanide, followed by deprotection (V.I. Torgov, V.N. Shibaev and N.K. Kochetkov, Bioorg. Khim., 1984, *10*, 946).

3-O-α-D-Mannopyranosyl-D-glucose, $[\alpha]_D$ +71.5° (c 2.6, water) (V.I. Torgov, V.N. Shibaev and N.K. Kochetkov, *loc. cit.*).

4-O-α-D-Mannopyranosyl-D-glucose, $[\alpha]_D$ +65.3° (c 2, water) (V.I. Torgov, V.N. Shibaev and N.K. Kochetkov, *loc. cit.*). Methyl α-glycoside, $[\alpha]_D$ +110.1° (c 3.2, water). α-Octaacetate, $[\alpha]_D$ +37.4° (c 2.5, chloroform).

6-O-α-D-Mannopyranosyl-D-glucose, $[\alpha]_D$ +68.2° (c 2.2, water). β-Octaacetate, $[\alpha]_D$ +40° (c 5.3, chloroform) (V.I. Targov, V.N. Shibaev and N.K. Kochetkov, *loc. cit.*).

3-O-α-D-Mannopyranosyl-D-talose. Methyl β-glycoside, $[\alpha]_D$ +48.6° (c 1.9, water). Methyl β-glycoside heptaacetate, $[\alpha]_D$ +12.1° (c 1.2, chloroform) (N.E. Nefant'ev *et al.*, Bioorg. Khim., 1988, *14*, 187).

3-O-β-D-Mannopyranosyl-D-talose. Methyl β-glycoside, $[\alpha]_D$ +40.5° (c 0.8, water) (N.E. Nefant'ev *et al.*, *loc. cit.*).

(i) KDO-containing disaccharides

2-Acetamido-2-deoxy-3-O-(3-deoxy-D-*manno*-2-octulopyranosylonic acid)-D-galactose. Methyl ester, $[\alpha]_D$ +90° (c 1.0, water) (J. Paquet and P. Sinaÿ, J. Am. Chem. Soc., 1984, *106*, 8313).

6-O-(3-Deoxy-D-*manno*-2-octulopyranosylonic acid)-D-glucose. Methyl ester α-octaacetate, $[\alpha]_D$ +105° (c 0.34, chloroform) (R.R. Schmidt and A. Esswein, Angew. Chem. Int. Ed. Engl. 1988, *27*, 1178).

(j) Non-reducing Disaccharides

β-D-Fructofuranosyl α-D-glucopyranoside, sucrose. Reviews on the chemistry of sucrose, including its conversion to a range of selectively substituted derivatives, have been published (L. Hough, Chem. Soc Rev., 1985, *14,* 357; L. Hough, Int. Sugar J., 1989, *91,* 23; T. Vydra, Chem.

Listy., 1989, *83*, 686; K.J. Parker, "Kirk Othmar Encycl. Chem. Technol.", 3rd Edn., 1983, *21*, 921; "Carbohydrates as Organic Raw Materials", ed. F.W. Lichtenthaler, Verlag Chemie, Weinheim, 1991). Conformation (K. Bock *et al.,* Carbohydr. Res.,1982, *100*, 63) and ^{1}H NMR (W.E. Hull, "Two-Dimensional NMR Brochure", Brucker Analytische Messtechnik, Karlsruhe, 1982). Sucrose, sweeteners and sweetness ("Progress in Sweeteners", ed. T.H. Grenby, Elsevier, 1989; "Sugarless, the Way Forward", ed. A.J. Rugg-Gunn, Elsevier, 1991). 6'-*O-tert*-Butyldiphenylsilyl glycoside, m.p. 192-195°, $[\alpha]_D$ +44° (*c* 1, methanol) (H. Karl *et al.,* Carbohydr. Res., 1982, *101*, 31). 6,6'-Dipalmityl glycoside, m.p. 109°, $[\alpha]_D$ +40° (*c* 1, chloroform) (S. Bottle *et al.,* J. Chem. Soc. Chem. Commun., 1984, 385).

β-D-Galactofuranosyl-β-D-galactofuranoside, m.p. 206-208°, $[\alpha]_D$ -150° (*c* 1, water) is prepared by controlled addition of water to penta-*O*-benzoyl-α,β-D-galactofuranose in the presence of tin (II) chloride followed by debenzoylation (C. Marino, O. Varela and R.M. de Lederkremer, Carbohydr. Res., 1989, *190*, 65).

α-D-Glucopyranosyl-α-D-glucopyranoside, α,α-trehalose has been reviewed (G.K. Lee, in "Developments in Food Carbohydrates-2", ed.C.K. Lee, Appl. Science Publ., London, 1980, p.1) as well as substrates and inhibitors of trehalase (J. Defaye *et al.,* in "Mechanisms of Saccharide Polymerization and Depolymerization", ed. J.J. Marshall, Academic Press, New York, 1980, p.331). 6,6'-Dipalmityl glycoside, m.p. 158.5-160°, $[\alpha]_D$ +78° (*c* 1, chloroform) (S. Bottle *et al.,* J. Chem. Soc. Chem. Commun., 1984, 385). 6'-*O*-Mycolate, m.p. 170-173°, $[\alpha]_D^{24}$ +45.5° (*c* 0.7,chloroform) (A. Liav *et al.,* Carbohydr. Res., 1984, *125*, 323). Partially protected trehaloses have been prepared (A. Lipták *et al.,* Carbohydr. Res., 1988, *175*, 241; A. Liav and M.B. Goren, *ibid*, 1980, *84*, 171; *ibid*, 1984, *125,* 323; J. Defaye *et al.,* Nouv. J. Chim., 1980, *4*, 59; T. Ogawa and M. Matsui, Tetrahedron, 1981, *37*, 2363; C. Vincent, Carbohydr. Res., 1989, *194*, 308). An enzymatic synthesis of α,α-trehalose from maltose (60% efficiency) has been reported (S. Murao *et al.,* Agric. Biol. Chem., 1985, *89,* 2113). Various deoxy and dideoxy derivatives of trehalose have been made by reductive desulphonylation of trehalose tosylates with lithium triethyl borohydride (H.H. Baer, M. Mekorska and F. Baucher, Carbohydr. Chem. 1985, *136*, 335). α,α-Trehalose has been converted by way of the 6,6'-ditrityl ether and its hexa-acetate to the 6,6'-dicarboxylic acid with Jones reagent (A. Liav and M.B. Goren, Carbohydr. Res., 1980, *84*, 171).

A review has been published on natural and synthetic trehalose esters (E. Lederer, Microbiol. Ser., 1984, *15*, 361). Aminodeoxy disaccharides related to trehalose have been reviewed (H.H. Baer, in "Trends in Synthetic Carbohydrate Chemistry", American Chemical Society, 1989, p.32). Regioselective syntheses of trehalose-containing trisaccharides involve the use of various glycohydrolases (K. Ajisaka and H. Fujimoto, Carbohydr. Res., 1990, *199*, 227). Orthoesterification of trehalose under kinetic control has been reviewed (M. Bouchra, P. Calinaud and J. Gelas, in "Trends in Synthetic Carbohydrate Chemistry," American Chemical Society, 1989, p.46).

β-D-Glucopyranosyl β-D-glucopyranoside, β,β-trehalose, m.p. 135-140°, $[\alpha]_D$ -40.2° (water) is prepared by reaction of 2,3,4,6-tetra-*O*-acetyl β-D-glucopyranose with 2,3,4,6-tetra-*O*-acetyl α-D-glucopyranoside trichloroacetimidate in the presence of boron trifluoride, followed by deacetylation (S.J. Cook, R.Khan and J.M. Brown, J. Carbohydr. Chem., 1984, *3*, 343).

β-D-Ribofuranosyl β-D-ribofuranoside, m.p. 154-155°, $[\alpha]_D$ -100° (*c* 0.487, water) is prepared from 1-*O*-acetyl-2,3,5-tri-*O*-benzoyl-β-D-ribofuranose by self-condensation in presence of boron trifluoride etherate and molecular sieves 3A, followed by removal of protecting groups (L.M. Lerner, Carbohydr. Res., 1990, *199*, 116). Hexaacetate, m.p. 107°, $[\alpha]_D$ -49.5° (*c* 0.536, chloroform).

(k) Nitrogen-containing disaccharides

(i) Disaccharides containing an amino-sugar and a neutral sugar residue

2-Acetamido-2-deoxy-3-*O*-(α-L-fucopyranosyl)-D-glucose. Methyl α-glycoside, m.p. 276-277°, $[\alpha]_D$ -19.8° (*c* 0.5, methanol) (S.S. Rana, J.J. Barlow and K.L. Matta, Carbohydr. Res., 1981, *91*, 140).

2-Acetamido-2-deoxy-3-*O*-(β-D-galactopyranosyl)-D-galactose. Methyl α-glycoside, m.p.228-230°, $[\alpha]_D$ +65° (*c* 0.2, water) is synthesised by halide ion-promoted glycosidation, using methanol and the disaccharide bromide derived from methyl 2-azido-3-*O*-(2,3,4,6-tetra-*O*-benzoyl-β-D-galactopyranosyl)-4,6-*O*-benzylidene-2-deoxy-1-thio-β-D-galactopyranoside. This derivative in turn was prepared by silver triflate-promoted condensation of monosaccharide derivatives (M. Haraldsson, H. Lönn and T. Norberg, Glycoconj. J., 1987, *4*, 225).

2-Acetamido-2-deoxy-4-*O*-(β-D-galactopyranosyl)-D-galactose, N-acetyllactosamine. A convenient large-scale synthesis has been reported involving addition of hydrogen cyanide to 1-*N*-benzyl-3-*O*-β-D-galactopyranosyl-D-arabinosylamine (J. Alais and A.Veyrières, Carbohydr. Res., 1981, *93*, 164) and a new synthesis by regioselective dibutylstannylene-mediated oxidation of methyl 3',4'-*O*-isopropylidene) α- and β-lactoside (A. Fernandez-Mayorales, Tetrahedron, 1988, *44*, 4877). Methyl α-glycoside, $[\alpha]_D$ +80° (*c* 0.5, water) (J. Dahmén *et al.,* Carbohydr. Res., 1985, *138*, 17). Methyl β-glycoside, $[\alpha]_D$ -25° (*c* 1.5, D_2O) (J. Dahmén *et al., loc. cit.*).

2-Acetamido-2-deoxy-6-*O*-(β-D-galactopyranosyl)-D-galactose, $[\alpha]_D$ +43.1° (*c* 1.1, water). *o*-Nitrophenyl α-glycoside, m.p. 253-254°, $[\alpha]_D$ +205.7° (*c* 0.5, dimethyl sulphoxide) (K.L. Matta, S.S. Rana and S.A. Abbas, Carbohydr. Res., 1984, *131*, 265). 2-Acetamido-2-deoxy-6-*O*-(β-D-galactopyranosyl)-D-galactose, $[\alpha]_D$ +26°, is also prepared by transfer of the galactopyranosyl residue from lactose using immobilised galactosidase (L. Hedbys *et al.*, Biochem. Biophys. Res. Commun., 1984, *123*, 8; P.O. Larsson *et al.*, Methods Enzymol., 1987, *136*, 230).

2-Acetamido-2-deoxy-*O*-(β-D-galactopyranosyl)-D-glucose. A mixture of isomers is prepared (yield 6-11%) by reverse hydrolytic enzyme activity (K. Ajisaka, H. Fujimoto and H. Nishida, Carbohydr. Res., 1988, *180*, 35).

2-Acetamido-2-deoxy-3-*O*-(β-D-glucopyranosyl)-D-galactose, *N*-acetylchondrosine. Dihydrate, m.p. 155-157°, $[\alpha]_D$ +47 (extrapolated) → 19° (*c* 1.07, water) (M.L. Wolfram and B.O. Juliano, J. Am. Chem. Soc., 1960, *82*, 1673) is prepared from chondroitin sulphate by heating with 90% DMSO-water (Y. Inoue and K. Nagasawa, Carbohydr. Res., 1981, *97*, 263) or by methanolysis, followed by borohydride reduction of the methyl ester (K. Blumberg and C.A. Bush, Anal. Biochem., 1982, *119*, 397).

2-Acetamido-2-deoxy-4-*O*-(α-D-glucopyranosyl)-D-glucopyranose. Hepta-*O*-benzyl-β-glycoside, $[\alpha]_D$ +91° (*c* 1.9, chloroform) is prepared from 1,3,6-tri-*O*-benzyl-2-acetamido-2-deoxy-α-D-glucopyranose as glycosyl acceptor and 2,3,4,6-tetra-*O*-benzyl-α-D-glucopyranose (1.3 equiv.), 4-nitrobenzenesulphonyl chloride (2.5 equiv.), and silver trifluoromethanesulphonate (2.5 equiv.) in dichloromethane, N,N-dimethylacetamide (2.5 equiv.) and triethylamine (2.5 equiv.) at -40° then 0° (S. Koto *et al.,* Carbohydr. Res., 1984, *130*, 73).

2-Acetamido-2-deoxy-6-*O*-(α-D-glucopyranosyl)-D-glucopyranose. Hepta-*O*-benzyl-β-glycoside, m.p. 179-181°, $[\alpha]_D$ +88° (*c* 0.8, chloroform) (S. Koto *et al., loc. cit.*).

2-*O*-(2-Acetamido-2-deoxy-β-D-glucopyranosyl)-α-D-mannopyranose. α-Heptaacetate, $[\alpha]_D$ -10° (*c* 0.8, chloroform) is prepared *via* glycosylation of methyl 3,4,6-tri-*O*-benzyl-α-D-mannopyranoside with 3,4,6-tri-*O*-acetyl-2-deoxy-2-phthalimido-β-D-glucopyranosyl bromide in the presence of silver trifluoromethanesulphonate, 2,4,6-trimethylpyridine and 4A molecular sieves in dichloromethane. The derived bromide is used to prepare a series of oligosaccharides containing the GlcNAc β(1→2)Man α sequence (S.H. Khan, S.A. Abbas and K.L. Matta, Carbohydr. Res., 1989, *193*, 125).

6-*O*-(2-Acetamido-2-deoxy-β-D-mannopyranosyl)-D-galactopyranose, monohydrate, m.p. 170° (dec. with sintering at 115°), $[\alpha]_D$ -14° (*c* 0.25, water) is prepared by a route involving glycosidation of tri-*O*-benzoyl-2-benzoyloxyimino-2-deoxy-α-D-*arabino*-hexopyranosyl bromide (E. Kaji *et al.,* Bull. Chem. Soc. Jpn., 1988, *61*, 1291).

4-*O*-(2-Acetamido-2-deoxy-β-D-mannopyranosyl)-D-glucopyranose. Methyl α-glycoside, semi-hydrate, m.p. 190° (dec.with sintering at 110°), $[\alpha]_D$ +45.2° (*c* 0.25, water) is prepared by a route involving glycosidation of tri-*O*-benzoyl-2-benzoyloxyimino-2-deoxy-α-D-*arabino*-hexopyranosyl bromide with the partially protected glucose, followed by hydroboration and deblocking (E. Kaji *et al., loc. cit.)*.

2-Acetamido-2-deoxy-3-*O*-(α-L-rhamnopyranosyl)-D-glucopyranose, $[\alpha]_D$ -30.5° (*c* 1, water) (S. Kamiya, S. Esaki and N. Shiba, Agric. Biol. Chem., 1987, *51*, 1195).

2-Acetamido-2-deoxy-4-*O*-(α-L-rhamnopyranosyl)-D-glucopyranose, $[\alpha]_D$ -7° (*c* 1, water) (S. Kamiya, S. Esaki and N. Shiba, *loc. cit.*).

2-Acetamido-2-deoxy-6-*O*-(α-L-rhamnopyranosyl)-D-glucopyranose, $[\alpha]_D$ +5° (*c* 1, water) (S. Kamiya, S. Esaki and N. Shiba, *loc. cit.*).

2-Acetamido-2-deoxy-6-*O*-(β-L-rhamnopyranosyl)-D-glucopyranose, $[\alpha]_D$ +53° (*c* 1, water) (S. Kamiya, S. Esaki and N. Shiba, *loc. cit.*).

(ii) Disaccharides containing two amino-sugar residues

2-Acetamido-3-*O*-(2-acetamido-2-deoxy-β-D-glucopyranosyl)-2-deoxy-D-galactose. Benzyl α-glycoside, m.p. 300° (dec.), $[\alpha]_D$ +131.4° (*c* 0.41, water) is prepared by glycosylation of benzyl 2-acetamido-4,6-*O*-benzylidene-2-deoxy-α-D-galactopyranoside with 3,4,6-tri-*O*-acetyl-2-deoxy-2-phthalimido-β-D-glucopyranosyl bromide in dichloromethane in the presence of silver triflate, 2,4,6-trimethylpyridine and molecular sieves, followed by removal of protecting groups (S.A. Abbas, J.J. Barlow and K.L. Matta, Carbohydr. Res., 1983, *112*, 201).

2-Acetamido-6-*O*-(2-acetamido-2-deoxy-β-D-glucopyranosyl)-2-deoxy-D-galactose. Benzyl α-glycoside, m.p. 290° (dec.), $[\alpha]_D$ +109.5° (*c* 0.52, water) is prepared by condensation of benzyl 2-acetamido-3-*O*-acetyl-2-deoxy-α-D-galactopyranoside with 2-methyl-(3,4,6-tri-*O*-acetyl-1,2-oxazoline in 1,2-dichloroethane at 70° in the presence of *p*-toluenesulphonic acid. The product is isolated as its peracetylated derivative which is finally deacetylated (S.A. Abbas, J.J. Barlow and K.L. Matta, *loc. cit.*).

2-Acetamido-4-*O*-(2-acetamido-2-deoxy-β-D-glucopyranosyl)-2-deoxy-D-glucose, di-N-acetyl chitobiose. Octaacetate is prepared in 31% yield by incubation of colloidal chitin for 30h at 50° with *Bacillus licheniformis* X-7u, acetylation and chromatography (S.-I. Nishimura *et al.*, Carbohydr. Res., 1989, *194*, 223).

2-Acetamido-4-*O*-(2-acetamido-2-deoxy-β-D-glucopyranosyl)-3-*O*-[(R)-1-carboxyethyl]-2-deoxy-D-glucose, 2-acetamido-2-deoxy-β-D-glucopyranosyl-(1→4)-N-acetylmuramic acid, m.p. 166-168° (softening at 164°), $[\alpha]_D$ +4° (water) (D. Kantoci, D. Keglevic and A.E. Derane, Carbohydr. Res., 1987, *162*, 227).

2-Acetamido-2-deoxy-6-*O*-(5-acetamido-3,5-dideoxy-D-*glycero*-α-D-*galacto*-2-nonulopyranosylonic acid)-D-galactose, 2-acetamido-2-deoxy-6-*O*-(N-acetyl-α-D-neuraminyl)-D-galactose. Allyl β-glycoside, $[\alpha]_D$ -6.5° (*c* 1.6, chloroform) is prepared by reacting allyl 2-azido-2-deoxy-β-D-galactopyranoside and methyl (5-acetamido-4,7,8,9-tetra-*O*-acetyl-3,5-dideoxy-D-*glycero*-β-D-*galacto*-2-nonulopyranosyl chloride)onate in the presence of silver triflate in tetrahydrofuran. The β(2→6)-linked disaccharide is formed in minor amount (H. Iijima and T. Ogawa, Carbohydr. Res., 1988, *172*, 183; 1989, *186*, 95).

3-*O*-(5-Acetamido-3,5-dideoxy-D-*glycero*-α-D-*galacto*-2-nonulopyranosylonic acid)-D-galactose, 3-*O*-(*N*-acetyl-α-D-neuraminyl)-D-galactose, $[\alpha]_D$ +23.1° (*c* 1.18, water) (T. Ogawa and M. Sugimoto, Carbohydr. Res., 1985, *135*, C5).

3-*O*-(*N*-acetyl-β-D-neuraminyl)-D-galactose, $[\alpha]_D$ +11.0° (*c* 0.30, water) (T. Ogawa and M. Sugimoto, *loc. cit.*).

6-*O*-(5-Acetamido-3,5-dideoxy-α-D-*glycero*-D-*galacto*-2-nonulopyranosylonic acid)-D-galactose, 6-*O*-(*N*-Acetyl-α-D-neuraminyl)-D-galactose. Methyl ester, $[\alpha]_D$ +5.4° (*c* 0.8, methanol) is prepared by condensation of methyl 5-acetamido-4,7,8,9-tetra-*O*-acetyl-2-chloro-2,3,5-trideoxy-β-D-*glycero*-D-*galacto*-2-nonulopyranosonate with benzyl 2,3,4-tri-*O*-benzyl-β-D-galactopyranoside, using silver salicylate as promoter, and removal of blocking groups. Methyl ester benzyl-β-glycoside m.p. 183-184°, $[\alpha]_D$ -1.3° (*c* 0.95, methanol) (A.Ya. Khorlim *et al.*, Carbohydr. Res., 1971, *19*, 272; D.J.M. Van Der Vleugel *et al.*, Carbohydr. Res., 1982, *104*, 221).

9-*O*-(*N*-acetyl-α-D-neuraminyl)-β-D-neuraminic acid. Potassium salt, m.p. 162-165°, $[\alpha]_D$ -14° (*c* 0.14, water) is prepared by glycosylation of the 2-*O*-allyl-7-*O*-benzyl-8,9-isopropylidene derivative methyl ester with pentaacetyl methyl ester 1-β-chloride in the presence of mercury(II) cyanide, bromide and 4A molecular sieves, followed by removal of protecting groups (C. Shimizu and K. Achiwa, Carbohydr. Res., 1987, *166*, 314).

(l) *Uronic Acids*

4-*O*-(α-D-Galactosyluronic acid)-D-galactose. Methyl ester methyl β-glycoside hexaacetate, m.p. 228.7-229.4°, $[\alpha]_D$ +95.89° (*c* 0.73, chloroform) is isolated after hydrolysis, with methanolic hydrogen chloride, of a fraction of the water-soluble polysaccharide of *P. amurense* (T. Fujiwara and K. Arai, Carbohydr. Res., 1982, *101*, 323).

(α-D-Glucopyranosyluronic acid)(α-D-glucopyranosiduronic acid), trehalose dicarboxylic acid, monohydrate, m.p. 140° (sintered), $[\alpha]_D$ +123° (*c* 2.0, water) (M.B. Goren and K.-S. Jiang, Carbohydr. Res., 1980, *79*, 225). Hexaacetate, m.p. 143-145°, $[\alpha]_D$ +157° (*c* 1.06, chloroform) (A. Liav and M.B. Goren, *ibid*, 1980, *84*, 171).

2-*O*-(4-*O*-Methyl-β-D-glucopyranosyluronic acid)-D-xylose. Methyl ester, m.p. 194-195.5° (from ethanol), $[\alpha]_D$ -12 (extrapolated) → -13 (3 min) → -14° (60 min, equil.) (P. Kováč, E. Petráková and P. Kočiš, *loc. cit.*).

2-*O*-(4-*O*-Methyl-α-D-glucopyranosyluronic acid)-D-xylose, $[\alpha]_D$ +104° (*c* 1.2, water). Methyl ester, m.p.174-176° (from ethanol), $[\alpha]_D$ +133 (extrapolated)→ +127(3.5min)→106° (90 min, equil.) (*c* 0.6, water) is obtained by reaction of methyl 2,3-di-*O*-benzyl-1-chloro-1-deoxy-4-*O*-methyl-α,β-D-glucopyranuronate, as glycosylating agent, with benzyl 3,4-di-*O*-benzyl-β-D-xylopyranoside in the presence of silver perchlorate and triethylamine in dichloromethane followed by hydrogenolysis. The β-linked product was also formed (α:β ratio 3:1) (P. Kováč, E. Petráková and P. Kočiš, Carbohydr. Res., 1981, *93,* 144).

4-*O*-α-L-Rhamnopyranosyl-D-glucopyranuronic acid, $[\alpha]_D$ -18.1° (*c* 0.99, water) is prepared by two approaches, rhamnosylation of benzyl 6-*O*-allyl-2,3-di-*O*-benzyl-β-D-glucopyranoside followed by deallylation and oxidation, or rhamnosylation of suitably protected D-glucuronic acid derivatives. Rhamnosylations were performed using 2,3,4-tri-*O*-benzyl-1-*O*-trichloroacetimidoyl-α-L-rhamnopyranose with trifluoromethanesulphonic acid catalysis. Methyl ester, $[\alpha]_D$ -7.4° (*c* 1.48, water) (P. Fügedi, J. Carbohydr. Chem., 1987, *6*, 377).

3. Trisaccharides and related compounds

A comprehensive survey of trisaccharide structures lists those synthesised from 1960 to 1986, with derivatives, physical properties and methods of synthesis (A. Lipták *et al.*, "CRC Handbook of Oligosaccharides," vol.1, CRC Press, Boca Raton, Florida, 1991); and literature references to trisaccharides are also listed annually (RSC Specialist Periodical Report, Carbohydate Chemistry, ed. R.J. Ferrier).

(a) Neutral sugar containing trisaccharides

O-β-D-Fructofuranosyl-(2→1)-β-D-fructofuranosyl α-D-glucopyranoside, 1-kestose. 2D NMR (proton and carbon) chemical shift assignments have been made (T.M. Calub *et al.,* Carbohydr. Res., 1990, *199*, 11).

Gal α(1→4)Gal β(1→4)Glc, globotriose, $[\alpha]_D$ +101° (water). Methyl β-glycoside, amorphous, $[\alpha]_D$ +65° (*c* 1, water) is prepared by galactosylation of 1,2,3,6-tetra-*O*-benzoyl-4-*O*-(2,3,6-tri-*O*-benzoyl-β-D-galactopyranosyl)-α-D-glucopyranose with 2,3,4,6-tetra-*O*-benzyl-α-D-galactopyranosyl chloride using silver triflate and 2,4,6-trimethylpyridine in toluene, followed by reaction of the derived α-bromide with methanol and removal of protecting groups (P.J. Garegg and H. Hultberg, Carbohydr.

Res., 1982, *110*, 261; J.-C. Jacquinet and P. Sinaÿ, Carbohydr. Res., 1985, *143*, 143).

Gal α(1→3)Man α(1→6)Man. Methyl α-glycoside, $[\alpha]_D$ +118° (*c* 1.1, water) is synthesised by reacting 2,3,4,6-tetra-*O*-benzoyl-α-D-mannopyranosyl bromide with methyl 2,3,4-tri-*O*-benzyl-α-D-mannopyranoside in dichloromethane in the presence of silver triflate. Hydrogenolysis of the (1→6)-linked disaccharide product, treatment with trimethylorthoacetate, then acetic anhydride followed by acid opening of the 2,3-orthoester gives methyl 2,4-di-*O*-acetyl-6-*O*-(2,3,4,6-tetra-*O*-benzoyl-α-D-mannopyranosyl)-α-D-mannopyranoside. This is glycosylated at OH-3 with 2,3,4,6-tetra-*O*-benzyl-D-galactopyranosyl bromide under halide-assisted conditions. Other glucose-, galactose-and mannose-containing branched trisaccharides are similarly prepared (P.J. Garegg, S. Oscarson and A.-K. Tidén, Carbohydr. Res., 1990, *200*, 475).

Glc α(1→2)Glc α(1→2)Glc , kojitriose. Monohydrate, m.p. 228-230° (dec.), $[\alpha]_D$ +150 (5 min) → +156° (*c* 1.7, water) is synthesised by condensation of 1,3,4,6-tetra-*O*-acetyl-α-D-glucopyranose with hepta-*O*-acetyl-α-kojibiosyl bromide in acetonitrile in the presence of mercury (II) cyanide and bromide, followed by removal of protecting groups. The 1,2-*trans* glycoside is also formed (K. Takeo, Carbohydr. Res., 1981, *88*, 158).

Glc α(1→3)Glc α(1→3)Glc, nigerotriose, $[\alpha]_D$ +182.7° (*c* 1.1, water) is prepared in approximately 10% yield by acetolysis of a D-glucan isolated from the fruit body of *haetiporus sulphureus* (K. Takeo and S. Matsuzaki, Carbohydr. Res., 1983, *113*, 281).

Glc β(1→4)Glc β(1→4)Glc, cellotriose, m.p. 206-208.5° (dec.) (aqueous ethanol), $[\alpha]_D$ +32.8 (3min) → +21.0° (*c* 4.2, water) is synthesised by condensation of 2,3,4,6-tetra-*O*-acetyl-α-D-glucopyranosyl bromide with hepta-*O*-benzyl-β-cellobioside in 1:1 benzene-nitromethane in the presence of mercury (II) cyanide (57h at 60°) followed by removal of protecting groups. Methyl β-glycoside, m.p. 253-255° (methanol-ethanol), $[\alpha]_D$ +68.8° (*c* 1.9, water). Methyl β-glycoside, m.p. 265-267°, (dec.) (aqueous ethanol), $[\alpha]_D$ -13.7° (*c* 3.4, water) (K. Takeo *et al.,* Carbohydr. Res., 1983, *121*, 163).

Glc β(1→3) [Glc β(1→6)]Glc, m.p. 183-188°, $[\alpha]_D$ -1.1° (*c* 1.5, water) is synthesised *via* one-stage β-glucosylation of benzyl 3-*O*-acetyl-2,4-di-*O*-benzyl-α-D-glucopyranoside with 2,3,4,6-tetra-*O*-benzyl-α-D-glucopyranose using a mixture of *p*-nitrobenzenesulphonyl chloride, silver triflate and triethylamine, followed by deacetylation. The disaccharide derivative is then subjected to the β-glucosylation, followed by catalytic hydrogenation. The α(1→3)β(1→6) isomer is formed in smaller amount (S. Koto *et al.,* Canad. J. Chem., 1981, *59*, 255; K. Takeo and S. Tei, Carbohydr. Res., 1986, *145*, 293). Methyl α-glycoside, $[\alpha]_D$ +28.5° (*c* 0.50, methanol) is synthesised from methyl 2,4-di-*O*-benzyl-α-D-glucopyranoside and 2,3,4,6-tetra-*O*-acetyl-α-D-glucopyranosyl bromide in 1,2-dichloroethane in the presence of mercury (II) bromide and molecular sieves 4A followed by deacetylation and hydrogenolysis (T. Ogawa and T. Kaburagi, Carbohydr. Res., 1982, *103*, 53).

Glc α(1→4) [Glc α(1→6)]Glc, $[\alpha]_D$ +139° (*c* 0.3, water) is synthesised starting from the glycosyl acceptor benzyl 4-*O*-allyl-2,3-di-*O*-benzyl-α-D-glucopyranoside and 2,3,4,6-tetra-*O*-benzyl-α-D-glucopyranose in the presence of 4-nitrobenzenesulphonyl chloride and silver triflate in dichloromethane with N,N-dimethylacetamide and triethylamine. The product disaccharide is then coupled with the same glycosyl donor after removal of the allyl group (S.Koto *et al.,* Carbohydr. Res., 1984, *130*, 73). Methyl β-glycoside, $[\alpha]_D$ +85.4° (water) (K. Bock and H. Pedersen, J. Carbohydr. Chem., 1984, *3*, 581).

Man α(1→2)Man α(1→3)Glc. Methyl α-glycoside, $[\alpha]_D$ +110° (water) (V. Bencano *et al.,* Rev. Cubana Farm., 1983, *17*, 36).

Man α(1→2)Man α(1→6)Man. Methyl β-glycoside, $[\alpha]_D$ +33° (*c* 1.2, water) (S.H. Khan, R.K. Jain and K.L. Matta, Carbohydr. Res., 1990, *207*, 57).

Man α(1→3) [Man α(1→6)]Man, $[\alpha]_D$ +111.0° (*c* 0.31, methanol) is synthesised by glycosylation of methyl 2-*O*-allyl-3-*O*-(2,3,4,6-tetra-*O*-acetyl-α-D-mannopyranosyl)-α-D-mannopyranoside with 2,3,4,6-tetra-*O*-acetyl-α-D-mannopyranosyl bromide in acetonitrile in the presence of mercury (II) bromide, cyanide and 4A molecular sieves, followed by removal of protecting groups. Methyl α-glycoside, $[\alpha]_D$ +111.0° (*c* 0.31, methanol) (F.M. Winnik *et al.,* Carbohydr. Res., 1982, *103*, 15). *p*-Trifluoracetamidophenyl β-glycoside, $[\alpha]_D$ +19° (*c* 1.4, methanol) (G.Ekborg and C.P.J. Glaudemans, Carbohydr. Res., 1985, *142*, 213).

Man α(1→6)Man α(1→6)Man. Methyl β-glycoside, amorphous, $[\alpha]_D$ +29° (*c* 0.6, water); 4-nitrophenyl β-glycoside, amorphous, $[\alpha]_D$ -17° (*c* 0.8, water) (S.H. Khan, R.K. Jain and K.L. Matta, *loc. cit.*).

Xyl β(1→4)Xyl β(1→4)Xyl, xylotriose m.p. 217-219°, $[\alpha]_D$ -47.7° (*c* 1, water) (J. Hirsch, P. Kovác and E. Petráková, Carbohydr. Res., 1982, *106*, 203) is synthesised using 1,2,3-tri-*O*-acetyl-4-*O*-benzyl-β-D-xylopyranose by way of the corresponding glycosyl bromide and the derived 1,2,3-tri-*O*-acetate (P.Kováč and J. Hirsch, *ibid*, 1981, *90*, C5). Removal of blocking groups left the methyl glycoside, m.p. 190-191°, $[\alpha]_D$ -80.5° (*c* 1, water) (P. Kováč, Chem. Zvesti, 1980, *34*, 234).

(b) Deoxysugar-containing trisaccharides

Ara*f* α(1→5) Ara*f* α(1→5)Ara. Benzyl α-*O*-glycoside heptaacetate, $[\alpha]_D$ +11° (chloroform) (K. Hatanaka and H. Kuzuhara, J. Carbohydr. Chem., 1985, *4*, 333).

L-Fuc α(1→2)Gal β(1→4)Glc, 2'-fucosyllactose, m.p. 230-231°, $[\alpha]_D$ -43→ - 48° (72h; *c* 0.47, water) (A. Fernandez-Mayoralas and M. Martin-Lomas, Carbohydr. Res., 1986, *154*, 93; S.A. Abbas, J.J. Barlow and K.L. Matta, Carbohydr. Res., 1981, *88*, 51).

L-Fuc α(1→3) [Gal β(1→4)]Glc, 3-fucosyllactose, $[\alpha]_D$ -43° (constant) (*c* 0.4, methanol) is synthesised from benzyl 2,6-di-*O*-benzyl-4-*O*-(2,3,4,6-tetra-*O*-benzyl-β-D-galactopyranosyl)-β-D-glucopyranoside and 2,3,4-tri-*O*-benzyl-α-L-fucopyranosyl bromide in the presence of mercury (II) bromide and molecular sieves 4A in dichloromethane followed by hydrogenolysis (A. Fernandez-Mayoralas and M. Martin-Lomas, Carbohydr. Res., 1986, *154*, 93).

Gal β(1→2) [Galβ(1→3)] L-Fuc. Methyl α-glycoside, m.p. 145-147°, $[\alpha]_D$ -26.6° (*c* 1, methanol) is prepared by glycosylation (with acetobromogalactose and mercury cyanide in nitromethane-benzene) of methyl 4-*O*-(2-nitrobenzoyl)-2-*O*-(tetra-*O*-acetyl-β-D-galactopyranosyl)-α-L-fucopyranoside which in turn is prepared by a procedure involving use of a photolabile *O*-(2-nitrobenzylidene) acetal blocking group (P.M. Collins and V.R.N. Munasinghe, *loc. cit.*).

Glc β(1→2) [Glcβ(1→3)]L-Fuc. Methyl α-glycoside, m.p. 143-145°, $[\alpha]_D$ -57.1° (*c* 0.2, methanol) is prepared by glycosylation (with acetobromoglucose and mercury cyanide in nitromethane-benzene) of methyl 4-*O*-(2-nitrobenzoyl)-2-*O*-(tetra-*O*-acetyl-β-D-glucopyranosyl)-α-L-fucopyranoside which in turn is prepared by a procedure involving use of a photolabile *O*-(2-nitrobenzylidene) acetal blocking group (P.M. Collins and V.R.N. Munasinghe, J. Chem. Soc. Perkin Trans. I, 1983, 921).

Glc α(1→4)Rha α(1→3)Glc. Methyl β-glycoside, $[\alpha]_D$ +63° (*c* 0.5, water).

Glc α(1→4)L-Rha α(1→3)Glc. Methyl β-glycoside, $[\alpha]_D$ +28° (*c* 0.6, water).

L-Rha α(1→2) [Glc β(1→3)]Gal, solatriose, m.p. 145-160° (foaming), 195-197°dec., $[\alpha]_D^{27}$ -7.2 (5 min)→ -4.5° (*c* 1.2, water) is prepared by condensation of 2,3,4,6-tetra-*O*-acetyl-α-D-glucopyranosyl bromide with benzyl 4,6-di-*O*-benzyl-2-*O*-(2,3,4-tri-*O*-benzyl-α-L-rhamnopyranosyl)-β-D-galactopyranoside in dichloromethane in the presence of silver triflate and tetramethylurea, followed by removal of protecting groups (K. Takeo, T. Fukatsa and K. Okushio, Carbohydr. Res., 1983, *121*, 328).

(c) Nitrogen-containing trisaccharides

Gal α(1→4)Gal β(1→4)Glc NAc, m.p. 179-180° (dec.) (methanol-acetone), $[\alpha]_D$ +68→+72° (*c* 0.36, methanol-water 9:1, 18h) is synthesised by condensation of benzyl 2-acetamido-3,6-di-*O*-benzyl-2-deoxy-4-*O*-(2,3,6-tri-*O*-benzyl-β-D-galactopyranosyl)-α-D-glucopyranoside with 2,3,4,6-tetra-*O*-benzyl-α-D-galactopyranosyl chloride in dichloromethane in the presence of *sym*-trimethylpyridine, silver triflate and molecular sieves 4A, followed by hydrogenolysis (P.H. Zollo, J.C. Jacquinet and P. Sinaÿ, Carbohydr. Res., 1983, *122*, 201; M.A. Nashed and L. Anderson, Carbohydr. Res., 1983, *114*, 43).

Gal β(1→3)GlcNAc β(1→3)Gal, m.p. 139-141° (dec.), $[\alpha]_D$ +13.6° (*c* 0.5, water) (R.U. Lemieux, S.Z. Abbas and B.Y. Chung, Canad. J. Chem., 1982, *60*, 68); m.p. 185-190°, $[\alpha]_D$ +19.8° (*c* 0.76, water) (C. Augé and A. Veyrières, J. Chem. Soc. Perkin I, 1977, 1343).

Gal β(1→3)GlcNAc β(1→6)Gal, m.p. 175-178° (dec.), $[\alpha]_D$ -0.9° (*c* 0.8, water) is synthesised by reaction of 4,6-di-*O*-acetyl-3-*O*-(tetra-*O*-acetyl-β-D-galactopyranosyl)-2-deoxy-2-phthalimido-α,β-D-glucopyranosyl chloride with 1,2:3,4-di-*O*-isopropylidene-α-D-galactopyranose in nitromethane in presence of silver triflate and *sym*-collidine at -15°, followed by removal of protecting groups (R.U. Lemieux, S.Z. Abbas and B.Y. Chung, *loc. cit.*).

GalNAc β(1→4)Gal β(1→4)Glc. Methyl β-glycoside, $[\alpha]_D$ -10.1° (*c* 0.8, methanol) is prepared by reaction of 4-*O*-acetyl-3,6-di-*O*-benzoyl-2-deoxy-2-

phthalimido-D-galactopyranosyl bromide with a 6'-benzylated derivative of methyl 4-*O*-β-D-galactopyranosyl-β-D-glucopyranoside followed by removal of protecting groups (H.-P. Wessel, T. Iversen and D.R. Bundle, Carbohydr. Res., 1984, *130*, 5).

Gal β(1→4)GlcNAc β(1→3)Gal, $[\alpha]_D$ +17.5° (*c* 0.5, water) (R.U. Lemieux, S.Z. Abbas and B.Y. Chung, *loc. cit.*). Methyl β-glycoside, $[\alpha]_D$ +4.5° (*c* 1.3, water) is prepared by glycosylation of methyl 3-*O*-(2-acetamido-3,6-di-*O*-benzyl-2-deoxy-β-D-glucopyranosyl)-2,4,6-tri-*O*-benzyl-β-D-galactopyranoside with 2,3,4,6-tetra-*O*-acetyl-α-D-galactopyranosyl bromide, catalysed by mercury (II) cyanide, followed by removal of protecting groups (R.K. Jain, S.A. Abbas and K.L. Matta, J. Carbohydr. Chem., 1988, *7*, 377).

Gal β(1→4)GlcNAc β(1→6)Gal, m.p. 155-159°, $[\alpha]_D$ +9.6° (*c* 0.5, water) is prepared by reaction of 3,6-di-*O*-acetyl-4-*O*-(tetra-*O*-acetyl-β-D-galactopyranosyl)-2-deoxy-2-phthalimido-α,β-D-glucopyranosyl chloride with 1,2:3,4-di-*O*-isopropylidene-α-D-galactopyranose in nitromethane in presence of silver triflae and *sym*-collidine at -15°, followed by removal of protecting groups (R.U. Lemieux, S.Z. Abbas and B.Y. Chung, Canad. J. Chem., 1982, *60*, 68).

Gal β(1→6)GlcNAc β(1→3)Gal. Methyl β-glycoside, $[\alpha]_D$ +10.76° (*c* 1.2, water) is prepared by glycosylation of methyl 3-*O*-(2-acetamido-3-*O*-acetyl-2-deoxy-β-D-glucopyranosyl)-2,4,6-tri-*O*-benzyl-β-D-galactopyranoside with 2,3,4,6-tetra-*O*-acetyl-α-D-galactopyranosyl bromide in the presence of mercury (II) cyanide, followed by removal of protecting groups (R.K. Jain, S.A. Abbas and K.L. Matta, *loc. cit.*).

GalNAc α(1→3)Gal β(1→3)GlcNAc, $[\alpha]_D$ +135.1° (*c* 1, water) is synthesised from benzyl 2-acetamido-3-*O*-(2-*O*-acetyl-4,6-*O*-benzylidene-β-D-galactopyranosyl)-4,6-*O*-benzylidene-2-deoxy-α-D-glucopyranoside by glycosylation with 2-azido-3,4,6-tri-*O*-benzyl-2-deoxy-β-D-galactopyranosyl chloride in dichloromethane in the presence of silver carbonate, silver perchlorate and molecular sieves 4A, followed by deprotection (N.V. Bovin, S.É. Zurabyan and A.Ya. Khorlin, Carbohydr. Res., 1983, *112*, 23).

GalNAc β(1→4)Gal β(1→4)Glc, gangliotriose, hygroscopic powder, m.p. 185-188°, $[\alpha]_D$ +30.3° (*c* 0.8, water), $[\alpha]_D$ +41° (*c* 0.5, methanol) (S. Sebesan and R.U. Lemieux, Canad.J.Chem., 1984, *62*, 644; H. Paulsen and M. Paal, Carbohydr. Res., 1985, *137*, 39; M. Sugimoto, T. Horisaki and T. Ogawa, Glycoconjugate J., 1985, *2*, 11). Methyl β-glycoside, $[\alpha]_D$

-10.1^o (methanol) (H.-P. Wessel. T. Iversen and D.R. Bundle, Carbohydr. Res., 1984, *130*, 5).

GalNAc β(1→6)Gal β(1→4)Glc. Benzyl β-glycoside, $[\alpha]_D$ -16° (water) (S.E. Zurabyan *et al.*, Bioorg. Khim., 1978, *4*, 928; H. Matsuda, H. Ishihara and S. Tejima, Chem. Pharm. Bull., 1979, *27*, 2564; T. Takamura, T. Chiba and S. Tejima, Chem. Pharm. Bull., 1981, *29* , 1027).

Man α(1→2)Man β(1→4)GlcNAc. α-Acetate, white powder, $[\alpha]_D$ +32.7^o (chloroform) (Y.Itoh and S. Tejima, Chem. Pharm. Bull., 1983, *31*, 1632).

Man α(1→3)Man β(1→4)GlcNAc, $[\alpha]_D$ +27.8^o (*c* 0.71, methanol) is prepared by glycosylation of 4-*O*-(6-O-acetyl-2,4-di-*O*-benzyl-β-D-mannopyranosyl)-1,6-anhydro-2-azido-3-*O*-benzyl-2-deoxy-β-D-glucopyranose with 2-*O*-acetyl-3,4,6-tri-*O*-benzyl-α-D-mannopyranosyl chloride in dichloromethane in the presence of silver triflate and tetramethylurea, followed by removal of protecting groups. α-Acetate, $[\alpha]_D$ +3.9^o (*c* 1.2, chloroform) (H. Paulsen and R. Lebuhn, Liebigs Ann. Chem., 1983, 1047).

Man β(1→4)GlcNAc β(1→4)GlcNAc, $[\alpha]_D$ +0.2^o (*c* 0.3, water) is prepared by glycosylation with the protected *N*-phthalimido-ManGlcN β-chloride in the presence of silver triflate and *sym*-collidine in dichloromethane, followed by removal of protecting groups (H. Paulsen and R. Lebuhn, Carbohydr. Res., 1984, *130*, 85).

NeuAc α(2→3)Gal β(1→4)Glc, $[\alpha]_D$ +19.2^o (*c* 1.53, water) is synthesised by employing the 1,2,3,6,2',6'-hexa-*O*-benzylated glycosyl acceptor and methyl (5-acetamido-4,7,8,9-tetra-*O*-acetyl-3,5-dideoxy-D-*glycero*-β-D-*galacto*-2-nonulopyranosyl chloride)onate as glycosyl donor in the presence of mercury (II) bromide and cyanide and powdered 4A molecular sieves in 1,2-dichloroethane to obtain a 1:2.1 mixture (18%) of the α and β anomeric products which are separated on a LiChroprep Si-60 column with 9:1 toluene-methanol before deprotection (T. Ogawa and M. Sugimoto, Carbohydr. Res., 1985, *135*, C5. Also: H. Paulsen and U. von Deessen, *ibid*, 1986, *146*, 147; K. Okamoto, T. Kondo and T. Goto, Tetrahedron Lett., 1986, *27*, 5233).

NeuAc α(2→6)Gal β(1→4)Glc, 6'-sialyllactose, $[\alpha]_D$ +5.6^o (water) (K. Furuhata *et al.*, Chem. Pharm. Bull., 1986, *34*, 2725).

4. Tetrasaccharides and higher oligosaccharides

Cyclomaltohexaose, α-cyclodextrin, cyclohexa-amylose, $[\alpha]_D$ +150.5° (*c* 1, water), chemistry has been reviewed (J. Szejtli, "Cyclodextrin Technology", Kluwer, Dordrecht, 1988). Hexakis (2,6-di-*O*-methyl)cyclomaltohexaose, m.p. 301-307° (from methanol), $[\alpha]_D$ +153° (*c* 1.1, water) (K. Takeo, *Carbohydr. Res., 1990,* 200, 481).

Cyclomaltoheptaose, β-cyclodextrin, cyclohepta-amylose, $[\alpha]_D$ +162.5° (*c* 1.1, water), chemistry has been reviewed (W. Saenger, Angew. Chem. Int. Ed. Engl., 1980, *19*, 344; J. Szejtli, "Cyclodextrins and their Inclusion Complexes," Akadémiai Kiadó, Budapest, 1982; derivatives, A.R. Croft and R.A. Bartsch, Tetrahedron, 1983, *39*, 1417; J. Szejtli, "Cyclodextrin Technology," Kluwer, Dordrecht, 1988; inclusion complexes, R.J. Clarke, J.H. Coates and S.F. Lincoln, Adv. Carbohydr. Chem. Biochem., 1988, *46*, 25). 6-Mono-*O*-tosylate, m.p.160-162° (dec.), $[\alpha]_D$ +118° (*c* 4, methyl sulphoxide) (J. Defaye *et al.,* Carbohydr. Res., 1989, *192*, 251). The total synthesis of cyclodextrins has been reviewed (Y. Takahashi and T. Ogawa, in "Trends in Synthetic Carbohydrate Chemistry," American Chemical Society, 1989, p.150). Heptakis(2,6-di-*O*-methyl)cyclomaltoheptaose, m.p. 299-305° (dec.) (from methanol), $[\alpha]_D$ +121° (*c* 1.1, chloroform), $[\alpha]_D$ +155.5° (*c* 1.1, water) (K. Takeo, *loc. cit.*). Heptabenzoate, m.p.134-136°, $[\alpha]_D$ -93° (*c* 1, chloroform) (C.M. Spencer, J.F. Stoddart and R. Zarzycki, J. Chem. Soc., Perkin Trans. 2, 1987, 1323; Y. Kubota *et al.,* Carbohydr. Res., 1989, *192*, 159).

Cyclomaltooctaose, γ-cyclodextrin, cycloocta-amylose, $[\alpha]_D$ +177° (*c* 1, water), chemistry has been reviewed (J. Szejtli, *loc. cit.*). Octakis (2,6-di-*O*-methyl)cyclomalto-octaose, m.p. 260-264° (dec.) (from ethanol), $[\alpha]_D$ +127° (*c* 1, chloroform), $[\alpha]_D$ +179.5° (*c* 1.3, water) (K. Takeo, *loc. cit.*).

L-Fuc α(1→2)Gal β(1→4) [L-Fuc α(1→3)]Glc, 3,2'-difucosyllactose, $[\alpha]_D$ -100° (*c* 0.5, water) (A. Fernandez-Mayoralas and M. Martin-Lomas, Carbohydr. Res., 1986, *154*, 93; K. Takeo and S. Tei, *ibid*, 1985, *141*, 159).

Gal β(1→4)GlcNAc β(1→3)Gal β(1→4)GlcNAc. Methyl β-glycoside, m.p. 259-261°, $[\alpha]_D$ -10° (*c* 1.02, water) is prepared by one of two block synthetic procedures for linear tetra-, hexa- and octa-saccharides of the poly-(*N*-acetyl-lactosamine) series. (J. Alais and A. Veyrières, Carbohydr. Res., 1990, *207,* 11).

Glc β(1→3)Glc β(1→3)[Glc β(1→6)]Glc, $[\alpha]_D$ -0.3° (*c* 0.50, water) is synthesised from benzyl 2,4-di-*O*-benzyl-6-*O*-(2,3,4,6-tetra-*O*-acetyl-β-D-glucopyranosyl)-α-D-glucopyranoside and 2,4,6-tri-*O*-acetyl-3-*O*-(2,3,4,6-tetra-*O*-acetyl-β-D-glucopyranosyl)-α-D-glucopyranosyl bromide in 1,2-dichloroethane in the presence of silver triflate and 4A molecular sieves at -10 to -15°, followed by removal of protecting groups (T. Ogawa and T. Kaburagi, Carbohydr. Res., 1982, *103*, 53).

Glc β(1→4)Glc β(1→4)Glc β(1→4)Glc, cellotetraose, m.p. 251-253° (dec.) (aqueous ethanol), $[\alpha]_D$ +7.1(3min)→ +17.1° (*c* 2.5, water) is prepared by condensation of benzyl 2,3,6-tri-*O*-benzyl-4-*O*-(2,3,6-tri-*O*-benzyl-β-D-glucopyranosyl)-β-D-glucopyranoside with hepta-*O*-acetyl-α-cellobiosyl bromide in benzene-nitromethane in the presence of mercury (II) cyanide at 65°, followed by removal of protecting groups (Takeo *et al.,* Carbohydr. Res., 1983, *121*, 163).

Glc β(1→6)[Glc β(1→3)]Glc β(1→6)Glc β(1→6)[Glc β(1→3)]Glc β(1→6)Glc β(1→6)[Glc β(1→3)]Glc β(1→6)Glc, $[\alpha]_D$ -17° (*c* 0.17, water), $[\alpha]_D$ -20° (*c* 0.6, DMSO) is prepared by a block synthesis using methyl 2-*O*-benzoyl-4-*O*-benzyl-3*O*-(2,3,4,6-tetra-*O*-benzoyl-β-D-glucopyranosyl)-6-*O*-(2,3,4-tri-*O*-benzoyl-6-*O*-chloracetyl-β-D-glucopyranosyl)-1-thio-β-D-glucopyranoside activated with methyl triflate in the presence of 4A molecular sieves in dichloromethane (W. Birberg *et al.,* J. Carbohydr. Chem., 1989, *8*, 47; P. Fügedi *et al.,* Carbohydr. Res., 1987, *164*, 297).

Glc α(1→4)Xyl α(1→4)Xyl α(1→4)Glc, $[\alpha]_D$ +157° (*c* 1.2, water) (K. Takeo *et al.,* Carbohydr. Res., 1990, *201*, 261).

NeuAc α(2→6)Gal β(1→4)GlcNAc β(1→2)Man, $[\alpha]_D$ -21° (*c* 1.0, water) (H. Paulsen and H. Tietz, Carbohydr. Res., 1985, *144*, 205; T. Kitajima *et al., ibid*, 1984, *127*, C1).

NeuAc β(2→6)Gal β(1→4)GlcNAc β(1→2)Man, $[\alpha]_D$ -25° (*c* 0.8, water) (H. Paulsen and H. Tietz, *loc. cit.;* T. Kitajima *et al., loc. cit.*).

Xyl β(1→4)Xyl β(1→4)Xyl β(1→4)Xyl, xylotetraose, is synthesised using 1,2,3-tri-*O*-acetyl-4-*O*-benzyl-β-D-xylopyranose by way of the corresponding glycosyl bromide and the derived 1,2,3-tri-*O*-acetate. After removal of blocking groups the tetrasaccharide was characterised as its deca-*O*-methyl derivative, m.p. 122-123°, $[\alpha]_D$ -87° (*c* 1, chloroform). Methyl β-*O*-glycoside, m.p. 238-239°, $[\alpha]_D$ -88° (*c* 1, water) (P. Kováč and J. Hirsch, *loc. cit.*)

5. Sulphur-containing oligosaccharides

Synthesis of thio-oligosaccharides, and their reactions with enzymes, have been reviewed (J. Defaye and J. Gelas, in "Studies in Natural Products Chemistry", vol. 8, ed. Atta-ur-Rahman, Elsevier, 1991, p.315).

4-*S*-β-D-Glucopyranosyl-4-thio-D-glucose, 4-thiocellobiose, $[\alpha]_D$ -16° (*c* 1, water) is prepared by reaction of methyl 2,3,6-tri-*O*-benzoyl-4-*O*-trifluoromethylsulphonyl-α-D-galactopyranose with the sodium salt of 1-thio-β-D-glucopyranose, followed by acetolysis of the resulting methyl β-glycoside. α-Octaacetate, m.p. 175-177°, $[\alpha]_D$ +23° (*c* 0.9, chloroform) (D. Rho *et al.,* J. Bacteriol., 1982, *149*, 47; J. Defaye, pers. commun.).

4-*O*-(2-Deoxy-6-*O*-sulpho-2-sulphoamino-α-D-glucopyranosyl)-L-idopyranosiduronic acid, heparin fragment. Methyl β-glycoside trisodium salt, $[\alpha]_D$ +94° (*c* 1.1, water) (T. Chiba *et al.,* Carbohydr. Res., 1988, *174*, 253).

2-Deoxy-4-*O*-(α-L-idopyranosyluronic acid)-6-*O*-sulpho-2-sulphoamino-D-glucose, heparin fragment. Methyl α-glycoside trisodium salt, $[\alpha]_D$ +22° (*c* 1, water) (T. Chiba *et al., loc. cit.*).

β-D-Fructofuranosyl 1-thio-α-D-glucopyranoside, 1-thiosucrose, $[\alpha]_D$ +26° (*c* 1.33, methanol) is prepared by Lewis acid (zirconium chloride) catalysed condensation between 2,3,4,6-tetra-*O*-acetyl-1-thio-α-D-glucopyranose and 1,3,4,6-tetra-*O*-benzyl-D-fructofuranose in dichloromethane, followed by liquid chromatography of the deprotected resulting isomeric disaccharides (J. Defaye *et al.,* Carbohydr. Res., 1984, *130*, 299).

4-*S*-β-D-Galactopyranosyl-4-thio-D-glucose, thiolactose, $[\alpha]_D$ +5→ -8° (*c* 1, water) (L.A. Reed and L. Goodman, Carbohydr. Res., 1981, *94, 91).*

4-*S*-β-D-Xylopyranosyl-4-thio-D-xylopyranose, 4-thioxylobiose, $[\alpha]_D$ -22.9° (*c* 2.6, methanol) is prepared by deacetylation of 2,3,4-tri-*O*-acetyl-1-*S*-acetyl-1-thio-β-D-xylopyranose with sodium methoxide and reaction of the resulting sodium 1-thiolate with crude 1,2,3-tri-*O*-benzoyl-4-*O*-trifluoromethylsulphonyl-β-L-arabinopyranose, followed by removal of protecting groups. (J. Defaye *et al.,* Carbohydr. Res., 1985, *139*, 123).

Second Supplements to the 2nd Edition of Rodd's Chemistry of Carbon Compounds, Vol. IE/F/G, edited by M. Sainsbury

Chapter 25

TETRAHYDRIC ALCOHOLS, THEIR ANALOGUES, DERIVATIVES AND OXIDATION PRODUCTS

R.A. HILL

1. Tetrahydric alcohols

(a) Tetritols

The production of erythritol by fermentation, its physical and chemical properties, metabolism and use in food products have been reviewed (T. Oda, M. Kusakabe and T. Sasaki, Shokukin Kogyo, 1989, **32**, 41; C.A., 1989, **111**, 38019). Osmotically-active erythritol has been detected in the internal solutes of xerophylic fungi by ^{13}C NMR spectroscopy (A.D. Hockin and R.S. Norton, J. Gen. Microbiol., 1983, **129**, 2915).The role of erythritol in the pathogenesis of abortion by *Brucella* has reviewed (A.I. De Diego, Rev. Med. Vet. (Buenos Aires), 1985, **66**, 389; C.A., 1987, **106**, 65079).

Erythitol can be prepared from arabinose, and threitol from xylose by decarbonylation reactions using a rhodium catalyst (M.A. Andrews, G.L. Gould and S.A. Klaeren, J. Org. Chem., 1989, **54**, 5257).

$$CHO\text{–}CHOH\text{–}CHOH\text{–}CHOH\text{–}CH_2OH \xrightarrow{(Ph_3P)_3RhCl} CH_2OH\text{–}CHOH\text{–}CHOH\text{–}CH_2OH$$

Erythritol, or threitol, when treated with acetic acid containing an acid catalyst produce mixtures of acetylated 1,4-anhydroerythritol, 1,4-anhydrothreitol, erythitol and threitol (A. Wisniewski *et al.*, Carbohydr. Res., 1983, **114**, 11). The relative and absolute stereochemistries of 1,2,3,4-tetrols can be assigned by Nakanishi's method, using circular dichroism, after a simple two-step derivatisation with exciton-coupling chromophores (W.T. Wiesler and K. Nakanishi, J. Am. Chem. Soc., 1989, **111**, 9205).

The conformations of erythritol, threitol and their acetates have been studied by ^{13}C-NMR (S.J. Angyal and R. Le Fur, Carbohydr. Res., 1980, **84**, 201). ^{1}H-NMR spectroscopy has been used to determine the conformations of erythritol and threitol in solution in D_2O (G.E. Hawkes and D. Lewis, J. Chem. Soc., Perkin 1, 1984, 2073). The conformation with a zig-zag planar carbon chain is favoured to the extent of 66% for erythritol and 56% for threitol.

(b) Derivatives of the tetritols

The complex of 1,4-di-*O*-methyl-L-threitol with Ti^{4+} has been studied by ^{1}H-NMR and ^{13}C-NMR giving an insight into the structure os this species in solution (P.G. Potvin *et al.*, Can. J. Chem., 1989, **67**, 1523).

Various benzyl-protected threitol derivatives have been prepared from dimethyl (*R,R*)-tartrate (S. Takano *et al.*, Synthesis, 1986, 811). These derivatives have been converted into the epoxides (1), (2) and (3). The epoxide (3) when treated with ammonia yields the aminoalcohol (4). The epoxide (1) has also been prepared by a different route from dimethyl (*R,R*)-tartrate (E. Hungerbuehler, D. Seebach and D. Wasmuth, Angew. Chem., Int. Ed., 1979, **18**, 958).

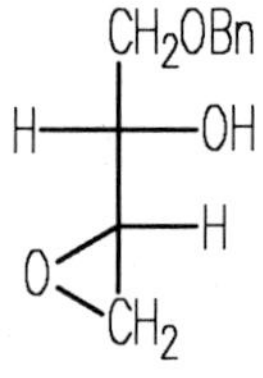

CH_2OBn / H—C—OH / H—C(O)—CH_2

(1)

CH_2OH / H—C—OH / H—C(O)—CH_2

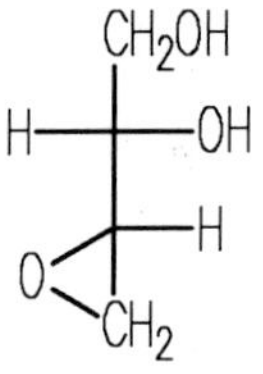

(2)

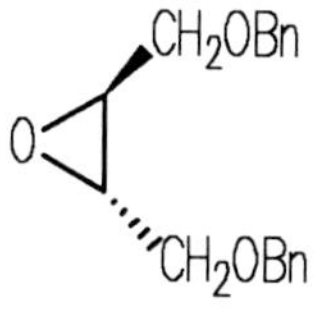

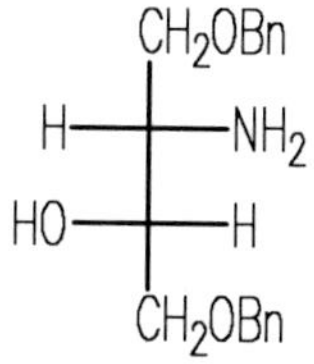

CH_2OBn O CH_2OBn

(3)

CH_2OBn H NH_2 HO H CH_2OBn

(4)

The diphosphinite ester of 1,4-anhydrothreitol (5), diphin, is a cheap chiral ligand that can be prepared from (*R,R*)-tartaric acid (W.R. Jackson and C.G. Lovel, Aust. J. Chem., 1982, **35**, 2069). Diphin has been used as a chiral ligand for rhodium in hydrogenation reactions and for palladium in hydrocyanation reactions.

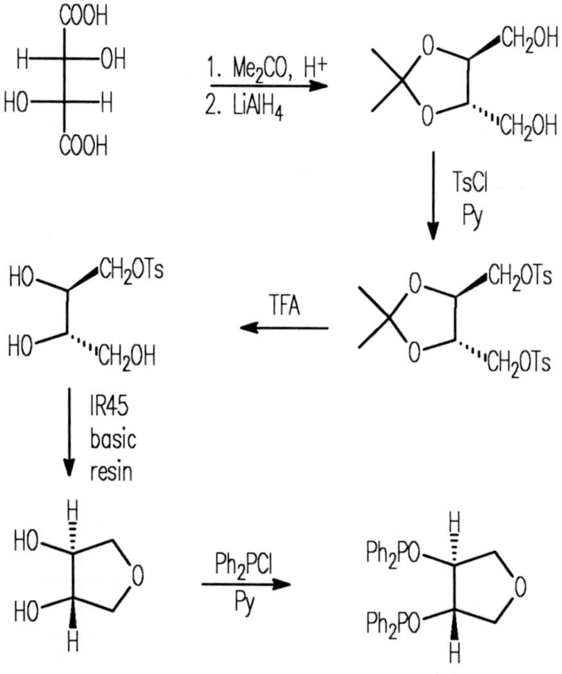

COOH H OH HO H COOH

1. Me_2CO, H^+
2. $LiAlH_4$

O CH_2OH O CH_2OH

TsCl
Py

O CH_2OTs O CH_2OTs

TFA

HO CH_2OTs HO CH_2OH

IR45
basic
resin

H HO O HO H

Ph_2PCl
Py

H Ph_2PO O Ph_2PO H

(5)

The optimum conditions for the hydolytic ring opening of 3,4-dichloro-1,2-epoxybutane to produce the dichlorodiol (6) involve using an acid ion exchange resin to initiate the reaction (E. Milchert and J. Myszkowski, Przem. Chem., 1983, **62**, 159).

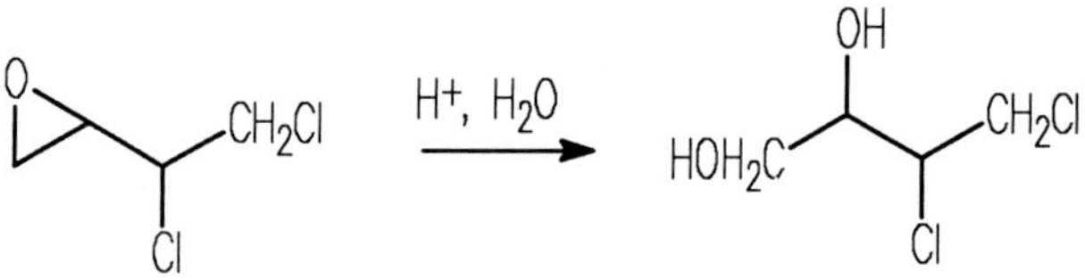
H+, H2O
OH
CH2Cl
HOH2C
Cl

(6)

(c) Homologous tetrahydric alcohols

The commercially-available adonitol (7) can be deoxygenated to provide *meso*-pentane-1,2,4,5-tetrol (Z.-F. Xie and K. Sakai, Chem. Pharm. Bull., 1989, **37**, 1650)

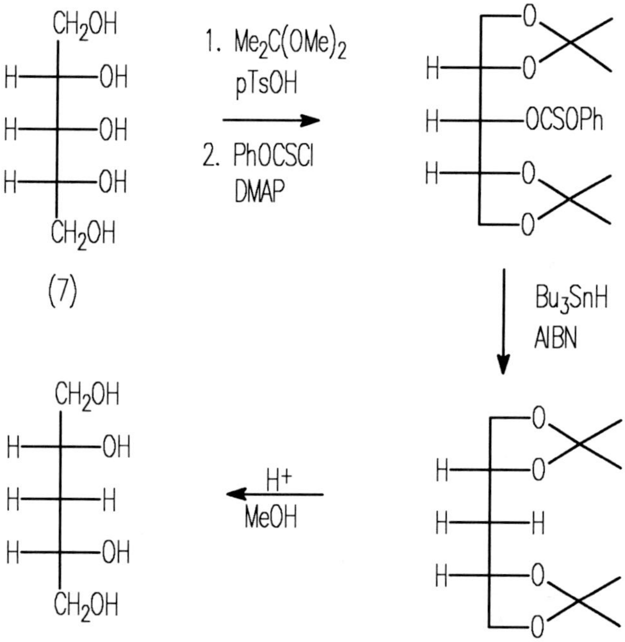
CH2OH
H
OH
(7)
1. Me2C(OMe)2
pTsOH
2. PhOCSCl
DMAP
OCSOPh
Bu3SnH
AIBN
H+
MeOH

The possible use of *meso*-2,3-dimercaptobutane-1,4-diol as a new arsenic antidote has been investigated (V.L. Boyd *et al.*, Chem. Res. Toxicol., 1989, **2**, 301). The complexes of the 1,4-dimercaptobutane-2,3-diols (8) and (9) with Cd^{2+} have been studied using ^{1}H-, ^{13}C- and ^{113}Cd-NMR spectroscopy (G.K. Carson, P.A.W. Dean and M.J. Stillman, Inorg. Chim. Acta, 1981, **56**, 59).

(8)

(9)

2. Trihydroxyaldehydes and trihydroxyketones

The glycolaldehyde derivative (10) undergoes a self aldol condensation to give mainly the L-erythrose derivative (11) (C.R. Noe, M. Knollmueller and P. Ettmayer, Liebig's Ann. Chem., 1989, 637).

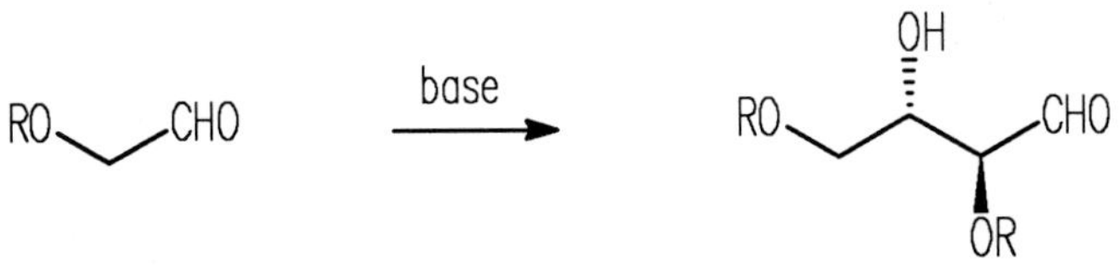

base

(10)

(11)

R =

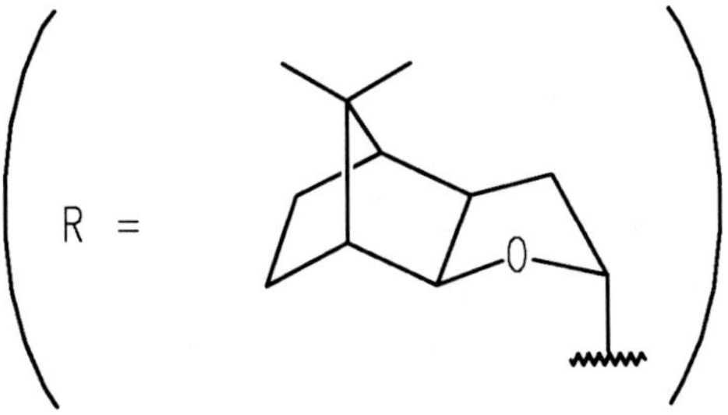

The anion of the sulphoxide (12) reacts with D-glyceraldehyde acetonide by *C*-alkylation to give a mixture of isomeric sulphoxides that can be separated. Reduction of the sulphoxides with lithium aluminium hydride, followed by hydrolysis of the intermediate product thioacetals gives erythrose and threose (J.A. Lopez Sastre, An. Quim., 1986, **82C**, 140).

O
SEt
SEt
(12)
+
CHO
O
O
BuLi
EtS
O
SEt
OH
H
O
O
$LiAlH_4$
EtS
SEt
OH
H
O
O
H^+, H_2O
CHO
HO—H
H—OH
CH_2OH
+
CHO
H—OH
H—OH
CH_2OH

L-Erythrose can be prepared in high yield from D-ribono-1,4-lactone (13) by sodium borohydride reduction of the acetonide (14) and sodium periodate cleavage of the resulting diol (R.H. Shah, Carbohydr. Res., 1986, **155**, 212).

HOH_2C
O
O
OH OH
(13)
Me_2CO
anhydr.
$CuSO_4$
HOH_2C
O
O
O O
(14)
$NaBH_4$
CH_2OH
HO—H
O—H
O—H
CH_2OH
$NaIO_4$
HO
O
O O
AcOH
H_2O
HO
O
OH OH

D-Erythrose and D-threose are produced by oxidation of D-arabinonic and D-xylonic acids, respectively (L.F. Sala *et al.*, An. Asoc. Quim. Argent., 1978, **66**, 57). D-Fructose is oxidatively degraded to D-erythrose with iron (III) chloride by photo-irradiation (K. Araki, M. Sakuma and S. Shiraishi, Chem. Lett., 1983, 665). Only the open-chain form of fructose absorbs ultra-violet light and produces, upon irradiation, a complex mixture of products, largely resulting from α-cleavage. One of the products has been identified as the 2,3-deoxy-2,3-di-*C*-hydroxymethyltetrose (15) (C. Triantaphylides, H.P. Schuchmann and C. Von Sonntag, Carbohydr. Res., 1982, **100**, 131).

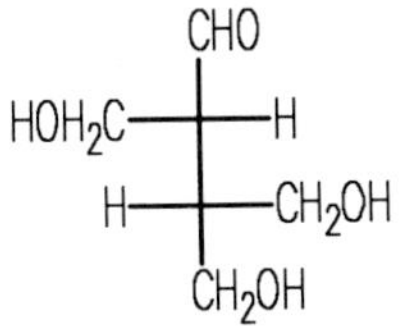

CHO
HOH_2C—C—H
H—C—CH_2OH
CH_2OH

(15)

[1-^{13}C,^{2}H]-Labelled erythrose and threose are produced by catalytic hydrogenation of the cyanohydrins of glyceraldehyde by deuterium gas (A.S. Serianni and R. Barker, Can. J. Chem., 1979, **57**, 3160).

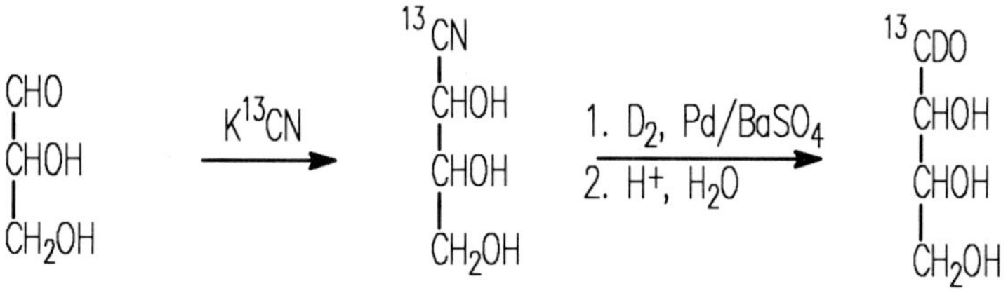

CHO–CHOH–CH_2OH $\xrightarrow{K^{13}CN}$ ^{13}CN–CHOH–CHOH–CH_2OH $\xrightarrow[2.\ H^+,\ H_2O]{1.\ D_2,\ Pd/BaSO_4}$ ^{13}CDO–CHOH–CHOH–CH_2OH

The labelled tetroses were used in a study of the kinetics of the interconversion of the α- and β-furanose forms using ^{1}H- and ^{13}C-NMR (A.S. Serianni *et al.*, J. Am. Chem. Soc., 1982, **104**, 4037).

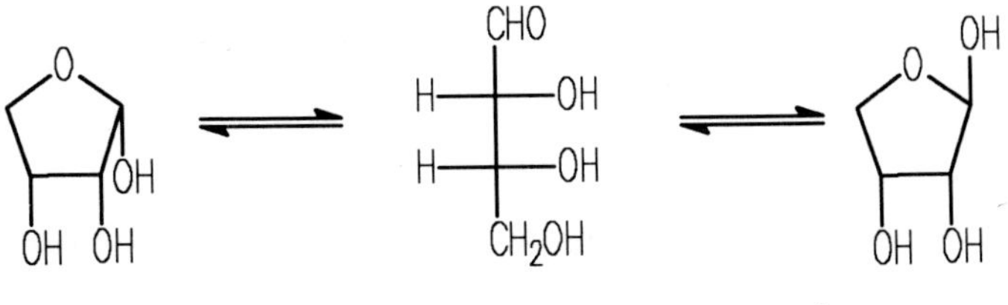

α-furanose ⇌ CHO / H—C—OH / H—C—OH / CH_2OH ⇌ β-furanose

α-furanose β-furanose

Swern oxidation of the protected threitol derivative (16) gives the threose derivative (17),a useful building block for L-sugars (T. Mukaiyama, K. Suzuki and T. Yamada, Chem. Lett. 1982, 929).

CH_2OBn CH_2OH O O → CH_2OBn CHO O O

(16) (17)

NMR spectroscopic studies have shown that a concentrated aqueous solution of erythrose-4-phosphate is composed of an equilibrium mixture of the monomeric aldehyde and hydrated aldehyde forms, which interconvert rapidly, together with a major contribution from 3 dimeric forms (C.C. Dale, J.K. MacLeod and J.F. Williams, Carbohydr. Res., 1981, **95**, 1). The conformational mobility of erythulose has been studied by measuring the temperature dependence of CD spectra in the region +40 to −140 °C (S. Bystricky, T. Sticzay and I. Tvaroska, Collect. Czech. Chem. Commun., 1980, **45**, 475).

Spinach leaves are a rich source of transketolase and this enzyme catalyses the condensation between hydroxypyruvate and glycolaldehyde to produce L-erythulose in high yield (J. Bolte, C. Demuynck and H. Samaki, Tetrahedron Lett., 1987, **28**, 5525).

CH_2OH–C=O–COOH + CHO–CH_2OH → CH_2OH–C=O–HO–C–H–CH_2OH

Aldol condensation of erythulose with glycolaldehyde gives *lyxo*-3-hexulose as a major product (S. Morgenlie, Acta Chem. Scand., 1987, **41B**, 745).

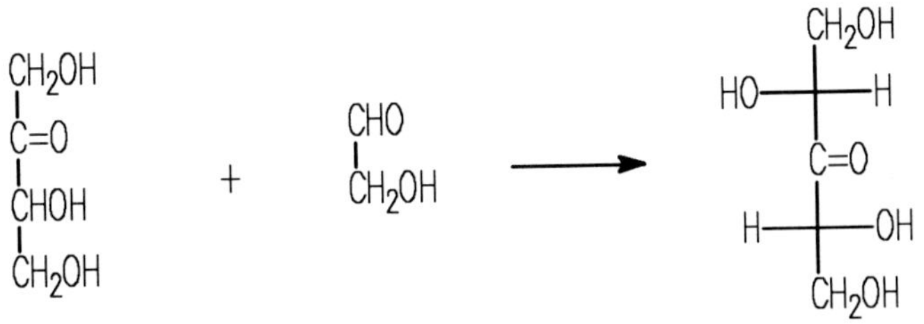

3. Dihydroxydiones and their derivatives

The useful synthetic intermediate (18) has been prepared from 1,4-anhydroerythritol by monoacetylation with ethyl orthoacetate, oxidation of the alcohol with pyridinium chlorochromate and bromination with *N*-bromosuccinimide. The intermediate (18) has been used in a synthesis of aflatoxin M_1 (G. Büchi *et al.*, J. Am. Chem. Soc., 1981, **103**, 3497)

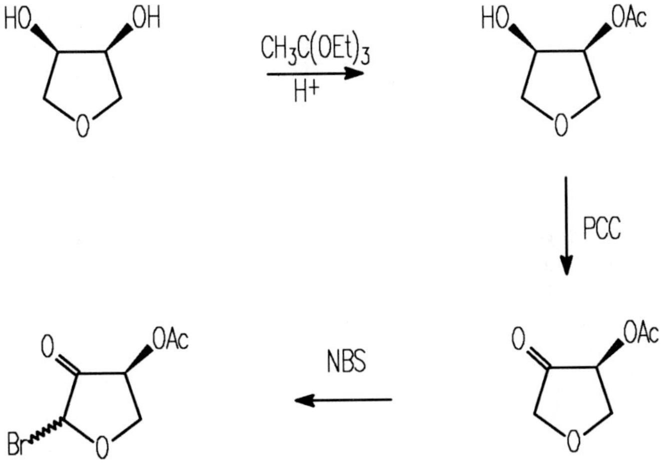

(18)

A crystal structure analysis of 1,4-dichlorobutane-2,3-dione has shown that the carbonyl groups are antiperiplanar and the chlorines are synperiplanar to the oxygen atoms in the solid state (B. Ducourant *et al.*, Acta Crystallogr., 1986, **42C**, 341)

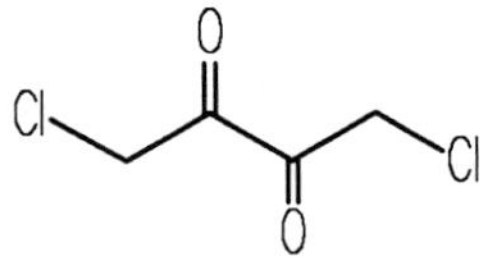

4. Trihydroxycarboxylic acids

The Kiliani reaction for the production of cyanohydrins from aldoses, for example, the reaction of D-glyceraldehyde with cyanide, has been studied using ^{13}C NMR spectroscopy (A.S. Serianni, H.A. Nunez and R. Barker, J. Org. Chem., 1980, **45**, 3329). The resultant cyanohydrins are hydrolysed *in situ* to the corresponding aldonic acids by a complex reaction pathway.

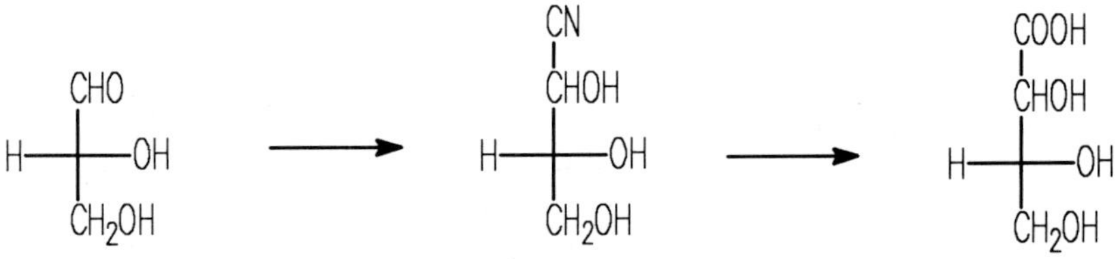

Both D-erythrono-1,4-lactone and D-threono-1,4-lactone have been prepared from the D-mannitol derivative (19) by sodium periodate cleavage, subsequent chain extension and deprotection (R. Herranz, An. Quim., 1987, **83C**, 318)

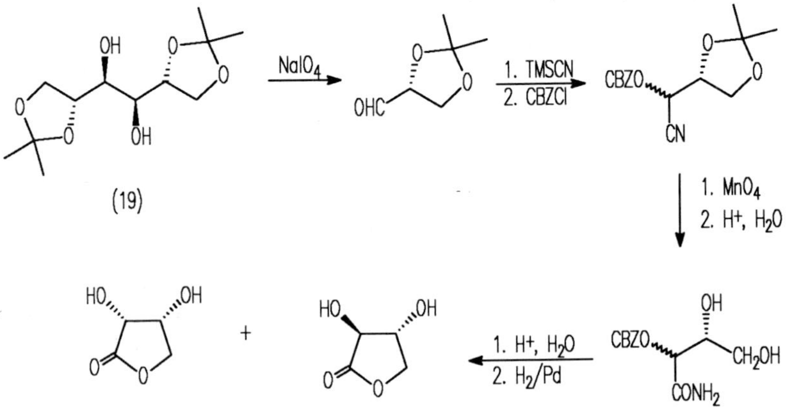

(19)

Enantioselective hydrolysis of the diacetate (20) using a lipase from *Pseudomonas fluorescens* , followed by oxidation and lactonisation, produces the useful chiral building block, 3(*S*)-hydroxy-5(*R*)-hydroxymethyl-2*H*-dihydrofuran-2-one (21) (Z.-F. Xie and K. Sakai, Chem. Pharm. Bull., 1989, **37**, 1650).

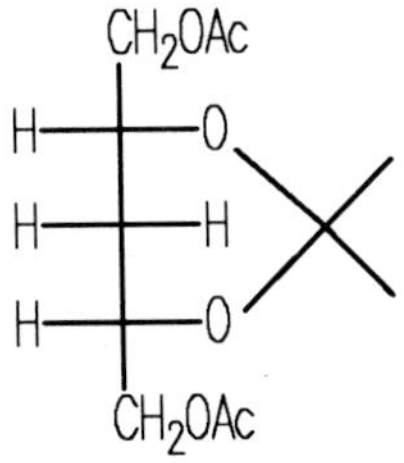

CH_2OAc H—O, H—H, H—O CH_2OAc

(20)

O, OH, O, CH_2OH

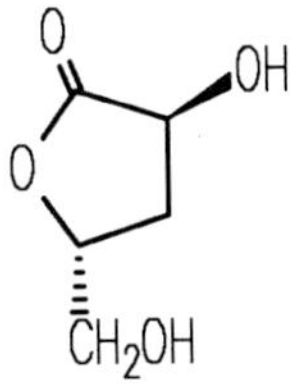

(21)

Threonic acid when treated with hydrogen bromide in acetic acid, followed by the addition of methanol produces the dibromo-ester (22), which yields the epoxide (23) through the reaction with potassium iodide and potassium acetate (P.S. Manchand *et al.*, J. Org. Chem., 1988, **53**, 5507).

COOH, H—OH, HO—H, CH_2OH $\xrightarrow[AcOH]{HBr}$ COOMe, H—OH, H—Br, CH_2Br $\xrightarrow[KOAc]{KI}$ COOMe, O, CH_2Br

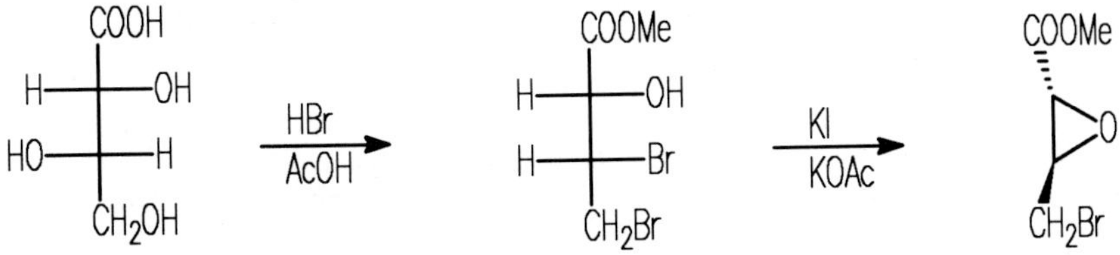

(22) (23)

The epoxide (24) is readily prepared from (*R,R*)-tartaric acid by treatment with hydrogen bromide in acetic acid and acetic anhydride to form a bromohydrin. Esterification is achieved by the addition of ethanol. The epoxide (25) is formed from the bromohydrin by treatment with triethylamine. Partial reduction of the diester (25) is performed using sodium borohydride at 0 °C producing the alcohol (24) in 45% overall yield from tartaric acid (P.S. Manchand *et al.*, J. Org. Chem., 1988, **53**, 5507).

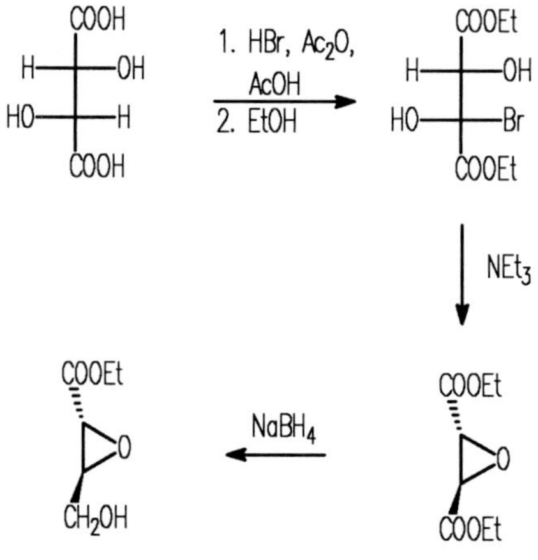

5. Dihydroxyalkanedicarboxylic acids

The crystal structure of (+)-tartaric acid has been studied at temperatures down to 20 K (J. Albertson, A. Oskarsson and K. Staahl, J. Appl. Crystallogr., 1979, **12**, 537). Chemical shift values in the ^{13}C-NMR spectra of several derivatives of tartaric acid have been studied and have been used to predict conformations (F.K. Velichko *et al.*, Org. Magn. Reson., 1980, **13**, 442; M. Ul Hasan, *ibid.*, 1980, **14**, 309; H.-J. Scneider and M. Lonsdorfer, *ibid.*, 1981, **16**, 133). The most stable conformation of (+)-tartaric acid in solution has been shown by ^{1}H- and ^{13}C-NMR spectroscopy to be that in which the carboxyl groups are are antiperiplanar (26) (J. Ascenso and V.M.S. Gil, Can. J. Chem., 1980, **58**, 1376). However, dimethyl tartrate adopts a conformation with the hydroxyl groups antiperiplanar (27), as shown by vibrational circular dichroism (C.N. Su and T.A. Keiderling, J. Am. Chem. Soc., 1980, **102**, 511).

The absolute configuration of (+)-tartaric acid has been confirmed as being (*R,R*). Tartaric acid was converted into the dithiol (28) and then into a dithioacetal derivative on which a Bijvoet X-ray analysis was performed (H. Buding, *et al.*, Angew. Chem. Int. Ed., 1985, **24**, 513).

COOH, H—OH, HO—H, COOH → 9 steps → CH_3, HS—H, H—SH, CH_3

(28)

Raman circular intensity differential spectra of some alkyl tartrates have been studied (H. Waki, S. Higuchi and S. Tanaka, Spectrochim. Acta, 1980, **36A**, 659) and the IR spectrum of tartaric acid has been interpreted (L.I. Kozhevina, L.G. Skryabina and Yu.K. Tselinskii, Zh. Prikl, Spektrosk., 1980, **23**, 1090). A study using ^{1}H-NMR and ^{13}C-NMR spectroscopy of the complexes of (*R,R*)- and *meso*-diisopropyl tartrates and tartramides with Ti^{4+} reveals details of their stuctures in solution (P.G. Potvin *et al.*, Can J. Chem., 1989, **67**, 1523). The X-ray crystal stuctures oxovanadium tartrate complexes $Rb[VO(tart)_2]$ and $Cs[VO(tart)_2]$ have been determined (J.T. Wrobleski and M.R. Thompson, Inorg. Chim. Acta, 1988, **150**, 269).

Dimethyl (*R,R*)-tartrate is formed by *syn*-hydroxylation of dimethyl fumarate. A high enantiomeric excess is achieved in the presence of chiral diamines such as (30) (E.J. Corey *et al.*, J. Am. Chem. Soc., 1989, **111**, 9243; K. Tomioka, M. Nakajima and K. Koga, *ibid.*, 1987, **109**, 6213; 1988, **110**, 988)

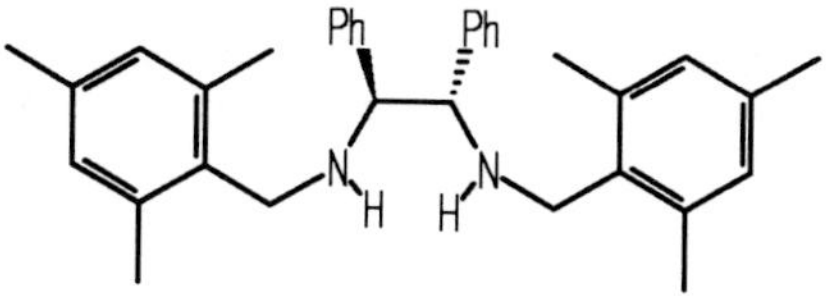

MeOOC–CH=CH–COOMe → OsO_4 → MeOOC–CH(OH)–CH(OH)–COOMe

Ph Ph N H H N

(30)

The β-lactam (31) has been synthesised from diethyl (*R,R*)-tartrate, the key step is the selective hydrolysis of one of the esters using pig liver esterase (PLE) (A. Gateau-Olesker, J. Cleophax and S.D. Gero, Tetrahedron Lett., 1986, **27**, 41).

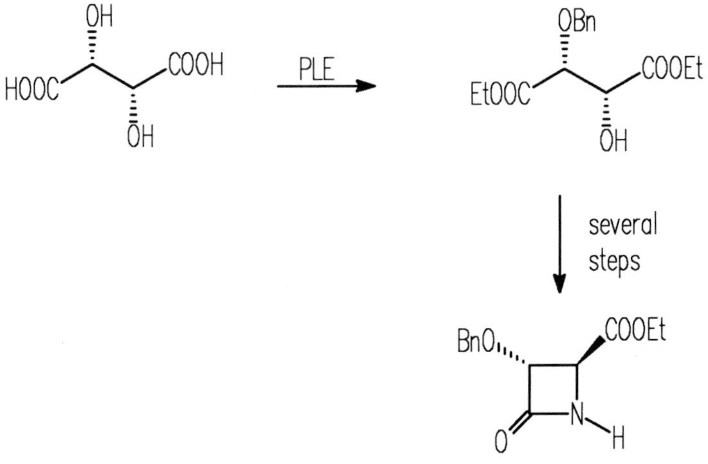

(31)

Probably one of the most significant advances in synthetic organic chemistry is the use of tartaric acid derivatives in the generation of new chiral centres. It is known that tartrates form complexes with metal ions. Early attempts used various metals, for example, molybdenum complexed with diisopropyl (+)-tartrate (DIPT) was employed in the asymmetric epoxidation of squalene. A low enantiomeric excess was claimed (K. Tani, M. Hanafusa and S. Otsuka, Tetrahedron Lett., 1979, 3017). However, a breakthough came with the introduction of the Sharpless epoxidation technique which uses DIPT or diethyl (+)-tartrate (DET), titanium tetraisopropyloxide and *tert*-butyl hydroperoxide (T. Katsuki and K.B. Sharpless, J. Am. Chem. Soc., 1980, **102**, 5974; J.G. Hill *et al.*, Org. Synth., 1985, **63**, 66). Enantiomeric excesses of the order of 95% have been achieved in the asymmetric epoxidation of allylic alcohols.

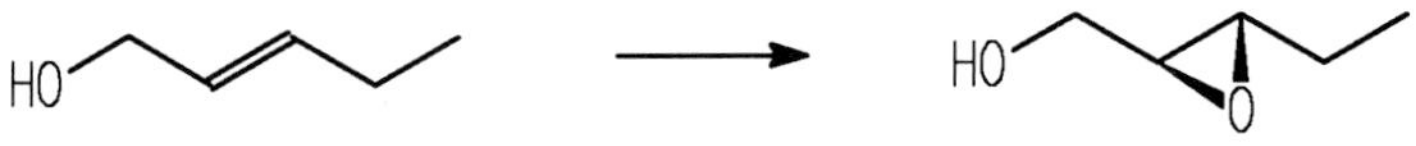

tetramethyl ester (F.R. Froczek, V.K. Gupta and G.R. Newkome, Acta Crystalogr. Sect. C, Cryst. Struct. Commun., 1983, **39**, 1113), show that these esters exist in the solid state in a conformation where the hydrogen atoms are in an antiperiplanar arrangement. 1,1,3,3-Tetracyanopropane (44) has been studied by X-ray analysis, ^{1}H- and ^{13}C-NMR, IR and mass spectrometry (R.A. Bell *et al.*, Can J. Chem., 1987, **65**, 261).

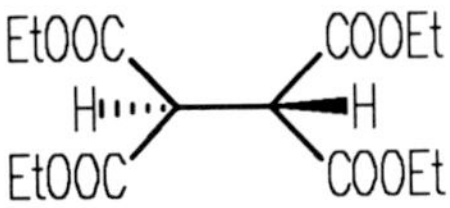

EtOOC COOEt H H EtOOC COOEt

(43)

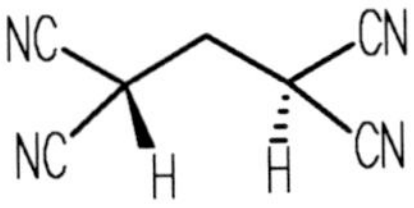

NC CN NC H H CN

(44)

(b) Alkenetetracarboxylic acids

The ^{13}C-NMR spectra of tetramethyl ethenetetracarboxylate (45) and tetracyanoethene (46) have been studied (M. Barfield, T, Gotoh and H.K. Hall Jr., Magn. Reson. Chem., 1985, **23**, 705). The crystal structure of tetramethyl ethenetetracarboxylate (45) shows that two opposite ester functions are twisted out of plane whereas dimethyl dicyanofumarate (47) is planar (H.K. Hall *et al.*, J. Polym. Sci., Polym. Chem. Ed., 1982, **20**, 361). The chemistry of tetracyanoethene and its complexes has been reviewed (H. Poradowska and K. Nowak, Wied. Chem., 1982, **36**, 649).

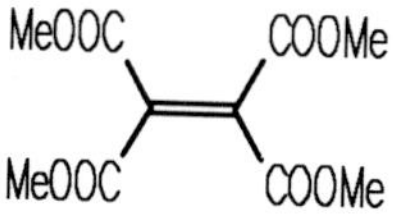

MeOOC COOMe MeOOC COOMe

(45)

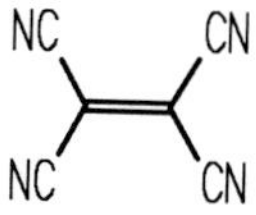

NC CN NC CN

(46)

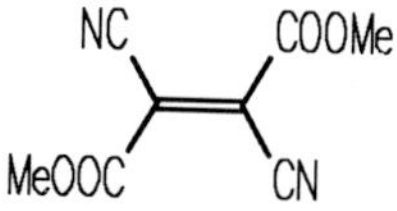

NC COOMe MeOOC CN

(47)

Guide to the Index

This index is constructed in a similar manner to the volume indexes of the first edition of the Chemistry of Carbon Compounds. However, to make the index easier to use, more descriptive entries have been made for the commonly occurring individual, and groups of chemicals.

The indexes cover primarily the chemical compounds mentioned in the text, and also include reactions and techniques, where named, and some sources of chemical compounds such as plant and animal species, oils, etc.

Chemical compounds have been indexed alphabetically under the names used by authors, editing being restricted to ensuring uniformity of entries under the same heading. In view of the alternative nomenclature that can often be used, a limited amount of cross-referencing has been done where it is considered to be helpful, but attention is particularly drawn to Convention 2 below.

For this and the succeeding volumes, the indexing conventions listed below have been adopted.

1. Alphabetisation

(a) A letter by letter alphabetical sequence is followed for entries, firstly for the main entry, followed by the descriptive entry.

(b) The following prefixes have not been counted for alphabetising:

n-	*o-*	*as-*	*meso-*	*C-*	*E-*
	m-	*sym-*	*cis-*	*O-*	*Z-*
	p-	*gem-*	*trans-*	*N-*	
	vic-			*S-*	
		lin-		*Bz-*	
				Py-	

Some prefixes and numbering have been omitted in the index, where they do not usefully contribute to the reference.

(c) The following prefixes have been alphabetised:

Allo	Epi	Neo
Anti	Hetero	Nor
Bis	Homo	Pseudo
Cyclo	Iso	

2. Cross references

In view of the many alternative trivial and systematic names for chemical compounds, the indexes should be searched under any alternative names which may be indicated in the main body of the text. Only a limited amount of cross-referencing has been carried out, where it is considered that it would be helpful to the user.

3. Derivatives

Simple derivatives are not normally indexed if they follow in the same short section of the text.

4. Collective and plural entries

In place of "– derivatives" the plural entry has normally been used. Plural entries have occasionally been used where compounds of the same name but differing numbering appear in the same section of the text.

5. Main entries

The main entry of the more common individual compounds is indicated by heavy type. Multiple entries, such as headings and sub-headings over several pages are shown by "–", e.g., 67–74, 137–139, etc.

Index

2-Acetamido-4-*O*-(2-acetamido-2-deoxy-β-D-glucopyranosyl)-3-*O*-[(*R*)-1-carboxyethyl]-2-deoxy-D-glucose, 487
2-Acetamido-3-*O*-(2-acetamido-2-deoxy-β-D-glucopyranosyl)-2-deoxy-D-galactose, 487
2-Acetamido-6-*O*-(2-acetamido-2-deoxy-β-D-glucopyranosyl)-2-deoxy-D-galactose, 487
2-Acetamido-4-*O*-(2-acetamido-2-deoxy-β-D-glucopyranosyl)-2-deoxy-D-glucose, 487
2-Acetamido-3-*O*-(2-acetamido-3,4,6-tri-*O*-acetyl-2-deoxy-α-D-galactopyranosyl)-1,4,6-tri-*O*-acetyl-2-deoxy-α-D-galactopyranose, X-ray crystallography, 469
1-Acetamido-1,1-benzamido-1-deoxy-D-glucitol, 265, 266
α-Acetamidocinnamic acid, enantioselective hydrogenation, 276
2-Acetamido-2-deoxy-6-*O*-(5-acetamido-3,5-dideoxy-D-*glycero*-α-D-*galacto*-2-nonulopyranosylonic acid)-D-galactose, 487
2-Acetamido-2-deoxy-3-*O*-(3-deoxy-D-*manno*-2-octulopyranosylonic acid)-D-galactose, 482
2-Acetamido-2-deoxy-3-*O*-(α-L-fucopyranosyl)-D-glucose, 484
2-Acetamido-2-deoxy-3-*O*-(β-D-galactopyranosyl)-D-galactose, 484
2-Acetamido-2-deoxy-4-*O*-(β-D-galactopyranosyl)-D-galactose, 485
2-Acetamido-2-deoxy-6-*O*-(β-D-galactopyranosyl)-D-galactose, 485
2-Acetamido-2-deoxy-*O*-(β-D-galactopyranosyl)-D-glucose, 485
2-Acetamido-2-deoxy-3-*O*-(β-D-glucopyranosyl)-D-galactose, 485
2-Acetamido-2-deoxy-4-*O*-(α-D-glucopyranosyl)-D-glucopyranose, 485
2-Acetamido-2-deoxy-6-*O*-(α-D-glucopyranosyl)-D-glucopyranose, 486
2-*O*-(2-Acetamido-2-deoxy-β-D-glucopyranosyl)-α-D-mannopyranose, 486
2-Acetamido-2-deoxy-D-glucose based oligosaccharides, 454
6-*O*-(2-Acetamido-2-deoxy-β-D-mannopyranosyl)-D-galactopyranose, 486
4-*O*-(2-Acetamido-2-deoxy-β-D-mannopyranosyl)-D-glucopyranose, 486
2-Acetamido-2-deoxy-3-*O*-(α-L-rhamnopyranosyl)-D-glucopyranose, 486
2-Acetamido-2-deoxy-4-*O*-(α-L-rhamnopyranosyl)-D-glucopyranose, 486
2-Acetamido-2-deoxy-6-*O*-(α-L-rhamnopyranosyl)-D-glucopyranose, 486
2-Acetamido-2-deoxy-6-*O*-(β-L-rhamnopyranosyl)-D-glucopyranose, 486
2-*O*-(5-Acetamido-3,5-dideoxy-D-*glycero*-α-D-*galacto*-2-nonulopyranosylonic acid)-D-galactose, 487
6-*O*-(5-Acetamido-3,5-dideoxy-α-D-*glycero*-D-*galacto*-2-nonulopyranosylonic acid)-D-galactose, 488
2-Acetamido-1,3,6-tri-*O*-acetyl-2-deoxy-4-*O*-(2,3,4,5-tetraacetyl-α-L-idopyranosyl)-α-D-glucopyranose, X-ray crystallography, 468
Acetophenone, asymmetric reduction, 185
4-Acetoxy-2-aminobutanoic acid (*O*-acetylhomoserine), 67
4-Acetoxy-2-bromo-2*H*-dihydrofuran-3-one, synthesis from 1,4-anhydroerythritol, 507
3-Acetoxy-4-cyano-3-methylbutanoic acid, 40
(*S*)-(−)-3-Acetoxy-4-cyano-3-methylbutanoic acid, 40
2-Acetoxy-3-deoxy-5,6:7,8-di-*O*-isopropylidene-D-manno-2-octenic acid γ-lactone, 187
2-Acetoxy-5,6-dihydro-2-pyrans, stereoselective alkylation, 226

2-(*N*-Acetylamino)propenoic acid, 54
N-Acetyl-L-aspartic acid, synthesis from ammonium fumarate, 83
N-Acetyl-2α-bromo-3β-hydroxy-neuraminic acid, glycosylation, 453
N-Acetyl-2α-chloro-3β-hydroxy-neuraminic acid, glycosylation, 453
(+)-*N*-Acetyldesalanylactinobolin, synthesis, 325, 326
Acetylglucals, reactions with phosphorodithioic acid, 258
2-Acetyl-1-*O*-hexadecyl-*sn*-glycero-3-phosphorylcholine, 136
2-Acetyl-4a,5,6,7,8,8a-hexahydro-4,4,7-trimethylbenz-1,3-oxathiane, 40
N-Acetylholacosamine, 240
O-Acetylhomoserine (4-acetoxy-2-aminobutanoic acid), 67
5-*O*-Acetyl-6-iodo-1,3:2,4-diethylidene-D-sorbitol, phosphorylation, 197
N-Acetyllactosamine, glycosylation, 455
–, synthesis, 455
N-Acetylneuraminic acid, homopolymers, 454
–, reaction with *N*-acetyllactosamine/2,6-sialyltransferase, 455, 456
3-*O*-(*N*-Acetyl-α-D-neuraminyl)-D-galactose, 487, 488
3-*O*-(*N*-Acetyl-β-D-neuraminyl)-D-galactose, 488
6-*O*-(*N*-Acetyl-α-D-neuraminyl)-D-galactose, 488
9-*O*-(*N*-Acetyl-α-D-neuraminyl)-β-D-neuraminic acid, 488
2-Acetyl-1-*O*-octadecyl-*sn*-glycero-3-phosphorylcholine, 136
2-Acetyl-1-oleoyl-*sn*-glycerol, 32
2-Acetyl-1-oleoylglycerophosphorylcholine, 32
N-Acetyl-L-ornithine, 77
O-Acetyloxymevalonolactone, 36
N-Acetyl-2-thioneuraminic acid, glycosylation, 453
N-Acetyl-L-tryptophan, synthesis from L-glutamic acid, 91
(+)-Acivicin, synthesis from *N*-glycosylnitrones, 311
–, – from *N*-glycosylnitrone/vinylglycine, 311
3-*O*-Acryloyl-1,2-*O*-isopropylidene-5-*O*-trimethylsilyl-α-D-xylofuranoside, cycloaddition with cyclopentadiene/titanium(IV) chloride, 300
Acrylphosphonate polymers, synthesis from diethyl diethoxyethylphosphonates and polyols, 11
N-Acylamino acids, *N*-chlorination, 51
(2*S*)-*N*-Acyl-α-amino acids, synthesis from *N*-acyldidehydro-α-amino acids, 273–275
2-(*N*-Acylamino)-2-alkenoate esters, synthesis from arylmethylene-5(4*H*)-oxazolones, 55
2-(*N*-Acylamino)-3-hydroxyalkanoic acids, dehydration, 50
N-Acylaspartate α-esters, oxidative β-decarboxylation, 84
1-*O*-Acyl-2-*O*-[bis(diethylamino)thionophosphonato]ethanediols, 128
1-*O*-Acyl-2-*O*-[bis(diethylamino)-α-tocopherol-6-yl thionophosphonato]-ethanediols, 128
N-Acyldehydroalanines, synthesis, 84
1-*O*-Acylethane-2-[*O*-bis(*N,N*-diethylamido)thionophosphonato]-stearate, 127
Acylisopropylidene glycerols, alkylation, 25
[^{14}C]-Acylphosphatidic acids, synthesis from silver dibenzylphosphate, 108
3-Acyltetramic acids, 393
S-Adenosyl-L-homocysteine, 62
Adiantum capillus veneris, 66
Adiantum pedatum, 96
Adonotol, 502
Aflatoxin M_1, 507
Agaricus hortensis, 98
Agrobacterium kumefaciens, 68
Alcohols, cyclic enediol phosphorylation, 110
Alditol acetals, synthesis, 190–193
Alditol allyl ethers, 199
Alditol-di-*O*-isopropylidene cyclic iminium salts, pyrolysis, 195
Alditol 2-haloethylidene acetals, 191
Alditols, anodic oxidation, 190
–, Ca(II) complexes, 180
–, conformation, 176, 177

–, cyclic phosphites, 195
–, dinuclear molybdate complexes, 180
–, enzymatic oxidation, 189
–, esterification, 193
–, gas chromatography, 181
–, high performance liquid chromatography, 181
–, ^{1}H NMR spectroscopy, 176, 177
–, lanthanide(III) complexes, 180
–, monoperfluoroalkyl ethers, 198
–, oxidation, 187–190
–, reactions with ammonium molybdate, 180
–, synthesis from aldoses (or ketoses), 167
–, – from formaldehyde/syn gas, 169, 170
–, – from glycosides, 168
–, – from sugars by enzymic reduction, 168
–, thiophosphorus derivatives, 195
–, use as chirons, 182
threo-Alditols, reactions with tungstate ion, 180
Alditol–hydroaluminate complexes, reductants for ketones, 186
Alditol–metal complexes, extended X-ray absorption fine structure spectroscopy, 180
Aldoses, electrolysis, 168
–, high performance liquid chromatography, 181
Alkanone oximes, enantioselective reduction to (*S*)-alkylamines, 282
Alkanones, asymmetric hydrosilylation, 281
–, enantioselective reduction by potassium glucoride, 284
–, – to (*S*)-alkanols, 282
Alkenamides, *N*-substitution by glycerol, 4
α-Alkenic acids, enantioselective hydrogenation, 280
α-Alkenones, enantioselective hydrogenation, 280
2-Alkenylmagnesium bromides, 292
Alkenyols, enantioselective reduction to (*R*)-allenic alcohols, 283
2-Alkoxy-2-aminoalkanoic acids, 49
β-Alkoxy-α-azidoalkanoate esters, 53
1-Alkoxy-1,4-cyclohexadienes, alkylation, 291
–, synthesis from phenyl-β-D-glucopyranoside, 291
2-Alkylcysteines, 61
Alkylethandiols, esterification with 2-bromoethylphosphonic acid, 131
Alkylglycooxyphospholipid analogues, 165
Alkylhalolysophospholipids, 135
Alkyllysophospholipids, radiolabelled, 135
1-Alkyl-(pentitol-1-yl)pyrroles, synthesis from 2-(alkylimino)-2-deoxyheptoses, 206
Alkylphospholipid analogues, 164
1-*O*-Alkyl-2-phosphonatoethanediol derivatives, 131
Alkyl tartrates, Raman circular intensity differential spectra, 511
2-Alkyltriacylglycerols, 25
L-Alloisoleucine, 47
(−)-Allosamizoline, synthesis from D-glucosamine, 318, 319
(−)-Alloyohimbane, synthesis from levoglucosenone, 325, 327, 328
Allyl alcohol, asymmetric epoxidation, 116
–, epoxidation, 2
–, synthesis from glycerol, 7
2-Allyl-3-benzyl-1-trityl-*sn*-glycerol, 22
Allylcholine phosphate, 105
3-Allyl-*sn*-glycerol, 22
Allylic alcohols, asymmetric epoxidation, 172
Amides, glycosylation with glycosyl halides, 447
Amidophospholipids, 135
2-Amido platelet activating factor, 148
Amino acid esters, *N*-chlorination, 51
α-Amino acid esters, synthesis from D-glucose sulphate esters, 294
(±)-Amino acids, resolution, 187
α-Amino acids, enantioselective synthesis, 276–278
–, synthesis from acetamidomalonates, 81, 82
D-α-Amino acids, synthesis, 298
R-α-Amino acids, synthesis from

didehydro-α-amino acids, 275
β-Amino acids, synthesis from acylaminomalonates, 81
–, – from allylsilanes (or allylstannanes)/ glycosylimines, 300
Aminoalditols, 203–208
–, high performance liquid chrommatography, 181
2-Amino-2-alkenic acids, synthesis from arylmethylene-5(4*H*)-oxazolones, 55, 56
2-Amino-2-alkenoate esters, synthesis from *N*-acyldialkoxyphosphorylglycine esters, 57
2-Amino-2-alkenoic acids, 50–57
–, synthesis from azidomalonic esters, 54
2-Amino-4-aminooxybutanoic acid (*O*-aminohomoserine), 66
7-Amino-3-(β-arabinofuranosyl)-pyrazolo[4,3-*d*]pyrimidine, 206
2-Amino-*N*-benzyloxycarbonyl-2-alkenoic acids, synthesis from 2-phenylaziridine-2-carboxylate esters, 54
2-Amino-3-benzylthio-2-methyl-propanoic acid, 63
2-Aminobutanoic acid, synthesis from methionine, 68
4-Aminobutanoic acid, synthesis from L-glutamic acid, 93
–, transaminase inhibitor, 93
4-Aminobutyrolactones, synthesis from aspartic acid, 84
3-Amino-4-carboxybutane sulphonic acid (homocysteic acid), 62
2-Amino-4-chloropentanoic acid, 77
Aminocyclitols, synthesis, 262–264
2-Amino-2-deoxy-D-galactitol, nitrosation and reductive deamination, 2
–, reaction with methanolic hydrogen chloride, 200
2-Amino-2-deoxy-D-glucitol, reaction with methanolic hydrogen chloride, 200
2-Amino-2-deoxypentitols, 172
2-Amino-2-deoxy-D-pentoses, 239
Aminodeoxy sugars, synthesis from 2,3-anhydro-α-D-ribopyranosides, 241, 242
Aminodialkanoic acids, 81
2-Amino-6-diazo-5-oxohexanoic acid, isolation from *Streptomyces ambofaciens*, 80
Aminodicarboxylic acids, 81–100
(*S*)-4-Amino-4,5-dihydrofuran-2-carboxylic acid, synthesis from L-glutamic acid, 93
2-Amino-4,5-dihydroxyadipic acid (2-amino-4,5-dihydroxyhexanedioic acid), 98
4-Amino-2,3-dihydroxybutanoic acid, 74
2-Amino-4,5-dihydroxyhexanedioic acid (2-amino-4,5-dihydroxyadipic acid), 98
4-Amino-2,3-dihydroxy-3-methyl-butanoic acid, isolation from carzinophilin, 74
2-Amino-1,1,2-ethanetricarboxylic acid (3-carboxyaspartic acid), 87
3-Amino-3-fluorobutanedioic acid (3-fluoroaspartic acid), 88
6-Aminoglucofuranoses, diazotisation, 244
3-Aminoglutamic acid (3-aminopentanedioic acid), 96
1-Amino-2-guanidinomethyl-cyclopropane-1-carboxylic acid (carnosadine), 94
O-Aminohomoserine (canaline), 66
2-Amino-3-hydroxybutanedioic acid (3-hydroxyaspartic acid), 86
2-Amino-2-hydroxy-1,2,4-butanetricarboxylic, isolation from *Caylusea abyssinica*, 74, 75
2-Amino-3-hydroxybutanoic acid (threonine), 67
2-Amino-4-hydroxybutanoic acid (homoserine), 66
4-Amino-2-hydroxybutanoic acid, 71
(*R*)-4-Amino-3-hydroxybutanoic acid, synthesis from D-arabinose, 72
(*S*)-4-Amino-3-hydroxybutanoic acid, synthesis from (*S*)-malic acid, 72
N'-(4-Amino-2-hydroxybutyl)lysine (hypusine), 78, 79
2-Amino-4-hydroxy-3-methylbutanoic acid (4-hydroxyvaline), 71
3-Amino-3-hydroxy-6-methylheptanoic acid (statine), 100

2-Amino-3-hydroxy-3-methylpentanedioic acid (3-hydroxy-4-methylglutamic acid), 97
2-Amino-4-hydroxy-4-methylpentanedioic acid (4-hydroxy-4-methylglutamic acid), 96
4-Amino-3-hydroxy-2-methylpentanoic acid, 76
2-Amino-4-hydroxy-3-methylpentanoic acid (4-hydroxyisoleucine), 76
(*S*)-5-Amino-4-hydroxypentanoic acid, synthesis from L-glutamic acid, 90, 91
2-Amino-3-hydroxypentanoic acid (β-hydroxynorvaline), 75
2-Amino-4-hydroxypentanoic acid (4-hydroxynorvaline), 75
3-Amino-2-hydroxy-4-phenylbutanoic acid, 72
3-Amino-2-hydroxy-3-phenylpropanoic acid (3-phenylisoserine), 59
3-Amino-4-hydroxypropanoic acid (isoserine), 59
4-Amino-3-isoxazolidine (cycloserine), 58
Aminomalonic acid (aminopropandioic acid), 81, 82
2-Amino-3-mercaptobutanoic acid (3-methylcysteine), 63
2-Amino-4-mercaptobutanoic acid (homocysteine), 62
2-Amino-3-mercapto-3-methylbutanoic acid (penicillamine), 64
2-Amino-3-mercapto-2-methylpropanoic acid (2-methylcysteine), 63
2-Amino-3-mercaptopropanoic acid (cysteine), 60, 61
2-Amino-3-methylbutanedioic acid (3-methylaspartic acid), 86
2-Amino-4-methylglutamic acid (4-methylglutamic acid), 94
2-Amino-4-methylheptanedioic acid (2-amino-4-methylpimelic acid), 98
2-Amino-4-methyl-4-hexenoic acid, isolation from *Aesculus californica*, 80
2-Amino-3-methyl-4-oxopentanoic acid, isolation from *Bacillus cereus*, 77
2-Amino-2-methylpentanedioic acid (2-methylglutamic acid), 94
2-Amino-3-methylpentanedioic acid (3-methylglutamic acid), 94
2-Amino-4-methylpentanedioic acid (4-methylglutamic acid), 94
4-Amino-6-methyl-2-piperidone, 79
2-Amino-2-methylpropanedioic acid (1-aminoisosuccinic acid), derivatives, 82
Aminomonosaccharides, synthesis, 230
–, – *via* Diels–Alder reaction, 235, 236
α-Aminonitriles, synthesis from 2,3,4,6-tetra-*O*-pivaloyl-β-D-galactopyranosylamine, 297
2-Aminooctanedioic acid (2-aminosuberic acid), 98
Aminooctodioses, 174
2-Aminopentanedioic acid (glutamic acid), 89
3-Aminopentanedioic acid (3-aminoglutamic acid), 94
α-Aminophosphonic acids, synthesis from (*Z*)-mannofuranosyl nitrones, 297
Aminopimelic acids, 2,6-disubstituted, 100
Aminopropandioic acid (aminomalonic acid), 81, 82
Amino pseudosugars, 264
2-Amino-3-selenylpropanoic acid (selenocysteine), 64
2-Aminosuberic acid (2-aminooctanedioic acid), 98
2-Amino sugars, glycosidation *via* 1,2-orthoesters, 443
–, – *via* oxazolines, 442, 443
3-Amino sugars, synthesis, 239, 240
D-Aminotransferase, 68
4-Aminovaline (2,4-diamino-3-methylbutanoic acid), 70
Amphomycin, 86
(*S*)-Anabasine, synthesis, 298
(+)-Anamarine, synthesis from D-glucose, 397
–, – from D-gulonolactone, 397, 398
Anhydrides, activation by 4-pyrrolidinopyridine, 119
Anhydroalditols, 199–202
–, synthesis *via* acyloxonium ions, 200
1,5-Anhydroalditols, trifluoroacetyl derivatives, gas chromatography, 181

1,5-Anhydro-D-alditols, synthesis from glycosylisothiocyanates, 201
2,5-Anhydroalditols, monotosylation, 200
1,5-Anhydro-L-arabinitol, partial *O*-tosylation, 200
1,5-Anhydro-4,6-*O*-benzylidene-D-galactitol, esterification, 193
1,3-Anhydro-2,4-*O*-benzylidene-D-glucitol, synthesis from 2,4-*O*-benzylidene-1,6-di-*O*-*p*-toluenesulphonyl-D-glucitol, 200
1,4-Anhydro-5-chloro-5-deoxypentitols, 200
1,4-Anhydroerythritol, acetyl derivatives, 500
1,4-Anhydro-5-*O*-hexadecyl-D-arabinitol, acylation, 193, 194
2,5-Anhydro-D-iditol, synthesis from 2,5-anhydro-D-mannitol, 201
1,4-Anhydropentitols, synthesis from D-arabinitol, 200
2,3-Anhydro-α-D-ribopyranosides, 241
1,4-Anhydrothreitol, acetyl derivatives, 500
1,4-Anhydrothreitol diphosphinite ester (diphin), 501
Anomeric effect, 268
Anthrylvinylphospholipids, synthesis from phosphatidylcholine, 5
6-*O*-α-D-Apiofuranosyl-D-glucose, 470
4-*O*-β-D-Apiofuranosyl-D-glucose, 470
6-*O*-β-D-Apiofuranosyl-D-glucose, 470
L-Arabinal 4,5-diacetate, reaction with sodium azide, 239
Arabinitol, stereochemistry, 177
–, use as a chiron, 184
(±)-Arabinitol, gas chromatographic resolution, 181
L-Arabinitol, cyclic phosphites, 195
D-Arabinitol monoacetonide, synthesis from D-glyceraldehyde acetonide, 172
2-*O*-α-L-Arabinofuranosyl-L-arabinofuranose, 470
3-*O*-α-L-Arabinofuranosyl-L-arabinofuranose, 471
4-*O*-β-L-Arabinofuranosyl-L-arabinose, 471
4-*O*-β-L-Arabinofuranosyl-D-glucose, 471
4-*O*-β-L-Arabinofuranosyl-D-xylose, 471
D-Arabinose, 260
D-Arabino-L-*xylo*-(tetra-*O*-acetyl)hexosulose-1,2-bis(acetylphenylhydrazone), 228
Arachidonyl platelet activating factor, 147
Archaebacteria, 30
Arthrinium phaeospermum, 87
Aryl alkyl ketones, enantioselective reduction by potassium glucoride, 284
(3*R*)-Aryl-β-amino acid esters, 2,3,4-tri-*O*-pivaloyl-α-D-arabinopyranosylimines, 300
(3*S*)-Aryl-β-amino acid esters, synthesis from silyl ketene acetals/galactosylimines, 298
(3*S*)-Aryl-β-amino acids, synthesis from silyl ketene acetals/galactosylimines, 300
Aryl bromides, cross coupling with *s*butylmagnesium bromide, 294
Arylideneserine esters, dehydration, 51
Asparagine, 88
α-Asparagine, 89
Aspartic acid (aminosuccinic acid), 83
–, metabolism in microorganisms, 66
–, preparation by Strecker synthesis, 84
–, synthesis from laevulinic acid, 83
1-[^{13}C]-Aspartic acid, 84
L-Aspartic acid, synthesis from diethyl malonate, 83
(*S*)-L-Aspartic acid derivatives, 84
Aspartic acid semialdehyde, 84
Aspartocin, 86
Aspergillus niger, 3
(+)-Asteltoxin C(1)-C(9)-segment, synthesis from (*R*)-isopropylidene glyceraldehyde, 399, 400
Astragalus sinicus, 66, 87
Asymmetric Diels–Alder reactions, 300–304
Asymmetric heterogenous hydrogenation, 280, 281
Asymmetric hydrosilylation, 281
Aureobasidium SW-45, 1
Avermectin A_{1a}, synthesis from monosaccharide precursors, 413–420
Avermectins, hexahydrobenzfuran component, synthesis from 1,4-

anhydrosorbitol, 353–357
–, oxahydrindene component, 341, 342, 353
–, synthesis by intramolecular Diels–Alder strategy, 341, 342
Azaserine (serine *O*-diazoacetate), 58
Azathiaphospholipids, 109
Azathiophospholipids, synthesis from 1-acetylthio-3-hydroxy-2-propanamide, 126
α-Azidoacylate esters, synthesis from α,β-dibromoesters, 53
α-Azidoalkanoic amides, 54
trans-2-Azidocyclohexanol, synthesis from cyclohexene, 186
2-Azido-2-deoxy-1-*O*-hexadexyl-*sn*-slycero-3-phosphorylcholine, synthesis from D-mannitol, 147
3′-Azido-3′-deoxythymidine, 19
2-Azidogalactosyl trichloroacetimidates, glycosidation reagents, 447
2-Azidoglucosyl trichloroacetimidates, glycosidation reagents, 447
α-Azidolactams, 54
2-Azidolactosyl trichloroacetimidates, glycosidation reagents, 447
Aziridino sugars, synthesis, 241, 242

Barker–Bourne rules, 193
Beckmann rearrangement, 3-bromolaevulinic acid oxime, 83
Benzaldehyde, reactions with lithium propionates, 287
(*Z*)-2-Benzamido-4-hydroxy-5-iodopent-4-enoic acid, 82
7*H*-Benzo[*h,i*]chrysen-7-one, synthesis from 1,2-benzanthraquinone, 9
N-Benzoyldaunosamine, synthesis, 236, 237
N^6-Benzoyllysine, 78
2-Benzyl-α-amino acids, 82
3-*O*-Benzyl-1,2-*O*-cyclohexylidene-α-D-glucofuramone/lithium aluminium hydride (Landor's reagent), 281, 282
S-Benzylcysteine, 61
Benzyl-2-deoxy-2-*C*-methylpentanopyranosides, ring-opening by organoaluminium reagents, 173
2-*O*-Benzylglyceraldehyde, 19
3-Benzyl-*sn*-glycerol, 22
3-*O*-Benzyl-*sn*-glycerol, synthesis from D-mannitol, 31
Benzyl 2,3-*O*-isopropylidene-β-L-arabinopyranoside acrylate, cycloaddition reactions, 300
(*R*)-*S*-Benzyl-2-methylcysteine, 61
O-Benzyl monosaccharides, selective de-*O*-benzylation, 266
3-Benzyl-2-myristoyl-1-stearoyl-*sn*-glycerol, 22
3-(*N*-Benzyloxycarbonyl)-α-asparagine, 89
N-Benzyloxycarbonyldimethoxyphosphorylglycine esters, synthesis from glyoxalic acid, 57
N-Benzyloxycarbonylisoglutamine, 98
N-Benzyloxycarbonyl-L-ornithine, 77
N-Benzyloxycarbonyl-L-proline methyl ester, synthesis from *N*-benzyloxycarbonyl-2-methyl-L-glutamic acid methyl ester, 94
2-Benzyl-1-stearoyl-*sn*-glycerol, synthesis from 3-alkyl-2-benzyl-*sn*-glycerol, 23
Benzyl 2,3,6-tri-*O*-benzyl-4-*O*-(but-1-en-3-yl)-α-D-glucopyranoside, addition to butyl glyoxalate, 304
(+)-Bicyclohumulenone, synthesis from D-mannitol, 344–346
Bilayer vesicles, 131, 132
1,1-Bis(benzamido)-1-deoxy-D-gluucitol, reaction with acetic anhydride/nitromethane, 265, 266
N,N-Bis(benzyloxycarbonyl)-L-ornithine, 77
(3*R*,4*R*)-3,4-Bis(diphenylphosphinoxy)-tetrahydropyran/rhodium(I), hydrogenation catalysts, 273
(−)-Bissetone, synthesis from 2-hydroxyglucal perbenzoate, 399, 401
1,2-Bis(vinyloxy)propane, synthesis from glycerol/ethene, 5
Bleomycin, 77
Boletus satanas, 75, 76
(+)-Bostrycin, synthesis, 307
Bovine protein C, 87
Branched cyclodextrins (cyclomalto-oligosaccharides), 457

Branched sugars, synthesis, 264
(+)-Brefeldin A, synthesis from D-mannitol, 337–339
Brevetoxin A, synthesis from 1,2:5,6-di-*O*-isopropylidene-α-D-glucofuranose, 379–383
2-Bromo-1-(bromomethyl)ethyl palmitate, 24, 25
3-Bromo-1,2-epoxypropane, 24
2-Bromoethyl phosphorodichloridate, 109
4-Bromoglutamic acids, syntheses from hydroxyglutamic acids, 90
3-Bromo-2-hydroxypropyl stearate, 24
3-Bromomethylenecyclohexene, alkylation agent for monocyanoacetamidomalonates, 81
3-Bromo-2-palmitoyl-1-stearoyloxypropane, synthesis from 2-bromo-1-(bromomethyl)ethyl palmitate/silver stearate, 25
Bromophenyldioxanyl aminobenzoate, 7
Bromo sugars, reactions with alkenes, 264
(−)-Bulgecinine, synthesis from D-glucose, 367, 368
–, – from D-glucuronic acid, 367
1,3-Butadienylglucosides, cycloadditions with methacrolein, 309
Butanetetracarboxylic acids, 516, 517
But-2-enetetracarboxylic acids, 517
4-(2-Butenyl)azetidinone, methanolysis, 236, 237
(*E*)-3-(*ᵗ*Butyldimethylsilyloxy)buta-1,3-dienyl-2,3,4-tri-*O*-acetylxylopyranosides, cycloaddition reactions, 305–307
1-*O*-(*ᵗ*Butyldiphenylsilyl)-3-*O*-benzyl-*sn*-glycerol, 31
1-*O*-(*ᵗ*Butyldiphenylsilyl)-2,3-di-*O*-[(3′*R*,*S*,7′*R*,11′*R*)-3′,7′,11′,15′-tetramethylhexadecyl-1-yl]-*sn*-glycerol, 32
1-*O*-(*ᵗ*Butyldiphenylsilyl)-2-*O*-[(3′*R*,*S*,7′*R*,11′*R*)-3′,7′,11′,15′-tetramethylhexadec-1-yl]-3-*O*-benzyl-*sn*-glycerol, 32
S-*ᵗ*Butylhomocysteine, 68
–, synthesis from methionine, 68
4-*ᵗ*Butylthiophenol, Michael addition to 2-cyclohexenone, 294

Calditol, isolation from thermoacidophilic archaebacteria, 175
–, synthesis from D-glyceraldehyde/(+)-(1*R*,4*R*)-7-oxabicyclo[2.2.1]hept-5-en-2-one, 175
(−)-Canadenosolide, synthesis *via* Wittig reaction, 220, 221
Canaline (*O*-aminohomoserine), 66
Canavalia ensiformis, 66
Candida albicans, 64
Caramelisation, 213
2-Carbamoylserine (serine carbamate), 57
Carbasugars, 258–264
2-Carboxyacylisopropylideneglycerols, 27
β-Carboxyaspartic acid, synthesis from di-*ᵗ*butyl malonate, 83
(*S*)-5-Carboxybutyrolactone, 90
5-Carboxyl-2-pyrrolidone, synthesis from L-glutamic acid metal salts, 93, 94
Cardiolipin, phosphono derivatives, 131
Carnitidine [2-hydroxy-3-(*N*,*N*,*N*-trimethylammonium)propanoate inner salt], 60
Carnitine, 72
Carnosadine (1-amino-2-guanidinomethylcyclopropane-1-carboxylic acid), 94
Carzinophilin, 74
(+)-Castanospermine, synthesis from D-glucose, 373, 374
Cellobiose, 478
–, ^{13}C NMR spectroscopy, 462
–, conformational analysis, 462, 463
–, spin–lattice relaxation studies, 462
Cellotetraose, 497
Cellotriose, 490
Cerebronic acid, synthesis from lignoceric acid, 83
Chestis glabra, 69
Chiral allyltitanium complexes, 292
Chiral epoxidation, 126
Chiral phosphate esters, synthesis from monosaccharide/1,3,2-oxazaphosphorinanes, 296
3-Chloro-D-alanine, 61

P-Chloro-1,3-dimethyl-1,2,3-diazaphosphacyclopentane, 128
2-Chloro-1,3-dimethyl-1,3-diazophosphacylcopentane, 128
4-Chloroglutamic acids, syntheses from hydroxyglutamic acids, 90
L-*threo*-β-Chloromalic acid, synthesis from chlorofumaric acid/fumarase, 514
2-Chloro-2-oxo-1,3,2-dioxaphospholane, 109, 157
2-Chloropropyl 4-(2-methylpropyl)phenyl ketone, reaction with glycerol/thionyl chloride, 9
Chlorothricolide, 18
Cholecalciferol-amidothiophospholipids, 132
Clavalanine, synthesis from D-xylose, 364, 366
Clostridium pasterianum, 68
Clostridium sticklandii, 80
Clostridium tetanomorphum, 86
Coconut oil, glycerolysis, 9
Colominic acid, acid hydrolysis, 454
(+)-Compactin, synthesis from D-glucose, 335–337
Corallocarpus epigaeuss, 89
Corey lactone, synthesis from isopropylidene-3-deoxy-α-D-glucofuranose, 316–319
Cortisol, 132
3-Crotonyl-4,4-dimethyl-1,3-oxazolidin-2-one, reaction with cyclopentadiene, 186
(−)-Cryptosporin, synthesis from L-fucal, 328, 330, 331
–, – from 1-nitro-L-fucal, 357, 358
Cyanomonosaccharides, 246–248
Cyanoselenylbenzylacetamidomalonates, 82
1-Cyanovinyl sugars, synthesis, 237, 238
Cyclic chlorophosphates, phosphorylation reagents, 109
Cyclic enediol phosphorylation, 110
Cyclic oxazaphospholidines, synthesis from 1,2-dipalmitoyl-*sn*-glycerol/(*R*)-2[*N*-(1-phenylethyl)amino]ethanol, 107
Cyclic phosphates, accelerated ring-opening by trimethylsilyltrifluoromethane sulphonate/amines, 109
α-Cyclodextrin, 496
–, X-ray crystallography, 469
β-Cyclodextrin, 496
–, X-ray crystallography, 469
γ-Cyclodextrin, 496
–, X-ray crystallography, 469
Cycloheptaamylose, 496
Cycloheptamycin, 75
Cyclohexaamylose, 496
Cyclomaltoheptaose, 496
–, X-ray crystallography, 469
Cyclomaltohexaose, 496
–, X-ray crystallography, 469
Cyclomaltooctoase, 496
–, X-ray crystallography, 469
Cyclooctaamylose, 496
Cyclopentanophospholipids, 127
Cyclopentenones, 19
Cyclopropane carboxylates, enantioselective synthesis from ethyl diazoacetate/imino monosaccharide–copper(II) complexes, 313
Cycloserine (4-amino-3-isoxazolidine), 58
Cysteine (2-amino-3-mercaptopropanoic acid), 60, 61
–, pyrolysis, 61
Cystine [3,3′-dithiobis(2-aminopropanoic acid)], 65
Cystine thiosulphinates, reduction by tris(dialkylamino)phosphines, 65

L-Daunosamine, synthesis from L-arabinose, 231, 232
Daunosamine bicyclic carba analogues, 234
Decitols, 174
Dehydroamino acids, catalytic hydrogenation, 185
Dehydroglutamic acid, synthesis from 2-ketoglutaric acid, 90
$\Delta^{5,7,9}$-Cholestatrien-3β-ol, coupling to phosphatidic acid, 132
Deltamethrin, 19
Deoxyalditols, 202, 203
–, synthesis, 172
–, – from formaldehyde/syn gas, 169
1-Deoxyalditols, synthesis, 202, 203

6-Deoxy-D-allofuranosyl bromides, synthesis, 251
4-Deoxycalditol, synthesis from D-glyceraldehyde/(+)-(1*R*,4*R*)-7-oxabicyclo[2.2.1]hept-5-en-2-one, 175
N-(6-Deoxy-1,2:3,4-di-*O*-cyclohexylidene-α-D-galactopyranosylidene)alanine methyl ester, alkylation, 291
2,3-Deoxy-2,3-di-*C*-hydroxymethyl-tetrose, synthesis from fructose, 505
2-Deoxy-2-ethylaminohexoses, reactions with 5,5-dimethyl-1,3-cyclohexanedione, 206
1′-Deoxy-1′-fluorosucrose, synthesis from 1-deoxy-1-flurorofructose, 456
6-Deoxyglucals, 265
2-Deoxyglucosyl dienes, reactions with *N*-phenylmaleimide, 305
2-Deoxy-D-glucuronic acid 3,6-lactone, synthesis, 240
2-Deoxyglycosidic orthoesters, 265
2-Deoxy-2-halogenoarabinofuranoses, synthesis from ribofuranoses, 248, 249
Deoxyhexosyldeoxyhexoses, 473, 474
Deoxyhexosylhexoses, 474–476
Deoxyhexosylpentoses, 472, 473
2-Deoxy-4-*O*-(α-L-idopyranosyluronic acid)-6-*O*-sulpho-2-sulphoamino-D-glucose, 498
3-Deoxy-3-iodo-D-glucose, 252
1-Deoxymannojirimycin, synthesis from methyl-α-D-mannopyranoside, 205
6-*O*-(3-Deoxy-D-*manno*-2-octulo-pyranosylonic acid)-D-glucose, 482
3-Deoxy-D-mannooctulosonic acid, synthesis from 1,2-anhydro-3,4:5,6-di-*O*-isopropylidene-D-mannitol, 187
(3-Deoxy-D-*manno*-2-octulosonic acid) glycosides, 452
Deoxynitroalditols, 208
1-Deoxynojirimycin, synthesis from tri-*O*-benzyl-6-bromopyranoside, 205
Deoxyribose, 228
2-Deoxy sugar α-dithiophosphates, 258
Deoxysugars, synthesis, 224
4-*O*-(2-Deoxy-6-*O*-sulpho-2-sulphoamino-α-D-glucopyranosyl)-L-idopyranosiduronic acid, 498
Dexamethasone, 132
Diacetone-2-keto-L-glunoic acid, synthesis from diacetone-L-sorbose, 190
2,2-Di(*N*-acetylamino)propionic acid, synthesis from pyruvic acid, 54
N,S-Diacetylcysteine, 61
Diacetyldeoxyrhamnal, iodination–fluorination by bis(*sym*-collidine)iodine tetrafluoroborate, 250
3,4-Di-*O*-acetyl-2,6-dibromo-2,6-dideoxy-α-D-glucopyranosyl bromide, glycosylation, 450
3,4-Di-*O*-acetyl-2,6-dibromo-2,6-dideoxy-α-D-mannopyranosyl bromide, glycosylation, 450
4′-*O*-Diacetylglyceryl-*N*-trimethyl-homoserine, 66
1,2-Diacyl-*sn*-glycerols, synthesis from acylphosphatidylcholines, 32
Dialdogalactopyranose diacetonide, homologation, 172, 173
3,3-Dialkylcysteines, synthesis from 2-(*N*-formylamino)enoate esters, 61
Dialkyl phosphoacetoins, hydrolysis, 110, 111
Dialkyl phospholipids, synthesis, 106
α,α-Diaminoalkanoic acids, 54
2,3-Diaminobutanoic acid, 69
2,4-Diaminobutanoic acid, 70
1,2-Diamino-1,2-dideoxy-D-glucitol, Pt(II) complexes, 179
2,6-Diaminoheptanedioic acid (2,6-diaminopimelic acid), 99
3,5-Diaminohexanoic acid, 79
2,6-Diamino-2-hydroxymethyl-heptanedioic acid (2,6-diamino-6-hydroxymethylpimelic acid), 99
2,6-Diamino-6-hydroxymethylpimelic acid (2,6-diamino-2-hydroxymethyl-heptanedioic acid), 99
2,4-Diamino-3-methylbutanoic acid (4-aminovaline), 70
2,5-Diaminopentanoic acid (ornithine), 77
2,6-Diaminopimelic acid (2,6-diaminoheptanedioic acid), 99
1,5-Dianhydroalditol benzylidene acetals, cleavage by butyllithium, 191

Dianhydroaldohexoses, 199
1,4:3,6-Dianhydro-2,5-dideoxy-2,5-bis(diphenylphosphino)-L-iditol/rhodium complexes, use as hydrogenation catalysts, 185
erythro-(*E*)-1,2:5,6-Dianhydro-3,4-dideoxy-3-hexenitol, 202
D-*threo*-(*E*)-1,2:5,6-Dianhydro-3,4-dideoxy-3-hexenitol, Payne epoxidation, 202
Dianhydromannitol (isomannide), 199
1,4:3,6-Dianhydro-D-mannitol/lithium aluminium hydride, enantioselective reductant for alkanones, 283
1,4:3,6-Dianhydromannopyranose, 199
Dianhydrosorbitol (isosorbide), 199
7*H*-Dibenzo[*a,k,l*]anthracen-7-one, synthesis from 1,2-benzanthraquinone, 9
13*H*-Dibenz[*a,d,e*]anthracen-13-one, synthesis from 1,2-benzanthraquinone, 9
N^3,N^5-Dibenzoyl-3,5-diaminohexanoic acid, 79
N^2,N^6-Dibenzoyllysine (lysuric acid), 78
1,6-Dibenzyl-2,5-ditosyl-3,4-dideoxy-D-*threo*-hexitol, 203
N,N-Dibenzylglycine esters, 49
(2*S*,3*S*,4*S*,5*S*)-Di-*O*-benzylidene-3,4-dihydroxy-2,5-bishydroxymethylthiolane, 211
1,3:4,6-Di-*O*-benzylidene-D-mannitol, oxidation by ruthenium tetraoxide, 188
–, reaction with tributyltin hydride, 211
1,3:4,6-Di-*O*-benzylidene-D-mannitol/lithium aluminium hydride, enantioselective reductant for alkanones, 283
2,5-Dibenzyl β-L-rhamnopyranose derivatives, synthesis from L-rhamnose, 230
2,3-Dibromopropyl-1,2,2,2-tetrafluoropropionate, 8
4,4-Dicarboxyprolines, 5-substituted, 94
1,4-Dichlorobutane-2,3-dione, crystal structure, 507
1,6-Dichloro-1,6-dideoxy-D-mannitol, reaction with tris(diethylamino)phosphine, 196
3,4-Dichloro-1,2-epoxide, hydrolytic ring opening, 501
1,2-Dichloro-1,1,3,3-tetraisopropyldisiloxane, reagent for selective protection of monosaccharides, 267
1,2:5,6-Di-*O*-cyclohexylidene-α-D-glucofuranose, reaction with arylsulphinyl chlorides, 295
1,2:5,6-Di-*O*-cyclohexylidene-α-D-glucofuranose/borohydride reagents, 285
Didehydro-α-amino acids, enantioselective hydrogenation catalysts, 273–278
–, homogenous hydrogenation, 273–278
1,2-Dideoxyalditols, synthesis, 202, 203
3-*O*-(2,6-Dideoxy-α-D-*arabino*-hexopyranosyl)-2,6-dideoxy-α-D-*lyxo*-hexopyranose, 473
3-*O*-(2,6-Dideoxy-β-D-*arabino*-hexopyranosyl)-2,6-dideoxy-α-D-*lyxo*-hexopyranose, 473
4-*O*-(2,6-Dideoxy-α-D-*arabino*-hexopyranosyl)-2,6-dideoxy-α-D-*lyxo*-hexopyranose, 473
4-*O*-(2,6-Dideoxy-β-D-*arabino*-hexopyranosyl)-2,6-dideoxy-α-D-*lyxo*-hexopyranose, 473
(*E*)-6,7-Dideoxy-1,2:3,4-di-*O*-isopropylene-α-D-*galacto*-oct-6-enopyranose, 174
1,4-Dideoxy-1,4-imino-D-arabinitol, 208
1,4-Dideoxy-1,4-imino-L-arabinitol, 208
1,4-Dideoxy-1,4-imino-D-lyxitol, 208
1,4-Dideoxy-1,4-iminopentitols, 208
Dideoxyribose, 228
3′-2′,3-Dideoxyuridine, 19
Diels–Alder reaction, acryldihydro-D-glucals, 300
–, glycosyldienes/alkyl glyoxalates, 304
Diethyl acetamidomalonate, alkylation, 81
Diethyl 2-amino-3-methylbutanedioate, 86
Diethyl benzamidomalonate, alkylation with 2-(trimethylsilyl)allyl bromide, 82
Diethyl malonate, alkylation by *N*-benzyloxycarbonyl-L-alanyl-2-chloroglycine methyl ester, 83

N,N-Diethyl-2,2,3,3,3-pentafluoropropylamine, 8
–, reactions with glycerol dihalohydrins, 8
(L)-Diethyl tartrate, glyceraldehyde equivalent, 18
Diethyl (+)-tartrate/tin complexes, reactions with allyl bromides/aldehydes, 513
Diethyl (+)-tartrate/titanium tetra-^{i}propyloxide/^{t}butylhydroperoxide, 512
1,2:3,4-Di-*O*-α-D-galactopyranoside, glycosidation, 445
Diglycerol, 5
2,6-Dihalotyrosines, 81
1,2-Di(6*Z*,9*Z*)-6,9-hexacosadienoylphospholipids, 136
3,4-Dihydro-(2*H*)-isoquinolines, 19
5,6-Dihydro-4-methyl-2-pyrone, 38
1,4-Dihydronicotinamide sugar pyranosides, enantioselective reductants for α,β-iminium salts and α-keto-esters, 286, 287
3,6-Dihydro-2*H*-1,2-oxazines, enantioselective synthesis from dienes/α-chloronitroso-D-ribose, 303
Dihydroxyacetone, reaction with hydrazine, 2
Dihydroxyalkanoic acids, 35–49
–, enzymic oxidation, 45
4,4′-Dihydroxybenzophenone derivatives, 133
Dihydroxybutanedioic acids, 510–515
Dihydroxybutanediones, 507
2,3-Dihydroxybutanoic acid, synthesis from crotonic acid, 49
1,2-Dihydroxy-3,4-dichlorobutane, synthesis from 3,4-dichloro-1,2-epoxide, 501
2,3-Dihydroxy-3-ethylpentanoic acid, synthesis from pentan-3-one, 48, 49
(*R*)-2,3-Dihydroxy-3-methylbutanoic acid, 47
2,3-Dihydroxy-3-methylheptanoic acid, synthesis from (*E*)-3-methylhept-2-enoic acid, 48
2,3-Dihydroxy-3-methylhexanoic acid, synthesis from (*E*)-3-methylhex-2-enoic acid, 48
2,3-Dihydroxy-3-methylpentanoic acid, enzymic oxidation, 45
2,3-Dihydroxypropyl *trans*-9-(4-methoxy-2,3,6-trimethylphenyl)-3,7-dimethyl-2,4,6,8-nonatetraenoate, synthesis from glycerol, 4
(2*R*,3*R*)-2,3-Dihydroxy-3-methylpentanoic acid, 47
1,2:5,6-Di-*O*-isopropylidene-α-D-allofuranose ester enolates, alkylation with alkyl iodides, 290
1,2:5,6-Di-*O*-isopropylidene-α-D-allofuranose/Grignard reagents, reactions with alkyl aryl ketones, 293
1,2:5,6-Di-*O*-isopropylidene-D-glucitol, synthesis from D-glucitol, 191
1,2:5,6-Di-*O*-isopropylidene-α-D-glucofuranose, glycosidation, 445
–, protonation agent for ester enolates, 295
1,2:5,6-Di-*O*-isopropylidene-α-D-glucofuranose/Grignard reagents, reactions with alkyl aryl ketones, 292
1,2:5,6-Di-*O*-isopropylidene-α-D-glucofuranose/sodium borohydride, enantioselective reductant for aryl alkyl ketones, 283, 284
1,2:5,6-Di-*O*-isopropylidene-α-D-glucofuranosyl alkoxycyclohexa-1,3-dienes, 310
1,2:5,6-Di-*O*-isopropylidene-D-mannitol, chiral promoter in Diels–Alder reactions, 186
–, 18-crown-6 complex, 186, 187
–, reaction with *N*-dichloromethylene-*N,N*-dimethylammonium chloride, 194
1,2:5,6-Diisopropylidenemannitol/lithium aluminium hydride, asymmetric reductant, 186
2,3:5,6-Di-*O*-isopropylidene-D-mannofuranosylnitrones, cycloadditions with methyl methacrylate, 310
1,2:3,4-Di-*O*-isopropylidenexylitol, reactions with perfluoroalkanoyl chlorides, 198
meso-2,3-Dimercaptobutane-1,4-diol

arsenic antidote, 502
1,4-Dimercaptobutane-2,3-diols, complexes with Cd^{2+} ion, 502
meso-Dimercaptosuccinic acid (succimer), antidote for heavy metal poisons, 516
Dimethyl aminomalonate, 81
Dimethyl chlorophosphate, 105
D-3,3-Dimethylcysteic acid, 64
3,3′-Dimethylcystine, 63
Dimethyl dicyanofumarate, structure and ^{13}C NMR spectrum, 517
Dimethyl dioxosuccinate, condensation with phenols/zinc chloride, 516
Dimethyl itaconate, enantioselective hydrogenation, 279
Dimethyl tartrate, conformations, 510
Dimethyl (*R,R*)-tartrate, synthesis from dimethyl fumarate, 511
1,4-Di-*O*-methyl-L-threitol, complex with Ti^{4+} ion, 500
N-(2,4-Dinitrophenyl)glutamine, 98
1,2-Diols, cleavage, 190
–, reactions with *N*-^{t}butyl-*N*′,*N*′,*N*″,*N*″-tetramethylguanidinium-3-iodoxybenzoate/thrichloroacetic acid, 190
Dioxobutanedicarboxylic acids, 516
Dipalmitoyl cyclopentanophosphoric acid, 127
1,2-Dipalmitoyl-*sn*-glycero-3-phosphatidylethanolamines, 108
1,2-Dipalmitoyl-*sn*-glycero-3-thiophosphocholine, synthesis from palmitin, 123
6′,6′-Dipalmityl glycoside, synthesis, 483
Diphenyl chlorophosphate, 105
6-*C*-(Diphenylphosphino)-1,2:3,4-di-*O*-isopropylidene-α-D-galactopyranose, 293
Dipin (1,4-anhydrothreitol diphosphinite ester), 501
(+)-Diplodiatoxin, synthesis from D-glucose, 331, 332
N,N-Di-^{i}propylmethylphosphonamidic chloride, 161
meso-Di-^{i}propyltartramide, complex with Ti^{4+} ion, 511
(*R,R*)-Di-^{i}propyltartramide, complex with Ti^{4+} ion, 511
(+)-Di-^{i}propyltartrate, 117
meso-Di-^{i}propyltartrate, complex with Ti^{4+} ion, 511
(*R,R*)-Di-^{i}propyltartrate, complex with Ti^{4+} ion, 511
(+)-Di-^{i}propyltartrate/metal complexes, asymmetric epoxidation, 512–514
Disaccharides, 437–498
–, ^{13}C NMR spectroscopy, 462, 463
–, ^1H NMR spectroscopy, 462, 463
–, nitrogen containing, 484–488
–, non-reducing, 482–484
–, synthesis, 439–458
α-Disaccharides, synthesis from glycosylthiopyridyls/sugar alcohols, 445
2,3-Di-*O*-[(*R,R,R*)-3′,7′,11′,15′-tetramethylhexadecyl]-*sn*-glycerol, 30
2,3-Di-*O*-[(3′*R,S*,7′*R*,11′*R*)-3′,7′,11′,15′-tetramethylhexadec-1-yl]-*sn*-glycerol, 32
(*R,R*)-1,2-Divinylglycol [(*R,R*)-1,5-hexadien-3,4-diol], synthesis from D-mannitol, 183
Dodecacarbonyltriruthenium/chiral phosphinites, enantioselective hydrogenation catalysts, 279
Dodecanoylglycerol, reaction with dimethyl disulphide, 26

Ecballium elaterium, 89
(−)-Echinosporin, synthesis from L-ascorbic acid, 341, 343, 344
Electron transport, in membranes, 134
Enantioselective aldol reactions, 287
Endolates, carboxylation, 25, 26
Enol ethers, enantioselective reduction, 282
1-Epicastanospermine, synthesis from D-glucose, 373, 374
5-Epidesosamine, synthesis, 235, 236
Epoxybenzoxocin sugars, 222, 223
β-(2,3-Epoxycyclohexyl)alanine, 81
Epoxy monosaccharides, ring-opening, 231, 232
2,3-Epoxypropyl stearate, 24
Epoxy sugars, reaction with diethyl-

aluminium cyanide, 246, 247
Erythritol, acetyl derivatives, 500
–, conformations, 500
–, occurrence, 499
–, reaction with acetic acid, 500
–, synthesis by fermentation, 499
–, – from arabinose, 499
–, – from glucose, 1
meso-Erythritol, cyclocondensation with 2-hydroxyketones, 7
D-Erythrono-1,4-lactone, synthesis from mannitol, 508
D-Erythrose, synthesis from D-arabinonic acid, 505
–, – from D-fructose, 505
–, – from glyceraldehyde, 503, 504
L-Erythrose, synthesis from D-ribono-1,4-lactone, 504
[1-^{13}C,^{2}H]-Erythrose, synthesis from glyceraldehyde, 505
Erythrose-4-phosphate monomer, equilibration with dimers, 506
Erythulose, reaction with glycolaldehyde, 506
L-Erythulose, synthesis from hydroxy-pyruvate/glycolaldehyde, 506
Escherichia coli, investigation of biochemical pathways, 269
Ester enolates, enantioselective protonation, 295
Ethanolaminophospholipids, 128
S-3-Ethoxycarbonylcysteine, 60
N-Ethoxycarbonylmethionine, 68
N-Ethoxycarbonyl-L-ornithine, 77
cis-4β-Ethoxycarbonylprolines, synthesis from *N*-benzylcarbonyl-L-glutamic acid, 94
Ethyl 2-amino-3-oxobutanoate (ethyl 2-aminoacetoacetate), 73
Ethylideneheptadienediol, reaction with glyceraldehyde derivatives, 19
Ethyl *N*-methoxycarbonyl-L-aspartate, reaction with benzene under Friedel–Crafts' conditions, 84
Ethyl octadecanimidate hydrochloride, 157
Eubacteria, 30
Eukaryotes, 30
Euphorbia pulcherrima, 70
Exo-anomeric effect, 268–271

Fecapentaene-12, 129
Fischer–Kiliani cyanohydrin synthesis, 174
Flavophospholipids, 134
Fluorescent membrane probes, 133
3-Fluoroaspartic acid, 88
β-Fluoromalic acid, synthesis from fluorooxaloacetic acid/malate dehydrogenase, 514
Fluorooxaloacetic acid, synthesis from dihydrofumaric acid, 515
–, – from fluoropyruvic acid, 515
1-Fluoro-2-thiophenyl monosaccharides, 451, 452
Formaldehyde, reductive condensation, 3
Formycin, 206
2-(*N*-Formylamino)-2-alkenoate esters, deformylation, 57
–, synthesis from carbonyl compounds, 56
(−)-Forskolin, synthesis from D-glucose, 331, 334, 335
O-β-D-Fructofuranosyl-(2→1)-β-D-fructofuranosyl α-D-glucopyranoside, 489
β-D-Fructofuranosyl-α-D-glucopyranoside, 482
–, chemistry, 483
6-*O*-Fructofuranosylsucrose monohydrate, X-ray crystallography, 469
β-D-Fructofuranosyl 1-thio-α-D-glucopyranoside, 498
D-Fructose, sweetness, 260
L-Fucitol, 188
2-*O*-α-L-Fucopyranosyl-D-galactose, 474
3-*O*-α-D-Fucopyranosyl-D-galactose, 475
3-*O*-α-L-Fucopyranosyl-D-galactose, 474
3-*O*-β-L-Fucopyranosyl-D-galactose, 475
4-*O*-α-L-Fucopyranosyl-D-galactose, 475
4-*O*-β-L-Fucopyranosyl-D-galactose, 475
2-*O*-α-L-Fucopyranosyl-D-glucose, 475
3-*O*-α-D-Fucopyranosyl-D-glucose, 475
3-*O*-α-L-Fucopyranosyl-D-glucose, 475
4-*O*-α-L-Fucopyranosyl-D-glucose, 475
4-*O*-β-L-Fucopyranosyl-D-glucose, 475
3-*O*-α-D-Fucopyranosyl-D-mannose, 475
3-*O*-α-L-Fucopyranosyl-D-mannose, 475

L-Fucose, synthesis from L-fucitol, 188
α-Fucose, IR spectrum, 271, 272
α-Furanose-β-furanose, interconversions of tetroses, 505
Furanoses, conformations, 271
Furoic acids, Birch reduction/ enantioselective protonation, 295
(−)-Fuscol, synthesis from D-mannitol, 357, 359, 360

Gabriel synthesis, 106
Galabiose, X-ray crystallography, 468
D-Galactitol, phosphorylation, 196, 197
Galactitol conformation, 176
Galactitol diisopropylidene acetals, 193
6-*O*-β-D-Galactofuranosyl-D-galactitol, 478
5-*O*-β-D-Galactofuranosyl-β-D-galactofuranose, 478
6-*O*-β-D-Galactofuranosyl-D-galactofuranose, 477
β-D-Galactofuranosyl-β-D-galactofuranoside, 483
–, chemistry, 483
3-*O*-β-D-Galactofuranosyl-D-mannopyranose, 480
6-*O*-β-D-Galactofuranosyl-D-mannopyranose, 480
Galactonic acid, synthesis from galactitol, 189
D-Galactopyranosides, analytical determination, 189
β-D-Galactopyranosyl disaccharides, 455
O-β-D-Galactopyranosyl-D-fructose, 479
4-*O*-β-D-Galactopyranosyl-L-fucose, 476
α-D-Galactopyranosyl-(1→4)-α,β-D-galactopyranose, X-ray crystallography, 468
2-*O*-α-D-Galactopyranosyl-D-galactose, 477
2-*O*-β-D-Galactopyranosyl-D-galactose, 477
3-*O*-α-D-Galactopyranosyl-D-galactose, 477
3-*O*-β-D-Galactopyranosyl-D-galactose, 478
4-*O*-α-D-Galactopyranosyl-D-galactose, 478
4-*O*-α-D-Galactopyranosyl-D-glucose, 480
4-*O*-β-D-Galactopyranosyl-D-glucose, 480
O-β-D-Galactopyranosyl-D-glucose, 479
2-*O*-α-D-Galactopyranosyl-D-mannose, 480
2-*O*-α-D-Galactopyranosyl-L-rhamnose, 476
4-*S*-β-D-Galactopyranosyl-4-thio-D-glucose, 498
Galactose, IR spectrum, 271, 272
Galactose oxidase, 189
α-Galactosidase, action on *o*- and *p*-nitrophenyl-α-galctosides, 457
N-Galactosyldehydropiperidines, synthesis from *N*-galactosylimines/ dienes, 304
Galactosylgalactoses, 477, 478
Galactosyltransferases/solid phase systems, 456
4-*O*-(α-D-Galactosyluronic acid)-D-galactose, 488
Galα(1→4)Galβ(1→4)GlcNAc, *O*-α-D-galactopyranosyl-(1→4)-*O*-β-D-galactopyranosyl-(1→4)-2-acetamido-2-deoxy-D-glucopyranose, 493
Galβ(1→3)GlcNAcβ(1→3)Gal, *O*-β-D-galactopyranosyl-(1→3)-*O*-(2-acetamido-2-deoxy-β-D-glucopyranosyl)-(1→3)-D-galactopyranose, 493
Galβ(1→3)GlcNAcβ(1→6)Gal, *O*-β-D-galactopyranosyl-(1→3)-*O*-(2-acetamido-2-deoxy-β-D-glucopyranosyl)-(1→6)-D-galactopyranose, 493
Galβ(1→4)GlcNAcβ(1→3)Gal, *O*-β-D-galactopyranosyl-(1→4)-*O*-(2-acetamido-2-deoxy-β-D-glucopyranosyl)-(1→3)-D-galactopyranose, 494
Galβ(1→4)GlcNAcβ(1→6)Gal, *O*-β-D-galactopyranosyl-(1→4)-*O*-(2-acetamido-2-deoxy-β-D-glucopyranosyl)-(1→6)-D-galactopyranose, 494
Galβ(1→6)GlcNAcβ(1→3)Gal, *O*-β-D-galactopyranosyl-(1→6)-*O*-(2-acetamido-2-deoxy-β-D-glucopyranosyl)-(1→3)-D-galactopyranose, 494

Galα(1→3)Manα(1→6)Man, *O*-α-D-Galactopyranosyl-(1→3)-*O*-α-D-mannopyranosyl-(1→6)-D-mannose, 490
GalNAcβ(1→4)Galβ(1→4)Glc, *O*-(2-acetamido-2-deoxy-β-D-galactopyranosyl)-(1→4)-*O*-β-D-galactopyranosyl-(1→4)-D-glucopyranose, 493
GalNAcβ(1→6)Galβ(1→4)Glc, *O*-(2-acetamido-2-deoxy-β-D-galactopyranosyl)-(1→6)-*O*-β-D-galactopyranosyl-(1→4)-D-glucopyranose, 495
GalNAcα(1→3)Galβ(1→3)GlcNAc, *O*-(2-acetamido-2-deoxy-α-D-galactopyranosyl)-(1→3)-*O*-β-D-galactopyranosyl-(1→3)-2-acetamido-2-deoxy-D-glucopyranose, 494
Gangliotriose, 494
Glcβ(1→3)[Glcβ(1→6)]Glc, *O*-β-D-glucopyranosyl-(1→3)-*O*-[α-D-glucopyranosyl-(1→6)]-D-glucopyranose, 490
Glcβ(1→3)Glcβ(1→3)[Glcβ(1→6)]Glc, *O*-β-D-glucopyranosyl-(1→3)-*O*-β-D-glucopyranosyl-(1→3)-*O*-[β-D-glucopyranosyl-(1→6)]-D-glucopyranose, 497
Glcβ(1→4)Glcβ(1→4)Glcβ(1→4)Glc, *O*-β-D-glucopyranosyl-(1→4)-*O*-β-D-glucopyranosyl-(1→4)-*O*-β-D-glucopyranosyl-(1→4)-D-glucopyranose, 497
Glcβ(1→6)[Glcβ(1→3)]Glcβ(1→6)-Glcβ(1→6)[Glcβ(1→3)]Glcβ(1→6)-Glcβ(1→6)[Glcβ(1→3)]Glcβ(1→6)Glc, *O*-β-D-glucopyranosyl-(1→6)-*O*-[β-D-glucopyranosyl-(1→3)]-*O*-β-D-glucopyranosyl-(1→6)-*O*-β-D-glucopyranosyl-(1→6)-*O*-[β-D-glucopyranosyl-(1→3)]-*O*-β-D-glucopyranosyl-(1→6)-*O*-β-D-glucopyranosyl-(1→6)-*O*-[β-D-glucopyranosyl-(1→3)]-*O*-β-D-glucopyranosyl-(1→6)-D-glucopyranose, 497
Glcα(1→4)L-Rhaα(1→3)Glc, *O*-α-D-glucopyranosyl-(1→4)-*O*-α-L-rhamnopyranosyl-(1→3)-D-glucopyranose, 493
Glcα(1→4)Rhaα(1→3)Glc, *O*-α-D-glucopyranosyl-(1→4)-*O*-α-D-rhamnopyranosyl-(1→3)-D-glucopyranose, 493
Glcα(1→4)Xylα(1→4)Xylα(1→4)Glc, *O*-α-D-glucopyranosyl-(1→4)-*O*-α-D-xylopyranosyl-(1→4)-*O*-α-D-xylopyranosyl-(1→4)-D-glucopyranose, 497
Globotriose, 489
D-Glucals, chlorination, 253
Glucitol, Cr(III) complexes, 180
D-Glucitol, conformation, 176
–, 18-crown-6 ethers, 186
D-Glucitol acetals, 191
D-Glucitol 2-haloethylidene acetals, 191
Glucitol–dextrin, metal complexes, 180
D-Glucobenzoxocins, synthesis, 222, 223
Glucocorticoid receptor, 132
Gluconobacter sp., oxidants for 3-methylpentane-1,3,5-triol, 44, 45
2-*O*-β-D-Glucopyranosyl-α-L-arabinopyranose, 472
2-*O*-β-D-Glucopyranosyl-β-L-arabinopyranose, 472
3-*O*-β-D-Glucopyranosyl-D-arabinopyranose, 472
1-(D-Glucopyranosyl)-1-deoxy-D-fructose, 18
α-D-Glucopyranosyl-(1→3)-β-D-fructofuranosyl-α-D-glucopyranose, X-ray crystallography, 469
3-*O*-β-D-Glucopyranosyl-D-fructose, 480
3-*O*-α-D-Glucopyranosyl-D-galactopyranose, 481
4-*O*-α-D-Glucopyranosyl-D-galactopyranose, 481
4-*O*-β-D-Glucopyranosyl-D-galactopyranose, 481
α-D-Glucopyranosyl-(1→6)-α-D-galactopyranosyl-(1→6)-α-D-glucopyranosyl-(1→2)-α-D-fructofuranose, X-ray crystallography, 469
2-*O*-α-D-Glucopyranosyl-D-galactose, 481
2-*O*-β-D-Glucopyranosyl-D-galactose, 481
3-*O*-β-D-Glucopyranosyl-D-galactose, 481
6-*O*-α-D-Glucopyranosyl-D-galactose, 481

6-*O*-β-D-Glucopyranosyl-D-galactose, 481
α-D-Glucopyranosyl-α-D-glucopyranoside, 483
β-D-Glucopyranosyl-α-D-glucopyranoside, chemistry, 483
β-D-Glucopyranosyl-β-D-glucopyranoside, 484
α-D-Glucopyranosyl-(1→6)-α-D-glucopyranosyl-(1→4)-α-D-glucopyranose, X-ray crystallography, 469
2-*O*-α-D-Glucopyranosyl-D-glucose, 478
4-*O*-α-D-Glucopyranosyl-D-glucose, 478
4-*O*-β-D-Glucopyranosyl-D-glucose, 478
6-*O*-α-D-Glucopyranosyl-D-glucose, 479
2-*O*-α-D-Glucopyranosyl-L-rhamnose, 477
2-*O*-β-D-Glucopyranosyl-L-rhamnose, 477
3-*O*-α-D-Glucopyranosyl-L-rhamnose, 477
3-*O*-β-D-Glucopyranosyl-L-rhamnose, 477
6-*S*-α-D-Glucopyranosyl-6-thiocyclomaltoheptaose, 453
6-*S*-β-D-Glucopyranosyl-6-thiocyclomaltoheptaose, 453
4-*S*-β-D-Glucopyranosyl-4-thio-D-glucose, 498
(α-D-Glucopyranosyluronic acid) (α-D-glucopyranosiduronic acid), 488
Glucoronic acid, Fe(II) complexes, 179
–, synthesis from aldoses, 168
–, – from glucose–fructose syrups, 168
D-[U-^{13}C]Glucose, 269
Glucose, IR spectrum, 271, 272
D-Glucose, Ferrier rearrangement, 262
Glucose acetals, reactions with sodium metaperiodate, 264
D-Glucose pentaacetate, reaction with thiophenol/zinc chloride, 257
β-Glucosyldiglycerides, hydroxyl group protection, 267
Glucosylglucoses, 478, 479
Glutamic acid (2-aminopentanedioic acid), 89
–, derivatives, 90
–, preparation from proteins, 90
–, synthesis, 90
3-[^{13}C]-Glutamic acid, synthesis, 90
L-Glutamic acid metal salts, 93, 94
L-Glutamic acid 4-semialdehyde, synthesis from *N*-acetyl-L-asparagine methyl ester, 91
Glutamine, synthesis from 2-ketoglutaric acid, 97
Glutamycin, 86
Glyceraldehyde, reaction with hydrazine, 2
D-Glyceraldehyde, cyanhydrin formation, 508
D-Glyceraldehyde acetonide, homologation, 172, 173
–, reactions with 1-trimethylsilylvinyl copper compounds, 172
–, – with the anion of di(ethylthio)methane monosulphoxide, 503
Glyceraldehyde isopropylidene acetal, reaction with ethyl triethoxyphosphonoacetate, 15
Glyceraldehydes, Knoevenagel reactions, 238, 239
Glycerides, transesterification, 2
Glycerol (1,2,3-trihydroxypropane), 1–33
–, methods of synthesis, 1–3
Glycerol acetonide, 11
Glycerol-1-benzyl ethers, 11
Glycerol 1,2-distearate, 24
Glycerol 1,3-distearate, 24
Glycerol 1,3-distearate-2-palmitate, synthesis from 2-bromo-1-(bromomethyl)ethyl palmitate/tris(decyl)methylammonium stearate, 25
1-Glycerol gallate, 5
Glycerols, ^{1}H NMR spectroscopy, 269, 270
–, reaction with chloro-(*N,N*-diisopropylamino)methoxyphosphine, 115
L-2-Glycerophosphate, 5
Glycerophospholipids, mixed chain derivatives, 116
Glycidol, hydrolysis, 2
(*S*)-Glycidol, 116
Glycidol benzyl ethers, 14
Glycidol derivatives, 116–118
Glycidyl arene sulphonates, nucleophilic ring-opening, 146
Glycocinnamoylspermidines, synthesis, 240
Glycolaldehyde derivatives, aldolisation

reactions, 503
Glycolipids, synthesis, 110
α-, β-D-Glycopyranosyl halides, isomerisation, 443–445
Glycosidation, imidate procedure, 447, 448
–, stereoselective, 437–458
–, *via* acetylglycal iodoniums, 449, 450
–, *via n*-pentenyl glycosides, 448, 449
β-Glycosidation, without neighbouring-group participation, 446, 447
Glycosides, silylation, 201
–, synthesis, 225
–, synthesis catalysed by halide ions, 443–445
Glycosyl acetates, reactions with thiolactic acid, 257
Glycosylalditols, conformation, 178
–, synthesis from formaldehyde/syn gas, 169
1,2-*trans*-Glycosylation, 440–443
Glycosylayanides, 246
Glycosyl bromides, synthesis from glucose, 251, 252
Glycosyl fluorides, 251
–, glycosidation reagents, 445
O-Glycosylhydroxylamines, synthesis from *O*-glycosyl-*N*-phthalimides, 241
Glycosyl-(1→4)-β-D-mannosamines, 446
Glycosyl-(1—4)-β-D-mannoses, 446
N-Glycosylnitrones, cycloadditions with alkenes, 310
–, synthesis from alkylglyoxalates, 310, 311
Glycosyloxyselenation, 452
Glycosylthiopyridyls, glycosidation reagents, 445
Glycosyltransferases, 454–458
Glyptal, laboratory preparation, 5
L-Gulose-*S*-phenylmonothiohemiacetal hexaacetates, 210
Gymnocladus sp., 97

Halichondrin B, synthesis from L-ascorbic acid/1,2:5,6-di-*O*-isopropylidene-α-D-glucofuranoside/2-deoxy-L-arabinose diethylthioacetal 4,5-acetonide, 420–432
Halichondrin B C(1)-C(15) segment, synthesis from D-ribose, 432–435
Halichondrin C(27)-C(38) segment, synthesis from D-galactal, 427
Halogeno monosaccharides, 248–254
Haptens, 154
Heptakis (2,6-di-*O*-methyl) cyclomaltoheptaose, 496
Heptanoid sugars, synthesis, 245
Heteropolysaccharides, synthesis from hetero-oligosaccharides, 443
Hexaalditols, ^{1}H NMR spectroscopy, 269
(*R,R*)-1,5-Hexadien-3,4-diol [(*R,R*)-1,2-divinylglycol], synthesis from D-mannitol, 183
Hexitols, oxidation by bromine/calcium carbonate, 188
–, synthesis *via* the Sharpless epoxidation procedure, 172
–, telluric acid complexes, 180
2,5-Hexodiuloses, 188
Hexoses, Pummerer rearrangement, 209
–, reaction with hydrazine, 203
Hexosyldeoxyhexoses, 476, 477
Hexosylhexoses, 477–482
Hexosylpentoses, 471, 472
lyxo-3-Hexulose, synthesis from erythulose/glyceraldehyde, 506
Hikosamine, isolation from hikimycin, 174
Homoallyl alcohols, enantioselective synthesis, 292
(2′*S*)-Homoallylamines, synthesis from β-L-fucosylimines, 300
–, – from β-D-galactosylimines, 300
Homoallylic alcohols, asymmetric synthesis, 513, 514
–, enantioselective synthesis from alkanals/chiral titanium complexes, 292
Homocysteic acid (3-amino-4-carboxybutane sulphonic acid), 62
Homocysteine (2-amino-4-mercaptobutanoic acid), 62
Homogeranic acid monosaccharide-ester, asymmetric cyclisation, 295
L-Homoglutamic acid, synthesis from *N*-acetyl-L-lysine ethyl ester, 90
Homoserine (2-amino-4-hydroxybutanoic acid), 66

Humicula langinosa, 194
α-Hydroxy-α-acetylaminopropionic acid, synthesis from pyruvic acid, 54
α-Hydroxyalkanals, synthesis from D-mannitol, 182
(*S*)-α-Hydroxyalkanoate esters, synthesis from α-oxoalkanoate esters, 285
β-Hydroxyamines, kinetic resolutions, 514
D-*threo*-β-Hydroxy-α-amino acids, synthesis from *N*-bis(silyl)glycine enolate, 289
3-Hydroxy-2-aminoalkanoic acids, dehydration, 50
–, synthesis from 2-amino-4-pentenoic acids, 50
3-Hydroxy-α-aminoalkanoic acids, synthesis, 49, 50
4-Hydroxy-2-aminoalkanoic acids, synthesis from 2-nitrosoalk-2-enoic acid amides, 50
(*S*)-3-Hydroxybutyrolactone, synthesis from L-aspartic acid, 84
Hydroxycarboxylic acids, 81–100
Hydroxydialkanoic acids, 81
3-Hydroxy-2,4-dimethylglutaric anhydride, 38
(1*S*,4*S*,5*R*)-4-Hydroxy-2,6-dioxabicyclo[3,3,0]octan-8-one, 199
5-Hydroxy-1,3-dioxane, 8
β-Hydroxyesters, enantioselective synthesis from, 287
3(*S*)-Hydroxy-5(*R*)-hydroxymethyl-2*H*-dihydrofuran-2-one, 509
4-Hydroxyisoleucine (2-amino-4-hydroxy-3-methylpentanoic acid), isolation from *Trigonella foenumgraecum*, 76
2-Hydroxyketones, cycloaddition with polyols, 7
Hydroxylysine, 78
5-Hydroxymethylbutrolactone, synthesis from glyceraldehyde isopropylidene acetal, 15
4-Hydroxymethyl-1,3-dioxane, 8
3-Hydroxy-4-methylglutamic acid (2-amino-3-hydroxy-3-methyl-pentanedioic acid), 97
4-Hydroxy-4-methylglutamic acid (2-amino-4-hydroxy-4-methyl-pentanedioic acid), 96
4-*O*-(3-*C*-Hydroxymethyl-β-D-*glycero*-tetrofuranosyl)-D-glucose, 470
3-Hydroxy-3-methylpentane-1,5-dioic acid, 36
4-Hydroxy-4-methylpentan-2-one, reaction with dimethyl phosphorochloridite, 245, 246
(*R*)-(–)-3-Hydroxy-3-methyl-4-phenylbutanonitrile, 40
4-Hydroxymethyl-2-^{i}propyl-1,3-dioxolane, synthesis from glycerol, 5
4-Hydroxymethyl-2-vinyl-1,3-dioxane, 8
cis-4-Hydroxymethyl-2-vinyl-1,3-dioxolane, synthesis from acrolein/glycerol, 8
4-Hydroxynorvaline (2-amino-4-hydroxy-pentanoic acid), 75
β-Hydroxynorvaline (2-amino-3-hydroxy-pentanoic acid), 75
N^6-Hydroxy-L-ornithine, synthesis from L-glutamic acid, 90
N^4-(4-Hydroxyphenyl)glutamine, 98
5-*C*-(Hydroxyphosphinyl)aldohexo-pyranoses, 244
(2*S*,4*R*)-4-Hydroxyproline, 72
3-Hydroxyprop-2-enylfuranoses, osmylation, 267
5-Hydroxy-2-^{i}propyl-1,3-dioxolane, synthesis from glycerol, 5
4-Hydroxyvaline (2-amino-4-hydroxy-3-methylbutanoic acid), 71
5-Hydroxy-2-vinyl-1,3-dioxane, 8
Hypusine [*N*′-(4-amino-2-hydroxybutyl)-lysine], 78
–, synthesis from *N*-benzyloxycarbonyl-L-lysine benzyl ester, 79

Ibuprofen, 10
L-Iditol biphosphine-rhodium catalysts, 186
"L-Iditol diepoxide", 182
L-Idobenzoxocins, synthesis, 222, 223
D-Idose, ^{13}C NMR spectroscopy, 270
Imidazolides, synthesis from fatty acids, 120
Immunoglobulins bonded to

phosphorylcholine, 135
Immunostimulants, 135
3-Iodo sugars, synthesis, 252
(*S*)-Ipsenol, synthesis *via* a chiral titanium enolate, 288
(−)-Isoavenenaciolide, synthesis, 397, 399, 400
α-Isocyanoacrylates, synthesis from carbonyl compounds/ diphenylphosphino(isocyanate)acetate, 57
Isoglutamine, 98
Isoleucine, biosynthesis, 45
Isomannide (dianhydromannitol), 199
Isomannide dinitrate, flash vacuum pyrolysis, 199
1,2-*O*-Isopropylidene-α-D-glucofuranose/ rutheniumtriphenylphine dichloride, enantioselective hydrogenation catalyst for α-alkenones, 280
1,2-Isopropylidene-*sn*-glycerol, synthesis from D-mannitol, 22
3,4-Isopropylidene-D-mannitol, 22, 203
Isoserine (3-amino-4-hydroxypropanoic acid), 59
–, synthesis from D-mannitol, 182
D-*myo*-Isositol triphosphate, synthesis, 261, 262
Isosorbide (dianhydrosorbitol), 199
–, catalytic oxidation, 199
Isosorbide-2,5-dinitrate, reduction, 208
Isosorbide-2-nitrate, 208
Isosorbide-5-nitrate, 208
(5′*R*)-Isoxazolidines, synthesis from D-ribosylhydroxyoximes, 310
(5*S*)-Isoxazolidines, synthesis from 2,3:5,6-di-*O*-isopropylidene-D-mannofuranosylnitrones methyl methacrylate, 310
Itacomic acid, enantioselective hydrogenation, 279

Kalanchoe daigremontiana, 71
2-Ketoglutaric acid, synthesis from lysine, 78
Ketones, asymmetric reduction to alcohols, 514
6-Ketose, X-ray crystallography, 469
Keto sugars, reactions with hydrogen cyanide, 246
Kijanolide, synthesis by intramolecular Diels–Alder strategy, 331, 333
Kiliani cyanohydrin synthesis, mechanism, 508
Kishi rules, for osmylation, 268
Kojitriose, 490

β-Lactams, syntheses from diethyl (*R,R*)-tartrate, 512
Lactarius quietus, 99
γ-Lactones, synthesis from glycerol, 11–16
Lactose, X-ray crystallography, 468
α-Lactose monohydrate, X-ray crystallography, 468
α-Laminaribose, X-ray crystallography, 468
Lanthionines, 65
Lathyrus odoratus, 75
Lecithin-cholesterol acyltransferase, 124
Leukotrienes, 147
(+)-Lineatin, synthesis from D-ribonolactone, 362, 363
1-*O*-Linoleoylethane-2-[*O*-bis(*N,N*-diethylamino)]thionophosphonate, 127, 128
Lipases, transesterification catalysts, 194
Lipid transfer, analysis with azobenzene derivatives, 133
Lipopolysaccharides, 188
Liposomes, quinone-functionalised, 134
Lotus tenius, 70
(−)-Lycoricidine, synthesis from D-glucose, 351–353
Lypoxygenase pathways, 119
Lysine (2,6-diaminohexanoic acid), 78
β-Lysine, fermentation, 79
Lysolecithins, esterification, 119
Lysophosphatidic acid, esterification, 121
Lysophospholipids, acylation, 120
–, hydrolysis, 155
–, synthesis, 108
2-Lysophospholipids, acylation, 118, 122
Lysuric acid (N^2,N^6-dibenzoyllysine), 78
D-Lyxitol peracetate, 172

Macrolide FK-506, synthesis from methyl β-D-galactopyranoside/tri-*O*-acetal-D-

glucal, 401–413
Macrolides FR-900520 and FR 900523, 410–412
Maltoheptaose-phosphorylase, X-ray crystallography, 469
Maltose, conformational analysis, 462, 463
β-Maltose octaacetate, X-ray crystallography, 468
(−)-Malyngolide, synthesis from 2,3-*O*-isopropylidene-D-apiose, 219, 220
Manβ(1→4)GlcNAcβ(1→4)GlcNAc, *O*-β-D-mannopyranosyl-(1→4)-*O*-(2-acetamido-2-deoxy-β-D-glucopyranosyl-(1→4)-2-acetamido-2-deoxy-D-glucopyranose, 495
Manα(1→2)Manβ(1→4)GlcNAc, *O*-α-D-mannopyranosyl-(1→2)-*O*-β-D-mannopyranosyl-(1→4)-2-acetamido-2-deoxy-D-glucopyranose, 495
Manα(1→3)Manβ(1→4)GlcNAc, *O*-α-D-mannopyranosyl-(1→3)-*O*-β-D-mannopyranosyl-(1→4)-2-acetamido-2-deoxy-D-glucopyranose, 495
Mannich reaction, 94
–, alkoxydienes, 298
Mannitol, phosphorylation, 195
–, synthesis from glucose–fructose syrups, 168
(±)-Mannitol, gas chromatographic resolution, 182
D-Mannitol, conformation, 176
–, 18-crown-6 ethers, 186
–, reaction with methanolic hydrogen chloride, 200
–, use as a chiron, 182
D-Mannitol acetals-titanium reagents, 186
Mannitol cyclic phosphites, chlorination, 196
D-Mannitol diacetonide, 11
Mannitol-1,6-dibenzoate, 192
"D-Mannitol diepoxide", 182
α-D-(1→4)-Mannobiose, X-ray crystallography, 468
β-D-Mannopyranopyranosides, synthesis from 6-*O*-acetyl-2,3,4-tri-*O*-benzyl-α-D-mannopyranosyl bromide, 446
2-*O*-α-D-Mannopyranosyl-D-glucose, 482
3-*O*-α-D-Mannopyranosyl-D-glucose, 482
4-*O*-α-D-Mannopyranosyl-D-glucose, 482
6-*O*-α-D-Mannopyranosyl-D-glucose, 482
2-*O*-α-D-Mannopyranosyl-D-mannose, 479
3-*O*-α-D-Mannopyranosyl-D-mannose, 479
6-*O*-α-D-Mannopyranosyl-D-mannose, 479
3-*O*-α-D-Mannopyranosyl-D-talose, 482
3-*O*-β-D-Mannopyranosyl-D-talose, 482
D-Mannose lactol, reaction with hydrazine, 205, 206
α-Mannosidase, isolation from jack bean, 457
(+)-Mannostatin A, synthesis, 346–348
Mannosylmannoses, 479
Meldrum's acid, 18
Melezitose, X-ray crystallography, 469
Mellitic acid, synthesis from glycerol, 3
Membrane fluidy, analysis by fluorescent anisotropy, 133
Membrane phospholipids, 119
Membrane probes, 132
2-Mercapto-1*H*-imidazo[4,5-*f*]quinoline, 7
Methanobacterium thermoautotrophicum, 32
Methionine (2-amino-4-methylbutanoic acid), 68
Methionine lyase, 68
Methionine sulphoxime [*S*-(3-amino-3-carboxypropyl)-*S*-methyl sulphoximine, 69
N^6-Methoxycarbonyllysine, 78
1-*O*-(2′-Methoxyhexadecyl)glycerol phosphatidylcholines, 135
1-Methoxy-3-(trimethylsiloxy)-1,3-butadiene, 91
Methyl α-acetamidocinnamate, enantioselective hydrogenation, 275
Methyl 2-(*N*-acetylamino)-2-alkenoates, 53
Methyl *N*-acetyl-α-L-vanucosamide, synthesis, 247
Methyl *N*-acylamino-α-alkoxyalkanoates, 51, 52
Methyl acylimino-α-alkoxyalkanoates, 52
Methyl α-acyloxy-γ-methylene-β-

tetronate, 18
Methylalkyl ethers, reactions with glycerol/methyloxirane, 10
α-Methyl-α-amino acids, synthesis from *N*-(6-deoxy-1,2:3,4-di-*O*-cyclohexylidene-α-D-galacto-pyranosylidene)alanine methyl ester, 291
N-Methylaminomalonic acid, 81
Methyl 2-amino-3-mercapto-2-methyl-propanoate, 63
Methyl 2,3-anhydro-5-deoxy-α-D-ribofuranoside, reaction with lithium dimethylcuprate, 230
Methyl 4,6-*O*-benzylidene-2,3-di-*O*-methyl-α-D-glucopyranoside, protonation agent for ester enolates, 295
Methyl 4,6-*O*-benzylidene-2,3-di-*O*-methyl-α-D-glucopyranoside/Grignard reagents, reactions with alkyl aryl ketones, 293
Methyl 4,6-*O*-benzylidene-β-glucopyranoside, glycosidation, 441
5-Methylcoumarins, 19
2-Methylcysteine (2-amino-3-mercapto-2-methylpropanoic acid), 63
3-Methylcysteine (2-amino-3-mercaptobutanoic acid), 63
Methyl 2,3-di-*O*-acetyl-α-D-galacto-pyranoside, Mitsunubo reaction with benzoic acid, 230
Methyl dichlorophosphate, 105
β-Methyleneglutamic acid, 81
(*S*)-4-(1-Methylethyl)-3-[(methyl-thio)acetyl]-2,5-oxazolidione, 73
Methyl D-galactopyranosiduronic acid, synthesis from methyl D-galacto-pyranoside, 189
Methyl α-L-garosamimide, 233, 234
Methyl α-D-glucopyranoside, reaction with diethylaminosulphur trifluoride, 251
2-*O*-(4-*O*-Methyl-α-D-gluco-pyranosyluronic acid)-D-xylose, 489
2-*O*-(4-*O*-Methyl-β-D-gluco-pyranosyluronic acid)-D-xylose, 488
2-Methylglutamic acid (2-amino-2-methylpentanedioic acid), 94
3-Methylglutamic acid (2-amino-3-methylpentanedioic acid), 94
4-Methylglutamic acid (2-amino-4-methylpentanedioic acid), 94
Methyl (9*R*)-hydroxy-(5*Z*,7*E*,11*E*,14*Z*)-eicosatetraenoate, 183
Methylidinoglycerol, 8
–, synthesis from glycerol, 8
3-Methyllanthione, synthesis from 3-methylcysteine, 61
β-Methyllinocosamine, synthesis from crotonaldehyde, 215–217
N-Methyllysine, 78
Methyl α-maltotrioside tetrahydrate, X-ray crystallography, 468
4-Methyl-2-oxoglutaric acid, 94
(*S*)-3-Methyl-2-oxopentanoic acid, biosynthesis from (2*R*,3*R*)-2,3-dihydroxy-3-methylpentanoic acid, 45
5-Methyl-4-oxo-3,6,8-trioxa-bicyclo[3.2.1]octane, synthesis from glycerol/2-methoxycarbonyl-2-methyl-1,3-dioxolane, 7
2-Methyloxy platelet activating factor, 153
3-Methylpentane-1,3,5-triol, bacterial oxidation, 44
–, oxidation by *Flavobacterium oxydans*, 45
Methyl phenylglyoxalate, enantioselective reduction, 286, 287
Methyl 6-*S*-phenyl-6-thio-α-D-gluco-pyranoside, 210
Methyl pyranosides, thiolation, 258
Methylselenylphenylalamines, synthesis, 82
Methyl-D-serine, 157
(–)-Methyl shikimate, synthesis from 1-deoxy-1-nitro-D-ribose, 360–362
2-Methylthioacylisopropylideneglycerols, 26, 27
Methyl triphenylphosphoranylidene acetate, reaction with glyceraldehyde, 19
Mevalonic acid, 35–45
Mevalonolactone, 35–45
–, synthesis from glutamic acid, 42
–, – from 3-hydroxy-3-methyl-4-

phenylbutanonitrile, 40
–, – from D-mannitol, 42
–, – from 3-methylpentane-1,3,5-triol, 44
[3-^{13}C]Mevalonolactone, synthesis from ethyl [3-^{13}C]acetoacetate, 36
Michaelis–Arbuzov reaction, 245
Microbial polysaccharides, synthesis, 443
Milbemycins, hexahydrobenzfuran component, synthesis from 1,4-anhydrosorbitol, 353–357
–, oxahydrindene component, 353
–, synthesis by intramolecular Diels–Alder strategy, 341, 342
Mitsunobu reaction, 140
Monoanhydroalditols, synthesis from pentitols, 200
Monolysocardiolipin, 164
Monosaccharide auxiliaries, asymmetric [3+2] cycloadditions, 310
Monosaccharide carbanions, enantioselective alkylation, 289
Monosaccharide chiral electrophiles, 296, 297
Monosaccharide cuprate reagents, 294
Monosaccharide leaving group, 294
Monosaccharide phosphine/rhodium(I), hydrogenation catalysts, 275
Monosaccharides, chemistry, 264–268
–, ^{13}C NMR spectroscopy, 269
–, conformations, 268, 269, 271
–, high performance chromatography, 213
–, ^{1}H NMR spectroscopy, 269
–, hydrogen bonding correlation with sweetness, 268, 269
–, IR spectra, 268
–, mass spectrometry, 269
–, methoxy-mercuration, 224
–, Raman spectra, 268
–, synthesis from oxiranes, 218–220
–, – using allyltins, 224, 225
–, – using Grignard reagents, 222–224
–, – *via* Diels–Alder reactions, 214–218
–, – *via* Wittig reactions, 220, 221
–, thermolysis, 213
–, unsaturated, 224
–, vibrational spectra, 268, 269
2-[^{2}H-]-Monosaccharides, 258
Monosaccharide thiocyanates, 258
Monosaccharide-*O*-thioformates, Diels–Alder reactions with 1,2-dithiooxalates, 304
(*S*)-Morpholino-1,1′-binaphthyl-2,2′-diamine, 38
Mucromonospora chalcea, 99
L-Mycarose, synthesis, 226, 227
Myobacterium rosens, 98
Myristic acid, 22
Myristoyl-*N*-benzoyloxycarbonyl-lysophosphatidylethanolamine, 121
2-Myristoyl-1-octadecyl-*sn*-glycerol, synthesis from 2-allyl-3-benzyl-*sn*-glycerol/octadecyl mesylate, 23
2-Myristoyl-1-stearoyl-*sn*-glycerol, 22, 23
3-Myristoyl-1-stearoyl-*sn*-glycerol, 23

Nakanishi's method, assignment of tetrol stereochemistry, 500
1,4-Naphthoquinone, cycloadditions with alkoxycyclohexa-1,3-dienes, 310
(*S*)-1-Naphthylphenyl sulphoxide, synthesis from 1-naphthylphenylsulphinyl chloride, 295
Nargenicins, synthesis from 1,7,9-decatrien-3-ones, 18
Neighbouring-group assisted, β-glycosidation procedures, 440–443
Neooxazolomycin, synthesis from (*Z*)-3-bromo-2-methyl-2-propenol/methyl 2,3-di-*O*-benzyl-α-D-glucopyranoside, 373, 375–379
NeuAcα(2→3)Galβ(1→4)Glc, *O*-(5-acetamido-3,5-dideoxy-D-*glycero*-α-D-*galacto*-2-nonulopyranosylonic acid)-(2→3)-*O*-β-D-galactopyranosyl-(1→4)-D-glucopyranose, 495
NeuAcα(2→6)Galβ(1→4)-GlcNAcβ(1→2)Man, *O*-(5-acetamido-3,5-dideoxy-D-*glycero*-α-D-*galacto*-2-nonulopyranosylonic acid)-(2→6)-*O*-β-D-galactopyranosyl-(1→4)-*O*-(2-acetamido-2-deoxy-β-D-glucopyranosyl)-(1→2)-D-mannopyranose, 497
NeuAcβ(2→6)Galβ(1→4)-GlcNAcβ(1→2)Man, *O*-(5-acetamido-3,5-dideoxy-D-*glycero*-β-D-*galacto*-2-nonulopyranosylonic acid)-

(2→6)-*O*-β-D-galactopyranosyl-(1→4)-*O*-(2-acetamido-2-deoxy-β-D-glucopyranosyl)-(1→2)-D-mannopyranose, 497
Neuraminic acids, glycosylation, 452, 453
Nigerotriose, 490
Nisin, antibiotic, 63
α-Nitroepoxy sugars, denitrification, 252
Nitroheptenitols, reaction with methyl acetoacetate, 208
Nojirimycin, synthesis from furan, 311
Nonadecanone, acetalisation with glycerol, 6
Nonitols, 174
Nonulosonic acids, 452, 453

Octadecanoylisopropylideneglycerol, reaction with dimethyl disulphide, 26
D-*erythro*-D-*galacto*-Octitol, 174
L-*erythro*-D-*galacto*-Octitol, 174
L-*threo*-D-*galacto*-Octitol, 174
L-*threo*-D-*gluco*-Octitol, synthesis from methyl (*E*)-2,3,4-tri-*O*-benzyl-6,7-dideoxy-α-D-*gluco*-octa-6-enopyranoside, 175
L-*threo*-D-*manno*-Octitol, synthesis from methyl (*E*)-2,3,4-tri-*O*-benzyl-6,7-dideoxy-α-D-*manno*-octa-6-enopyranoside, 175
Octitols, 174
Octulosonic acids, 452, 453
Okadaic acid, synthesis from pyranoid glycals, 383–393
1-*O*-Oleoylethane-2-[*O*-bis(*N,N*-diethylamino)]thionophosphonate, 127, 128
2-Oleoyl-1-stearoyl-*sn*-glycero-3-phosphocholine, synthesis from 2-benzyl-1-stearoyl-*sn*-glycerol, 23
Oligonucleotides, synthesis from dipalmitoyl- *sn*-glycerol, 112
Oligosaccharides, 437–498
–, chromatography, 463–467
–, ^{13}C NMR spectroscopy, 458–463
–, containing sulphur, 498
–, enzyme-catalysed synthesis, 454–458
–, enzymic release from glycoprotein, 458
–, – from proteoglycans, 458
–, fast atom bombardment mass spectrometry, 464–467
–, ^{1}H–^{1}H correlated spectroscopy, 459
–, Hassel–Ottar effects, 461
–, ^{1}H NMR spectroscopy, 458–463
–, interproton distances, 462
–, IR spectroscopy, 469, 470
–, mass spectrometry, 463–467
–, NOE spectroscopy, 460, 461
–, X-ray crystallography, 467–469
–, solution conformations, 460–462
–, synthesis, 225, 439–458
–, – by reverse hydrolysis, 457
–, – *via* Diels–Alder reactions, 304
Ornithine (2,5-diaminopentanoic acid), 77
Ornithine biosynthesis, 90
L-Ornithine methyl ester, 77
Oxaloacetic acid, synthesis from fluorofumaric acid/fumarase, 514
Oxamino sugars, synthesis from *O*-glycosyl-*N*-phthalimides, 241
1,3,2-Oxazaphosphorinanes/monosaccharides, ring opening, 295
Oxiranemethanol-^{t}butyldiphenylsilyl ethers, 118
2-Oxoalkanoic acids, synthesis through enzymic oxidation of dihydroxyalkanoic acids, 45
3-Oxo[3-^{13}C]butyl acetate, 36
Oxovanadium tartrate complexes, X-ray crystal structures, 511

Palmaria palmata, 62
Pandanus veitchii, 96
Paniculide B, synthesis from D-glucose, 348, 349
α-Panose, X-ray crystallography, 469
Pantolactone, 132
Penicillamine (2-amino-3-mercapto-3-methylbutanoic acid), 64
Penicillum farinosa, 2
L-Penta-*O*-acetylarabinitol, 210
Pentaglycerol, 5
meso-Pentane-1,2,4,5-tetrol, synthesis from adonitol, 502
Pentiol, oxidation by bromine/calcium carbonate, 188
Pentiols, dehydration, 199, 200
Pentitols, reactions with tosyl chloride/

pyridine, 200
–, synthesis from 2-alkoxy-3-(1,3-dioxolan-4-yl)-3-silyloxy-1-propenes, 174
–, – from divinylmethanol, 171
–, – *via* the Sharpless epoxidation procedure, 172
Pentoses, reaction with hydrazine, 203
Pentosylhexoses, 471
Pentosylpentoses, 470, 471
Pepstatin A, hydrolysis, 100
Peptidoglycans, 99
Peracetyl monosaccharides, selective 1-*O*-deacetylation, 266
Peracetylpyranosyl bromides, reaction with mercuric cyanide, 228, 229
Per(alkoxybicyclophosphoranes), synthesis from polyols/bicyclophosphorane, 3
Perbenzoylgluculosyl bromide, exchange with iodide ion, 446
Permethyloligosaccharides, mass spectrometry, 464–467
Phenacylglycines, synthesis, 84
Phenols, condensation with dimethyl dioxosuccinate/zinc chloride, 516
2-Phenyl-α-amino acids, 82
N-Phenylazomethines, enantioselective reduction to (*S*)-*N*-phenylamines, 282
Phenyl 4,6-*O*-(*R*)-benzylidene-2,3-*O*-bis(diphenylphosphino)-β-D-glucopyranoiside/rhodium(I), hydrogenation catalysts, 273
3-Phenylisoserine (3-amino-2-hydroxy-3-phenylpropanoic acid), 59
(*R*)-(+)-Phenylmethylcarbinol, synthesis from acetaphenone, 186
2-Phenyl-1,3-oxazolines, synthesis from aspartic acid, 84
β-Phenylthio-α-azidoalkanoate esters, 53
6-*S*-Phenyl-6-thio-6-deoxyalditol, 209
6-*S*-Phenyl-6-thio-D-glucitol pentaacetate, 210
Phosphamonosaccharides, 243–246
Phosphatidic acid, 105
–, reaction with *N*-(β-hydroxy-methyl)carbazole/2,4,6-trisisopropylbenzenesulphonyl chloride, 111
Phosphatidylcholines, 105, 113, 122
–, fluorescent labelled, 111
–, radiolabelled, 120, 121
Phosphatidylethanolamine nitroxides, spin labelled, 111
Phosphatidylethanolamines, 113, 120, 122
–, fluorescent labelled, 111
–, radiolabelled, 120
–, synthesis from *N*-tritylaminoethyl-phosphoric acid, 108
Phosphatidylserines, 112, 113, 122, 124
Phosphodiesters, protection by 1,1,1-trichloro-2-methylpropylation, 114
–, synthesis, 106
–, – from phosphomonoesters, 111
Phosphoester bond formation, from halophosphate reagents, 105, 108
Phospholipase A_1, 155
Phospholipase A_2, 124, 150, 155
–, inhibitors, 154–156
Phospholipase C, 124, 150, 154
Phospholipase D, 122, 148, 154
Phospholipid analogues, 123–136
Phospholipid conjugates, 110
Phospholipid ribosome/iron porphyrin complexes, 135
Phospholipids, 101–166
–, acylation, 119
–, antitumour properties, 109
–, bibliography, 101–104
–, cyclopropyl derivatives, 119
–, 2-diazirinylphenoxy derivatives, 133
–, 2-diazocyclopentadienylcarbonyl derivative, 133
–, ether linked, 116, 117
–, from sponges, 135
–, *P*-labelled, 107
–, membrane-spanning probes, 132
–, synthesis, 104, 105
–, – by phosphoramidite methodology, 128
–, – from 1,2-isopropylidene-*sn*-glycerol, 121
–, – from methallyl alcohol, 127
–, transesterifaication, 122
–, 4-(trifluoromethyl)diazirinylphenyl derivatives, 133
Phospholipid–pantoic acid aggregates,

132
Phospholiposterols, membrane-forming properties, 132
Phosphomonoesters, synthesis, 106
Phosphonolipids, 129–131
Phosphonophosphatidic acid, 129
Phosphonophosphatidylcholine, 129
Phosphorylative coupling methods, 112
Phosphoryl-*N*-methylethanolamines, synthesis from *P*-chloro-1,3-dimethyl-1,2,3-diazaphosphacyclopentane/alcohols, 128
N-Phthaloyl-4-bromo-D-glutamic acid dimethyl ester, 91
(+)-Phyllanthocin, synthesis from isopropylidene-α-D-glucofuranose, 323–325
Phytanol [(3*R*,*S*,7*R*,11*R*)-3,7,11,15-tetramethylhexadecan-1-ol], 31
L-Pipecolic acid, synthesis from L-lysine, 78
Pipecolic acid derivatives, synthesis from L-asparagine, 88
Piperidinolactones, biosynthesis from glyceraldehyde equivalents, 16
(−)-Pipitzol, synthesis from mannose, 321–323
Pisum sativum, 67
4-*O*-Pivaloyldihydro-L-rhamnal, 303
Plasmalogen quinones, 134
Plasmalogens, 129
Platelet activating factor, 25, 130, 135–154
–, agonists and antagonists, 134, 145–154
–, constrained analogues, 151, 152
–, modification of glycerol backbone, 149
–, non-constrained analogues, 150
–, 1-octadecylthio derivative, 146
–, structural variants, 145–154
–, synthesis from epichlorohydrin, 143
–, – from glycidylarene sulphonates, 143
–, – from 1,2-isopropylidene-*sn*-glycerol, 140
–, – from D-mannitol, 137–142
–, – from 3-octadecyl-2-myristoyl-*sn*-glycero-1-phosphocholine, 144
–, – from tartaric acids, 142
Polyalkoxyaldehydes, synthesis from glyceraldehyde derivatives and 2-(trimethylsilyl)thiazole, 18
Poly-*O*-benzylpolyols, de-*O*-benzylation, 3
Polyglycerols, synthesis from polyols, 5
Polyhydric alcohols, 167
Polyhydroxylated pyrrolidines, 211
–, synthesis from D-mannitol, 207
Polyhydroxylated tetrahydrofurans, 211
Polyhydroxylated thiolanes, 211
Poly-*N*-neuraminic acid, acid hydrolysis, 454
Polyol, Mn(II) complexes, 179
Polyols, borate complexes, 178, 179
–, carboalkoxylation, 195
–, Cu(II) complexes, 180
–, metal complexation and chelation, 178
–, oxidation by manganese(III) pyrophosphate, 189
–, – by ruthenium tetraoxide, 189
–, – by vanadium(V) compounds, 189
–, sequestering properties, 179
–, stereochemistry, 178
Polyols trifluoroacetyl derivatives, gas chromatography, 181
Poly(oxypropylene)polyols, synthesis from glycerol/propylene oxide, 8
Polysaccharides, ^{13}C NMR spectroscopy, 463
–, degradation to oligosaccharides, 453, 454
–, solvolysis, 453, 454
Potassium 9-*O*-(1,2:5,6-di-*O*-isopropylidene-α-D-glucofuranosyl)-9-boratobicyclo[3,3,1]nonane, 285
Potassium glucoride, 285
Prochloron didemnii, 96
L-Proline, synthesis from L-glutamic acid, 93
S-2-Propyloxycarbonyl-L-cysteine, 61
(−)-Prostaglandin E_2, synthesis, 346, 347
Prostaglandins, 147
–, synthesis from carbohydrates, 315–319, 346, 347
Protein kinase C, 32, 33
Proteins, glycosidation, 438
Proteus norganic, 78
Pseudofructopyranoses, sweetness, 260
Pseudo-α-L-mannopyranose, synthesis from D-ribose, 259

Pseudomonas putida, 61
Pseudosugars, 258–264
"P-sugars", 243–246
Pullulanase, action on cyclodextrins, 457
Pyranoid glycals, reactions with methanol/palladium(II) acetate, 265
Pyranoses, synthesis from ethoxyacetaldehyde, 227, 228
4-Pyrrolidinopyridine, activation of anhydrides, 119
–, acylation mediator, 119

Quinones, membrane-bound, 134

Rapamycin, synthesis, 348, 350, 351
Reseda luteola, 96
Retro Claisen condensations, 3
5-*O*-α-L-Rhamnopyranosyl-α-L-arabinofuranose, 473
2-*O*-α-L-Rhamnopyranosyl-L-arabinopyranose, 472
3-*O*-α-L-Rhamnopyranosyl-L-arabinopyranose, 472
2-*O*-α-L-Rhamnopyranosyl-L-arabinose, 472
6-*O*-α-L-Rhamnopyranosyl-D-galactose, 475
6-*O*-β-L-Rhamnopyranosyl-D-galactose, 476
4-*O*-α-L-Rhamnopyranosyl-D-glucopyranuronic acid, 489
2-*O*-β-L-Rhamnopyranosyl-D-glucose, 476
3-*O*-α-D-Rhamnopyranosyl-D-glucose, 476
3-*O*-α-L-Rhamnopyranosyl-D-glucose, 476
6-*O*-β-L-Rhamnopyranosyl-D-glucose, 476
2-*O*-β-L-Rhamnopyranosyl-L-rhamnopyranose, 473
2-*O*-α-D-Rhamnopyranosyl-L-rhamnose, 474
3-*O*-α-D-Rhamnopyranosyl-L-rhamnose, 474
3-*O*-β-L-Rhamnopyranosyl-L-rhamnose, 474
4-*O*-β-L-Rhamnopyranosyl-L-rhamnose, 474
3-*O*-α-D-Rhamnopyranosyl-D-talose, 476
3-*O*-α-L-Rhamnopyranosyl-D-talose, 476
3-*O*-β-D-Rhamnopyranosyl-D-talose, 476
Rhizobium bacteria, 70
Rhizopus japonicus, 194
L-Rhodinose, synthesis by 1,3-dipolar cycloaddition reaction, 214
Rhodium(I) dicyclooctene chloride, hydrogenation catalyst, 276
Rhodium(I)/monosaccharide phosphine complexes, enantioselective hydrogenation catalysts, 273
Rhodium(I)/monosaccharide phosphinites, enantioselective hydrogenation catalysts, 278, 279
Ribitol, cyclic phosphites, 195
–, monoacetonide, synthesis from D-glyceraldehyde acetonide, 172
Ribitol peracetate, 172
D-Ribofuranose, pseudorotation, 271
β-D-Ribofuranosyl-β-D-ribofuranoside, 484
2-*O*-β-D-Ribofuranosyl-D-ribose, 471
3-*O*-β-D-Ribofuranosyl-D-ribose, 471
5-*O*-β-D-Ribofuranosyl-D-ribose, 471
Robinobiose, 475
L-Rodinose, synthesis from (*S*)-glycidylsulphide, 218, 219
(−)-Rosmarinecine, synthesis from 2-amino-D-xylofuranose, 367, 369, 370

Saccharomyces cerevisiae, 2
Saccharpine dehydrogenase, 78
Sagittaria pygmacea, 78
Salicylchlorophosphite, 112
Salmonella typhimurium, 45, 47
β-Santalene, synthesis from D-mannitol, 328, 329
Selenocysteine (3-selenylalanine), 64
–, synthesis, 64
Selenohomocysteine, synthesis from *O*-acetylserine, 64
Serine carbamate (2-carbamoylserine), 57
Serine *O*-diazoacetate (azaserine), 58
Sharpless, asymmetric epoxidation, 171, 172, 174
Sharpless epoxidation, 11
–, allylic alcohols, 512–514
–, phenylalkenes, 513
Sharpless epoxidation procedure, 186, 203

(–)-Shikimic acid, synthesis from D-mannose, 360, 361
2,6-Sialyltransferase, 455, 456
(–)-Silphiperfolene, synthesis from monosaccharide precursor, 318, 320, 321
–, – from (*R*)-(+)-pulegone, 318, 320, 321
Solatriose, 493
Sorbitol, Fe(II) complexes, 179
–, Sn(II) complexes, 178
–, synthesis from glucose, 168
–, – from sugars by enzymic reduction, 168
–, transesterification by methyl fatty acid esters, 194
–, W(VI) complexes, 179
Sorbitol linoleic acid esters, 194
(–)-Specionin, synthesis from D-xylal, 339–341
Spirotetronic acids, 18
Squalene, asymmetric epoxidation, 512
Stachyose, X-ray crystallography, 469
Statine (3-amino-3-hydroxy-6-methyl-heptanoic acid), 100
(3*S*,4*S*)-Statine, synthesis from *N*-benzyloxycarbonyl-L-leucinal/(–)-gabaculine, 73
1-*O*-Stearoylethane-2-[*O*-bis(*N,N*-diethylamino)]thionophosphonate, 127, 128
Strecker synthesis, chiral version, 297
Streptococcus lactis, 77
(+)-Streptolic acid, synthesis from streptolydigin, 393–396
–, – from tri-*O*-acetal-D-glucal, 393–397
Streptomyces aureus, 98
Streptomyces griseosporus, 75, 77
Streptomyces pyogenes, 98
Streptomyces sp., 87
–, metabolites, 58
Styrene, hydroformylation, 186
Subtilin, antibiotic, 63
Succimer (*meso*-dimercaptosuccinic acid), antidote for heavy metal poisons, 516
Sucrose, chemistry, 482, 483
Sucrose octaacetate, X-ray crystallography, 468
Sucrose synthetase, 456
(–)-Swainsonine, synthesis from 3-amino-3-deoxy-α-D-mannopyranoside, 370, 372, 373
–, – from D-mannose, 370, 371

(+)-Tartaric acid, absolute configuration, 511
–, conformations, 510
–, crystal structure, 510
(+)-Tartaric acid dithioacetal derivative, crystal structure, 511
Tartrate esters/allylboronates, reactions with aldehydes, 513
Tetraalditols, ^{1}H NMR spectroscopy, 269
O-(Tetra-*O*-benzyl-D-glycopyranoside) trichloroacetimidate, reactions with sugar alcohols, 448
2,3,4,6-Tetra-*O*-benzyl-1-mercapto-1-deoxy-D-glucitol, 211
2,3,4,6-Tetra-*O*-benzyl-1-*O*-tosyl-D-glucitol, 211
Tetracyanoethene, chemical reactions, 517
–, ^{13}C NMR spectrum, 517
1,1,3,3-Tetracyanopropane, structure and spectroscopy, 517
Tetraethyl 1,1,2,2-ethane-tetracarboxylate, crystal structure, 516, 517
Tetraglycerol, 5
Tetrahydric alcohols, 499–502
cis-Tetrahydroactinidiolide, synthesis from homogeranic acid, 295
Tetrahydropyrans, conformattion, 270, 271
2,3,4,5-Tetrahydroxypentyladenine, 206
2,3,4,5-Tetrahydroxypentylcytosine, 206
2,3,4,5-Tetrahydroxypentyluracil, 206
Tetramethyl 1,1,2,2-ethane-tetracarboxylate, crystal structure, 517
Tetramethyl 1,1,2,2-ethene-tetracarboxylate, structure and ^{13}C NMR spectrum, 517
N,N,N′,N′-Tetramethylguanidinium-*syn*-4-nitrobenzaloximate, 139
(3*R,S*,7*R*,11*R*)-3,7,11,15-Tetramethyl-hexadecan-1-yl triflate, 31, 32
2,3,4,6-Tetra-*O*-pivaloyl-β-D-galacto-

pyranosylamine, 297
–, auxiliary in Ugi-four-component-condensations, 298
Tetra-^{i}propylmethylenebisphosphonate, reaction with isopropylidene-*sn*-glycerol, 129
Tetrasaccharides, 496, 497
Tetritols, 499–502
Tetrols, synthesis *via* the Sharpless epoxidation procedure, 171
1,2,3,4-Tetrols, stereochemistry, 500
1,2,3,5-Tetrols, synthesis from benzyl 2-deoxy-2-*C*-methylpentanopyranosides, 173
Tetronolide, synthesis by intramolecular Diels–Alder strategy, 331, 333
Tetroses, synthesis from ethoxyacetaldehyde, 227, 228
Tetrosylhexoses, 470
Thalidomide, metabolism, 98
5-Thia-D-allose, synthesis, 256
5-Thia-D-altrose, synthesis, 256, 257
5-Thia-D-glucose, synthesis, 255, 256
5-Thia-D-mannose, natural occurrence, 255
Thia monosaccharides, 255–258
2-Thia platelet activating factor, 153
1-Thio-2-acetimido platelet activating factor, 153
Thioalditols, 209
Thioalkylphospholipid analogues, 164
sn-1-Thioalkylphospholipids, 163
Thiocellobiose, 498
4-Thiocellobiose, 498
1-Thio-α-D-glucopyranose, synthesis, 257
1-Thioglycerol, cyclocondensation with 2-hydroxyketones, 7
Thioglycerophospholipids, 126
Thioglycosides, conversion into bromides, 445
–, synthesis, 453
β-Thioglycosides, glycosidation agents, 442
1-Thioglycosyl acetates, synthesis from glycosyl acetates, 257
Thiolactose, 498
Thiomalto-oligosaccharides, 453
Thio monosaccharides, 255–258
Thiophenyl α-D-glycosides, reactions with diethylaminosulphur trifluoride, 451, 452
Thiophospholipid-1-β-D-arabinofuranosylcytosine conjugates, 126
Thiophospholipids, 123–127
–, synthesis from 1,2-dimercaptopropanol, 126
–, – from thiophosphates, 125
1-Thiosucrose, 498
4-Thioxylobiose, 498
Threitol, acetyl derivatives, 500
–, benzyl derivatives, epoxidation, 500
–, –, synthesis from dimethyl (*R,R*)-tartrate, 500
–, conformations, 500
–, reaction with acetic acid, 500
–, synthesis from xylose, 499
Threitol derivatives, Swern oxidation, 506
Threo-3-hydroxy-L-glutamic acid, synthesis from L-serine, 91
Threo-4-hydroxy-D-glutamic acid, 91
Threonic acid, reaction with hydrogen bromide, 509
Threonine (2-amino-3-hydroxybutanoic acid), 67
D-Threono-1,4-lactone, synthesis from mannitol, 508
D-Threose, synthesis from glyceraldehyde, 503, 504
–, – from D-xylonic acid, 505
D-(+)-Threose, synthesis from D-(+)-threitol, 189, 190
[1-^{13}C,^{2}H]-Threose, synthesis from glyceraldehyde, 505
L-Threose acetonide, homologation, 172, 173
Thyroglobulin, 154
(±)-1-*O*-Tosyl-2,3:4,5-di-*O*-isopropylideneribitol, 206
Transglycosylation, 457
Transketolase, isolation from spinach, 506
Transphosphatidylation, 108, 122
α,α-Trehalose, 483
–, chemistry and synthesis, 483
β,β-Trehalose, 484
Trehalose esters, 484

3,4,6-Tri-*O*-acetyl-2-azido-2-deoxy-α-D-galactopyranosyl bromide, 241
Tri-*O*-acetyl-1,2-*O*-(1-cyanoethylidene)-α-D-galactopyranose, isomerisation, 228
3,4,6-Tri-*O*-acetyl-D-galactal, chlorination–iodination, 253
Triacylglycerols, self condensation, 27
Trianhydroglucitol, 202
1,3:2,5:4,6-Trianhydrohexitols, monotosylation, 200
2,4,6-Tri-^{t}butylphenyldithiophosphorane, reaction with glycerol, 11
Trichothecene skeleton, synthesis, 364, 365
Tridecanoylglycerols, 27
3-*O*-Trifluoroacetyl-α-D-daunosamine dienes, use in the synthesis of anthracycline derivatives, 309
Trifluoroacetyl sugars, reactions with cyanide ion, 246
Triglycerides, synthesis from 1,3-dibromopropan-2-ol, 23
Triglycerol, 5
Trihydroxybutanals, 503–506
Trihydroxybutanones, 503–506
Trihydroxyoctadecadienoic acid, 18
Trihydroxyoctadecenoic acid, 18
2(*S*),3(*R*),4(*R*),5(*S*)-Trihydroxypipecolic acid, synthesis from D-glucuronic acid, 367, 369
1,2,3-Trihydroxypropane (glycerol), 1–33
1,2:3,4:5,6-Tri-*O*-isopropylidene-galactitol, synthesis from galactitol, 193
Trimethylsilylketene acetals, alkylation, 27, 29
O-Trimethylsilylketene acetals, synthesis from ester enolates, 27
(*E*)-3-Trimethylsilyloxybuta-1,3-dienyl-2,3,4,6-tetra-*O*-acetyl-β-D-glucopyranoside, reactions with dienophiles, 305
Trimethylsilyl trifluoromethane-sulphonate, catalyst for glycosidation, 441, 442
2,3,4-Tri-*O*-pivaloyl-α-D-arabino-pyranosylamine, 300
–, auxiliary in Ugi-four-component-condensations, 298
Trisaccharides, 489–492
–, containing deoxysugars, 492, 493
–, containing nitrogen, 493–495
Tris(trimethylsilyl)phosphite, addition to (*Z*)-mannofuranosyl nitrones, 297
Tritylglycidol, 126
N-Tritylphosphatidylethanolamine, 120

Ugi-four-component-condensations, 298
Uloses, 188
Ulosyl bromides, synthesis, 250, 251
α,β-Unsaturated acids, dihydroxylation, 47
Uronic acids, 488, 489

Valine, biosynthesis, 45
L-Valylglycine dioxopiperazine, 84
Viehe's salt (*N*-dichloromethylene-*N,N*-dimethylammonium chloride), 194
L-Vinylglycine, synthesis from L-glutamic acid, 82
Vliegenthart's reporter groups, 459, 460

Xylitol, cyclic phosphites, 195
–, synthesis from aldoses, 168
–, – from cyclopentadiene, 171
–, – from formaldehyde, 169
Xylitol pentaacetate, 171
Xylitol peracetate, 172
4-*S*-β-D-Xylopyranosyl-4-thio-D-xylo-pyranose, 498
3-*O*-β-D-Xylopyranosyl-D-xylopyranose, 471
D-Xylose, 260
L-(−)-Xylose, synthesis from xylitol, 189, 190
Xylotetraose, 497
Xylβ(1→4)Xylβ(1→4)Xylβ(1→4)Xyl, *O*-β-D-Xylopyranosyl-(1→4)-*O*-β-D-xylo-pyranosyl-(1→4)-*O*-β-D-xylopyranosyl-(1→4)-D-xylopyranose, 497